Carbohydrate Chemistry:

State of the Art and Challenges for Drug Development

An Overview on Structure,
Biological Roles, Synthetic Methods
and Application as Therapeutics

Carbohydrate Chemistry:

State of the Art and Challenges for Drug Development

An Overview on Structure, Biological Roles, Synthetic Methods and Application as Therapeutics

Edited by

Laura Cipolla

University of Milano-Bicocca, Italy

Imperial College Press

Published by

Imperial College Press
57 Shelton Street
Covent Garden
London WC2H 9HE

Distributed by

World Scientific Publishing Co. Pte. Ltd.
5 Toh Tuck Link, Singapore 596224
USA office: 27 Warren Street, Suite 401-402, Hackensack, NJ 07601
UK office: 57 Shelton Street, Covent Garden, London WC2H 9HE

Library of Congress Cataloging-in-Publication Data
Cipolla, Laura, 1968–
Carbohydrate chemistry : state of the art and challenges for drug development / by Laura Cipolla, University of Milano-Bicocca, Italy.
pages cm
Includes bibliographical references and index.
ISBN 978-1-78326-719-4 (hardcover : alk. paper)
1. Carbohydrates--Synthesis. 2. Pharmaceutical chemistry. 3. Drug development. I. Title.
QP701.C525 2015
612'.01578--dc23
2015005247

British Library Cataloguing-in-Publication Data
A catalogue record for this book is available from the British Library.

Typeset by Stallion Press
Email: enquiries@stallionpress.com

Printed in Singapore

Preface

Glycans are universal and essential components to life, recurring in different forms, such as protein and lipid conjugates, large structural polysaccharides, and secreted mucus components. The structural complexity and the synthetic challenges facing glycans have hampered both the study of their biologically relevant yet multifaceted roles, and the application of carbohydrates in drug development. However, in very recent years, new synthetic techniques flanked by the growing knowledge about carbohydrate involvement in physiological and pathological states has spurred renewed interest in the chemistry, biology and therapeutic potentialities of carbohydrates. This book will give an overview of key aspects of carbohydrate biology and chemistry that are fundamental for the design of novel therapeutics. Several topics will be outlined, in order to put forward to the reader how complex but simultaneously intriguing and challenging carbohydrates are.

Part 1 of the book will give a concise introduction on glycans found in living systems and their role in physiological conditions. Part 2 of the book is focused on the most recent techniques that can be used for structural elucidation and for the investigation of the biological role of carbohydrates, including their interactions with specific proteins and receptors. There is still much to be discovered, and the deep knowledge of structure and biological activity is a pre-requisite for the design of carbohydrate-based therapeutics. One of the main drawbacks in the use of carbohydrate building blocks for the design and development of carbohydrate-based therapeutics is their synthesis. However, in the last few years several efforts have been made in order to develop efficient synthetic methodologies, including automated techniques that give a big relief to this issue. We are still far from a fully automated synthesis, as we know it, for peptides or oligonucleotides.

Part 3 of the book will give some outlines about the synthetic up-to-date methodologies that carbohydrate chemists have developed. Part 4 will give emphasis to relevant cases and opportunities for the development of therapeutics based on carbohydrate structures. Due to the great involvement of glycans in biological systems, different pathologies can take advantage of carbohydrate chemistry. Finally, a concluding chapter will give examples of commercially available drugs based on carbohydrates and their mimics.

Carbohydrates have very different structures and play a plethora of biological functions. For this reason, not all the aspects of carbohydrate chemistry, biology and therapeutics have been included in this book. For example, glycosaminoglycans (GAGs) are an important class of complex sugars abundant in the extracellular matrix (ECM); for this reason, they bring a large amount of information eliciting a large number of biological responses that are relevant, for example, in tissue development and regeneration. They have not been cited in this book. Also, glycosphingolipid-related metabolic disorders and frontiers therapies have been omitted, as well as chemoselective and bio-orthogonal strategies for complex glycoconjugate synthesis or advanced techniques for protein and cell surface glycan engineering.

I hope that this book will give a general overview of the richness of glycans in terms of development of therapeutics and advancement of our knowledge in the medical fields.

Laura Cipolla
June 2015

Contents

Acronyms and Abbreviations

6-Acetamido-6-deoxycastanospermine	CASNAc
2-Acetamido-2-deoxynojirimycin	NJNAc
2-Acetamido-1,2-dideoxynojirimycin	DNJNAc
2-Acetamido-1,4-imino-1,2,4-trideoxy-D-arabinitol	DABNAc
2-Acetamido-1,4-imino-1,2,4-trideoxy-L-arabinitol	LABNAc
Acetate	Ac
Acetic anhydride	Ac_2O
Acid α-glucosidase	GAA
Acquired immunodeficiency syndrome	AIDS
Adenosine triphosphate	ATP
Alanine	Ala
Aldehyde	–CHO
Alkyl	C–C
17-Allylamino-17-demethoxygeldanamycin	17-AAG
Allyl methacrylate	AMA
Alpha-D-mannose-(1→2)-α-D-mannose-(1→6)-D-*myo*-inositol	PIM_2
Alpha-2,8-sialyltransferase	ST8Sia
Amine	RNH_2
Amino	NH_2
2-Aminoacridone	AMAC
Aminobenzamide	AB
8-Amino-8-deoxy-*manno*-oct-2-ulosonic acid	Kdo8N
2-Amino-ethanol	EtN
2-Amino-*N*-(2-aminoethyl)benzamide	AEAB
Amyothrophic lateral sclerosis	ALS

Antibody	Ab
Antibody-dependent cellular cytotoxicity	ADCC
Antigen 85	Ag85
Antigen presenting cell	APC
Apoptosis-inducing factor	AIF
Arabinofuranosyl	Ara*f*
Arabinogalactan	AG
Arabinosamine	Ara*p*4N
Arabinosyltransferase	AraT
Arginine	Arg
Aryl hydrocarbon receptor nuclear translocator	ArnT
Asialoglycoprotein receptor	ASGPR
Asparagine	Asn
Atom transfer radical polymerization	ATRP
ATP-binding cassette	ABC
Azide	N_3
6-Azido-6-deoxy-α,α-trehalose	ADT
Azidohomoalanine	Aha
Azidophenyl lactamine	AzPhLac
β-D-arabinofuranosyl-1-monophosphoryldecaprenol	DPA
β1–2-*N*-acetylglucosaminyltransferase I	GnT-I
β1–2-*N*-acetylglucosaminyltransferase II	GnT-II
Baby Hamster Kidney	BHK21
Bacillus Calmette–Guerin	BCG
Bacillus circulans chitinase A1	BcChiA1
Bacterial carbohydrate structure database	BCSDB
B-cell lymphoma 2	Bcl-2
Benzoate	Bz
(Benzotriazol-1-yloxy)tripyrrolidinophosphonium hexafluorophosphate	PyBOP
Benzyl	Bn
β-galactosyltransferase 4	β3Gal T4
2,2-Bis(methylol)propionic acid	Bis-MPA
β-*N*-acetyl-galactosaminyltransferase	β4GalNAc T1
Borane	BH_3

Bovine serum albumin	BSA
Bovine viral diarrhea virus	BVDV
Calcium	Ca^{2+}
Calcium-dependent type lectin	C-type lectin
Capillary electrophoresis	CE
Capsular polysaccharide	CPS
Carbohydrate-Active Enzymes	CAZy
Carbohydrate binding protein	CBP
Carbohydrate recognition domain	CRD
Carbon	C
Carbon isotope with mass number 13	^{13}C
Carbonate hydroxyapatite	CHA
Carbon nanotube	CNT
Carbobenzyloxy	Cbz
Cationic antimicrobial peptide	CAMP
Cationic peroxidase of peanut	CPRx
Central nervous system	CNS
Centre national de la recherche scientifique	CNRS
Ceramide	Cer
Chinese Hamster Ovary	CHO
Chitinase	GH18
Chloride	Cl
Chondroitin sulfate	CS
Chronic kidney disease	CKD
Chronic obstructive pulmonary disease	COPD
Cluster of differentiation	CD
Cluster of differentiation 14	CD14
Cluster of differentiation 82	CD82
Cluster of differentiation 95	CD95
Cobalt	Co
Collisionally-activated dissociation	CAD
Collision-induced dissociation	CID
Colorectal adenocarcinoma	Caco
Community-acquired pneumonia	CAP
Complement-dependent cytotoxicity	CDC

Complementary deoxyribonucleic acid	cDNA
Complex Carbohydrate Structure Database	CCSD
Consortium for Functional Glycomics	CFG
Copper	Cu
Copper (I)-catalysed azide–alkyne cycloaddition	CuAAC
Core 1 β1–3 galactosyltransferase	C1GalT-1 (or T synthase)
Core 2 β1–6-*N*-acetylglucosaminyltransferase	C2GnT
Core 3 β1–3-*N*-acetylglucosaminetransferase	C3GnT
C-type lectin domain family 4, member C	CLEC4C
Cyclodextrin	CD
Cyclodextrin glucanotransferase	CGTase
Cysteine	Cys
Cytidine-5′-monophospho-*N*-acetylneuraminic acid	CMP-Neu5Ac
Cytoplasmic linker protein	CLIPR
Cytosine monophosphate	CMP
Cytotoxic T-lymphocyte	CTL
Damage-associated molecular pattern	DAMP
Deaminated neuraminic acid (2-keto-3-deoxy-nonulosonic acid)	Kdn
Decaprenyl phosphate	DP
Decaprenylphosphoryl-D-2-keto erythro pentose reductase	DprE2
Decaprenylphosphoryl-β-D-ribose-2′-epimerase	DprE
Decaprenylphosphoryl 5-phosphoribose	DPPR
Decaprenylphosphoryl ribose	DPR
Degree of polymerization	DP
Dendritic cell	DC
Dendritic cell-specific intercellular adhesion molecule-3-grabbing non-integrin	DC-SIGN
Dengue virus	DENV
1-Deoxygalactonojirimycin	DGJ
6-Deoxy-L-mannose	L-Rha
1-Deoxynojirimycin	DNJ
Dermatan sulfate	DS

Detergent-resistant membrane	DRM
D-*glycero*-D-*manno* heptose residues	D,D-Hep
D-*glycero*-D-*talo*-oct-2-ulosonic acid	Ko
2,3-Diamino-2,3-dideoxy-D-glucopyranose	GlcpN3N
1,4-Dideoxy-1,4-imino-D-arabinitol	DAB
2,5-Dideoxy-2,5-imino-D-mannitol	D-DMDP
2,5-Dideoxy-1,4-imino-L-mannitol	L-DMDP
1,4-Dideoxy-1,4-imino-L-arabinitol	LAB
Diisopropylethylamine	DIEA
Dimethylacetamide	DMAc
17-Dimethylaminoethylamino-17-demethoxygeldanamycin	17-DMAG
Dimethylamino pyridine	DMAP
Dimethyl dioxirane	DMDO
Dimethylformamide	DMF
Dimethyl sulfoxide	DMSO
Diphtheria toxin/toxoid	DT
D-Isofagomine	D-IFG
Dissociation constant	K_d
Dolichol phosphate	Dol-P
Dolichol pyrophosphate *N*-acetylglucosamine	Dol-P-P-GlcNAc
Double quantum filter correlation spectroscopy	DQF-COSY
Ethylenediaminetetraacetic acid	EDTA
Electron capture dissociation	ECD
Electron detachment dissociation	EDD
Electron excitation dissociation	EED
Electron ionization	EI
Electron transfer dissociation	ETD
Electrospray Ionization mass spectrometry	ESI MS
Endo-β-acetylglucosaminidase	GH85, or ENGase
Endoglycosidase H	Endo-H
Endoplasmic reticulum	ER
Enzyme-linked immunosorbent assay	ELISA

Enzyme replacement therapy	ERT
Epidermal growth factor	EGF
Epidermal growth factor receptor	EGFR
Erythropoietin	EPO
European Union	EU
Expert Protein Analysis System	ExPASy
Extensible markup language	XML
Extensive drug resistant	XDR
Extracellular matrix	ECM
Extracellular signal-regulated kinase	Erk
Fibroblast growth factor	FGF
Field Asymmetric Ion Mobility Spectrometry	FAIMS
Field desorption-mass spectrometry	FD-MS
Fluorenylmethoxycarbonyl	Fmoc
Fluorine isotope 19	^{19}F
Focal adhesion kinase	FAK
Food and Drug Administration	FDA
Fourier transform infrared	FT-IR
Fourier transform ion cyclotron resonance	FT-ICR
Free radical polymerization	FRP
Fucose	Fuc
Gadolinium	Gd
Galactose	Gal
α-Galactosidase A	α-Gal A
Galactosylceramide	Galβ1-Cer
Galactosyltransferase	GalT
Galacturonic acid	Gal*p*A
Galβ1–4GlcNAcβ1–3	poly-*N*-acetyllactosamine
GalNAcβ1–4GlcNAc-	LacdiNAc
Gas chromatography–mass spectrometry	GC–MS
Gas liquid chromatography mass spectrometry	GLC-MS
Glucocerebrosidase	prGCD or GCase

Glucosamine	GlcN
Glucose	Glc
Glucuronic acid	GlcA
Glutamate	Glu
Glycine	Gly
Glycomics ontology	GlycoO
Glyconanoparticle	GNP
Glycopeptide N-Glycosidase F	PNGase F
Glycosaminoglycan	GAG
Glycoside hydrolase	GH
Glycosphingolipid	GPI or GSL
Glycosylation-dependent cell adhesion molecule	GLYCAM
Glycosylphosphoinositide	GPI
Glycosyltransferase	GT
Gold–sulfur	Au–S
Good manufacturing practice	GMP
Granulocyte colony stimulating factor	G-CSF
Granulocyte/macrophage colony stimulating factor	GM-CSF
Growth factor receptor-bound protein 2	Grb2
Guanine nucleotide-binding protein	G protein
Guanosine triphosphatase	GTPase
Haemophilus influenza	*H. influenza*
H. influenza type A	Hia
H. influenza type B	Hib or HiB
H. influenza type B: first generation	HbPV
Heat shock protein 90	Hsp90
Henrietta Lacks	HeLa
Heparan sulfate	HS
Hepatitis B virus	HBV
Hepatitis C virus	HCV
Hepatocyte growth factor	HGF
Heteronuclear multiple bond correlation	HMBC
Heteronuclear NOE spectroscopy	HOESY
Heteronuclear single quantum coherence	HSQC
Hexaarabinofuranosyl	Ara_6

Hexosaminidase	GH20
Hexosaminidase A	Hex A
Hexosaminidase B	Hex B
Hexosaminidase S	Hex S
Hexuronic acid	HexA
High-capacity ion trap	HCT
High-field asymmetric wave form ion mobility spectrometry	FAIMS
High-molecular-weight adhesin	HMW1
High performance	HP
High-performance/pressure liquid chromatography	HPLC
High-throughput screening	HTS
Histamine	His
Homopropargylglycine	Hpg
Human colon tumor	HCT116
Human Embryo Kidney 293	HEK293
Human epidermal growth factor receptor 2	HER-2
Human fibrosarcoma	HT1080
Human galectin	hGal
Human immunodeficiency virus	HIV
Human leukocyte antigen	HLA
Human lymphoma	Namalwa
Human papilloma virus	HPV
Human serum albumin	HSA
Hyaluronic acid	HA
Hyaluronidases	GH52
Hydrochloride acid	HCl
Hydrogen	H
Hydrophilic interaction liquid chromatography	HILIC
Hydroxyapatite	HA
Hydroxyl	OH
Iduronic acid	IdoA
Imaging mass spectrometry	IMS
Immunoglobulin	Ig

Immunoglobulin G	IgG
Immunoglobulin M	IgM
Indium tin oxide	ITO
Infrared	IR
Infrared multiphoton dissociation	IRMPD
Inhibitory constant	K_i
Inositol-1-phosphate	IP
Insulin receptor	IR
Interferon-α2b	IFN-α2b
Interferon-γ	IFN-γ
International Union of Pure and Applied Chemistry	IUPAC
Intravenous immunoglobulin	IVIG
Intrinsic reaction coordinates	IRC
Invasive ductal breast carcinoma	IDC
Iron	Fe
Isoleucine	Ile
Isothermal titration microcalorimetry	ITC
Japanese Consortium for Glycobiology and Glycotechnology Database	JCGGDB
Kartogenin	KGN
KEGG Chemical Function	KCF
Keratan sulfate	KS
2-Keto-3-deoxy octulosonic acid	KDO
Keyhole limpet hemocyanin	KLH
Knock-out	KO
Kyoto Encyclopedia of Genes and Genomes	KEGG
Lactosylceramide	LacCer
Leaving group	LG
Leucine	Leu
Levulinoate	Lev
Lewis x	Le^x
Lewis y	Le^y

Ligand place exchange	LPE
L-Isofagomine	L-IFG
Linear notation for unique description of carbohydrate sequences	LINUCS
Lipid peroxidise	Lpx
Lipoarabinomannan	LAM
Lipomannan	LM
Lipooligosaccharide	LOS
Lipopolysaccharide	LPS
Lipopolysaccharide transporter	*lpt*
Liquid chromatography	LC
Liquid Crystal on Silicon	L-COS
Liquid extraction surface analysis	LESA
Logarithmic acid dissociation constant	pKa
Lophozozymus pictor toxin	Lptx
LPS binding protein	LBP
Lysosomal storage disorder	LSD
Madin-Darby canine kidney	MDCK
Magnesium	Mg^{2+}
Magnetic resonance imaging	MRI
Major histocompatibility complex	MHC
Maltose-C-phosphonate	MCP
Maltose-1-phosphate	M1P
Manganese	Mn
1,2-Manobioside	Man-α-1,2-Man
Man-6-phosphoethanolamine	Man-6-P-EthN
Mannose	Man
Mannose 6-phosphate	M6P
Mannosylated bovine serum albumin	Man-BSA
Mannosylated LAM	ManLAM
Mannosyltransferase	ManT
Mass spectrometry	MS

Matrix-Assisted Laser Desorption/Ionization – Time of Flight	MALDI–TOF
Median effective concentration	EC_{50}
Median inhibitory concentration	IC_{50}
Meningitidis Vaccine Project	MVP
Meningococcal X	Men X
Merkel cell polyomavirus	MCPyV
Meta-chloroperoxybenzoic acid	*m*CPBA
Methanol	MeOH
Methyl ethyl ketone	MEK
Methyl ketone	MK
Methyl trifluoromethanesulfonate	MeOTf
4-Methylumbelliferyl-β-D-*N*-acetylglucosamine	MUG
Michaelis constant	K_M
Michigan Cancer Foundation-7	MCF-7
Minimal inhibitory concentration	MIC
Minimum information about a microarray experiment	MIAME
Minimum information required for a glycomics experiment	MIRAGE
Mitogen-activated protein kinase	MAPK
Molecular dynamics	MD
Molecular mechanic	MM
Molecular weight	MW
Monoclonal antibody	mAb
Mucin 1	MUC1
Mucopolysaccharidose	MPS
MUG-6-sulfate	MUGS
Multi drug resistant	MDR
Multiple antigen glycopeptides	MAG
Multiple reaction monitoring	MRM
Mycobacterium tuberculosis	*MTB*
Mycolyl arabinogalactan	mAG
Mycolyl arabinogalactan peptidoglycan	mAGP
Myelin-associated glycoprotein	MAG
Myeloid differentiation factor 2	MD-2

N-acetyl-2,6-dideoxy-D-glucosamine	D-QuiNAc
N-acetylglucosamine	GlcNAc
N-acetyl-glucosamine-thiazoline	NGT
N-acetylglucosaminyltransferase	GlcNAcT
N-acetyllactosamine	LacNAc
N-acetylneuraminic acid	Neu5Ac or NANA
Nanoporous gold	NPG
Nanotube field-effect transistor	NTFET
National Institute for Bioprocessing Research and Training	NIBRT
Native chemical ligation	NCL
N-benzyl	*N*-Bn
N-chlorosuccinimide	–NCS
Neisseria meningitidis	*N. meningitidis*
Nerve growth factor	NGF
Neuraminidase	NA
N-glycolylneuraminic acid	Neu5Gc or NGNA
N-(2-hydroxypropyl) methacrylamide	HPMA
N-hydroxysuccinimide	NHS
Nicotinamide adenine dinucleotide	NAD^+
N-iodosuccinimide	NIS
Nitroxide-mediated polymerization	NMP
Nogo receptor	NgR
Nojirimycin	NJ
Non-reducing glucosamine	Glc*p*N II
Nuclear magnetic resonance	NMR
Nuclear overhauser effect spectroscopy	NOESY
O-benzoxazolyl	OBox
O-GlcNAc transferase	OGT
O-glycoPEGylated coagulation factor VIII	FVIII
Oligodendrocyte myelin glycoprotein	OMgp
Oligosaccharide	OS
Oligosaccharyltransferase	OST

6-*O*-methylglucose lipopolysaccharide	MGLP
3-*O*-methylmannose polysaccharide	MMP
3-*O*-methyltransferase	OMT
O-*N*-acetylgalactosamine	*O*-GalNAc
O-*N*-acetylglucosamine	*O*-GlcNAc
One-dimensional	1D
Osteoarthritic	OA
Outer membrane protein complex	OMPC
Parallel artificial membrane permeability assay	PAMPA
Paramagnetic relaxation enhancements	PRE
Pathogen-associated molecular pattern	PAMP
Pattern recognition receptor	PRR
Pentafluorophenyl acrylate	PFPA
Peptide-*N*-glycosidase	PNGase
Peptide nucleic acid	PNA
Peptidoglycan	PG
Peripheral blood mononuclear cell	PBMC
Permeability transition pore complex	PTP
Pharmacological chaperone	PC
Phenolic glycolipid	PGL
Phosphate	P
Phosphate-activated glutaminase	Pag
Phosphatidylinositol	PI
Phosphatidylinositol 3-kinase	PI3K
Phosphotidylinositol mannoside	PIM
Phosphoethanolamine	PEtN
Phosphoethanolamine transferase	EptA
Phosphoinositol	PI
5-Phosphoribose-1-diphosphate	PRPP
5-Phospho-ribosyltransferase	DPPR synthase
Phosphorus-31	^{31}P
Phthiocerol dimycocerosate	PDIM
Pivaloyl	Piv

Pivaloyl chloride	PivCl
Plasma protein binding	PPB
Platelet-derived growth factor	PDGF
p-Nitrophenyl	PNP
Poisson-Boltzmann surface area	PBSA
Poliovirus protein	PV
Poly(ε-caprolactone)	PCL
Polyethylene glycol	PEG
Polyethylene terephthalate	PET
Polymer of ribose, ribitol, and phosphate	PRP
Polymethylated polysaccharides	PMPS
Poly-*N*-acetyl glucosamine	PNAG
Polypeptide-*N*-acetylgalactosaminyltransferase	ppGalNAcT
Polyprenolphosphomannose	PPM
Polypropylene	PP
Poly vinylidene difluoride	PVDF
Porous graphitized carbon	PGC
Porous polymers synthesized within high internal phase emulsions	polyHIPE
Post-translational modification	PTM
Potassium hydroxide	KOH
Proline	Pro
Protecting	P
Protein Data Bank	PDB
Protein Informatics Group	PIG
Protein kinase B	Akt
Proton	^{1}H
Pseudo contact shift	PCS
Pseudomonas aeruginosa lectin I	PA-IL
Quadrupole time-of-flight	QTOF
Quantum dot	QD
Quantum mechanics	QM
Ras homolog gene family member A	RhoA
Reactive oxygen species	ROS

Receptor tyrosine kinase	RTK
Recombinant human GAA	rhGAA
Recombinant protein	r-protein
Reducing-end-xylose releasing exo-oligoxylanase	REX
Reducing glucosamine	Glc*p*N I
Regioselectively addressable functionalized templates	RAFTs
Relative reactivity value	RRV
Residual dipolar coupling	RDC
Reverse-phase	RP
Reversible addition–fragmentation chain transfer	RAFT
Rhamnosyltransferase	RhamT
Ribonuclease B	RNase B
Ribonucleic acid	RNA
Ribosomal ribonucleic acid	rRNA
Ricinus communis agglutinin	RCA_{120}
Ring-opening metathesis polymerization	ROMP
Rotating-frame overhauser effect spectroscopy	ROESY
Rough-form lipopolysaccharide	R-LPS
Salmonella typhi	*S. typhi*
Saturation transfer difference	STD
S-benzoxazolyl	SBox
Scanning electron microscope	SEM
Selected reaction monitoring	SRM
Self-assembled monolayer	SAM
Serine	Ser
Severe combined immunodeficiency	SCID
Sialic acid-binding immunoglobulin-type lectin	Siglec
Sialyl Lewis x	sLe^x
Sialyl Tn	sTn
Silver trifluoromethanesulfonate	AgOTf
Simian polyomavirus 40	SV40
Single-electron transfer living radical polymerization	SET-LRP

Single reaction monitoring	SRM
Single-walled carbon nanotube	SWNT
Small cell lung cancer	SCLC
Smooth-form lipopolysaccharide	S-LPS
Sodium borohydride	$NaBH_4$
Sodium cyanoborohydride	$NaCNBH_3$
Sodium dodecyl sulphate polyacrylamide gel electrophoresis	SDS-PAGE
Sodium–glucose cotransporter type 2	SGLT-2
Sodium hydroxide	NaOH
Sodium methoxide	NaOMe
Solid-phase peptide synthesis	SPPS
Solvent exposed-amines	SEA
Specificity constant	k_{cat}/K_M
Stage-specific embryonic antigen	SSEA
Streptococcus pneumonia	*S. pneumonia*
Structure–activity relationship	SAR
Substrate reduction therapy	SRT
Sugar–amino acid–nucleoside	SAAN
Surface plasmon resonance	SPR
Surface plasmon resonance imaging	SPRi
Surface-Tethered Iterative Carbohydrate Synthesis	STICS
Swiss Institute of Bioinformatics	SIB
Tetanus toxoid/toxin	TT or TTox
Tetraazocyclododecane triacetic	DO_3A
Tetrabutylammonium fluoride	TBAF, Bu_4NF
Tert-butyldimethylsilyl	TBS
Tert-butyldiphenylsilyl	TBDPS
Tetrahydrofuran	THF
Three-dimensional	3D
Threonine	Thr
Thin-layer chromatography	TLC
Thiol	SH
Thomsen–Friedenreich	TF
Thymidine diphosphate	dTDP

Toll-like receptor 4	TLR4
Total correlated spectroscopy	TOCSY
Transferred NOE	trNOE
Transferred NOE spectroscopy	trNOESY
Transglycosylase	TG
Transglycosylation/hydrolysis balance	T/H
Trans-Golgi network	TGN
Transition metal-mediated living radical polymerization	TMM-LRP
Transverse relaxation optimized spectroscopy	TROSY
Trehalose dimycolate	TDM
Trehalose monomycolate	TMM
Trehalose synthase	TreS
Trichloroacetamide	TCANH
Trichloroacetyl	TCA
Triethylsilane	Et_3SiH
Trifluoroacetic acid	TFA
Trifluoroacetic anhydride	TFAA
Trifluoromethanesulfonate	Bu_2BOTf
Trifluoromethanesulfonic acid	TfOH
Trimethylsilyl trifluoromethanesulfonate	TMSOTf
Tripalmitoyl-S-glycerylcysteinylserine	Pam_3Cys
Tropaeolum majus	TmNXG1
Tumor associated carbohydrate antigen	TACA
Tumor necrosis factor	TNF
Two-dimensional	2D
Tyrosine	Tyr
UDP galactopyranose mutase	UGM
UDP-glucose:glycoprotein glucosyltransferase 1	UGT1 or UGGT
Ultra performance liquid chromatography	UPLC
Ultraviolet	UV
Ultraviolet-visible	UV-Vis
Undecaprenyl phosphate	Und-P
United Kingdom	UK

United States	US
University of Oxford Format	UOXF
Uridine 5c-diphosphate galactofuranose	UDP-Gal*f*
Uridine 5c-diphosphate galactopyranose	UDP-Galp
Uridine diphosphate	UDP
Uridine diphosphate glucose	UDP-Glc
Uridine monophosphate	UMP
Valine	Val
Vascular endothelial growth factor	VEGF
Voltage-dependent anion channel	VDAC
World Health Organization	WHO
X-ray photoelectron spectroscopy	XPS
Xyloglucan *endo*-transglycosylase	XET
Xylose	Xyl
Zinc	Zn
Zona pellucida	ZP

Part 1:
Structure and Biological Function of Glycoconjugates

Chapter 1

Glycoproteins: Chemical Features and Biological Roles

Yanhong Li, Xixuan Li and Xi Chen

Department of Chemistry, University of California-Davis, USA

1.1 Introduction

Glycoproteins consisting of one or more glycan chains covalently linked to a protein were first discovered by Albert Neuberger in 1938 [1]. In 1960, it was clear that the glycans were covalently linked, via the anomeric carbon at the reducing end, to mainly two types of amino acid residues on proteins to form two major types of protein glycans. Linkage to the nitrogen in the amide side chain of L-asparagine forms an *N*-glycosyl bond for protein *N*-glycans and linkage to the hydroxyl group of L-serine or L-threonine forms an *O*-glycosyl bond in protein *O*-glycans.

Recently, it has been found that in addition to *N*-glycosylated proteins in eukaryotic systems, some bacteria also have *N*-glycosylated proteins, although the structures of bacterial *N*-glycans differ from those in eukaryotic systems [2]. There are also some differences in the *N*-glycosylation site consensus sequences between the two systems [3]. Furthermore, the surface (S)-layer proteins in archaea can also be *N*-glycosylated [4].

For eukaryotic *O*-glycans, the major ones are protein *O*-*N*-acetylgalactosamine (*O*-GalNAc) glycans or mucin-type *O*-glycans. More recently, protein *O*-*N*-acetylglucosamine (*O*-GlcNAc) modification was discovered. In addition, other protein *O*-glycans have been identified, including containing glycoproteins mannose (Man) or fucose (Fuc) linked to the hydroxyl group at threonine (Thr) and serine (Ser). *O*-Glycosylation has also been found at the hydroxyl group of tyrosine, hydroxyproline, and hydroxylysine [5,6]. Proteoglycans containing one or more glycosaminoglycan (GAG) polysaccharide chains linked via a xylose (Xyl) to the hydroxyl group at the serine residue of the protein backbone are a unique group of *O*-glycosylated proteins [7,8]. *O*-Glycosylation has also been found at the tyrosine, serine, and threonine residues on the S-layer glycoproteins of archaea and bacteria [9,10].

Carbon (C)-glycosylation with an example of mannose linked to tryptophan, phosopho (P)-glycosylation with GlcNAc/Man/Fuc/Xyl-1-phosphate linked to the hydroxyl group at serine, and glypiation with Man-6-phosphoethanolamine (Man-6-P-EthN) linked to the carboxyl (C)-terminus of the proteins in glycosphingolipid (GPI)-anchored proteins have also been identified [11]. More recently, sulphur (S)-linked glycopeptide antibiotics (sublancin, thurandacin A and B, and glycocin F) with a glucose (Glc) or a GlcNAc linked to the cysteine (Cys) residue have been identified [12,13].

In higher eukaryotes, more than 50% of all proteins are glycosylated [14]. Glycans can modulate the biological functions of a glycoprotein, making a protein more soluble, protecting a protein against hydrolysis via protease activity, and reducing or preventing aggregation of a protein, covering the antigenic site of a protein, and altering the orientation of a protein on the cell surface. The glycans of a glycoprotein also play important roles on a protein's localization and trafficking, biological half-life of a protein, as well as cell–cell interactions [15].

This chapter describes the chemical features of different types of glycoproteins with focus on *N*- and *O*-linked glycans in glycoproteins. Their biosynthesis, biological roles, production, applications in drug development, and advances on glycan modification of glycoproteins for developing drugs with improved functions are also discussed briefly.

1.2 Glycoproteins and Biosynthesis

1.2.1 *N-Linked glycoproteins*

N-Linked glycoproteins were traditionally considered unique for eukaryotic systems. It was not until more recently that their presences in bacteria and archaea have become recognized. *N*-Glycans in eukaryotes share some common features and a general core structure. They consist of common monosaccharide units. However, their structures in bacteria and archaea are more diverse and contain both common and rare monosaccharide building blocks [16].

1.2.1.1 *N-Linked glycoproteins in eukaryotic systems*

N-Glycans in eukaryotic cells have a common $Man_3GlcNAc_2$ core linked to the asparagine (Asn) residue in the Asn-X-Ser/Thr sequons of proteins (where X is an amino acid other than proline) and can be classified to one of three types: High-mannose, complex, and hybrid (Fig. 1.1). They are presented on many secreted and membrane-bound glycoproteins [17].

In *Saccharomyces cerevisiae* and vertebrates, N-linked glycoproteins are initially synthesized on the cytoplasmic side of the

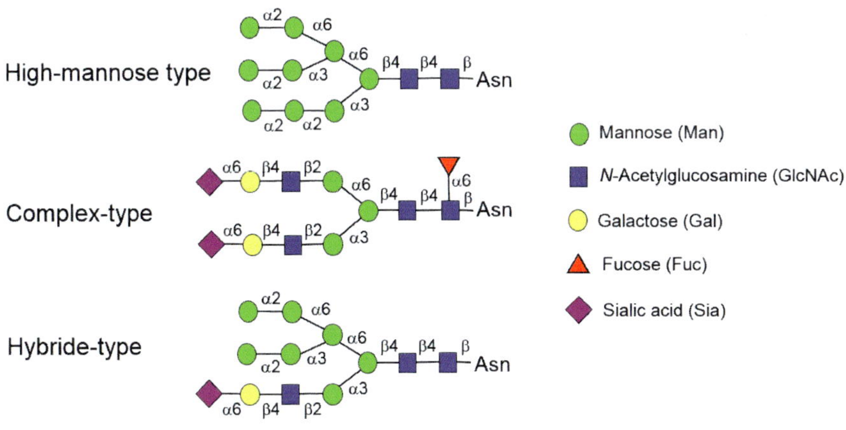

Figure 1.1 Representative high-mannose, complex, and hybrid type *N*-glycans.

endoplasmic reticulum (ER) membrane, starting with the transfer of GlcNAc-phosphate (P) from nucleotide-activated sugar donor uridine diphosphate (UDP)-GlcNAc to the ER membrane-anchored molecule dolichol phosphate (Dol-P). The dolichol pyrophosphate *N*-acetylglucosamine (Dol-P-P-GlcNAc) formed is further processed by glycosyltransferases. One GlcNAc and five mannose residues are subsequently added from UDP-GlcNAc and GDP-Man, respectively, generating $Man_5GlcNAc_2$-P-P-Dol. This sugar chain is flipped into the luminal side of the ER membrane and extended by the transfer of four mannose residues from Dol-P-Man and three glucose residues from Dol-P-Glc. The 14-sugar precursor ($Glc_3Man_9GlcNAc_2$) is then transferred *en bloc* from the dolichol carrier to the Asn-X-Ser/Thr sequon (where X is an amino acid other than proline) of a polypeptide chain by the oligosaccharyltransferase (OST) in the ER lumen. The attached glycan chain is sequentially trimmed by various glycosidases and glycosyltransferases. In the ER, α-glucosidase I and II remove the terminal α1–2-linked glucose and the penultimate α1–3-linked glucose, respectively, to produce $Glc_1Man_9GlcNAc_2$, which binds to chaperons calnexin and calreticulin in the ER to help protein folding [18]. Further cleavage of an additional α1–3-linked glucose catalyzed by the α-glucosidase II produces $Man_9GlcNAc_2$. Most glycoproteins have another mannose cleaved off by α-mannosidase I to form $Man_8GlcNAc_2$ before entering the Golgi apparatus [19].

$Man_5GlcNAc_2$, the crucial intermediate in the conversion of high-mannose type *N*-glycan to complex or hybrid type, is produced in the *cis*-Golgi by α1–2-mannosidases IA, IB, and IC. In some single-cell organisms such as yeast, $Man_8GlcNAc_2$ is added with a mannose residue rather than being truncated, giving rise to high-mannose type *N*-glycans. In the *medial*-Golgi, *N*-acetylglucosaminyltransferase (GlcNAcT)-I transfers GlcNAc from UDP-GlcNAc to the α1–3 arm of $Man_5GlcNAc_2$ in a β1–2 linkage, triggering the action of α-mannosidase II, which removes the terminal α1–3 and α1–6 mannose residues of the α1–6 arm to form the precursor of complex-type *N*-glycans, $GlcNAcMan_3GlcNAc_2$. In complex-type *N*-glycans, GlcNAcT-II adds another GlcNAc to the α1–6 arm of $GlcNAcMan_3GlcNAc_2$ via the β1–2 linkage. GlcNAc residues can be further attached to the core mannose by other GlcNAcTs. For example, GlcNAcT-IV transfers a GlcNAc to C-4 of the α1–3 linked mannose and GlcNAcT-V links a GlcNAc to C-6 of the α1–6 linked mannose, forming complex *N*-glycans with three and four branches [19]. Besides, GlcNAcT-III can catalyze the addition of a bisecting GlcNAc to the β1–4 linked mannose.

Most of the further modifications on N glycans occur in the *trans*-Golgi. There are many possible terminal glycosylation pathways. For mammalian *N*-glycans, a core fucose is added to the innermost asparagine-linked GlcNAc residue through an α1–6 linkage. The sugar chain is elongated by β-linked galactose (Gal) residues to the C3 or the C4 of the GlcNAc, generating type-1 Galβ1–3GlcNAc or type-2 *N*-acetyllactosamine (LacNAc) structures. Tandem repeats of LacNAc are also common. The antennae can be capped with α-linked sialic acid to facilitate the interaction of glycans with other molecules. Besides those mentioned above, there are other glycosyltransferases and glycosidases which would act on the glycan chains. The activities of these enzymes depend on the physiological condition, cellular environment, the nature of proteins and many other factors, which have great influence on the glycosylation pattern. *N*-glycans, with a huge diversity in their composition, length and linkage, play an important role in protein folding, quality control, trafficking, stability, and many other functions [19].

1.2.1.2 *N-Linked glycoproteins in bacteria*

The first bacterial *N*-glycan identified was from *Campylobacter jejuni*. It has a heptasaccharide GalNAcα1–4GalNAcα1–4(Glcβ1–3) GalNAcα1–4GalNAcα1–4GalNAcα1–3diNAcBac β-linked to the Asn residue of the protein, where diNAcBac is di-*N*-acetylbacillosamine [20]. Its biosynthesis shares some similarity to that of the eukaryotic systems. It is proposed that the monosaccharides for *N*-glycosylation are sequentially assembled onto undecaprenyl phosphate (Und-P) at the cytoplasmic face of the inner membrane. The five glycosyltransferases (PglC, PglA, PglJ, PglH, and PglI) that catalyze these reactions are encoded in the *pgl* gene cluster. Then, the formed lipid-linked oligosaccharide (LLO) is flipped to the periplasmic face catalyzed by flippase PglK. PglB, serving as the oligosaccharyltransferase (OST), catalyzes the transfer of the oligosaccharide component from the lipid carrier onto the acceptor protein. *C. jejuni* requires an extended *N*-glycosylation consensus sequence of Asp/Glu-X_1-Asn-X_2-Ser/Thr, where X_1 and X_2 represent any amino acid except proline. There is no further glycan modification after glycan *en bloc* transfer from the glycolipid. Another bacterial *N*-glycan biosynthesis pathway reported was in *Haemophilus influenzae* [21]. *H. influenza* high-molecular-weight adhesin (HMW1) is *N*-glycosylated in the cytoplasm by the glycosyltransferase, HMW1C. Monohexoses and dihexoses are added sequentially to the eukaryotic sequon Asn-X-Ser/Thr, rather than being transferred *en bloc* from the lipid carrier. *N*-Glycosylation has not been reported in Gram-positive bacteria yet [22].

1.2.1.3 *N-Linked glycoproteins in archaea*

N-Glycosylation patterns in archaea can be very distinct among species. In *Haloferax volcanii*, glycoproteins are modified with a pentasaccharide, which consists of a hexose, two hexuronic acids, a methyl ester of hexuronic acid, and a mannose [23]. Four monosaccharide units are assembled onto Dol-P by glycosyltransferases AglJ, AglG, AglI and AglE, while mannose is attached to a different Dol-P by AglD. Both types of Dol-P are translocated across the plasma membrane and attached to the accepting protein, but in different manners and orders. The mechanism by which Dol-P with tetrasaccharide is flipped is still unknown and the transfer to protein is catalyzed by an OST, AglB. The translocation of Dol-P with the mannose involves AglR and the transfer of mannose by AglS occurs after the tetrasaccharide be bound to target protein. The *N*-glycan chain of *Methanococcus voltae* contains a trisaccharide with or without a 220 kD or 260 kD sugar moiety [23]. The GlcNAc, di-*N*-acetylglucuronic acid and *N*-acetylmannuronic acid are transferred by AglH, AglK and AglA, respectively, in sequence.

1.2.2 *O-Linked glycoproteins*

O-Linked glycoproteins have long been recognized in eukaryotic, bacterial, and archaeal systems. *N*-glycans in eukaryotic systems have more predictable structures, due to the *en bloc* transfer of oligosaccharides to form a common precursor leading to a shared core structure, and are easier to analyze, due to the availability of specific hydrolyzing enzymes, such as endoglycosidases and peptide-*N*-glycosidases (PNGases). On the contrary, the major eukaryotic *O*-glycans (mucin-type *O*-glycans or *O*-GalNAc glycans) are more diverse, due to the addition of individual monosaccharides catalyzed by diverse glycosyltransferases, which may or may not share or compete with the same acceptor substrates. Also, *O*-glycans are challenging for analysis, due to the lack of suitable hydrolyzing enzymes and selective chemical hydrolyzing approaches [24].

1.2.2.1 *O-Linked glycoproteins in eukaryotic systems*

O-Linked glycoproteins in eukaryotic systems are mainly of the *O*-GalNAc mucin-type. More recently, *O*-GlcNAc modification has drawn increasing attention. The presence and the importance of other types of *O*-glycans in eukaryotic systems are less well appreciated, but data are emerging to support their important roles in biology [25].

1.2.2.1.1 *O*-GalNAc glycans

O-Glycans α-linked to the hydroxyl group of serine or threonine residues via a GalNAc residue are called *O*-GalNAc glycans. *O*-GalNAc glycans are found in many glycoproteins such as interleukin-2, erythropoietin, CD34, etc. Mucins, a kind of glycoprotein found on the surface of epithelial cells, are highly *O*-GalNAc glycosylated and, therefore,*O*-GalNAc glycans are also called mucin *O*-glycans. Hundreds of *O*-GalNAc glycans are attached to the variable number of tandem repeat (VNTR) region of mucins. This region is rich in serine and threonine residues which can serve as potential *O*-glycosylation sites. *O*-GalNAc glycans are heterogeneous in mucins, with distinct lengths and compositions [26].

The biosynthesis of *O*-GalNAc glycans are initialized by the transfer of GalNAc from UDP-GalNAc to the acceptor site of the protein, catalyzed by a family of polypeptide-*N*-acetylgalactosaminyltransferases (ppGalNAcT). Different sugars are then added to form different *O*-GalNAc core structures. The simplest *O*-GalNAc glycan is Tn antigen, with solely a GalNAc residue attached to the serine or threonine of the peptides or proteins (Fig. 1.2). Tn and sialylated Tn (Neu5Acα2–6GalNAcα-*O*-Ser/Thr) antigens are often found in mucins of cancer cells rather than normal cells. A specific glycosyltransferase core 1 β1–3 galactosyltransferase (C1GalT-1, or also called T synthase) that transfers a galactose from UDP-Gal to Tn antigen in a β1–3 linkage is responsible for the synthesis of T antigen (core 1). This enzyme needs to be exported from the ER to the Golgi with the help of a molecular chaperone, Cosmc. Sialylation of T antigens in which sialic acid is α2–3 linked to the galactose and α2–6 linked to the GalNAc are common in glycoproteins. Core 2 *O*-GalNAc glycans have a GlcNAc β1–6 linked to the GalNAc of core 1, which is catalyzed by core 2 β1–6-*N*-acetylglucosaminyltransferase (C2GnT).

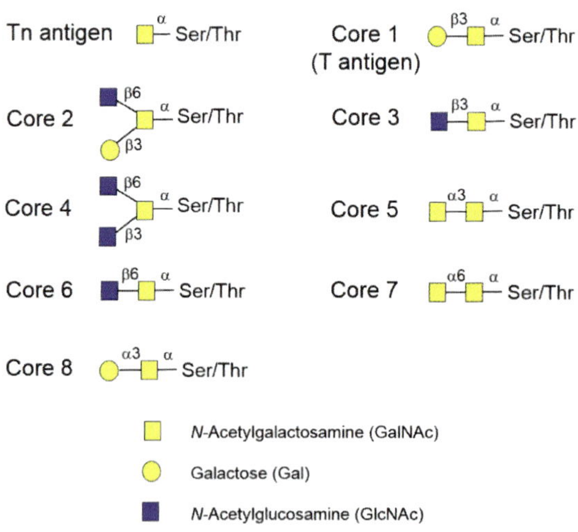

Figure 1.2 Representative *O*-GalNAc glycans.

An abnormal amount of core 2 glycans are associated with tumor progression. In the synthesis of core 3, the transfer of GlcNAc to GalNAc is catalyzed by core 3 β1–3-*N*-acetylglucosaminetransferase (C3GnT). Core 4 *O*-glycan has another GlcNAc β1–6 linked to core 3 and M-type C2GnT-2 is the enzyme catalyzing the synthesis of this type of glycan. Core 1 through core 4 are the most prevalent *O*-GalNAc structures, and core 5 through core 8 are the less common ones. Complex *O*-GalNAc glycans can be synthesized by extending core 1 and core 2 *O*-GalNAc glycans with repeated Galβ1–4GlcNAcβ1–3 (poly-*N*-acetyllactosamine), GalNAcβ1–4GlcNAc- (LacdiNAc), or Galβ1–3GlcNAc- sequences. Cores 1–4 and core 6 glycans can be elongated with ABO or Lewis blood group antigens. The termini of *O*-GalNAc glycans may be added with fucose, galactose, GlcNAc, GalNAc, sialic acid, or sulfate which leads to *O*-GalNAc diversity in nature.

1.2.2.1.2 *O*-GlcNAc glycans

N-Acetylglucosamine β-linked to serine or threonine in proteins is called protein *O*-β-GlcNAc or *O*-GlcNAc modification. Protein *O*-GlcNAcylation occurs only in the nucleus and cytoplasm, and the GlcNAc is the only moiety attached, which means that no other sugar residues will be added to form complex structures. It was reported initially by Gerald W. Hart and colleagues in 1984 [27]. Many intracellular proteins including transcription factors, cytoskeletal proteins, metabolic enzymes [28], and other proteins have now been found with *O*-GlcNAc-modification [29]. Aberrant *O*-GlcNAcylation has also been related to cancer [30]. Under the competing effects of *O*-GlcNAc transferase (OGT) and *O*-GlcNAcase, *O*-GlcNAc is attached to and removed from the protein several times at different rates. Cells in different phases of cell cycle have various levels of *O*-GlcNAcylation. *O*-GlcNAc glycans have been found in many multicellular organisms such as insects, plants, mammals and fungi, but not in yeast. On the other hand, *O*-α-GlcNAc is common in cell surface and secreted glycoproteins from protozoa, but the existence of *O*-β-GlcNAc is still under investigation [31].

1.2.2.1.3 Other O-linked glycans

In addition to protein *O*-GalNAc glycans and *O*-GlcNAc-modifications, other *O*-glycans linked to the Ser/Thr residues of proteins, including mannose, glucose, galactose, and fucose, have been identified. These are found in glycoproteins present in the nucleus and cytoplasm. In addition to the hydroxyl group at threonine and serine, *O*-glycosylation has also been found at the hydroxyl group of tyrosine, hydroxyproline, and hydroxylysine [5]. Vertebrate and invertebrate collagens have hydroxylysine residues with galactose β-linked to them. Nevertheless, threonine and serine are the primary *O*-glycosylated residues for glycoproteins in the secretory pathway. Other than *O*-GlcNAc modification of threonine or serine residues in cytoplasmic and nucleus *O*-glycosylated proteins, *O*-glycosylation on the tyrosine and hydroxyproline have been identified for cytoplasmic *O*-glycosylated proteins [5, 6]. For example, the cell walls of plants are rich in glycoproteins with hydroxyproline residues that are converted from proline by prolyl hydroxylases in the ER and *O*-glycosylated in the ER and the Golgi [32]. This process is initiated by the attachment of an arabinose or galactose residue, followed by further arabinosylation or galactosylation. These *O*-glycoproteins are involved in signalling, development, cell proliferation, structural role, etc. Another unique example is glycogenin, which is present in dimers, and one subunit of the glycogenin dimer catalyzes the addition of glucose residues to the hydroxyl group at tyrosine of the adjacent subunit to form oligo-glucose primers to allow the downstream of glucogen synthase-catalyzed formation of large polysaccharides of glucose residues ($MW = 1 \times 10^6$ to 2×10^9, glycogens containing α1–4-linked glucan with 10% α–6-branch linkages) on the protein [33].

Proteoglycans are another type of *O*-glycosylated proteins. They are glycoconjugates composed of a protein central core and one or more GAG chains. GAGs are polysaccharides that consist of repeating disaccharide building blocks consisting of an *N*-acetylhexosamine (GalNAc or GlcNAc) and a uronic acid (glucuronic acid (GlcA) or iduronic acid (IdoA)) or a galactose with multiple carbohydrate postglycosylational modifications (modifications on the carbohydrates after the formation of glycosidic bonds) [34]. Other than hyaluronans, or hyaluronic acids (HAs), which exist as free unattached polysaccharides, and keratin sulfates, which are 6-*O*-sulfated polyLacNAc attached to both *N*- or *O*-glycans, GAGs including chondroitin sulfate (CS), dermatan sulfate (DS), keratan sulfate (KS), heparan sulfate (HS), and heparin are all covalently attached to the serine residue in the protein core through a xylose via a tetrasaccharide GlcAβ1–3Galβ1–3Galβ1–4Xylβ-linked to *O*-Ser [7,8]. Some proteoglycans can have both GAGs and more common *O*- and *N*-linked glycans.

1.2.2.2 *O-Linked glycoproteins in bacteria*

O-Glycosylation has been found in both Gram-negative and Gram-positive bacteria. There are OST-dependent and independent pathways. The components of *O*-glycans vary between species and even strains. In OST-dependent pathways, several glycosyltransferases attach the monosaccharides to an undecaprenyl phosphate (Und-P) on the inner face of the plasma membrane. Then the lipid-linked glycan is flipped to the periplasmic face and transferred to the acceptor proteins by an OST. For instance, the Galβ1–4Galα1–3diNAcBac β-structure of *Neisseria gonorrhoeae* pilin is assembled by PglB, PglA and PglE onto Und-P, flipped by PglF to the periplasm and transferred by PglO onto the serine residue of pilin [35]. In an alternative pathway, PglH adds a glucose, rather than two galactose residues to diNAcBac. In OST-independent O-glycosylation, glycosyltransferases add monosaccharides to proteins in the cytoplasm and the resulted glycosylated proteins are transported to the outer membrane or secreted by the flagellum [36]. S-Layer glycoproteins with long glycans of disaccharide to hexasaccharide repeats linked to Ser/Thr or Tyr have been identified from numerous bacteria [37].

1.2.2.3 *O-Linked glycoproteins in archaea*

In addition to *N*-glycosylation, archaeal proteins can also be *O*-glycosylated [38,39]. In the S-layer glycoproteins of both *Halobacterium salinarum* and *H. volcanii*, Thr-rich regions adjacent to the predicted membrane-spanning domain of the protein are modified with galactose-glucose disaccharides [40,41]. The *O*-glycosylation pathway in archaea is underexplored.

1.3 Glycoprotein Functions

Glycoproteins play important biological roles. They can serve as structural components (e.g. in cell wall), lubricants, protective agents [42], transport molecules (carriers), inhibitors, hormones, and enzymes. *N*-Glycans on glycoproteins affect the folding, stability, solubility, function, and the activity of certain proteins [43–46]. Heavily *O*-glycosylated mucins produced by epithelial tissues in most animals form viscous gels serving functions from lubrication to cell signalling to forming chemical barriers [47]. Many glycoproteins have both *N*-glycans and *O*-GalNAc glycans. *N*-Glycans and *O*-GalNAc glycans in eukaryotic glycoproteins may have different glycan-protein linkages and core glycan structures, but often share the same terminal oligosaccharide sequences and are often detected by carbohydrate-binding proteins (especially plant lectins) and carbohydrate-specific antibodies commonly used for functional studies. The terminal sequences of *N*- and *O*-GalNAc-glycans often serve as recognition molecules and sites for cell attachment [48,49].

During the glycoprotein *N*-glycan biosynthesis, the structures of *N*-glycans play a key role in quality control of protein folding in the ER. For example, $Glc_1Man_9GlcNAc_2$ on the glycoproteins allow for good interactions with chaperons calnexin and calreticulin in the ER, which help proper protein folding. Improperly folded $Man_9GlcNAc_2$-containing glycoproteins can be re-glucosylated, catalyzed by UDP-glucose:glycoprotein glucosyltransferase (UGT1 or UGGT) in the Golgi to allow a second chance of the glycoprotein to interact with chaperons to help folding [18]. Correctly folded glycoproteins eventually leave the ER to continue its maturation in the Golgi, while incorrectly folded glycoprotein is degraded [18,50].

N-Glycan structures on glycoproteins can also be used as a mechanism to direct their sorting and transport. For example, modification of soluble acid hydrolases by mannose 6-phosphate (M6P) on their

N-glycans allows their recognition by M6P receptors in the Golgi complex, directing their transport to endosomes and lysosomes [51].

The important roles of glycans on mature proteins are being increasingly recognized. Several examples are described here. *N*-Glycans were shown to enhance glycoprotein folding *in vitro* [52–54]. More recently, the core trisaccharide Manβ1-4GlcNAcβ1-4GlcNAcβ or $ManGlcNAc_2$ was found to be the major component that contributes to the enhanced folding of glycoproteins [44].

The major cationic peroxidase of peanut (CPRx) is a glycoprotein with three *N*-linked glycans at Asn_{60} (*N*-60), Asn_{144} (*N*-144) and Asn_{185} (*N*-185). Site-directed mutagenesis studies showed that mutation of *N*-60 or *N*-144 decreased the specific activity of the protein, mutation of *N*-185 decreased the thermal stability of the enzyme, and mutation of any of the three sites influenced the rate of protein folding [55].

The zona pellucida (ZP), a multi-glycoprotein-containing matrix surrounding the mammalian oocyte, plays an important role in sperm–egg binding and induction of acrosome reaction. In humans, there are four glycoproteins in the ZP matrix designated as ZP1, ZP2, ZP3 and ZP4 (*Zp4* pseudogene in mouse). ZP2 binds to the acrosome-reacted spermatozoa as a secondary sperm receptor. Both ZP3 and ZP4 are involved in sperm–egg binding and mediate induction of the acrosome reaction. Glycosylation is not critical for binding, but it is essential for induction of acrosomal exocytosis. In humans, *N*-linked glycosylation is more critical for the human ZP3/ZP4 mediated induction of acrosomal exocytosis [56]. However, in mouse, instead of *N*-linked glycosylation, *O*-linked glycosylation of ZP3 is critical [57]. *N*- and *O*-Glycans linked to human ZP were found to be predominantly terminated with sialyl Lewis x (sLe^x) tetrasaccharide or sLe^x–Le^x heptasaccharide and are believed to be important for sperm–egg binding [58,59].

Immunoglobulins (Igs) are Y-shaped glycoproteins secreted by B lymphocytes as the major molecules during an adaptive immune response. Both Fab and Fc portions of all immunoglobulin G (IgG) subclasses are *N*-glycosylated, with the Fc region having a single *N*-linked glycosylation site at conserved Asn_{297} (*N*-297) in each heavy chain [60]. The *N*-glycan structures at *N*-297 are diverse, including monosialylation, no disialylation, little bisecting GlcNAc, a high chance of core fucose, and heterogeneity of galactose residues with G0, G1, and G2 forms [61]. The influence of the biological functions of IgGs by their *N*-linked glycoforms in the Fc is well documented. An increase in the number of galactose residues at the terminus of the Fc *N*-glycans enhances C1q binding to the IgG, resulting in an enhanced complement-dependent cytotoxicity (CDC) without affecting the antibody-dependent cellular cytotoxicity (ADCC) activity [62]. Removal of the core fucose enhanced ADCC activity both *in vitro* [63] and *in vivo* [64]. In addition, the galactose content of human Fc *N*-glycans of IgGs reversely correlated with disease progression in rheumatoid arthritis and other autoimmune diseases [65]. The anti-inflammatory activity of intravenous immunoglobulin (IVIG) was influenced by the α2–6-sialic acid residues linked to the terminal galactose at the Fc *N*-glycans [66].

Mucins are high molecular weight glycoproteins with many *O*-GalNAc glycans [67,68]. In normal healthy tissue, short glycan structures such as core 1 T (Galβ1–3GalNAcα-) and Tn (GalNAcα-) epitopes, as well as their sialylated glycans, including sialyl T (sT) and sialyl Tn (sTn) antigens, are almost absent. However, these antigens have been detected in a number of carcinomas. In addition, the presence of T, Tn, sT, and sTn antigens in human cancers correlates with cancer progression, metastasis, and poor clinical prognosis. For example, in human breast cancer cells, expression of sTn on mucin MUC1 has been associated with reduced cell adhesion and increased cell migration [69].

In general, glycoproteins are responsible for a wide diversity of biological functions and represent a promising source for disease biomarkers and molecular targets for drug development.

1.4 Glycoproteins as Therapeutics

The vast majority of the therapeutic proteins (over 70%) are glycoproteins with *N*-glycan and less often with *O*-glycans. Erythropoietin (EPO, Epogen®; Amgen) is a glycoprotein with both *N*- and

O-glycans used for the treatment of chronic kidney disease (CKD) patients on hemodialysis [70]. Glycosylation is a very critical modification of therapeutic proteins, known to significantly modulate yield, bioactivity, solubility, stability against proteolysis, immunogenicity, and clearance rate from circulation [71–73].

Among glycoprotein therapeutics, antibody-based drugs are the largest and fastest growing category and have a wide range of applications, particularly in cancer, immune disorders, and infectious disease [74]. For example, the top five therapeutic proteins in 2010 sales are four mAbs and one antibody (IgG1Fc)-derived fusion protein, namely Etanercept (Enbrel®), bevacizumab (Avastin®), rituximab (Rituxan® and MabThera®), adalimumab (Humira®), and infliximab (Remicade®) [75]. In 2012, an antibody drug, Poteligeo (mogamulizumab), developed by Tokyo-based Kyowa Hakko Kirin, was approved in Japan to treat a form of T-cell lymphoma, a rare cancer. On November 1, 2013, Gazyva (obinutuzumab, GA101; Genentech), a fully humanized monoclonal antibody, directed against CD20-positive B cells to treat chronic lymphocytic leukemia, was approved by the Food and Drug Administration (FDA) in the United States. It was the first glyco-engineered antibody drug to reach the Western market. Compared to Rituxan (rituximab, also from Genentech), the core fucose on the *N*-glycans in the conserved *N*-glycosylation site in Fc is removed in Gazyva, leading to high ADCC and CDC activity [76].

1.5 Glycoprotein Production

Therapeutic glycoproteins have been mainly prepared from human or animal sources [72]. Considering the purity, potential contamination issues, availability, and the rapid increase in demand, the recombinant expression systems such as mammalian cell lines, insects, plants, yeasts and fungi, and bacteria have been developed for producing therapeutic glycoproteins with the assistant of genetic engineering tools. All expression systems have their challenges and may produce glycoproteins with glycan structures differing from those in human native glycoproteins.

1.5.1 *Mammalian expression systems*

Mammalian cell expression systems, especially Chinese Hamster Ovary (CHO) cell lines, are the preferred host to produce glycoproteins harboring human-like complex glycosylation at an industrial scale [72,77]. Currently, the great majority of FDA approved therapeutic glycoproteins and most of the recombinant protein (r-protein) therapeutics under development are produced in CHO cells. Some monoclonal antibodies and Fc-fusion proteins of licensed therapeutics are also produced in mouse myeloma cells (NS0, SP2/0) and hybridomas [78]. Other animal and human cell lines used for production include Baby Hamster Kidney (BHK21), human fibrosarcoma (HT1080), human lymphoma (Namalwa) and Human Embryo Kidney 293 (HEK293) [72]. These cell lines offer several advantages, including low risk for the transmission of the major human viruses, high protein yield, and robustness towards pH, temperature, oxygen level, and pressure variations [79]. They are also able to produce glycoproteins with *N*-glycans similar to those found on human proteins [80]. Nevertheless, CHO cells can present the non-human *N*-glycolylneuraminic acid (Neu5Gc) and the non-human α-Gal epitope (Fig. 1.3) to glycoproteins [81], and such glycoproteins can be cleared by anti-Neu5Gc antibodies and anti-Gal antibodies naturally present in human serum [82,83]. Another drawback is that glycoproteins produced in CHO cells display inherent glycan heterogeneity, resulting in a mixture of molecules with varying efficacy profiles [84]. Even with human cell expression systems, Neu5Gc can still be introduced through cell culture media and cause potential complications [76,85].

In addition to cell lines, transgenic animals have been developed as hosts for glycoprotein production. Several glycoproteins, including monoclonal antibodies, have been expressed [86]. Nevertheless, transgenic animals are also capable of incorporating Neu5Gc into glycan structures and introducing α-Gal epitopes, which are recognized by specific human natural antibodies [85,87].

1.5.2 *Insect cell expression systems*

Insect cell lines are alternative expression hosts for efficient production of glycoproteins [88]. Major insect cell lines used are *Spodoptera frugiperda* SF9 or SF21, and *Trichoplusia ni* BTI 5B1-4 (High Five™). Insect cells are capable of performing post-translational modifications (PTMS), such as *N*-glycosylation. However, recombinant proteins obtained from the insect cells contain glycans that differ from those in human glycoproteins. Insect cell *N*-glycans lack complex-type glycans and contain high mannose and paucimannose glycans [77,89]. Some also contain potentially immunogenic core α1–3-fucose attached to the innermost GlcNAc residue, in contrast to the core α1–6-fucose residue in human (Fig. 1.3) [90]. Furthermore, insect cell *N*-glycans lack galactose and terminal sialic acid residues [91], which are very important sugars for the biological functions of some glycoproteins. Therapeutic glycoprotein drugs require a high degree of sialylation of their *N*-glycans for a better circulatory half-life that results in greater

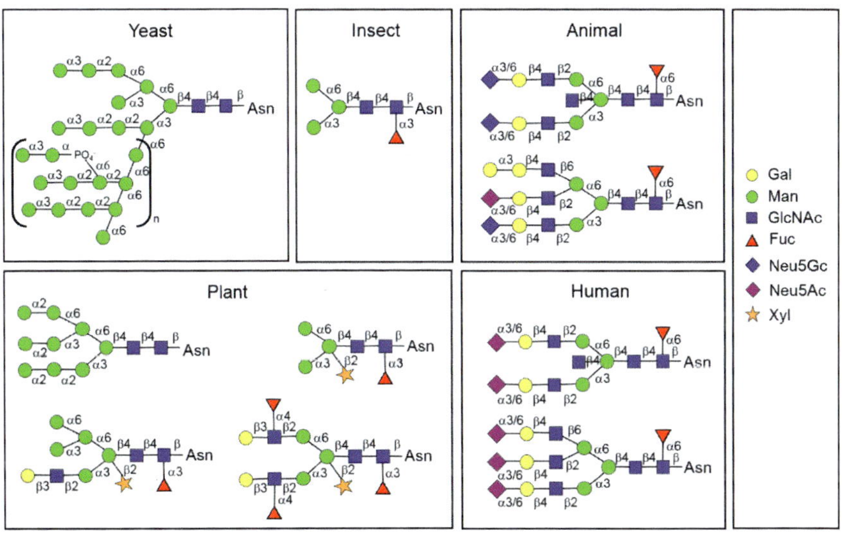

Figure 1.3 Non-human glycan structures in available glycoprotein expression systems [70].

efficacy. Owing to the poor half-life associated with these *N*-glycans, insect cell lines are not suitable for the expression of glycoproteins requiring long serum half-life for the therapeutic purposes. These major limitations may be overcome by glycoengineering approaches to allow improved *N*-glycan processing and to generate human-type complex glycosylation [80].

1.5.3 *Plant-based expression systems*

Plants have been established as alternative hosts to mammalian cell lines for glycoprotein production. Plant-based expression systems have several advantages, such as easy to scale-up to industrial scale with a relatively low cost [92], as well as providing glycosylated and pathogen-free therapeutics. Some edible oral vaccines are produced from the genetically modified plants [93,94]. Plant glycan structures are not always immunogenic and may not hinder FDA-approval and commercialization [95]. However, plant-derived glycoproteins lack the bisecting GlcNAc residue, β1–4-galactose residues, sialic acids, and core α1–6-fucose residues commonly found in human glycoproteins. In contrast, plant glycoproteins may contain potentially immunogenic β1–2-xylose and core α1–3-fucose residues (Fig. 1.3) [96]. The potential allergic responses to these sugars could limit the development of plants for the production of therapeutic glycoproteins. Significant progresses have been achieved in the removal of enzymatic pathways which can generate immunogenic sugars in glycoproteins to produce humanized glycosylation. For example, transgenic plants have been developed to incorporate different components of mammalian glycosylation machinery, allowing the production of multi-antennary and α2–6-sialylated *N*-glycosylation to some extent [92, 97–99]. Currently, a few plant-derived therapeutic proteins have been approved in Europe for topical use in human and several products are in clinical trials, including interferon alpha (Locteron1) produced in Lemna minor and glucocerebrosidase (prGCD) in carrot cells [100]. It is obvious that plant-based expression systems are gaining acceptance to produce therapeutic glycoproteins.

1.5.4 *Yeast expression systems*

Yeasts provide an appealing alternative for the production of recombinant glycoproteins. The yeast expression systems are attractive as they are more robust than mammalian cells and have well-characterized glycosylation machineries. They provide high titers (usually in the g/L ranges) and are easily adaptable to fermentation processes in simple and cost-effective culture media [72,101]. Although humans and yeasts share the same *N*-glycan core structure ($Man_3GlcNAc_2$) (Fig. 1.3), the glycans produced by yeast cells may differ from the native protein and contain *N*-linked high mannose glycans which are immunogenic and have poor pharmacokinetic and pharmacodynamic properties in human. Therefore, the wide type yeast expression systems are limited to produce proteins that do not require *N*-glycosylation for therapeutic efficacy. For example, many approved non-glycosylated therapeutic proteins are currently produced in *S. cerevisiae* and the methylotrophics, *Hansenula polymorpha* and *Pichia pastoris*, including insulin and analogs, growth hormone, hirudin (a leech-derived anticoagulant) and albumin [72]. Other drawbacks of yeasts are the low to absent amounts of fucose and complete lack of terminal sialic acids [102]. In addition, yeasts cannot perform tyrosine *O*-sulfation, PTM that occurs in higher eukaryotes and in mammals [103]. Recently, the yeasts are successfully genetically engineered to produce human-like *N*-linked glycans, in which genes responsible for yeast high mannose glycans, e.g. *och1*, are disrupted and a series of glycosidases and glycosyltransferases are introduced. For example, although the first fully functional IgG was reported in 1999, use of the *P. pastoris* expression system to produce IgG with human-like *N*-linked glycans was not successful until recently [104]. The yeast *S. cerevisiae* was also successfully genetically modified to provide protein *O*-fucosylation capabilities [105], a key modification essential for the bioactivity of many epidermal growth factor (EGF) domain-containing proteins, such as Notch and urokinase-type plasminogen activator [106].

1.5.5 *Bacterial expression systems*

Glycoproteins containing human glycans may be produced by a combined method involving the engineering and functional transfer of the required glycosylation machinery into bacterial systems to express glycosylated proteins followed by *in vitro* glycan modification [107]. For example, glycoproteins containing a *C. jejuni* heptasaccharide were obtained from the *C. jejuni* glycosylation machinery transferred into recombinant *Escherichia coli* cells. The bacterial glycans were then trimmed and remodelled *in vitro* by enzymatic transglycosylation [108]. There are many advantages of using well-understood bacterial systems such as *E. coli* for producing human glycoproteins. The published genome sequence and the well-studied metabolic pathways make it easy to be genetically modified [109]. The high-titer production, the shorter fermentation time, and the reduced viral contamination issues are also vital advantages. Moreover, bacteria are less sensitive to glycosylation changes; that is, the bacteria do not die upon glycosylation [110]. Therefore, bacteria would be ideal hosts compared to eukaryotes where hundreds of enzymes are involved. However, many issues may have to be solved to make the *in vivo* glycoengineering approach successful; for example, whether the yields of humanized glycoproteins produced from bacteria can compete with those obtained from engineered eukaryotic systems. The specificity of the glycosyltransferases may change when they are expressed in bacterial systems. Moreover, due to the fact that some glycosyltransferases physically interacting with each other *in vivo* to efficiently produce oligo- and polysaccharides [111], some glycosyltransferases expressed in bacteria may show impaired activity or specificity in the absence of their partner enzymes [107]. Therefore, additional improvements are needed to establish a cost-efficient, reliable system to produce humanized glycoproteins in *E. coli* or other bacteria.

1.6 Advances on Glycan Engineering of Glycoproteins

Significant progress has been made over the past decade to overcome the current limitations of non-mammalian expression systems to achieve the production of human-like glycosylation patterns [72,80]. Several glycoengineering approaches have been made to reduce or eliminate the contamination of the non-human glycan epitopes produced by the non-human mammalian cells, including an anti-sense ribonucleic acid (RNA) strategy [112], changing of culture conditions [113], and feeding CHO cells with the human sialic acid, Neu5Ac [85]. There are an increasing number of therapeutic glycoproteins with modified glycans to alter their therapeutic properties, such as the licensed products darbepoetin a (Aranesp®), a variant of erythropoietin, and the enzyme Cerezyme® (imiglucerase) [114]. In addition, the glycan-defined glycoproteins are urgently needed in both structure–function relationship studies and therapeutic applications.

Three approaches have been developed to engineer glycosylation, including alteration of glycosylation sites on proteins, enzymatic and chemoenzymatic modification of glycans *in vitro*, and manipulation of cellular glycosylation machinery [115].

Alteration of glycosylation sites on proteins is mainly done by inserting *N*-glycosylation consensus sequences (Asn-X-Ser/Thr) to various positions of proteins to introduce more *N*-glycosylation sites. For example, adding extra *N*-linked glycosylation sites to recombinant erythropoietin (EPO; darbepoietin α) by glycoengineering increased its *in vivo* activity and prolonged its duration of action [116]. Site-specific glycosylation of recombinant proteins can also be obtained by chemoselective ligation between bio-orthogonally tagged proteins and glycans [117,118]. The free cysteine residue in the expressed protein as a tag can be selectively modified with a thio-reactive functional group, which is pre-attached in a sugar moiety to perform a site-specific glyco-conjugation, in which cysteine can be introduced by site-directed mutagenesis. This method was successfully used to selectively introduce sugar chains at the conserved *N*-glycosylation sites (Asn-297) of human IgG-Fc to test the effects of the glycosylation on the functions of antibody [119]. In addition to the cysteine residue, a series of novel functionalized unnatural amino acid residues with a bioorthogonal chemical handle, such as those with an azide-, an aldehyde-, an alkyne-, or an alkene group, can also be introduced to proteins through genetic manipulation. The method has significantly expanded the scope and diversity of the "tag and modify" strategy [115].

Chemoenzymatic glycosylation of natural and recombinant glycoproteins *in vitro* provides an attractive approach to produce glycan-defined glycoforms. For example, coagulation factor IX produced by CHO cells carries glycans with α2–3-linked sialic acid. However, human plasma-derived coagulation factor IX has α2–6-linked sialic acids as the terminal residues. Using a strategy of desialylation followed by resialylation with specific sialyltransferases, such as Pd2–6ST [120], the CHO-cell derived sialylation pattern was converted into a human-like sialylation pattern *in vitro* [121]. Another example of this approach is the glycan remodelling of ribonuclease B (RNase B) to a homogeneous glycoform with an *N*-linked sLe^x moiety. RNase B with high-mannose *N*-glycans was treated with Endoglycosidase H (Endo-H) to form GlcNAc-RNase B. The galactose, sialic acid, and fucose were then subsequently added in sequence by glycosylation reactions catalyzed by β1–4-galactosyltransferase, α2–3-sialyltransferase, and α1–3-fucosyltransferase, respectively, to produce a ribonuclease with a novel glycoform [122]. An alternative way to sequential sugar chain extension is the *en bloc* transfer of pre-assembled large oligosaccharides to the protein in a single step by endo-β-*N*-acetylglucosaminidase (ENGase) or mutants-catalyzed transglycosylation [115]. Site-directed enzymatic *O*-glycoPEGylation of nonglycosylated polypeptides produced by *E. coli* using glycosyltransferases has also been developed. The process includes enzymatic GalNAc glycosylation at specific Ser/Thr residues in proteins followed by enzymatic transfer of sialic acid conjugated polyethylene glycol (PEG) to the GalNAc residues [123]. The strategy has been used for modifying important clinically used glycoprotein drugs, including granulocyte colony stimulating factor (G-CSF), interferon-α2b (IFN-α2b), and granulocyte/macrophage colony stimulating factor (GM-CSF). Recently, *O*-glycoPEGylated coagulation factor VIII (FVIII) (a therapeutic for treating hemophilia A, an inherited bleeding disorder caused by deficiency or dysfunction of FVIII) was shown to have the same efficacy but prolonged effect in animal models compared to the native FVIII [124]. More recently, adding polysaccharide to the *N*-glycans of therapeutic proteins has been shown to be an effective approach to increase the serum half-life of the glycoproteins [125].

Manipulation of cellular glycosylation machinery has also been developed. Knockout mutagenesis and overexpression of certain glycoprocessing enzymes can redirect glycosylation. For example, in order to produce therapeutic IgG with human-like *N*-linked glycans in yeast, the *N*-glycan biosynthetic pathway in *P. pastoris* was glycoengineered to mimic human *N*-glycan synthesis by eliminating the genes for producing yeast high-mannose glycans and introducing a series of glycosidases and glycosyltransferases, such as α1–2-mannosidase I, mannosidase II, β1–2-*N*-acetylglucosaminyltransferase I (GnT-I), β1–2-*N*-acetylglucosaminyltransferase II (GnT-II), and β1–4-galactosyltransferase for producing human *N*-glycans [126–128]. Using specific inhibitors to block selected enzymes in the biosynthesis pathway can also generate simplified and/or more uniformed glycoforms. For example, *N*-butyl deoxynojirimycin inhibits the activities of α-glucosidases I and II in the ER, resulting in the glycoprotein carrying the full-length *N*-glycan precursor ($Glc_3Man_9GlcNAc_2$-protein) [115]. Kifunensine has been used successfully as an α-mannosidase I inhibitor for producing *N*-glycosylated proteins with high-mannose type *N*-glycans [129,130]. Moreover, overexpression of certain enzymes in the host system can also change glycosylation profiles and enrich the production of desired glycoforms. CHO cells transfected with human α2–6-sialyltransferase complementary deoxyribonucleic acid (cDNA) increases the sialylation of recombinant glycoproteins, such as interferon-γ (IFN-γ) [131].

1.7 Conclusions and Perspectives

Glycoproteins are proteins containing one or more covalently linked glycan chains, which exist widely in eukaryotic systems, as well as in some bacteria and archaea. More than 20 different types of sugar-amino acid linkages are known and new linkages continue to emerge. Other than the essential roles of *N*-glycans in quality control, folding, and sorting of eukaryotic glycoproteins, their importance in influencing the stability, solubility, function, and the activity of glycoproteins is increasingly realized. *O*-Linked glycoproteins in eukaryotic systems are mainly of *O*-GalNAc mucin-type. Unlike *N*-glycans in eukaryotic systems, the structures of *O*-glycans are more diverse, more challenging for analysis, and are less understood. Despite their differences in the core structures, *N*- and *O*-glycans in eukaryotes often share similar terminal glycan structures and are common recognition molecules that are essential for molecular interactions. Glycoproteins are major therapeutic proteins. Establishing cost-efficient and robust systems to produce humanized therapeutic glycoproteins with improved functions and homogenous forms at industrial scale is urgently needed. Technologies in glycoprotein production and analysis, chemoenzymatic synthesis, genetic manipulation, and chemical and biological approaches are advancing. This will lead to a greater availability of glycoproteins with homogenous, structurally-defined, and tailor-made glycans. This will also lead to a better understanding of the roles of glycoproteins and specific functions of glycans on glycoproteins at the molecular level. We expect to see continuous emerging of new generation glycoprotein therapeutics with increased stability and improved efficacy.

Acknowledgments

The authors appreciate the funding supports from the National Science Foundation grant CHE-1300449 and National Institutes of Health grants R01HD065122 and R01GM094523. X.C. is a Camille Dreyfus Teacher-Scholar and a UC-Davis Chancellor's Fellow.

References

1. Neuberger, A. (1938) *Biochem. J.* **32**, 1435–1451.
2. Dell, A., Galadari, A., Sastre, F., Hitchen, P. (2010) *Int. J. Microbiol.* **2010**, 148178–148191.

3. Kowarik, M., Young, N. M., Numao, S., Schulz, B. L., Hug, I., Callewaert, N., Mills, D. C., Watson, D. C., Hernandez, M., Kelly, J. F., Wacker, M., Aebi, M. (2006) *EMBO J.* **25**, 1957–1966.
4. Yurist-Doutsch, S., Chaban, B., VanDyke, D. J., Jarrell, K. F., Eichler, J. (2008) *Mol. Microbiol.* **68**, 1079–1084.
5. Hart, G. W., West, C. M. (2009) In *Essentials of Glycobiology*, 2nd ed. Varki, A., Cummings, R. D., Esko, J. D., Freeze, H. H., Stanley, P., Bertozzi, C. R., Hart, G. W., Etzler, M. E. Eds. Cold Spring Harbor (New York). pp. 249–262.
6. Wells, L., Vosseller, K., Hart, G. W. (2001) *Science* **291**, 2376–2378.
7. Esko, J. D., Kimata, K., Lindahl, U. (2009) In *Essentials of Glycobiology*, 2nd ed. Varki, A., Cummings, R. D., Esko, J. D., Freeze, H. H., Stanley, P., Bertozzi, C. R., Hart, G. W., Etzler, M. E. Eds. Cold Spring Harbor (New York). pp. 229–248.
8. Sasisekharan, R., Raman, R., Prabhakar, V. (2006) *Annu. Rev. Biomed. Eng.* **8**, 181–231.
9. Esko, J. D., Doering, T. L., Raetz, C. R. H. (2009) In *Essentials of Glycobiology*, 2nd ed. Varki, A., Cummings, R. D., Esko, J. D., Freeze, H. H., Stanley, P., Bertozzi, C. R., Hart, G. W., Etzler, M. E. Eds. Cold Spring Harbor (New York). pp. 293–307.
10. Schaffer, C., Messner, P. (2001) *Biochimie* **83**, 591–599.
11. Spiro, R. G. (2002) *Glycobiology* **12**, 43R–56R.
12. Oman, T. J., Boettcher, J. M., Wang, H., Okalibe, X. N., van der Donk, W. A. (2011) *Nat. Chem. Biol.* **7**, 78–80.
13. Wang, H., Oman, T. J., Zhang, R., Garcia De Gonzalo, C. V., Zhang, Q., van der Donk, W. A. (2014) *J. Am. Chem. Soc.* **136**, 84–87.
14. Apweiler, R., Hermjakob, H., Sharon, N. (1999) *Biochim. Biophys. Acta* **1473**, 4–8.
15. Varki, A., Lowe, J. B. (2009) In *Essentials of Glycobiology*, 2nd ed. Varki, A., Cummings, R. D., Esko, J. D., Freeze, H. H., Stanley, P., Bertozzi, C. R., Hart, G. W., Etzler, M. E. Eds. Cold Spring Harbor (New York). pp. 75–88.
16. Schwarz, F., Aebi, M. (2011) *Curr. Opin. Struct. Biol.* **21**, 576–582.
17. Stanley, P., Schachter, H., Taniguchi, N. (2009) In *Essentials of Glycobiology*, 2nd ed. Varki, A., Cummings, R. D., Esko, J. D., Freeze, H. H., Stanley, P., Bertozzi, C. R., Hart, G. W., Etzler, M. E. Eds. Cold Spring Harbor (New York). pp. 101–114.
18. Aebi, M., Bernasconi, R., Clerc, S., Molinari, M. (2010) *Trends Biochem. Sci.* **35**, 74–82.

19. Stanley, P., Schachter, H., Taniguchi, N. (2009) In *Essentials of Glycobiology*, 2nd ed. Varki, A., Cummings, R. D., Esko, J. D., Freeze, H. H., Stanley, P., Bertozzi, C. R., Hart, G. W., Etzler, M. E. Eds. Cold Spring Harbor (New York). pp. 101–114.
20. Young, N. M., Brisson, J.-R., Kelly, J., Watson, D. C., Tessier, L., Lanthier, P. H., Jarrell, H. C., Cadotte, N., Michael, F. S., Aberg, E. (2002) *J. Biol. Chem.* **277**, 42530–42539.
21. Gross, J., Grass, S., Davis, A. E., Gilmore-Erdmann, P., Townsend, R. R., Geme, J. W. S. (2008) *J. Biol. Chem.* **283**, 26010–26015.
22. Dell, A., Galadari, A., Sastre, F., Hitchen, P. (2010) *Int. J. Microbiol.* **2010**, 148178.
23. Eichler, J. (2013) *Nat. Rev. Microbiol.* **11**, 151–156.
24. Marino, K., Bones, J., Kattla, J. J., Rudd, P. M. (2010 *Nat. Chem. Biol.* **6**, 713–723.
25. Dell, A., Galadari, A., Sastre, F., Hitchen, P. (2010) *Int. J. Microbiol.* **2010**, 148178.
26. Brockhausen, I., Schachter, H., Stanley, P. (2009) In *Essentials of Glycobiology*, 2nd ed. Varki, A., Cummings, R. D., Esko, J. D., Freeze, H. H., Stanley, P., Bertozzi, C. R., Hart, G. W., Etzler, M. E. Eds. Cold Spring Harbor (New York). pp. 115–128.
27. Torres, C. R., Hart, G. W. (1984) *J. Biol. Chem.* **259**, 3308–3317.
28. Yi, W., Clark, P. M., Mason, D. E., Keenan, M. C., Hill, C., Goddard, W. A., 3rd, Peters, E. C., Driggers, E. M., Hsieh-Wilson, L. C. (2012) *Science* **337**, 975–980.
29. Comer, F. I., Hart, G. W. (2000) *J. Biol. Chem.* **275**, 29179–29182.
30. Slawson, C., Hart, G. W. (2011) *Nat. Rev. Cancer* **11**, 678–684.
31. Hart, G. W., Akimoto, Y. (2009) In *Essentials of Glycobiology*, 2nd ed. Varki, A., Cummings, R. D., Esko, J. D., Freeze, H. H., Stanley, P., Bertozzi, C. R., Hart, G. W., Etzler, M. E. Eds. Cold Spring Harbor (New York). pp. 263–279.
32. Sommer-Knudsen, J., Bacic, A., Clarke, A. E. (1998) *Phytochemistry* **47**, 483–497.
33. Lomako, J., Lomako, W. M., Whelan, W. J. (2004) *Biochim. Biophys. Acta* **1673**, 45–55.
34. Yu, H., Chen, X. (2007) *Org. Biomol. Chem.* **5**, 865–872.
35. Hartley, M. D., Morrison, M. J., Aas, F. E., Børud, B., Koomey, M., Imperiali, B. (2011) *Biochemistry* **50**, 4936–4948.
36. Iwashkiw, J. A., Vozza, N. F., Kinsella, R. L., Feldman, M. F. (2013) *Mol. Microbiol.* Vol. 89: 14–28.

37. Messner, P., Steiner, K., Zarschler, K., Schaffer, C. (2008) *Carbohydr. Res.* **343**, 1934–1951.
38. Abu-Qarn, M., Eichler, J., Sharon, N. (2008) *Curr. Opin. Struct. Biol.* **18**, 544–550.
39. Calo, D., Kaminski, L., Eichler, J. (2010) *Glycobiology* **20**, 1065–1076.
40. Mescher, M. F., Hansen, U., Strominger, J. L. (1976) *J. Biol. Chem.* **251**, 7289–7294.
41. Sumper, M., Berg, E., Mengele, R., Strobel, I. (1990) *J. Bacteriol.* **172**, 7111–7118.
42. Kundra, R., Kornfeld, S. (1999) *J. Biol. Chem.* **274**, 31039–31046.
43. Shental-Bechor, D., Levy, Y. (2008) *Proc. Natl. Acad. Sci. USA* **105**, 8256–8261.
44. Hanson, S. R., Culyba, E. K., Hsu, T. L., Wong, C. H., Kelly, J. W., Powers, E. T. (2009) *Proc. Natl. Acad. Sci. USA* **106**, 3131–3136.
45. Culyba, E. K., Price, J. L., Hanson, S. R., Dhar, A., Wong, C. H., Gruebele, M., Powers, E. T., Kelly, J. W. (2011) *Science* **331**, 571–575.
46. Skropeta, D. (2009) *Biorg. Med. Chem.* **17**, 2645–2653.
47. Marin, F., Luquet, G., Marie, B., Medakovic, D. (2008) *Curr. Top. Dev. Biol.* **80**, 209–276.
48. Spiro, R. G. (1963) *New Engl. J. Med.* **269**, 616–621.
49. Murray, R. K., Granner, D. K., Rodwell, V. W. (2006) *Harper's Illustrated Biochemistry* 27th ed. McGraw-Hill (New York).
50. Helenius, A., Aebi, M. (2004) *Annu. Rev. Biochem.* **73**, 1019–1049.
51. Helenius, A., Aebi, M. (2001) *Science* **291**, 2364–2369.
52. O'Connor, S. E., Pohlmann, J., Imperiali, B., Saskiawan, I., Yamamoto, K. (2001) *J. Am. Chem. Soc.* **123**, 6187–6188.
53. Wormald, M. R., Dwek, R. A. (1999) *Structure* **7**, R155–160.
54. Mitra, N., Sinha, S., Ramya, T. N., Surolia, A. (2006) *Trends Biochem. Sci.* **31**, 156–163.
55. Lige, B., Ma, S., van Huystee, R. B. (2001) *Arch. Biochem. Biophys.* **386**, 17–24.
56. Gupta, S. K., Bansal, P., Ganguly, A., Bhandari, B., Chakrabarti, K. (2009) *J. Reprod. Immunol.* **83**, 50–55.
57. Wassarman, P. M., Litscher, E. S. (2008) *Int. J. Dev. Biol.* **52**, 665–676.
58. Pang, P. C., Chiu, P. C., Lee, C. L., Chang, L. Y., Panico, M., Morris, H. R., Haslam, S. M., Khoo, K. H., Clark, G. F., Yeung, W. S., Dell, A. (2011) *Science* **333**, 1761–1764.
59. Clark, G. F. (2013) *Hum. Reprod.* **28**, 566–577.

60. Xue, J., Zhu, L. P., Wei, Q. (2013) *Glycoconjugate J.* **30**, 735–745.
61. Dwek, R. A., Lellouch, A. C., Wormald, M. R. (1995) *J. Anat.* **187 (Pt 2)**, 279–292.
62. Hodoniczky, J., Zheng, Y. Z., James, D. C. (2005) *Biotechnol. Prog.* **21**, 1644–1652.
63. Shields, R. L., Lai, J., Keck, R., O'Connell, L. Y., Hong, K., Meng, Y. G., Weikert, S. H., Presta, L. G. (2002) *J. Biol. Chem.* **277**, 26733–26740.
64. Nimmerjahn, F., Ravetch, J. V. (2005) *Science* **310**, 1510–1512.
65. Parekh, R. B., Dwek, R. A., Sutton, B. J., Fernandes, D. L., Leung, A., Stanworth, D., Rademacher, T. W., Mizuochi, T., Taniguchi, T., Matsuta, K., Takeuchi, F., Nagano, Y., Miyamoto, T., Kobata, A. (1985) *Nature* **316**, 452–457.
66. Samuelsson, A., Towers, T. L., Ravetch, J. V. (2001) *Science* **291**, 484–486.
67. Kannagi, R. (1997) *Glycoconjugate J.* **14**, 577–584.
68. Hollingsworth, M. A., Swanson, B. J. (2004) *Nat. Rev. Cancer* **4**, 45–60.
69. Julien, S., Adriaenssens, E., Ottenberg, K., Furlan, A., Courtand, G., Vercoutter-Edouart, A. S., Hanisch, F. G., Delannoy, P., Le Bourhis, X. (2006) *Glycobiology* **16**, 54–64.
70. Jelkmann, W. (2013) *Transfus. Med. Hemother.* **40**, 302–309.
71. Arnold, J. N., Wormald, M. R., Sim, R. B., Rudd, P. M., Dwek, R. A. (2007) *Annu. Rev. Immunol.* **25**, 21–50.
72. Durocher, Y., Butler, M. (2009) *Curr. Opin. Biotechnol.* **20**, 700–707.
73. Higgins, E. (2010) *Glycoconjugate J.* **27**, 211–225.
74. Reichert, J. M. (2010) *mAbs* **2**, 695–700.
75. Dimitrov, D. S. (2012) *Methods Mol. Biol.* **899**, 1–26.
76. Ratner, M. (2014) *Nat. Biotechnol.* **32**, 6–7.
77. Ghaderi, D., Zhang, M., Hurtado-Ziola, N., Varki, A. (2012) *Biotechnol. Genet. Eng. Rev.* **28**, 147–175.
78. Beck, A., Wagner-Rousset, E., Bussat, M. C., Lokteff, M., Klinguer-Hamour, C., Haeuw, J. F., Goetsch, L., Wurch, T., Van Dorsselaer, A., Corvaia, N. (2008) *Curr. Pharm. Biotechnol.* **9**, 482–501.
79. Wurm, F. M. (2007) *Kidney Blood Press. Res.* **30 Suppl 1**, 6–8.
80. Jacobs, P. P., Callewaert, N. (2009) *Curr. Mol. Med.* **9**, 774–800.
81. Bosques, C. J., Collins, B. E., Meador, J. W., 3rd, Sarvaiya, H., Murphy, J. L., Dellorusso, G., Bulik, D. A., Hsu, I. H., Washburn, N., Sipsey, S. F., Myette, J. R., Raman, R., Shriver, Z., Sasisekharan, R., Venkataraman, G. (2010) *Nat. Biotechnol.* **28**, 1153–1156.
82. Zhu, A., Hurst, R. (2002) *Xenotransplantation* **9**, 376–381.
83. Galili, U. (2013) *Immunology* **140**, 1–11.

84. Sethuraman, N., Stadheim, T. A. (2006) *Curr. Opin. Biotechnol.* **17**, 341–346.
85. Ghaderi, D., Taylor, R. E., Padler-Karavani, V., Diaz, S., Varki, A. (2010) *Nat. Biotechnol.* **28**, 863–867.
86. Houdebine, L. M. (2002) *Curr. Opin. Biotechnol.* **13**, 625–629.
87. Tangvoranuntakul, P., Gagneux, P., Diaz, S., Bardor, M., Varki, N., Varki, A., Muchmore, E. (2003) *Proc. Natl. Acad. Sci. USA* **100**, 12045–12050.
88. Kost, T. A., Condreay, J. P., Jarvis, D. L. (2005) *Nat. Biotechnol.* **23**, 567–575.
89. Kim, Y. K., Shin, H. S., Tomiya, N., Lee, Y. C., Betenbaugh, M. J., Cha, H. J. (2005) *Biotechnol. Bioeng.* **92**, 452–461.
90. Staudacher, E., Altmann, F., Wilson, I. B., Marz, L. (1999) *Biochim. Biophys. Acta* **1473**, 216–236.
91. Hillar, A., Jarvis, D. L. (2010) *Glycobiology* **20**, 1323–1330.
92. Gomord, V., Fitchette, A. C., Menu-Bouaouiche, L., Saint-Jore-Dupas, C., Plasson, C., Michaud, D., Faye, L. (2010) *Plant Biotechnol. J.* **8**, 564–587.
93. Streatfield, S. J. (2005) *Expert Rev. Vaccines* **4**, 591–601.
94. Davoodi-Semiromi, A., Samson, N., Daniell, H. (2009) *Hum. Vaccin.* **5**, 488–493.
95. Boothe, J., Nykiforuk, C., Shen, Y., Zaplachinski, S., Szarka, S., Kuhlman, P., Murray, E., Morck, D., Moloney, M. M. (2010) *Plant Biotechnol. J.* **8**, 588–606.
96. Gomord, V., Faye, L. (2004) *Curr. Opin. Plant Biol.* **7**, 171–181.
97. Castilho, A., Strasser, R., Stadlmann, J., Grass, J., Jez, J., Gattinger, P., Kunert, R., Quendler, H., Pabst, M., Leonard, R., Altmann, F., Steinkellner, H. (2010) *J. Biol. Chem.* **285**, 15923–15930.
98. Castilho, A., Gattinger, P., Grass, J., Jez, J., Pabst, M., Altmann, F., Gorfer, M., Strasser, R., Steinkellner, H. (2011) *Glycobiology* **21**, 813–823.
99. Nagels, B., Van Damme, E. J., Pabst, M., Callewaert, N., Weterings, K. (2011) *Plant Physiol.* **155**, 1103–1112.
100. Karg, S. R., Kallio, P. T. (2009) *Biotechnol. Adv.* **27**, 879–894.
101. Costa, A. R., Rodrigues, M. E., Henriques, M., Oliveira, R., Azeredo, J. (2013) *Crit. Rev. Biotechnol.* **34**, 77–88.
102. Cummings, R. D., Doering, T. L. (2009) In *Essentials of Glycobiology*, 2nd ed. Varki, A., Cummings, R. D., Esko, J. D., Freeze, H. H., Stanley, P., Bertozzi, C. R., Hart, G. W., Etzler, M. E. Eds. Cold Spring Harbor (New York). pp. 309–320.

103. Moore, K. L. (2003) *J. Biol. Chem.* **278**, 24243–24246.
104. Ha, S., Wang, Y., Rustandi, R. R. (2011) *mAbs* **3**, 453–460.
105. Chigira, Y., Oka, T., Okajima, T., Jigami, Y. (2008) *Glycobiology* **18**, 303–314.
106. Okajima, T., Matsuura, A., Matsuda, T. (2008) *J. Biochem.* **144**, 1–6.
107. Hug, I., Feldman, M. F. (2011) *Glycobiology* **21**, 138–151.
108. Schwarz, F., Huang, W., Li, C., Schulz, B. L., Lizak, C., Palumbo, A., Numao, S., Neri, D., Aebi, M., Wang, L. X. (2010) *Nat. Chem. Biol.* **6**, 264–266.
109. Baneyx, F. (1999) *Curr. Opin. Biotechnol.* **10**, 411–421.
110. Pandhal, J., Wright, P. C. (2010) *Biotechnol. Lett.* **32**, 1189–1198.
111. Kos, V., Whitfield, C. (2010) *J. Biol. Chem.* **285**, 19668–19687.
112. Chenu, S., Gregoire, A., Malykh, Y., Visvikis, A., Monaco, L., Shaw, L., Schauer, R., Marc, A., Goergen, J. L. (2003) *Biochim. Biophys. Acta* **1622**, 133–144.
113. Borys, M. C., Dalal, N. G., Abu-Absi, N. R., Khattak, S. F., Jing, Y., Xing, Z., Li, Z. J. (2010) *Biotechnol. Bioeng.* **105**, 1048–1057.
114. Birch, J. R., Onakunle, Y. (2005) *Methods Mol. Biol.* **308**, 1–16.
115. Wang, L. X., Lomino, J. V. (2012) *ACS Chem. Biol.* **7**, 110–122.
116. Elliott, S., Lorenzini, T., Asher, S., Aoki, K., Brankow, D., Buck, L., Busse, L., Chang, D., Fuller, J., Grant, J., Hernday, N., Hokum, M., Hu, S., Knudten, A., Levin, N., Komorowski, R., Martin, F., Navarro, R., Osslund, T., Rogers, G., Rogers, N., Trail, G., Egrie, J. (2003) *Nat. Biotechnol.* **21**, 414–421.
117. Gamblin, D. P., Scanlan, E. M., Davis, B. G. (2009) *Chem. Rev.* **109**, 131–163.
118. Chalker, J. M., Bernardes, G. J., Davis, B. G. (2011) *Acc. Chem. Res.* **44**, 730–741.
119. Watt, G. M., Lund, J., Levens, M., Kolli, V. S., Jefferis, R., Boons, G. J. (2003) *Chem. Biol.* **10**, 807–814.
120. Sun, M., Li, Y., Chokhawala, H. A., Henning, R., Chen, X. (2008) *Biotechnol. Lett.* **30**, 671–676.
121. Fischer, B. E., Dorner, F. (1998) *Thromb. Res.* **89**, 147–150.
122. Witte, K., Sears, P., Martin, R., Wong, C. H. (1997) *J. Am. Chem. Soc.* **119**, 2114–2118.
123. DeFrees, S., Wang, Z. G., Xing, R., Scott, A. E., Wang, J., Zopf, D., Gouty, D. L., Sjoberg, E. R., Panneerselvam, K., Brinkman-Van der Linden, E. C., Bayer, R. J., Tarp, M. A., Clausen, H. (2006) *Glycobiology* **16**, 833–843.

124. Stennicke, H. R., Kjalke, M., Karpf, D. M., Balling, K. W., Johansen, P. B., Elm, T., Ovlisen, K., Moller, F., Holmberg, H. L., Gudme, C. N., Persson, E., Hilden, I., Pelzer, H., Rahbek-Nielsen, H., Jespersgaard, C., Bogsnes, A., Pedersen, A. A., Kristensen, A. K., Peschke, B., Kappers, W., Rode, F., Thim, L., Tranholm, M., Ezban, M., Olsen, E. H., Bjorn, S. E. (2013) *Blood* **121**, 2108–2116.
125. Lindhout, T., Iqbal, U., Willis, L. M., Reid, A. N., Li, J., Liu, X., Moreno, M., Wakarchuk, W. W. (2011) *Proc. Natl. Acad. Sci. USA* **108**, 7397–7402.
126. Hamilton, S. R., Davidson, R. C., Sethuraman, N., Nett, J. H., Jiang, Y., Rios, S., Bobrowicz, P., Stadheim, T. A., Li, H., Choi, B. K., Hopkins, D., Wischnewski, H., Roser, J., Mitchell, T., Strawbridge, R. R., Hoopes, J., Wildt, S., Gerngross, T. U. (2006) *Science* **313**, 1441–1443.
127. Li, H., Sethuraman, N., Stadheim, T. A., Zha, D., Prinz, B., Ballew, N., Bobrowicz, P., Choi, B. K., Cook, W. J., Cukan, M., Houston-Cummings, N. R., Davidson, R., Gong, B., Hamilton, S. R., Hoopes, J. P., Jiang, Y., Kim, N., Mansfield, R., Nett, J. H., Rios, S., Strawbridge, R., Wildt, S., Gerngross, T. U. (2006) *Nat. Biotechnol.* **24**, 210–215.
128. Jacobs, P. P., Geysens, S., Vervecken, W., Contreras, R., Callewaert, N. (2009) *Nat. Protoc.* **4**, 58–70.
129. Zhou, Q., Shankara, S., Roy, A., Qiu, H., Estes, S., McVie-Wylie, A., Culm-Merdek, K., Park, A., Pan, C., Edmunds, T. (2008) *Biotechnol. Bioeng.* **99**, 652–665.
130. Chang, V. T., Crispin, M., Aricescu, A. R., Harvey, D. J., Nettleship, J. E., Fennelly, J. A., Yu, C., Boles, K. S., Evans, E. J., Stuart, D. I., Dwek, R. A., Jones, E. Y., Owens, R. J., Davis, S. J. (2007) *Structure* **15**, 267–273.
131. Bragonzi, A., Distefano, G., Buckberry, L. D., Acerbis, G., Foglieni, C., Lamotte, D., Campi, G., Marc, A., Soria, M. R., Jenkins, N., Monaco, L. (2000) *Biochim. Biophys. Acta* **1474**, 273–282.

Chapter 2
Gangliosphingolipids: Structure and Biological Roles

Sophie Groux-Degroote, Yann Guérardel and Philippe Delannoy

Structural and Functional Glycobiology Unit,
University of Sciences and Technologies of Lille, France

2.1 Introduction

Glycosphingolipids (GSLs) are a type of glycolipids composed of a carbohydrate chain linked to a ceramide lipid moiety ubiquitously expressed in animal cells. They are essential compounds of the plasma membranes where they interact with other sphingolipids and cholesterol, forming lipid rafts. Within GSLs, sialylated species, also known as gangliosides, were demonstrated to be central molecules involved in cell adhesion and recognition processes, in the modulation of signal transduction pathways, and are also well-characterized receptors for toxins and pathogens. These different functions are mainly supported by the glycan moiety and changes in the structure of gangliosides may occur under pathological conditions including cancers [1].

2.2 Structure and Analysis of Gangliosides

Gangliosphingolipids, also called gangliosides, are a complex family of GSLs characterized by the presence of one or several sialic acid

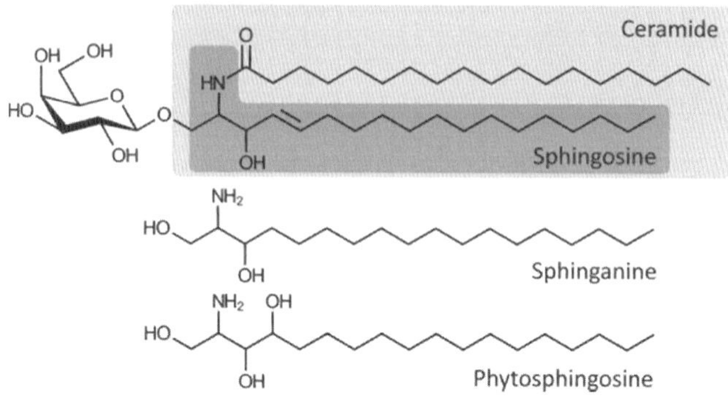

Figure 2.1 General structure of galactosylceramides with different sphingoid bases.

residues in their carbohydrate moieties. The lipid moiety of vertebrate GSLs consists of either a sphingoid or a ceramide, which is a sphingoid base linked to a fatty acid through an amide bond. Sphingoids are long-chain 2-amino-hydroxy-alkans. The most abundant representative of sphingoid bases in vertebrates is (2S, 3R, 4E)-2-amino-4-octadecen-1,3-diol, also called 4E-sphingenine or, in short, sphingosine (Fig. 2.1). Structural variations of sphingosine may also occur through differential hydroxylation or saturation of the alkane chain, as observed in sphinganine (dihydrosphingosine) or phytosphyngosine (4-hydroxysphinganine). A simplified nomenclature similar to that of fatty acids is used for sphingoid bases. It takes into account the number of carbons, the number of hydroxyl groups ('d' for two and 't' for three), and the number of insaturations, such as in d18:1 for sphingosine and t18:0 for phytosphingosine. Altogether, the nomenclature of ceramide part-juxtaposes those of sphyngoid bases and fatty acids, such as in d18:1-C16:0 for a ceramide made of sphingosine and palmitic acid [5].

However, most of the structural variability of gangliosides and GSLs is borne by the carbohydrate domain that exhibits a staggering structural diversity. As a result, GSLs are classified in a number of series defined by their monosaccharide sequences that include (iso)globo, (iso)ganglio, (neo)lacto, (neo)gala and muco for the most abundant series in vertebrates (Table 2.1). Some of these series

Table 2.1 Major series of glycosphingolipids in animals.

Series	Structure	Abbreviation
globo/isoglobo	Galα1–4Galβ1–4/3Glcβ1-Cer	Gb/iGb
lacto/neolacto	Galβ1–3/4GlcNAcβ1–3Galβ1–4Glcβ1-Cer	Lc/nLc
Ganglio/isoganglio	GalNAcβ1–4/3Galβ1–4Glcβ1-Cer	Gg/iGg
Muco	Galβ1–4Galβ1–4Glcβ1-Cer	Mc
Gala	Galα1–4Glcβ1-Cer	Ga

appear to be very specific to selected groups of organisms, such as the arthro, mollu, schisto or spirometo series. From these series, hundreds of GSLs have been identified from all of the animal kingdom [2]. It is noteworthy that the vast majority of GSLs derive from a glucosylceramide scaffold, leaving only a small number of molecules built upon a galactoceramide scaffold in the Gala series. Among those, galactoceramide and its sulphated derivative are the major glycolipids in the human brain where their presence is essential to the function of myelin [3].

From a historical perspective, it should be noted that the main structural features of GSLs were established long before those of protein-linked glycans, including *N*-linked and *O*-linked glycans. An obvious explanation to this is the fact that glycolipids could be isolated from complex mixtures of molecules purely by extraction and chromatography methods without having prior knowledge of their composition and without relying on chemical or enzymatic degradation processes. Indeed, the precise structure of brain sphingosine was established by Carter and collaborators in 1947 [4], whereas a homogeneous sphingoglycolipid fraction identified as a ganglioside was isolated from bovine brain in 1942, whose exact composition was established in 1951 by Klenk [5]. Similarly, the presence of sulphated glycolipids was suggested as early as 1878 and initially identified by Blix in 1933 as 6-*O*-sulfated galactosylceramide (Galβ1-Cer), before being finally corrected as 3-*O*-sulfated Galβ1-Cer by Yamakawa *et al.* in 1962 [6]. This is to compare to the structural elucidation of mucin-type *O*-linked glycans in the late sixties [7] or *N*-glycans in the mid-seventies [8].

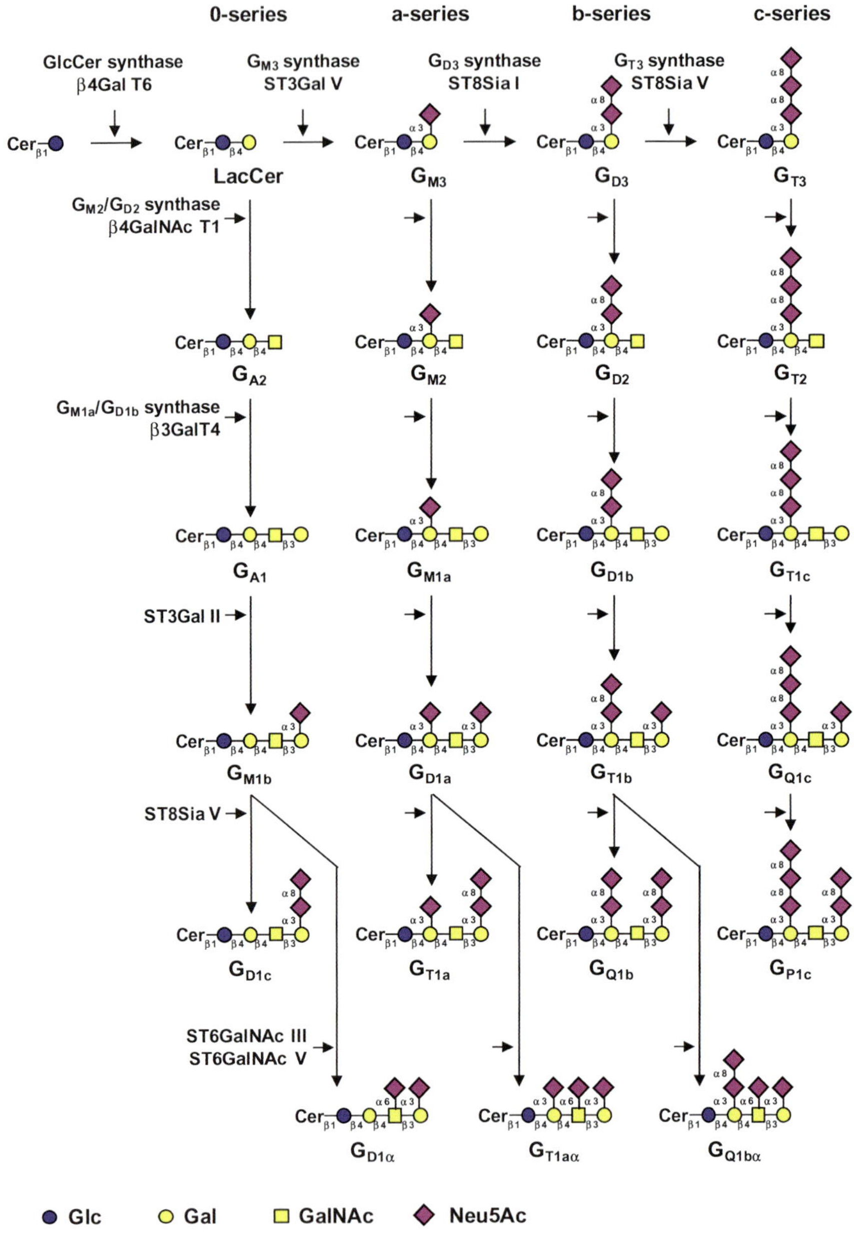

Figure 2.2 Biosynthetic pathway of gangliosides. Gangliosides are synthesized by the stepwise addition of monosaccharides to ceramide. Ceramide (Cer) is the acceptor for uridine diphosphate glucose (UDP-Glc): ceramide β-glucosyltransferase.

Figure 2.2 (*Continued*) Extension of glucosylceramide (GlcCer) occurs by the action of lactosylceramide (LacCer) synthase to make lactosylceramide. The action of ST3Gal V (G_{M3} synthase), ST8Sia I (G_{D3} synthase) and ST8Sia V (G_{T3} synthase) leads to the biosynthesis of the precursors of a-, b- and c-series gangliosides, respectively. The 0-series gangliosides are directly synthesized from LacCer. Elongation is performed by the sequential action of β-*N*-acetyl-galactosaminyltransferase (β4GalNAc T1), β-galactosyltransferase (β3Gal T4) and sialyltransferases (ST3Gal II and ST8Sia V). α-gangliosides derive from the action of sialyltransferase ST6GalNAc III or V on G_{M1b}, G_{D1a} or G_{T1b}. Adapted from [9].

Although represented in vast majority by ganglio-series (derived from GalNAcβ1–4Galβ1–4Glcβ1-Cer), gangliosides may also derive from other series, such as (neo)lacto-, and (iso)globo-series. Ganglio-series gangliosides are a heterogeneous family of gangliosides, in which carbohydrate sequence varies according to the length of the core sequence and the number and positions of sialic acid residues (Fig. 2.2). The core limit sequence is the tetrasaccharide, Galβ1–3GalNAcβ1–4Galβ1–4Glcβ, but the di-, tri-, and tetrasaccharide versions usually coexist within a single cell type, whereas the total number of substituting sialic acids varies from zero (which should formally not appear as gangliosides) to five.

In the simplified nomenclature established by Svennerholm [10], a letter is first added to G depending on the sialic acid number (A = 0; M = 1; D = 2; T = 3; Q = 4; P = 5), then a number depending on the length of the core sequence (1 for tetra-, 2 for tri-, and 3 for disaccharides established according to their mobilities on thin-layer chromatography (TLC)). Finally, a, b, c, or α may be added to differentiate the sialylation patterns (Fig. 2.2). The sialic acids associated to gangliosides are either *N*-acetylneuraminic acid (Neu5Ac) or *N*-glycolylneuraminic acid (Neu5Gc), or even deaminated neuraminic acid (Kdn) as observed, for example, in rainbow trout testis [11]. Finally, ganglio-series gangliosides may be further substituted by fucose residues [12], *O*-acetyl, or sulphate groups [13]. Although observed in small quantities within the cell membranes isolated from most animal tissues, gangliosides are concentrated in the neural system. Indeed, they represent the predominant form of glycosylation of mammalian neural tissue where gangliosides account for about 75% of the total linked sialic acids. Out of those, only four gangliosides, G_{M1}, G_{D1a}, G_{D1b}, and G_{T1b}, constitute the vast majority of the total gangliosides in mammalian and bird brains [14]. Recent studies report that these four compounds are differentially distributed within mice brains [15]. Whereas G_{D1b} and G_{T1b} are widely expressed throughout mice brains, G_{M1} is predominately present in white matter.

2.3 Organization of Gangliosides in Membranes: Structure and Functions of the Plasma Membrane Lipid Rafts

The model of the fluid mosaic of Singer and Nicholson [16], in which proteins and lipids diffuse within a considered homogenous membrane lipid bilayer, has considerably evolved in these last few decades. Because of their tightly packed backbone, GSLs have the ability to associate preferentially with other sphingolipids and cholesterol and segregate from classical membrane glycerophospholipids, forming microdomains called lipid rafts [17]. These domains were first termed "glycolipid-enriched domains" [18] and later "detergent-resistant membranes (DRMs)", based on their insolubility in non-ionic detergents at 4°C, which is the basis for most techniques used for lipid raft purification.

A variety of proteins have been shown to be raft-associated, such as GSL-anchored proteins in the extracellular leaflet [19], heterotrimeric G-proteins and src-family kinases related to signal transduction in the cytoplasmic leaflet, cytoskeletal components [20], receptor tyrosine kinases (RTKs), tetraspanins [21], but also adhesion molecules, such as integrins [22]. Lipid rafts are therefore highly dynamic structures containing specific lipids and proteins and can be seen as critical signalling platforms for many cellular functions, such as cell adhesion, growth, and migration (reviewed in [23]). GSLs, and especially gangliosides present in lipid rafts, can interact laterally with other molecules in their own membrane or with molecules present on other cells, and are therefore key molecules driving or modulating recognition and signalling events [24].

Lipid rafts have been widely studied in the plasma membrane of cells, but are thought to originate from the Golgi apparatus, where GSLs are synthesized, and it has been shown that polarized epithelial cells use these domains as sorting platforms for the delivery of proteins from the *trans*-Golgi network (TGN) to the apical plasma membrane [17,25]. Interestingly, polarized hepatic cells exhibit an indirect transport of GSL-anchored proteins to the apical cell surface, which are transported from the TGN to the basolateral plasma membrane, where they are then endocytosed and transcytosed to the apical surface. This sorting pathway is also dependent on GSLs and cholesterol, which indicates that lipid raft-dependent protein sorting does not depend on the polarized state of a cell [26].

Distinct types of membrane rafts enriched in different gangliosides have been shown to co-exist in the apical plasma membrane domain. In Madin–Darby canine kidney (MDCK) cells, G_{M1} was found to co-localize with raft-associated prominin-1 on microvilli, whereas G_{M3} was segregated and localized in the planar region. However, overlapping localization of G_{M1} or G_{M3} and prominin-1 was observed in the primary cilium [27]. Moreover, gangliosides G_{M1} and G_{D3} do not co-localize in rat primary cerebellar granule neurones and in membrane rafts isolated from rat cerebellum, suggesting that the ganglioside composition of lipid rafts is heterogenous, although the underlying cellular mechanisms remain unclear [28].

2.4 Gangliosides and Neurogenesis

Many studies have focused on the role of gangliosides in the development and functions of the nervous system, where they are expressed at very high levels, especially G_{M1}, G_{D1a}, G_{D1b} and G_{T1b}. These are the main species present in the brain of humans and most mammals. The addition of gangliosides on cultured neurons or neuroblastoma cells indicated a clear role of these molecules in neuronal development, with effects on neuritogenesis, neurite outgrowth, axonal development, and even in nervous system repair [29]. In particular, exogenous G_{M1} promoted neuritogenesis in mouse neuroblastoma cells [30] and increased neurite outgrowth from primary peripheral and central neurons [31].

The functions of specific gangliosides were later highlighted by studies using knock-out (KO) mice for specific glycosyltransferases involved in their biosynthesis (for review [32]). Except for the KO mice of glucosyl-ceramide synthase that exhibits early embryonic lethality, all other KO mice exhibit almost normal brain development at birth [33]. Nevertheless, most mice lacking glycosyltransferases involved in the biosynthesis of complex GSLs exhibit progressive degenerative defects in the central nervous system with aging, suggesting that GSLs play roles in the maintenance and repair of the nervous system. Interestingly, the KO mice of G_{M2}/G_{D2} synthase do not express any of the major complex gangliosides (G_{M1}, G_{D1a}, G_{D1b}, or G_{T1b}) in the adult brain, but these GSLs species are replaced by comparable concentration of gangliosides G_{M3} and G_{D3}. With aging, these KO mice displayed axon degeneration and demyelination, and destroyed neurons were detected. At one year of age, they were severely affected, dragging their hind limbs and having severe motor problems. These defects result in part from altered interactions of the myelin-associated glycoprotein (MAG), a sialic acid binding protein from the sialic acid-binding immunoglobulin-type lectins (Siglec) family called Siglec 4, with specific gangliosides of the axon membrane. It has been shown that the binding of MAG to G_{D1a} and G_{T1b} present on the axon cell membrane is necessary to stabilize myelin-axon interactions [34]. Genetic disruption of MAG or its ganglioside ligands results in similar destabilization of axons and myelin [35]. The intracellular signalling pathways triggered by MAG–ganglioside interactions and involved in axon-myelin stability remain to be established. GSLs are also involved in axon-myelin interactions at nodes of Ranvier [36], which explains alterations in motor nerve conduction velocity in the G_{M2}/G_{D2} synthase KO mice.

Following spinal cord injuries, axons are usually unable to regenerate due, in part, to a number of growth inhibitory factors that are present at the site of injury. The three myelin-associated inhibitors of axon growth: MAG, oligodendrocyte myelin glycoprotein (OMgp), and Nogo-A bind to a single receptor/co-receptor complex corresponding to the Nogo receptor (NgR1), LINGO-1, and either the p75 or TROY co-receptors [37–40]. It was first shown that MAG–G_{D1a}–G_{T1b} gangliosides interactions, via MAG-induced clustering of these GSLs, are involved in axon outgrowth inhibition at the site of lesion. The inhibition of axon outgrowth is reversed by sialidase treatment, inhibition of ganglioside synthesis, or by antibodies against G_{D1a} and G_{T1b} [41]. Intracellular signalling pathways leading to axon outgrowth inhibition involves the activation of the small guanosine triphosphatase (GTPase), ras homolog gene family member A (RhoA) [42]. Interestingly, both NgRs and gangliosides are lipid raft-localized molecules. Indeed, Saha and co-workers recently demonstrated that gangliosides facilitate the interactions of NgR1 with its co-receptors LINGO-1, p75, and the myelin-associated inhibitor Nogo-A, the carbohydrate moiety of G_{T1b} being directly involved in NgR1/LINGO-1 interactions [43].

In order to further elucidate the function of specific GSLs in neuron maintenance, double KO mice lacking the G_{M2}/G_{D2} and the G_{D3} synthase genes were generated, leading to the sole expression of G_{M3} [44]. The mutant mice displayed severe neurodegeneration shortly after birth, in particular, peripheral nerve degeneration, which induced reduced sensory function and skin over-scratching and lesions. However, the link between ganglioside defect and neurodegeneration remains unclear. In the double KO mice, many genes involved in inflammation and immune response are up-regulated, including complements and its related genes [45], and it was demonstrated that neurodegeneration found in mutant mice was largely due to the activation of complements [32]. The deficiency in the synthesis of GSLs in KO mice modifies the organization and function of lipid rafts. Lipid raft analysis in the cerebellum tissues of mutant mice revealed different disorders, depending on the type of GSL deficiency [45]. The localization and organization of CD55 and CD59, GPI-anchored complement regulating proteins, are modified, resulting in an increased sensitivity to complements, inflammation reactions, and neurodegeneration [45]. The identification of the molecules and signalling pathways involved in ganglioside-induced inflammatory reactions due to complement deposition could be important for the treatment of neurodegenerative diseases such as Alzheimer's disease.

2.5 Gangliosides as Ligands for Pathogens

Gangliosides serve as receptors for a large set of pathogens, including viruses, bacteria, and toxins, that can only act on cells that express the appropriate gangliosides on the outer surface of their plasma membranes. One of the best examples is the binding of cholera toxin to G_{M1}. The adenylate cyclase-activating enterotoxin of *Vibrio cholerae* contains six L subunits that bind to G_{M1} at the plasma membrane of targeted cells before the penetration of the H subunit that mediates the toxic action [46]. It was also shown that *Clostridium botulinum* neurotoxins bind to nerve endings to block acetylcholine release by recognizing G_{D1b} and G_{T1b} as functional receptors [47]. On the other hand, molecular mimicry between gangliosides on the human nerves and sialylated lipooligosaccharides (LOS) on the cell envelope of bacteria can induce a cross-reactive immune response, resulting in autoimmune-driven nerve damage. As an example, the *Campylobacter jejuni* cell envelope contains G_{M1}-like and G_{D1a}-like LOS that induce the production of anti-G_{M1} or anti-G_{D1a} immunoglobulin G (IgG) antibodies in infected patients, responsible for the development of Guillain–Barré syndrome, a post-infectious autoimmune neuropathy [48]. Gangliosides are also used by many human and animal viruses as receptors for cell entry, including noro-, rota-, parvo-, polyoma-, and paramyxoviruses [49]. Different gangliosides have been found to be involved in rotavirus cell entry: G_{M3} and G_{D1a} for porcine; G_{M3}, G_{M2}, and G_{D1a} for simian; and G_{M1} for human rotaviruses [50]. G_{D1a} and G_{T1b} were identified as receptors for murine polyomavirus, whereas simian polyomavirus 40 (SV40) was shown to bind to G_{M1} [51]. G_{D1b} and G_{T1b} serve as receptors for human polyomavirus, BkPyV [52], and G_{T1b} was also identified as a receptor for Merkel cell polyomavirus (MCPyV) [53]. Attachment to gangliosides is mediated by receptor-binding proteins of the viral envelope or exposed at the surface of non-enveloped viruses. Variations in the receptor specificity are important determinants for host susceptibility, tissue tropism, pathogenicity, and transmissibility of these viruses [49].

2.6 Role of Gangliosides in RTKs Signalling Regulation

Several RTKs, including receptors for epidermal growth factor (EGF), fibroblast growth factor (FGF), platelet-derived growth factor (PDGF), nerve growth factor (NGF), hepatocyte growth factor (HGF) and insulin, were demonstrated to be regulated by gangliosides [24,54–57]. RTKs are key membrane proteins involved in the control of many cellular processes, such as survival, proliferation, differentiation, migration, and invasion. They share a similar structural organization comprising an extracellular domain that contains the ligand binding site, a unique transmembrane domain, and a cytoplasmic region containing the tyrosine kinase activity [54]. RTKs are associated to lipid rafts and are activated by the binding of their specific ligand that induces receptor dimerization and intracellular domain autophosphorylation. The role of gangliosides in fine regulation of RTKs signalling was first clearly demonstrated by the addition of exogenous gangliosides in cell culture medium and by the modulation of specific enzymes involved in their biosynthesis. Depending on the receptor, gangliosides can negatively or positively regulate RTKs by inhibiting the ligand-induced dimerization and autophosphorylation of the receptor, or by activating receptor signalling with or without ligand binding [55,58]. Basically, a-series gangliosides, such as G_{M3} or G_{M1}, are usually negative regulators of RTKs, whereas disialogangliosides including G_{D2}, G_{D1a} or G_{D1b}, mostly activate RTK-mediated signal transduction [54,57,59] (Fig. 2.3). In parallel, the regulation of RTKs by gangliosides is highly dependent on the carbohydrate moiety of gangliosides, as shown for the c-Met receptor, which is constitutively activated by G_{D2} in a ligand-independent manner, whereas G_{D3} shows no effect [60,61]. However, the molecular mechanisms by which gangliosides regulate RTKs remain poorly understood. Direct interactions between the carbohydrate moiety of gangliosides and RTKs have been clearly identified as demonstrated for G_{M3} that inhibits EGF

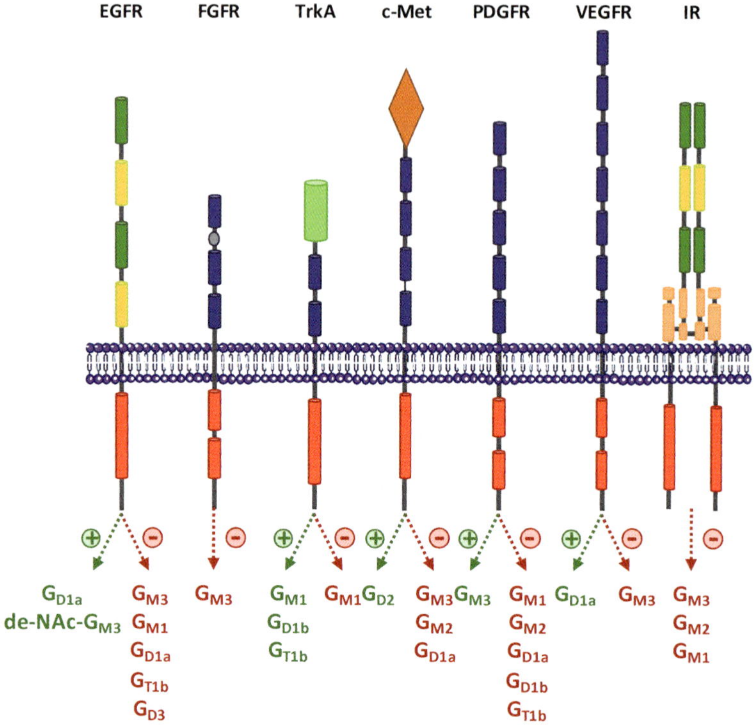

Figure 2.3 Regulation of RTKs by gangliosides. Gangliosides can either inhibit (red arrows) or activate (green arrows) RTKs signalling, depending on the expression pattern, cell type and experimental conditions of the ganglioside. Three different mechanisms can be involved: ganglioside/ligand interactions such as FGF/G_{M1} interaction, the regulation of receptor dimerization as for G_{M3} with epidermal growth factor receptor (EGFR), or the regulation of RTKs activity due to the localization inside lipid rafts as in the case for G_{M3} with insulin receptor. Adapted and updated from [54, 57].

receptor signalling by interacting with terminal *N*-acetylglucosamine (GlcNAc) residues of receptor *N*-glycans [62,63]. Regulation of gangliosides can also involve the dissociation of RTKs from lipid rafts, resulting in a reduced phosphorylation of the receptors as it has been demonstrated for insulin receptors (IRs). G_{M3} interacts with a lysine residue localized above the transmembrane domain of the IR beta-subunit and induces the dissociation of the IR-caveolin-1 complex, which is essential for insulin signalling [64]. Other indirect interactions of gangliosides could also regulate RTKs, such as interactions with integrins and tetraspanins that interact with RTKs and gangliosides. G_{M3} and G_{M2} can form calcium (Ca^{2+}) dependent heterodimers that specifically interact with tetraspanin cluster of differentiation 82 (CD82), thus impairing the *trans*-phosphorylation of the c-Met receptor, the recruiting of growth factor receptor-bound protein 2 (Grb2), and the activation of phosphatidylinositol 3-kinase/protein kinase B (PI3K/Akt) and methyl ethyl ketone/extracellular signal-regulated kinase (MEK/Erk) pathways [65,66].

2.7 Gangliosides and Apoptosis

GSLs, particularly disialyloganglioside G_{D3}, have been involved in mitochondrial damage and apoptosis. G_{D3} mediates different cellular responses, such as proliferation, differentiation or apoptosis, depending on the cell type, tissues, G_{D3}, and other gangliosides contents. G_{D3} was first implicated in apoptosis in lymphoid and myeloid tumor cells [67]. In these cells, the activation of the cluster of differentiation 95 (CD95)/Fas death receptor induced G_{D3} synthesis and transient accumulation, and subsequent disruption of mitochondrial transmembrane potential and apoptosis [67]. Acidic sphingomyelinase was shown to be required for G_{D3} ganglioside accumulation in response to Fas cross-linking and efficient apoptosis in lymphoid cells [68]. Importantly, lipid raft microdomains at the plasma membrane are clearly involved in CD95/Fas-mediated apoptosis, and the association of specific gangliosides, such as G_{D3} with cytoskeletal elements, such as the cytoplasmic peripheral membrane protein ezrin, is necessary to trigger CD95/Fas-mediated apoptosis [69].

Studies on the trafficking of G_{D3} to mitochondria in response to tumor necrosis factor (TNF) in rat hepatocytes showed that G_{D3} is normally confined to the plasma membrane and the endosomal/Golgi network. Raft associated G_{D3} can proceed from the cell plasma membrane (and/or from the TGN) to the mitochondria via a microtubule-dependent mechanism. In lymphoblastoid T cells, this transport may be regulated by cytoplasmic linker protein (CLIPR)-59, a new CLIP-170-related protein involved in the regulation of microtubule dynamics [70]. CLIPR-59 is a microtubule binding protein, but is also lipid raft associated by a double palmytoylation on two cysteine residues in its carboxyl (C)-terminal domain [71]. Interestingly, CLIPR-59 is also associated with TGN membranes and could be involved in G_{D3} trafficking from the TGN. After its transport to the mitochondria through interactions with mitochondrial raft-like microdomains, G_{D3} may trigger specific events involved in the apoptogenic program, including the opening of the permeability transition pore (PTP) complex, mitochondria hyperpolarization and depolarization, outer membrane disruption, and subsequent release of apoptogenic factors, such as cytochrome c, apoptosis-inducing factor (AIF), and mitochondrial caspases [72]. At the mitochondrial level, G_{D3} would be involved in PTP opening by allowing the formation of a multimolecular complex that includes the mitochondrial porin channel (also called the voltage-dependent anion channel or VDAC-1), B-cell lymphoma 2 (Bcl-2) family proteins and fission proteins, such as h-Fis1 [73].

Although apoptosis could be induced by G_{M1} in neuronal cells and by G_{M3} in thymocytes and in proliferating neuronal cells [74], most data available show that G_{D3} is the major player in programmed cell death. Apoptosis can be induced by direct exposure of some cell types to exogenously added G_{D3}, or by over-expression of the G_{D3} synthase, alpha-2,8-sialyltransferase (ST8Sia) I. Other gangliosides, such as G_{D1a} and G_{M3}, but also ceramide itself, had no effect on apoptosis induction on isolated liver mitochondria [75]. Interestingly, it has been shown that 9-*O*-acetylation of G_{D3} suppresses its pro-apoptotic activity [76]. 9-*O*-acetylated gangliosides are expressed at high levels in tumors of neuro-ectodermal origin such as melanoma, where they play roles in the growth and migration of cells [77]. Solid tumors also shed gangliosides into the microenvironment in greater quantity than healthy tissues and this phenomenon could promote tumor development by inhibiting the immune response. Thus, G_{D3} released by human melanoma inhibits the differentiation and induces apoptosis of dendritic cells [78] in a caspase-dependent manner involving direct interaction of G_{D3}

with mitochondria as well as reactive oxygen species (ROS) production [79]. *O*-acetylation of sialic acids affects their recognition by lectins present on pathogens or immune cells, such as Siglecs, and could be a factor regulating sialic acid-dependent interactions. 9-*O*-acetyl-G_{D3} could therefore modify the pro-apoptotic activity of G_{D3}, either by affecting its transport to the mitochondria, or its interactions with mitochondrial components. Although the molecular mechanism that links aberrant levels of mitochondrial Ca^{2+} with apoptosis are not fully understood, a new mechanism of Ca^{2+}-mediated apoptotic signalling was recently described: ganglioside G_{M1} accumulation in the lipid rafts of mitochondria-associated ER membranes modifies Ca^{2+} transport across membranes, inducing dissipation of mitochondrial membrane potential, release of apoptogenic factors, and mitochondria-mediated apoptosis of neuronal cells [80]. Specific GSLs may therefore be considered as potential therapeutic targets to attenuate neuronal apoptosis in neurodegenerative disorders caused by excessive apoptosis.

2.8 Gangliosides and Cancer

Complex gangliosides from b- and c-series are over-expressed in several types of cancer including melanoma, brain tumors, small cell lung carcinoma and breast cancer [9]. Whereas mostly absent in healthy human melanocytes, G_{D3} is highly expressed in melanoma primary tumors and cell lines [81, 82] and has been suggested to play a central role in the maintenance of malignancy of melanoma cells [83]. G_{D3} induces the activation of adaptor proteins such as paxillin, p130Cas, and focal adhesion kinase (FAK), that promote cell growth, migration, and invasive properties of melanoma cells [84, 85]. G_{D3} is also involved in the attachment of melanoma cells to extracellular matrix proteins, including collagen type I. Within lipid rafts, G_{D3} forms clusters with β1 integrin and interacts with collagen type I, enhancing integrin-linked kinase/Akt signalling pathways and malignant properties of melanoma cells [86–88]. Melanomas are also enriched in de-*N*-acetyl-G_{M3}, G_{D2}, 9-*O*-acetyl-G_{D3} (CDw60), and 9-*O*-acetyl-G_{D2} [77,89–91], with de-*N*-acetyl-G_{M3} being a marker of metastatic melanoma and promoting cell migration and invasion [92].

G_{D3} is also over-expressed in brain tumors [93–96] and involved in the proliferative and invasive capacities of tumor cells. In neuroblastoma cells, the inhibition of G_{D3} expression reduces cell migration and metastatic potential [97]. Similarly, the expression of G_{D3} increases tumorigenicity and invasion of glioma cells, whereas anti-G_{D3} monoclonal antibodies (mAbs) inhibit tumor growth [98,99]. On the contrary, the proliferation of human neuroblastoma cells is reduced in the presence of G_{M3} G_{M1}, G_{D1a}, and G_{T1b} via the inhibition of EGFR phosphorylation [100]. G_{M3} also inhibits vascular endothelial growth factor (VEGF)-induced angiogenesis and brain tumors expressing high levels of G_{M3} are less vascularized and grow slower than tumors that express low levels of G_{M3} [101].

Complex gangliosides from b-series including G_{D2} are re-expressed in small cell lung tumors (SCLC) [102]. G_{D2} was shown to be involved in the acquisition of the malignant phenotype of SCLC, and G_{D2} expression is sufficient to increase cell growth and invasion of small cell lung carcinoma cell line SK-LC-17 cells [102]. In parallel, inhibition of G_{D2} by small interfering RNAs reduces cell proliferation and tumor growth in severe combined immunodeficiency (SCID) mice [103] and anti-G_{D2} mAb inhibits cell growth and induces apoptosis of SCLC cells by a mechanism dependent on MEK/Erk and p38/mitogen-activated protein kinase (MAPK) signalling pathways [102,104]. G_{D3}, 9-*O*-acetyl-G_{D3} and 9-*O*-acetyl-G_{T3} are over-expressed in about 50% of invasive ductal breast carcinoma (IDC) [105] and *N*-glycolyl-G_{M3} is also detected in 100% of grade II IDC [106]. Recently, G_{D2} and G_{D3} were shown to be greatly increased in breast cancer stem cells, and involved in mammosphere formation and cell motility [107,108].

2.9 Concluding Remarks

Gangliosides are important compounds of the plasma membranes involved in a large set of cellular functions. These different functions are mainly supported by the glycan moiety of gangliosides, although the diversity of the sphingoid bases and the amine-linked fatty acids could also be involved in ganglioside specific cellular roles. The recent progress in GSL structure determination based on mass spectrometry approaches will allow more detailed investigations into the nature of ganglioside glycan structure in tissues and cells, as well as their changes that occur under physio-pathological conditions. Such studies will allow to highlight subtle changes in ganglioside glycans and to define more clearly the mechanisms of recognition that should involve specific lectins that remain to be identified.

Acknowledgments

This work was supported by the University of Sciences and Technologies of Lille, the Association pour la Recherche sur le Cancer et le comité du Pas-de-Calais de La Ligue contre le Cancer.

References

1. Bobowski, M., Cazet, A., Steenackers, A., Delannoy, P. (2012) *Carbohydr. Chem.* **37**, 1–20.
2. Yu, R. K., Yanagisawa, M., Ariga, T. (2007) In *Comprehensive Glycoscience.* Kamerling, H. Ed. Elsevier (Oxford). pp. 73–122.
3. Boggs, J. M. (2014) *Adv. Neurobiol.* **9**, 263–291.
4. Carter, H. E., Glick, F. J., Norris, W. P., Phillips, G. E. (1947) *J. Biol. Chem.* **170**, 285–294.
5. Hakomori, S. (1983) In *Sphingolipid Biochemistry — Handbook of Lipid Research.* Kanfer, J.N., Hakomori, S.I. Eds. Plenum Press (New York). pp. 1–165.
6. Yamakawa, T., Kiso, N., Handa, S., Makita, A., Yokoyama, S. (1962) *J. Biochem.* **52**, 226–227.
7. Thomas, D. B., Winzler, R. J. (1969) *J. Biol. Chem.* **244**, 5943–5946.
8. Montreuil, J. (1975) *Pure Appl. Chem.* **42**, 431–477.
9. Bobowski, M., Cazet, A., Steenackers, A., Delannoy, P. (2012) *Carbohydr. Chem.* **37**, 1–20.
10. Svennerholm, L. (1980) *Adv. Exp. Med. Biol.* **125**, 11.

11. Song, Y., Kitajima, K., Inoue, S., Khoo, K. H., Morris, H. R., Dell, A., Inoue, Y. (1995) *Glycobiology* **5**, 207–218.
12. Ariga, T., Kobayashi, K., Kuroda, Y., Yu, R. K., Suzuki, M., Kitagawa, H., Inagaki, F., Miyatake, T. (1987) *J. Biol. Chem.* **262**, 14146–14153.
13. Tadano-Aritomi, K., Kubo, H., Ireland, P., Hikita, T., Ishizuka, I. (1998) *Glycobiology* **8**, 341–350.
14. Tettamanti, G., Bonali, F., Marchesini, S., Zambotti, V. (1973) *Biochim. Biophys. Acta* **296**, 160–170.
15. Vajn, K., Viljetić, B., Degmečić, I. V., Schnaar, R. L., Heffer, M. (2013) *PLoS One* **8**, e75720.
16. Singer, S. J., Nicolson, G. L. (1972) *Science* **175**, 720–731.
17. Simons, K., van Meer, G. (1988) *Biochemistry* **27**, 6197–6202.
18. Fra, A. M., Williamson, E., Simons, K., Parton, R. G. (1994) *J. Biol. Chem.* **269**, 30745–30748.
19. Brown, D. A., Rose, J. K. (1992) *Cell* **68**, 533–544.
20. Kusumi, A., Fujiwara, T. K., Chadda, R., Xie, M., Tsunoyama, T. A., Kalay, Z., Kasai, R. S., Suzuki, K. G. (2012) *Annu. Rev. Cell Dev. Biol.* **28**, 215–250.
21. Cherukuri, A., Shoham, T., Sohn, H. W., Levy, S., Brooks, S., Carter, R., Pierce, S. K. (2004) *J. Immunol.* **172**, 370–380.
22. Wang, C., Yoo, Y., Fan, H., Kim, E., Guan, K. L., Guan, J. L. (2010) *J. Biol. Chem.* **285**, 29398–29405.
23. Head, B. P., Patel, H. H., Insel, P. A. (2014) *Biochim. Biophys. Acta* **1838**, 532–545.
24. Regina Todeschini, A., Hakomori, S. I. (2008) *Biochim. Biophys. Acta* **1780**, 421–433.
25. Schuck, S., Simons, K. (2004) *J. Cell Sci.* **117**, 5955–5964.
26. Nyasae, L. K., Hubbard, A. L., Tuma, P. L. (2003) *Mol. Biol. Cell.* **14**, 2689–2705.
27. Janich, P., Corbeil, D. (2007) *FEBS Lett.* **581**, 1783–1787.
28. Vyas, K. A., Pateln H. V., Vyasn A. A., Schnaarn R. L. (2001) *Biol. Chem.* **382**, 241–250.
29. Skaper, S. D., Leon, A., Toffano, G. (1989) *Mol. Neurobiol.* **3**, 173–199.
30. Facci, L., Leon, A., Toffano, G., Sonnino, S., Ghidoni, R., Tettamanti, G. (1984) *J. Neurochem.* **42**, 299–305.
31. Skaper, S. D., Katoh-Semba, R., Varon, S. (1985) *Brain Res.* **355**, 19–26.
32. Furukawa, K., Ohmi, Y., Ohkawa, Y., Tokuda, N., Kondo, Y., Tajima, O. (2011) *Neurochem. Res.* **36**, 1578–1586.

33. Proia, R. L. (2003) *Philos. Trans. R. Soc. Lond. B. Biol. Sci.* **358**, 879–883.
34. Schnaar, R. L. (2010) *FEBS Lett.* **584**, 1741–1747.
35. Pan, B., Fromholt, S. E., Hess, E. J., Crawford, T. O., Griffin, J. W., Sheikh, K. A., Schnaar, R. L. (2005) *Exp. Neurol.* **195**, 208–217.
36. Susuki, K., Baba, H., Tohyama, K., Kanai, K., Kuwabara, S., Hirata, K., Furukawa, K., Furukawa, K., Rasband, M. N., Yuki, N. (2007) *Glia* **55**, 746–757.
37. Fournier, A. E., GrandPre, T., Strittmatter, S. M. (2001) *Nature* **409**, 341–346.
38. Park, J. B., Yiu, G., Kaneko, S., Wang, J., Chang, J., He, X. L., Garcia, K. C., He, Z. A. (2005) *Neuron* **45**, 345–351.
39. Shao, Z., Browning, J. L., Lee, X., Scott, M. L., Shulga-Morskaya, S., Allaire, N., Thill, G., Levesque, M., Sah, D., McCoy, J. M., Murray, B., Jung, V., Pepinsky, R. B., Mi, S. (2005) *Neuron* **45**, 353–359.
40. McDonald, C. L., Bandtlow, C., Reindl, M. (2011) *Curr. Med. Chem.* **18**, 234–244.
41. Lehmann, H. C., Lopez, P. H., Zhang, G., Ngyuen, T., Zhang, J., Kieseier, B. C., Mori, S., Sheikh, K. A. (2007) *J. Neurosci.* **27**, 27–34.
42. Zhang, G., Lehmann, H. C., Manoharan, S., Hashmi, M., Shim, S., Ming, G. L., Schnaar, R. L., Lopez, P. H., Bogdanova, N., Sheikh, K. A. (2011) *J. Neurosci.* **31**, 1664–1675.
43. Saha, N., Kolev, M. V., Semavina, M., Himanen, J., Nikolov, D. B. (2011) *Biochem. Biophys. Res. Commun.* **413**, 92–97.
44. Inoue, M., Fujii, Y., Furukawa, K., Okada, M., Okumura, K., Hayakawa, T., Furukawa, K., Sugiura, Y. (2002) *J. Biol. Chem.* **277**, 29881–29888.
45. Ohmi, Y., Tajima, O., Ohkawa, Y., Mori, A., Sugiura, Y., Furukawa, K. (2009) *Proc. Natl. Acad. Sci. USA* **106**, 22405–22410.
46. Holmgren, J., Svennerholm, A. M. (1977) *J. Infect. Dis.* **136**, S105–S112.
47. Tsukamoto, K., Kohda, T., Mukamoto, M., Takeuchi, K., Ihara, H., Saito, M., Kozaki, S. (2005) *J. Biol. Chem.* **280**, 35164–35171.
48. Yuki, N. (2010) *Methods Mol. Biol.* **600**, 51–65.
49. Matrosovich, M., Herrler, G., Klenk, H. D. (2013) *Top Curr. Chem.* **Jul 20**, Epub ahead of print.
50. Taube, S., Jiang, M., Wobus, C. E. (2010) *Viruses* **2**, 1011–1049.
51. Tsai, B., Gilbert, J. M., Stehle, T., Lencer, W., Benjamin, T. L., Rapoport, T. A. (2003) *EMBO J.* **22**, 4346–4355.
52. Low, J. A., Magnuson, B., Tsai, B., Imperiale, M. J. (2006) *J. Virol.* **80**, 1361–1366.
53. Erickson, K. D., Garcea, R. L., Tsai, B. (2009) *J. Virol.* **83**, 10275–10279.

54. Bremer, E. G., Hakomori, S. I. (1984) *Adv. Exp. Med. Biol.* **174**, 381–394.
55. Hakomori, S. I., Igarashi, Y. (1995) *J. Biochem.* **118**, 1091–1103.
56. Hakomori, S. I. (2002) *Proc. Natl. Acad. Sci. USA* **99**, 225–232.
57. Miljan, E. A., Bremer, E. G. (2002) *Sci. STKE.* **2002**, re15.
58. Kaucic, K., Liu, Y., Ladisch, S. (2006) *Methods Enzymol.* **417**, 168–185.
59. Furukawa, K., Ohkawa, Y., Yamauchi, Y., Hamamura, K., Ohmi, Y. (2012) *J. Biochem.* **151**, 573–578.
60. Cazet, A., Lefebvre, J., Adriaenssens, E., Julien, S., Bobowski, M., Grigoriadis, A., Tutt, A., Tulasne, D., Le Bourhis, X., Delannoy, P. (2010) *Mol. Cancer Res.* **8**, 1526–1535.
61. Cazet, A., Bobowski, M., Rombouts, Y., Lefebvre, J., Steenackers, A., Popa, I., Guérardel, Y., Le Bourhis, X., Tulasne, D., Delannoy, P. (2012) *Glycobiology* **22**, 806–816.
62. Huang, X., Li, Y., Zhang, J., Xu, Y., Tian, Y., Ma, K. (2013) *J. Cell. Biochem.* **114**, 1616–1624.
63. Guan, F., Handa, K., Hakomori, S. I. (2011) *Neurochem. Res.* **36**, 1645–1653.
64. Kabayama, K., Sato, T., Saito, K., Loberto, N., Prinetti, A., Sonnino, S., Kinjo, M., Igarashi, Y., Inokuchi, J. (2007) *Proc. Natl. Acad. Sci. USA* **104**, 13678–13683.
65. Todeschini, A. R., Dos Santos, J. N., Handa, K., Hakomori, S. I. (2007) *J. Biol. Chem.* **282**, 8123–8133.
66. Todeschini, A. R., Dos Santos, J. N., Handa, K., Hakomori, S. I. (2008) *Proc. Natl. Acad. Sci. USA* **105**, 1925–1930.
67. De Maria, R., Lenti, L., Malisan, F., d'Agostino, F., Tomassini, B., Zeuner, A., Rippo, M. R., Testi, R. (1997) *Science* **277**, 1652–1655.
68. De Maria, R., Rippo, M. R., Schuchman, E. H., Testi, R. (1998) *J. Exp. Med.* **187**, 897–902.
69. Giammarioli, A. M., Garofalo, T., Sorice, M., Misasi, R., Gambardella, L., Gradini, R., Fais, S., Pavan, A., Malorni, W. (2001) *FEBS Lett.* **506**, 45–50.
70. Sorice, M., Matarrese, P., Manganelli, V., Tinari, A., Giammarioli, A. M., Mattei, V., Misasi, R., Garofalo, T., Malorni, W. (2010) *PLoS One* **5**, e8567.
71. Lallemand-Breitenbach, V., Quesnoit, M., Braun, V., El Marjou, A., Poüs, C., Goud, B., Perez, F. (2004) *J. Biol. Chem.* **279**, 41168–41178.
72. Susin, S. A., Lorenzo, H. K., Zamzami, N., Marzo, I., Brenner, C., Larochette, N., Prévost, M. C., Alzari, P. M., Kroemer, G. (1999) *J. Exp. Med.* **189**, 381–394.

73. Ouasti, S., Matarrese, P., Paddon, R., Khosravi-Far, R., Sorice, M., Tinari, A., Malorni, W., Degli Esposti, M. (2007) *Cell Death Differ.* **14**, 453–461.
74. Nakatsuji, Y., Miller, R. H. (2001) *Exp. Neurol.* **168**, 290–299.
75. Rippo, M. R., Malisan, F., Ravagnan, L., Tomassini, B., Condo, I., Costantini, P., Susin, S. A., Rufini, A., Todaro, M., Kroemer, G., Testi, R. (2000) *FASEB J.* **14**, 2047–2054.
76. Malisan, F., Franchi, L., Tomassini, B., Ventura, N., Condò, I., Rippo, M. R., Rufini, A., Liberati, L., Nachtigall, C., Kniep, B., Testi, R. (2002) *J. Exp. Med.* **196**, 1535–1541.
77. Kohla, G., Stockfleth, E., Schauer, R. (2002) *Neurochem. Res.* **27**, 583–592.
78. Péguet-Navarro, J., Sportouch, M., Popa, I., Berthier, O., Schmitt, D., Portoukalian, J. (2003) *J. Immunol.* **170**, 3488–3494.
79. Bennaceur, K., Popa, I., Chapman, J. A., Migdal, C., Péguet-Navarro, J., Touraine, J. L., Portoukalian, J. (2009) *Glycobiology* **19**, 576–582.
80. Sano, R., Annunziata, I., Patterson, A., Moshiach, S., Gomero, E., Opferman, J., Forte, M., d'Azzo A. (2009) *Mol. Cell* **36**, 500–511.
81. Portoukalian, J., Zwingelstein, G., Doré, J. F. (1979) *Eur. J. Biochem.* **94**, 19–23.
82. Thampoe, I. J., Furukawa, K., Vellvé, E., Lloyd, K. O. (1989) *Cancer Res.* **49**, 6258–6264.
83. Birklé, S., Gao, L., Zeng, G., Yu, R. K. (2000) *J. Neurochem.* **74**, 547–554.
84. Hamamura, K., Furukawa, K., Hayashi, T., Hattori, T., Nakano, J., Nakashima, H., Okuda, T., Mizutani, H., Hattori, H., Ueda, M., Urano, T., Lloyd, K. O., Furukawa, K. (2005) *Proc. Natl. Acad. Sci. USA* **102**, 11041–11046.
85. Hamamura, K., Tsuji, M., Ohkawa, Y., Nakashima, H., Miyazaki, S., Urano, T., Yamamoto, N., Ueda, M., Furukawa, K., Furukawa, K. (2008) *Biochim. Biophys. Acta* **1780**, 513–519.
86. Ohkawa, Y., Miyazaki, S., Hamamura, K., Kambe, M., Miyata, M., Tajima, O., Ohmi, Y., Yamauchi, Y., Furukawa, K., Furukawa, K. (2010) *J. Biol. Chem.* **285**, 27213–27223.
87. Birklé, S., Zeng, G., Gao, L., Yu, R. K., Aubry, J. (2003) *Biochimie* **85**, 455–463.
88. Kuphal, S., Bauer, R., Bosserhoff, A. K. (2005) *Cancer Metastasis Rev.* **24**, 195–222.
89. Cheresh, D. A., Reisfeld, R. A., Varki, A. P. (1984) *Science* **225**, 844–846.
90. Popa, I., Pons, A., Mariller, C., Tai, T., Zanetta, J. P., Thomas, L., Portoukalian, J. (2007) *Glycobiology* **17**, 367–373.

91. Chammas, R., Sonnenburg, J. L., Watson, N. E., Tai, T., Farquhar, M. G., Varki, N. M., Varki, A. (1999) *Cancer Res.* **59**, 1337–1346.
92. Liu, J. W., Sun, P., Yan, Q., Paller, A. S., Gerami, P., Ho, N., Vashi, N., Le Poole, I. C., Wang, X. Q. (2009) *Cancer Res.* **69**, 8662–8669.
93. Ruan, S., Raj, B. K., Lloyd, K. O. (1999) *J. Neurochem.* **72**, 514–521.
94. Ladisch, S., Chang, F., Li, R., Cogen, P., Johnson, D. (1997) *Cancer Lett.* **120**, 71–78.
95. Berra, B., Gaini, S. M., Riboni, L. (1985) *Int. J. Cancer* **36**, 363–366.
96. Shinoura, N., Dohi, T., Kondo, T., Yoshioka, M., Takakura, K., Oshima, M. (1992) *Neurosurgery* **31**, 541–549.
97. Zeng, G., Gao, L., Birklé, S., Yu, R. K. (2000) *Cancer Res.* **60**, 6670–6676.
98. Sottocornola, E., Colombo, I., Vergani, V., Taraboletti, G., Berra, B. (1998) *Invasion Metastasis* **18**, 142–154.
99. Hedberg, K. M., Dellheden, B., Wikstrand, C. J., Fredman, P. (2000) *Glycoconj. J.* **17**, 717–726.
100. Mirkin, B. L., Clark, S. H., Zhang, C. (2002) *Cell Prolif.* **35**, 105–115.
101. Seyfried, T. N., Mukherjee, P. (2010) *J. Oncol.* **2010**, 961243.
102. Yoshida, S., Fukumoto, S., Kawaguchi, H., Sato, S., Ueda, R., Furukawa, K. (2001) *Cancer Res.* **61**, 4244–4252.
103. Ko, K, Furukawa, K, Takahashi, T, Urano, T, Sanai, Y, Nagino, M, Nimura, Y, Furukawa, K. (2006) *Oncogene* **25**, 6924–6935.
104. Aixinjueluo, W., Furukawa, K., Zhang, Q., Hamamura, K., Tokuda, N., Yoshida, S., Ueda, R., Furukawa, K. (2005) *J. Biol. Chem.* **280**, 29828–29836.
105 Marquina, G., Waki, H., Fernandez, L. E., Kon, K., Carr, A., Valiente, O., Perez, R., Ando, S. (1996) *Cancer Res.* **56**, 5165–5171.
106. Oliva, J. P., Valdés, Z., Casacó, A., Pimentel, G., González, J., Alvarez, I., Osorio, M., Velazco, M., Figueroa, M., Ortiz, R., Escobar, X., Orozco, M., Cruz, J., Franco, S., Díaz, M., Roque, L., Carr, A., Vázquez, A. M., Mateos, C., Rubio, M. C., Pérez, R., Fernández, L. E. (2006) *Breast Cancer Res. Treat.* **96**, 115–121.
107. Liang, Y. J., Ding, Y., Levery, S. B., Lobaton, M., Handa, K., Hakomori, S. I. (2013) *Proc. Natl. Acad. Sci. USA* **110**, 4968–4973.
108. Battula, V. L., Shi, Y., Evans, K. W., Wang, R. Y., Spaeth, E. L., Jacamo, R. O., Guerra, R., Sahin, A. A., Marini, F. C., Hortobagyi, G., Mani, S. A., Andreeff, M. (2012) *J. Clin. Invest.* **122**, 2066–2078.

Chapter 3

Bacterial Lipopolysaccharides: An Overview of Their Structure, Biosynthesis and Immunological Activity

Flaviana Di Lorenzo, Alba Silipo, Rosa Lanzetta, Michelangelo Parrilli and Antonio Molinaro

Department of Chemical Sciences, University of Naples Federico II, Via Cintia 4, 80126, Naples, Italy

3.1 Introduction

The lipopolysaccharide (LPS) macromolecule, also known as the "endotoxin", is the major component of the outer membrane (OM) of almost all Gram-negative bacteria and of some cyanobacteria [1–3], constituting approximately 75% of the outer surface. The historical term "endotoxin" (from the Greek "endo" meaning "within"), indicating the capability to activate the immune system, derives from work on *Vibrio cholerae* [4,5] executed in the year 1892 by Richard Pfeiffer, a disciple of Robert Koch, who postulated that the toxic shock reactions in guinea pigs treated with cell-lysates of *V. cholerae* were caused by a principle attached to the bacterial cells [4,5], unlike the previously known exotoxins that were secreted by bacteria. The clear proof of this hypothesis was given only in 1923 by Florence Seibert, one of the most important pioneers in pyrogen research [5]. Her work attempted to identify the bioactive material contaminating the infusion fluids, and the pharmaceutical drugs led her to give further evidence of the endobacterial nature of these "contaminants" that she used to call "her little blue devil", due to the difficulties in eradicating them [5]. The current term "lipopolysaccharide" was introduced in 1943 by Murray Shear, highlighting that endotoxins mainly consist of carbohydrates and lipids [5], although it took several years to reach the complete elucidation of the structure of the molecule; this was obtained in 1952 by Otto Lüderitz and Otto Westphal, who were the first to isolate the endotoxic material in sufficiently pure amounts required for structural studies [5]. Their protocol [6] of LPS extraction from bacterial cells is currently used in the endotoxin research field. The opportunity to obtain large amounts of purified LPS finally allowed studies to help define the moieties responsible for their endotoxic and antigenic properties [5]. Indeed, in the sixties, it was demonstrated that the lipidic portion of the LPS is the endotoxic principle of the entire macromolecule and this was further confirmed by Tetsuo Shiba, in the eighties, synthetically preparing the first free lipid A [7]. Within the last three decades, LPSs have been isolated from a large number of Gram-negative bacteria and the structural details of each have been characterized with advanced techniques, such as nuclear magnetic resonance (NMR) spectroscopy, high-performance liquid chromatography (HPLC) and mass spectrometry. At the same time, extensive studies have been executed that have been aimed at the biological elucidation of LPS activity, in relation to its role as an elicitor of host innate immune systems.

The aim of this chapter is to present the main information related to the structural aspects of the LPS molecule, illustrating the common and uncommon features found in several microorganisms and showing the most used procedure to extract, purify and characterize LPS structure. A large part of this chapter will also be dedicated to describing the biosynthesis and the biological role of LPSs and their structure-dependent activity.

3.2 The LPS Molecule: An Overview

LPSs are heat-stable amphiphilic molecules indispensable for the viability and survival of Gram-negative bacteria, as they heavily contribute to the structural integrity of the OM and to the protection of the bacterial cell envelope [8]. Being intercalated into the OM, they form hydrophobic interactions with the phospholipid bilayer, composing the inner layer, giving rise to the typical asymmetry of the OM of Gram-negative bacteria [8]. The highly ordered structure and low fluidity of the LPS monolayer, stabilized by electrostatic interactions between divalent cations (as calcium (Ca^{2+}) and magnesium (Mg^{2+})) and negatively charged groups present on LPS molecules, are responsible for the increase of permeability to hydrophobic compounds and to higher molecular weight hydrophilic compounds, but also for the bacterial resistance to external stress factors. Indeed, most of the commonly used antibiotics directed against Gram-negative bacteria, such as polymyxin B, are able to destabilize the above-mentioned ionic interactions, leading to the disruption of membrane integrity [9]. In addition, since they are exposed to the external environment, LPS molecules participate in crucial mechanisms of host-bacterium interactions such as colonization, virulence in the case of pathogen and opportunistic bacteria, adhesion and symbiosis [10]. Among all these activities, LPSs have been shown to be the most potent immunostimulant molecules, playing a key role in the pathogenesis of Gram-negative infections, triggering the immune system in a wide range of eukaryotic organisms ranging from insects to humans [8]. LPSs belonging to different bacterial species possess different structures, but also, the LPS of an individual bacterial strain is not a single molecule possessing a specific chemical structure, but rather, a blend of various molecules characterized by an intrinsic size and structural heterogeneity [11–12]. Furthermore, bacteria of the same species producing diverse LPS molecules under different growth conditions were also found [12]. Despite the high structural heterogeneity, all LPSs conform, for the most part, to a common basic architecture composed of well-conserved domains in which three distinct regions can be distinguished (Fig. 3.1), since they are encoded by different gene clusters. Indeed, they consist of a polysaccharide (known as *O*-antigen, *O*-side chain or *O*-specific polysaccharide) characterized by a highly variable structure, which constitutes the chemical

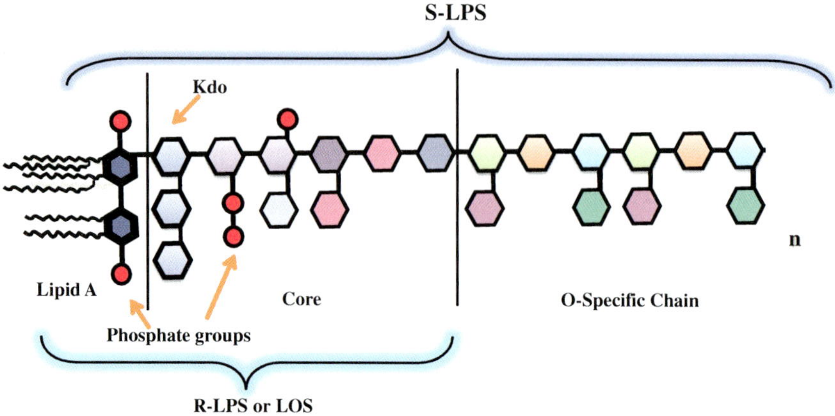

Figure 3.1 Scheme of the general chemical structure of lipopolysaccharides.

basis for the serological classification of bacterial strains, covalently linked to an oligosaccharide part (core), in turn, linked to a glycolipid portion (lipid A) that is the most conserved region of the entire macromolecule, even among different species belonging to the same genus [13]. The lipid A moiety is embedded in the outer leaflet of the OM, whereas the sugar chain is oriented outwards [12]. The *O*-polysaccharide chain is not ubiquitous, as it can be lacking or partly truncated in some Gram-negative strains. Bacterial colonies can be catalogued by gross colony morphology as rough or smooth on the basis of the occurrence of the *O*-side chain, being absent in the former and present in the latter [14]. The terminology currently used to designate the different LPS types, namely with or without the *O*-polysaccharide portion, is S-LPS or R-LPS (or lipooligosaccharide, LOS), respectively [14].

3.3 The Endotoxic Centre of LPSs: The Lipid A Moiety

From the first structural characterization of lipid A by Westphal and Lüderitz until experiments by Galanos on solubilized free lipid A moieties [14–17], it was clearly demonstrated that lipid A represents the hydrophobic and endotoxic principle of the LPS molecule, since it acts as a potent stimulator of the host innate immune system, being recognized by the toll-like receptor 4 (TLR4)/myeloid differentiation factor 2 (MD-2) complex. It possesses a highly conservative structure consisting of a backbone of β-(1→6)-linked D-glucosamine disaccharides substituted with a number of amide- and ester-linked 3-hydroxyl fatty acids at the positions C-2 and C-2′, and C-3 and C-3′, respectively. The acyl groups that are directly linked to the sugar backbone are defined as primary. Some of the primary fatty acids are further acylated at the hydroxyl groups by secondary acyl chains. Furthermore, the sugar backbone is generally α-phosphorylated at position O-1 of the reducing glucosamine (Glc*p*N I) and at position O-4′ of the non-reducing glucosamine (Glc*p*N II) [18]. The first lipid A structural characterizations were from *Escherichia coli* and *Salmonella enterica* LPSs in 1983. *E. coli* LPS lipid A (Fig. 3.2) is considered to date the prototype that other LPS structures are often compared to; it is built up of the typical bisphosphorylated disaccharide backbone of D-Glc*p*Ns [P→4-β-D-Glc*p*N-(1→6)-α-D-Glc*p*N-1→P], acylated at position 2 and 3 of both Glc*p*Ns by four C14:0 (3-OH) (Fig. 3.2). The primary fatty acids located on the Glc*p*N II were both esterified at their hydroxyl group by two secondary fatty acids; the *N*-linked C14:0 (3-OH) was esterified by a C12:0, while the *O*-linked C14:0 (3-OH) by a C14:0 (Fig. 3.2). Despite their general architecture, lipid A moieties also present a sort of microheterogeneity due to the presence of subtle chemical differences depending on a wide number of factors including bacterial adaptation, incomplete biosynthesis, changing of environment, presence of external stimuli and chemical modifications resulting from the procedures used for lipid A extraction from bacterial cells. Microheterogeneity has been observed in the acylation (number, type and distribution of acyl chains) and phosphorylation patterns, and less commonly, also in the disaccharide backbone, as in the case of a number of bacterial species, such as *Aquifex pyrophilus* [19], *Brucella abortus* [20], *Bacteriovorax stolpii* [21], *Caulobacter crescentus* [22], *Mesorhizobium huakuii* [23], *Bradyrhizobium elkanii* [24], *Bartonella henselae* [25], *Legionella pneumophila* [26], whose glucosamine (GlcN) residues may be

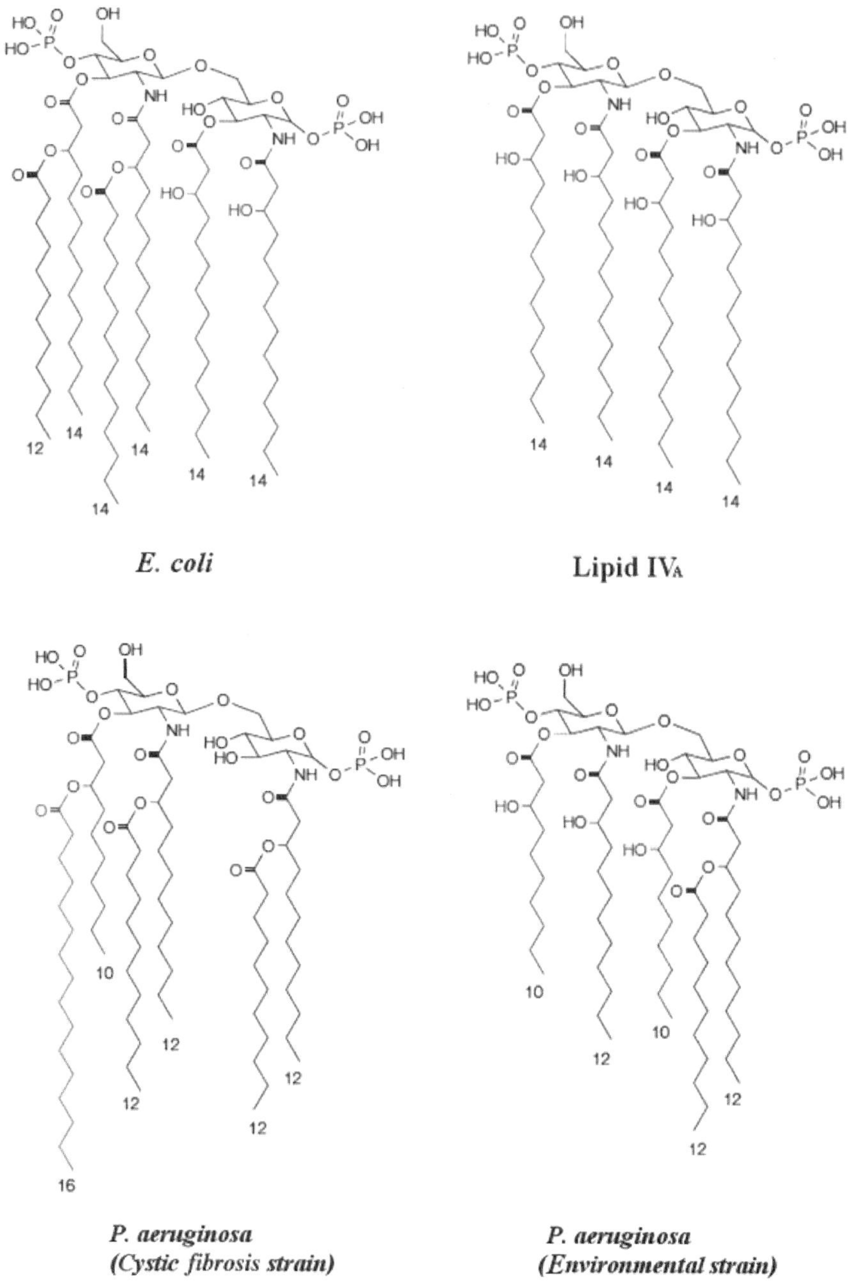

Figure 3.2 Lipid A chemical structures showing their diversity in Gram-negative bacteria, and also among the same species.

replaced with 2,3-diamino-2,3-dideoxy-D-glucopyranose (Glc*p*N3N) residues. Moreover, phosphate groups can be substituted by further phosphate groups, producing a pyrophosphate, but also by other polar substituents, such as 4-amino-4-deoxy-L-arabinopyranose (arabinosamine, Ara*p*4N) and 2-amino-ethanol groups (EtN), or by acid residues, such as galacturonic acid (Gal*p*A). Phosphate groups can be absent as in the case of *Bdellovibrio bacteriovorus* LPS lipid A that is characterized by the replacement of phosphate groups with two mannose residues, generating a totally neutral lipid A [27]. Exceptionally, the lipid A of the marine bacterium *Loktanella rosea* has a very unusual structure, as the molecule is non-phosphorylated, both Glc*p*N residues are β-linked, and the Glc*p*N I forms with α-Gal*p*A a unique mixed trehalose-like structure [28]. Concerning the acylation pattern, fatty acids can be attached to the disaccharide backbone either symmetrically (3 + 3, e.g. *Neisseria meningitides*) or asymmetrically (4 + 2, e.g. *E. coli*). Finally, the following are some structural features that are less frequently presented in lipid A fatty acids: methyl branching, variations in functional and hydroxyl groups, chain lengths up to 28 carbon atoms and odd numbered carbon chains.

Changes in lipid A structure that are dependent on environmental or growth conditions have been found in several bacterial species [29]; for example, lipid A from *E. coli* can undergo structural changes as the addition of palmitate residues in the presence of antimicrobial peptides and low concentration of Mg^{2+}. It has been found that in other species, additional enzymes that are not present in *E. coli* are responsible for further structural modifications. For instance, the presence of late functioning phosphatases in *Rhizobium etli, Rhizobium leguminosarum* and *Francisella tularensis* induces the synthesis of partially or completely dephosphorylated lipid A species [30]. It is worth noting that some bacterial species, such as the cystic fibrosis opportunistic pathogens, *Burkholderia* [31], were found to constitutively synthesize a lipid A substituted by L-Ara4N residues, that are, in physiological conditions, positively charged elements whose presence reduces the net charge on the surface of the outer membrane inferred by LPS molecules, increasing bacterial resistance against cationic antimicrobial compounds, such as polymyxin B or cationic antimicrobial peptides (CAMP) [31]. Moreover, a very frequent substitution, mainly involving bacteria that live in extreme habitats (termed extremophiles), is that of galacturonic acid residues that can be attached either on the Kdo (3-deoxy-D-manno-octulosonic acid) moiety or on the lipid A glucosamine-backbone [28]. Even the most subtle variation in chemical structure of lipid A can affect the LPS molecule bioactivity, since this latter is strictly structure-related; indeed, the lipid A intrinsic conformation is responsible for its agonistic and antagonistic activity [32–38]. Exposition to tiny and balanced amounts of agonistic lipid A prepares the immune system to protect the host against further infection, while uncontrolled and excessive exposure (or excessive responses) can lead to uncontrolled inflammatory responses that can induce septic shock and cellular death.

3.3.1 *Lipid A structure elucidation*

The lipid fraction is obtained by precipitation after mild acid hydrolysis of LPSs (or LOSs). The structural approach makes extensive use of mass spectrometry analysis by Matrix-Assisted Laser Desorption/Ionization-Time of Flight (MALDI-TOF) mass spectrometry and Electrospray Ionization mass spectrometry (ESI-MS) techniques both on the native and selectively degraded lipid A fractions. The MALDI-TOF and ESI-MS data offer insights into the number of lipid A species present in the fraction, the presence of polar heads and the distribution of acyl residues on each Glc*p*N units of the disaccharide backbone [39]. Through the chemical and spectroscopical study of several lipid A moieties and LPSs, novel general approaches to gain structural insights on such molecules have been developed. In particular, combining MALDI-MS or ESI-MS and selective chemical degradation by ammonium hydroxide hydrolysis, a useful methodology to obtain the secondary fatty acid distribution has been developed, which is one of most demanding issues in the structural determination of lipid A [40–42]. These approaches must be supported by the classical chemical analyses of the lipid A, which imply the removal of ester-linked fatty acid moieties by mild alkaline treatment, or complete removal by strong alkaline treatment and gas liquid chromatography mass spectrometry (GLC-MS) analyses of fatty acids as methyl ester derivatives [43–44]. The NMR study of LPSs is less frequently used, but is equally important [45–47]. The amphiphilic nature of lipid A entails a very poor solubility of the sample in any solvent system; however, when possible, it is invaluable in obtaining diagnostic information. NMR is highly useful for studies on the saccharide backbone, such as the anomeric configuration of the carbohydrate residues and their sequence, and the occurrence and position of the polar heads [45–47].

3.4 The Oligosaccharide Portion of LPSs: The Core Moiety

The core oligosaccharide moiety is a complex component of the LPS molecule, since it can be characterized by up to fifteen monosaccharides, which can be organized giving either a linear or a branched structure [48]. In the core LPS portion, two different regions can be distinguished on the basis of the monosaccharide composition: inner core and outer core [48]. The inner core region is directly linked to the lipid A, is well conserved and consists of uncommon sugar residues, such as heptoses (L-*glycero*-D-*manno*-heptose and D-*glycero*-D-*manno*-heptose) and Kdo (3-deoxy-D-*manno*-octulosonic acid) [49]; this latter is considered a diagnostic marker for all Gram-negative bacteria, which connects the core oligosaccharide to the lipid A backbone with an α-configured ketosidic linkage in almost every LPS investigated to date [50]. The only two exceptions in which Kdo is not the first residue of the inner core regards *Acinetobacter* [51] and *Shewanella* [52]: In the former, Kdo is replaced by its C-3 hydroxy-derivative, the D-*glycero*-D-*talo*-oct-2-ulosonic acid (Ko) [51], whereas in the latter it is replaced by its C-8 amino-derivative, the 8-amino-8-deoxy-*manno*-oct-2-ul-osonic acid (Kdo8N) [52]. Interestingly, in *Xanthomonas campestris* pv. *campestris*, a galacturonyl phosphate substituent linked to position

O4 of Kdo has been identified [54]. Frequently, inner core structures containing other peculiar residues, such as D-*glycero*-D-*manno*-heptose residues (D,D-Hep), have also been described [53]. The L-*glycero*-D-*manno*-heptose residues are often decorated with charged substituents like phosphate, pyrophosphate, Ara*p*4N or uronic acids, as found in bacteria belonging to the *Pseudomonas* genus and many other species. Particularly, the Ara*p*4N residue is always present as a terminal non-reducing monosaccharide of the trisaccharide Ara*p*4N-(1→8)-α-Ko-(2→4)-α-Kdo in *Burkholderia* LPSs. It is speculated that all the substituents bearing a positively charged free amino group, such as Ara*p*4N residues, might play a role in pathogenesis since they reduce the negatively charged surface on the outer membrane, rendering it positively charged or in an isoelectric state that, in turn, confers resistance to antibiotic compounds and antimicrobial peptides [12,31]. Moreover, all the negatively charged residues found in both inner and outer core regions seem to maintain an association with divalent cations required for membrane integrity [33].

The outer core region is the most exposed portion, often branched, and is characterized by a higher structural variability than the inner core region. It is typically composed of common hexoses, such as glucose, galactose, *N*-acetyl glucosamine and *N*-acetyl galactosamine and it may also contain residues such as 6-deoxy-L-mannose (L-Rha) and *N*-acetyl-2,6-dideoxy-D-glucosamine (D-QuiNAc). It is possible to find peculiarity also in this region, although more commonly; for example, in *Shewanella* LPSs, a new kind of glycosydic linkage has been found that involves an open chain acetal linkage of a glucosamine residue that is present as a non cyclic carbonyl form [52]. The plant pathogen *Xanthomonas campestris* pv. *campestris* was shown to possess a strong accumulation of negatively charged groups in the lipid A-inner core region and has a number of novel features, including a galacturonyl phosphate attached at a 3-deoxy-D-*manno*-oct-2-ulosonic acid residue and a unique phosphoramide group [54]. Furthermore, in *Rhizobium*, *Acinetobacter* and *Agrobacterium* strains, the 3-deoxy-D-*lyxo*-hept-2-ulosaric acid has been identified [51,55]. The core oligosaccharide moiety together with the lipid A portion is the structural motive common to all LPSs; this suggests that these domains are indispensable for Gram-negative bacterial life. Indeed, until a short while ago, it was believed that mutants of Gram-negative bacteria without a minimal core oligosaccharide (OS) structure were not viable, but mutants synthesising only lipid A have been recently isolated [49–50]. However, in bacteria producing LOSs, as in many *Burkholderia* strains or in *Nesseria meningitidis, Nesseria gonorrhoeae* and some chronic strains of *Pseudomonas aeruginosa*, the core moiety results in the most exposed portion of the LPS molecule; thus, it is recognized by host adaptive immune system components and it accounts for serological specificity [29,56]. A further consequence of core moiety exposition to the environment is that both bacterial viability and resistance against antibiotic compounds are strictly related to its chemical structure. Some of the bacteria that possess a LOS and not a complete LPS are nasty microbes, such as *N. meningitidis* and *H. influenzae* that are both human pathogens responsible for meningitis. LOS produced by the first is the antigenic determinant of the macromolecule and contains residues, such as *N*-acetylneuraminic acid (NeuAc) or L-fucose (L-Fuc), which is a reminder of human glycosphingolipids [29,56]. This mimicry of host cellular surface components is responsible for the increase of pathogen resistance against phagocytosis and host bactericidal activity escaping the attack from the host complement membrane attack complex [29,57].

3.4.1 *Core oligosaccharide structure elucidation*

The target of the structural elucidation of a LOS core oligosaccharide is generally much more complex than that of the *O*-chain because of the higher monosaccharide number and the presence of labile groups and other substituents that are often present in non stoichiometric amounts. In addition, the acid hydrolysis (Fig. 3.3) used to remove the lipid A from LOS occurs by hydrolysis of Kdo that, in its reducing form, usually appears in different anomeric and cyclic forms, thus determining an "artificial" increase of

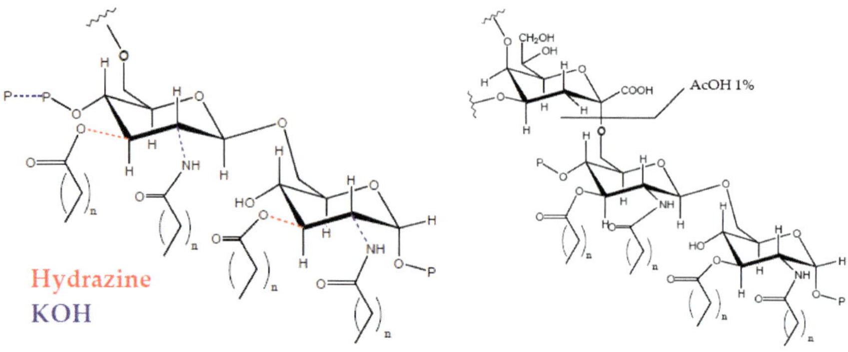

Figure 3.3 Left panel: Lipid A after treatment with hydrazine (red dotted lines) and potassium hydroxide (KOH) (blue dotted lines). Right panel: Schematic representation of the cleavage after mild acid hydrolysis of LOS/LPS.

the core heterogeneity. Therefore, LOS samples are generally treated under alkaline conditions (Fig. 3.3) that do not cleave the Kdo linkage, thereby allowing the analysis of the complete sugar backbone, including the two Glc*p*N of lipid A. Notably the LOS sample should first be treated under mild alkaline conditions in order to selectively remove the lipid A ester-bound fatty acids (Fig. 3.3). This more soluble *O*-deacylated LOS is *N*-deacylated (Fig. 3.3) under stronger alkaline conditions, affording the completely deacylated backbone oligosaccharide, which represents the intact sugar skeleton of LOS suitable for sequence determination by NMR analysis [58]. The direct combined de-*O*- and de-*N*-acylation is usually not performed in order to avoid the formation of alkali resistant unsaturated *N*-linked fatty acids. In order to assign phosphate groups and their possible further attached substituents, one- and two-dimensional (1D and 2D) phosphorus-31 (^{31}P)-NMR experiments are also carried out. Since alkaline treatment frequently determines the loss of base-labile functions, the oligosaccharide moiety released under acid conditions is also studied as a complementary and informative approach. More recently, the full NMR study of the *O*-deacylated sample has been carried out in a water solution containing a denaturing agent [59]. Every step of the selective degradation is usually accompanied by MALDI mass spectrometry analysis, which is a very valuable and complementary approach in the study of such molecules. Our collaborators have pioneered the development of new MALDI mass spectrometry approaches through the usage of new methods of preparation and experimental study. It is worth emphasizing that when such experiments are carried out on the intact sample, they are often of very high value as they yield a complete "picture" of the intact endotoxin without any loss of structural information [42].

3.5 The Antigenic Determinant of LPSs: The *O*-Side Chain Moiety

The *O*-chain polysaccharide is the most variable portion of the LPS, and also within bacteria belonging to the same genus. For most Gram-negative bacteria, it consists of up to 50 identical repeating oligosaccharide units, each consisting of two to eight different glycosyl residues (heteroglycans) or, in some bacteria such as *L. pneumophila* [60], of identical sugars (homoglycans). A single bacteria produces LPSs with *O*-chains characterized by a wide range of lengths as a result of incomplete synthesis of the polysaccharide chain [12]; this different degree of polymerization is responsible for the ladder-like pattern, showed by sodium dodecyl sulfate polyacrylamide gel electrophoresis (SDS-PAGE) [61], typical of a smooth LPS. The high structural variability of the *O*-polysaccharide is ascribable to the large number of sugar residues (in both pyranose and furanose rings, anomeric and absolute configurations) that can build up the repeating units, as well as to the glycosydic sequence and to the presence of non-carbohydrate substituents, such as phosphate, amino acids, sulfate, acetyl or formamide groups, which often present in a non-stoichiometric fashion [62–65]. Some examples of peculiar monosaccharides that can characterize the LPS *O*-chain portion are the C_{12} branched-chain sugar caryophyllose [66–67] (3,6,10-trideoxy-4-*C*-[(R)-1-hydroxy-ethyl]-D-*erythro*-D-*gulo*-decose) (Fig. 3.4) and the caryose [66–67] (4,8-cyclo-3,9-dideoxy-L-*erythro*-D-*ido*-nonose), the only carbocyclic monosaccharide found in nature so far. Both of these sugars were isolated from the plant pathogen *Burkholderia caryophylli*, responsible for the wilting of carnation [66–67]. It is worth outlining the unique LPS *O*-chain structure of the nitrogen-fixing soil bacterium *Bradyrhizobium* sp. BTAi1, characterized by a bicyclic monosaccharide, the bradyrhizose (Fig. 3.4) [68]. On the other hand, an example of particular substituents that can decorate the *O*-chain is the 3-hydroxy-2,3-dimethyl-5-oxoprolyl group found linked to the QuiNAc residue of the LPS *O*-chain moiety from the newly identified cystic fibrosis pathogen *Pandoraea pulmonicola* (Fig. 3.4) [69]. The function of these substituents is frequently unknown, although it can be speculated that bacteria can modify their LPSs to mask themselves to the host immune system [63–64]. The classical example of this "stealth-mechanism" involves the *O*-chain moiety from the human pathogen *Helicobacter pylori*, which expresses Lewis antigenic epitopes (mainly Le^x and Le^y) essential for bacterial survival in the acidic environment of the human stomach; this arises from

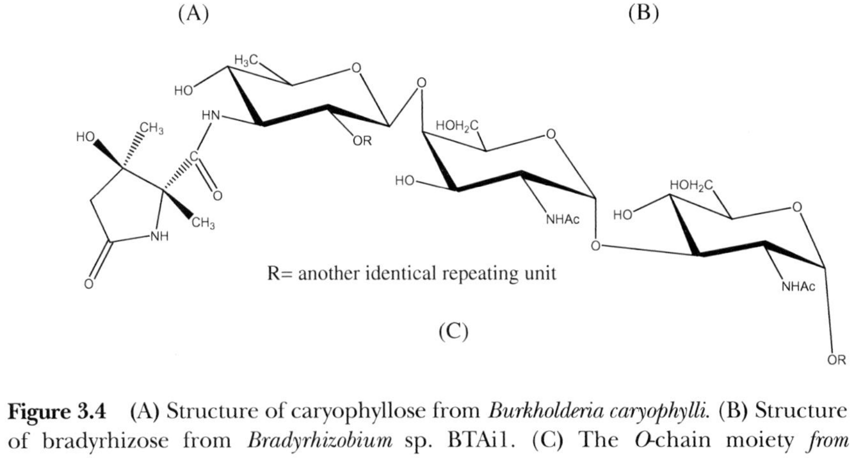

Figure 3.4 (A) Structure of caryophyllose from *Burkholderia caryophylli.* (B) Structure of bradyrhizose from *Bradyrhizobium* sp. BTAi1. (C) The *O*-chain moiety *from Pandoraea pulmonicola.*

the observation that *H. pylori* possessing rough LPSs is characterized by a strongly reduced colonizing ability [70]. The *O*-chain provides a plethora of biological activities, among which it is responsible for the serological specificity, and is used for bacterial serotyping, and the protection of bacteria from host immune defenses, as described above. Moreover, it is involved in interactions of bacteria with plants and bacteriophages, and it constitutes a physical barrier necessary for bacterial viability that prevents cellular dehydration and promotes the adhesion of colonies. Taking into account that the *O*-chain is the major antigen targeted by host adaptive responses, the structural elucidation of this LPS moiety is crucial, since it contributes to a better understanding of the mechanisms of the pathogenesis of infectious diseases, and it also provides pivotal information to develop novel vaccines and diagnostic reagents [71]. The synthesis of the *O*-chain starts with the *O*-unit that is first assembled on a lipid carrier and then polymerized; on the contrary, homoglycans and part of the heteroglycans with disaccharide *O*-units are synthesized via sequential transfer of single sugar residues to the growing chain [71]. Recently, it has been reported that the first monosaccharide of the *O*-unit is a derivative of a 2-amino-2,6-dideoxy-D-hexose (QuiN, FucN, QuiN4N, FucN4N) or a 2-amino-2-deoxy-D-hexose (GlcN, GalN) in the D-gluco or D-galacto configuration [71]. Furthermore, many LPSs were shown to have non-repetitive domains, which are thought to be a signal for cessation of the *O*-chain synthesis; for example, the incorporation of an *O*-methylated sugar residue allows for the termination of the *O*-polysaccharide chain at a specific sugar residue rather than at any residue [71]. Finally, it is possible to find a non-repetitive domain between the *O*-chain and the core oligosaccharide, termed a primer, such as a 2-*N*-acetylamino sugar whose transfer to a lipid carrier initiates the *O*-antigen synthesis [71].

3.5.1 *O-chain polysaccharide structure elucidation*

The *O*-chain of LPSs, recovered in supernatants after mild acid treatment, usually consists of a regular long chain polysaccharide with the core oligosaccharide still at the reducing end of the chain, but since this oligosaccharide moiety represents a small percentage of the whole polymer, it is generally barely detectable from the chemical and spectroscopical point of view. *O*-chain structure determination aims to define the composition in monosaccharides, the relative, the anomeric and the absolute configurations of each monosaccharide, their ring sizes, the linkage of each residue and their sequence within the repeating unit. In some cases, the non-stoichiometric presence of some substituents can determine heterogeneity and hide the structural regularity [72].

The monosaccharide composition is achieved by the classical analytic approaches consisting of acid hydrolysis and/or methanolysis followed by further chemical modification, such as acetylation, to perform GLC-MS analyses. The reaction sequence includes methylation, acid hydrolysis, reduction and acetylation, and these steps are carried out to determine the glycoside substitution sites and the ring sizes of each monosaccharide [73]. For the absolute configurations, GLC analysis of convenient glycoside derivatives are usually exploited [74]. The anomeric configurations and the sequence of the monosaccharide residues are determined by 1D and 2D proton (^{1}H) and carbon (^{13}C)-NMR experiments (double quantum filter correlation spectroscopy (DQF-COSY), total correlated spectroscopy (TOCSY), nuclear overhauser effect spectroscopy (NOESY), rotating-frame overhauser effect spectroscopy (ROESY), ^{1}H,^{13}C-heteronuclear single quantum coherence (HSQC), ^{1}H,^{13}C-heteronuclear multiple bond correlation (HMBC). In detail, TOCSY and DQF-COSY spectra allow for the correct identification and assignment of all ring protons, and on the basis of these data, the assignment for all ^{13}C resonances usually follows from the analysis of a ^{1}H,^{13}C-HSQC spectrum. The intra-residue pattern of dipolar correlations in the 2D NOESY or ROESY spectra gives the proof for specific configurations. Information concerning the sequence of the monosac-charides within the oligosaccharide (or the repeating unit of the polysaccharide) are then deduced from the observation of the inter-residual dipolar correlations in 2D NOESY and ROESY spectra and from the existence of scalar long range correlations in the $^{1}H,^{13}C$-HMBC spectrum, among the ^{1}H and ^{13}C nuclei involved in glycosidic linkages.

3.6 Biosynthesis of the LPS Molecule

The biosynthesis of LPSs has been extensively investigated, since it represents an attractive target in the development of new antibiotics and in the synthesis of molecules that can reduce or eliminate bacterial virulence and, moreover, that can work as antibiotic adjuvant potentiating antibiotic effects. The three structural domains of the LPS molecule are synthesized through different biosynthetical pathways that take place in the cytoplasm. Since endotoxins are distributed on the bacterial outer membrane, the cytoplasmic synthesis implies their subsequent export to the surface of bacteria cells. Herein, after the assembling of lipid A and core oligosaccharides in the cytoplasm, nascent LPSs are translocated across the inner membrane. The *O*-chain is ligated to the core-lipid A moiety only at the periplasmic side of the membrane, thus concluding LPS synthesis [12]. The first step of the biosynthetic pathway of LPSs is the synthesis of Kdo_2-lipid A [12,75] that has been well characterized in *E. coli* and shown to be mediated by nine enzymes (Table 3.1) highly conserved among almost all Gram-negative bacteria. The initial reactions are represented by the acylation of the small molecule UDP-GlcNAc, followed by the addition of two 3-hydroxy (OH) acyl chains to the positions 2 and 3 of this UDP-GlcNAc, forming the UDP-diacyl-GlcN catalyzed by the enzymes lipid peroxidise (Lpx) A, C and D [76]. In detail, *E. coli* LpxA is involved in the transfer of a C14 hydroxyacyl chain to the 3-OH group of UDP-GlcNAc, thus forming an ester linkage. The capability of LpxA to incorporate only C14 fatty acid chains led to consider the active site of this enzyme as possessing a "hydrocarbon ruler" able to work with a β-hydroxymyristate [C14:0(3-OH)] at a rate two orders of magnitude faster than C12 or C16 chains [77], as demonstrated by the structure of *E. coli* lipid A. On the contrary, the "ruler" of LpxA from *Pseudomonas aeruginosa* is set on acyl chains of 10 carbon atoms [78].

Table 3.1. Enzymes involved in *E. coli* lipid A biosynthesis.

Enzyme	Function
LpxA	Fatty acylation of UDP-GlcNAc.
LpxC	Deacetylation of UDP-3-*O*-(acyl)-GlcNAc.
LpxD	Addition of a second R-3-hydroxymyristate chain to make UDP-2,3-diacyl-GlcN.
LpxH	Cleavage of the pyrophosphate linkage of UDP-2,3-diacyl-GlcN to form lipid X.
LpxB	Allowing condensation between UDP-2,3-diacyl-GlcN and lipid X.
LpxK	Phosphorylation of the position 4′ of the disaccharide 1-phosphate to form lipid IV_A.
KdtA	Addition of two Kdo residues to the position 6′ of lipid IV_A.
LpxL	Addition of a secondary lauroyl fatty acid chain to the non-reducing glucosamine residue.
LpxM	Addition of a secondary myristoyl fatty acid chain to the non-reducing glucosamine residue.

It was demonstrated that a single point mutation is required to convert the "hydrocarbon ruler" of *E. coli* to a 10 carbon atom enzyme [77]. LpxC is a zinc metallo-enzyme that acts as a deacetylase of the UDP-3-*O*-acyl-GlcNAc to produce UDP-3-*O*-acylglucosamine [79–82]. Since LpxC is encoded from a single-copy gene present in several Gram-negative bacteria, and displays no sequence similarity to other deacetylases, it represents a promising target for development of broad-spectrum Gram-negative novel antibiotics aimed at interrupting lipid A biosynthesis [82–85]. The deacylation allows the biosynthesis equilibrium moving towards the incorporation of the second acyl chain on the UDP-3-*O*-acylglucosamine by *Lophozozymus* pictor toxin (LPTX) D to form UDP-2,3-diacylglucosamine [86]. Thereafter, the pyrophosphate bond of UDP-2,3-diacylglucosamine is cleaved by the highly selective pyrophosphatase LpxH to form 2,3-diacylglucosamine-1-phosphate (lipid X) and uridine monophosphate (UMP) [87]. The disaccharide synthase LpxB is responsible for the subsequent condensation of one molecule of lipid X with one molecule of its precursor, the UDP-2,3-diacylglucosamine [88–89].

Both LpxB and LpxH are peripheral membrane proteins on the contrary of the previous step-enzymes and the subsequent LpxK, KdtA, LpxL and LpxM enzymes, which are integral proteins of the inner membrane [76]. LpxK is a specific kinase that phosphorylates the position 4′ of the disaccharide-1-P to form the precursor lipid IV_A (Fig. 3.2) [90–91]. This latter is of great interest since it was observed that on mouse cells it acts as a potent agonist on the TLR4/MD-2 receptorial complex, whereas on human cells, it plays as an antagonist underlying a species-specific difference in the recognition of lipid IV_A by the receptor [37]. *E. coli* KdtA acts as a bifunctional enzyme transferring two Kdo residues to the 6′ position of the non-reducing lipid IV_A glucosamine, by using the sugar nucleotide cytosine monophosphate (CMP)-Kdo as the donor [76,92]. Previous studies have showed that KdtA homologs can transfer from one to four Kdo residues to the nascent LPS molecule, but it is not possible to determine whether this Kdo transferase acts as a mono-, bi- or even trifunctional enzyme by bioinformatic analyses. The capability to transfer more than two Kdo residues was only observed to date in *Chlamydia* species, with *C. psittacae* capable of transferring up to four Kdo units [82,93–95]. Subsequent steps in LPS biosynthesis involve LpxL and LpxM enzymes that are responsible for the addition of a secondary lauroyl residue and myristoyl residues, respectively, to the non-reducing glucosamine residue, producing the Kdo_2-lipid A [96]. Interestingly, both *E. coli* transferases of secondary acyl chains require the presence of the Kdo disaccharide for their activity [82], and in non-thermophilic bacteria, they need the occurrence of a phosphate group linked to the Kdo; this latter phosphate group is transferred on to lipid A by the enzyme KdkA [97].

E. coli presents one additional acyltransferase homologous to LpxL, known as LpxP, which is activated under cold shock conditions (12°C), introducing a palmitoleate (C16:1) in place of a laurate (C12:0), thus incorporating an unsaturated acyl chain into the lipid A structure, which increases membrane fluidity necessary at low temperatures [98].

In parallel to the conserved biosynthetic pathway, other enzymes able to modulate the number of acyl chains of a given lipid A species have been described. In particular, the phosphate-activated glutaminase (Pag) P enzyme is responsible for the increase in fatty acid number [99–100], whereas LpxR and PagL are involved in the reduction of acyl chain numbers [101–102]. All three proteins are immersed in the bacterial outer membrane. Furthermore, in the periplasm or outer membrane, several other modifications in the lipid A structure can occur in order to confer an advantage to the bacteria in escaping the host innate immune system. For example, the decoration of the phosphate groups with Ara*p*4N by the enzyme aryl hydrocarbon receptor nuclear translocator (ArnT) is typical of several species, among which *Burkholderia*, *E. coli*, *Pseudomonas* and *Salmonella* species involve Ara*p*4N in their resistance to the most commonly used cationic antimicrobial peptides [103]. The transferase ArnT employs an undecaprenyl-linked Ara*p*4N as the donor substrate [104], preferentially modifying the 4′-phosphate group of lipid A on the periplasmic side of the inner membrane [105]. Recently, it was demonstrated that in the *Burkholderia* genus, ArnT is involved in the transfer of Ara*p*N residues both on the lipid A and on the core oligosaccharide moieties [31]. In other bacterial species, phosphoethanolamine transferase (EptA) is responsible for the transfer of a phosphoethanolamine (PEtN) on to the lipid A backbone; the incorporation of this zwitterionic residue may be advantageous in low pH conditions when the 2-amino-ethanol group (EtN) is protonated [82, 106].

Once the biosynthesis of the Kdo_2-lipid A is complete, the molecule acts as an acceptor on which the core oligosaccharide is assembled; this process of assembly provides sequential glycosyl transfers, catalyzed by specific membrane-associated glycosyltransferases that utilize nucleotide sugar precursors [12]. As opposed to the highly conserved lipid A structure, the core oligosaccharide shows some variations and is assembled on Kdo_2-lipid A at the cytoplasmic surface of the inner membrane. Genes involved in inner core synthesis, as well as the outer core assembly, are well studied in *Salmonella*, *E. coli* and *K. pneumonia* due to their location as a cluster on the chromosome. In particular, in *E. coli* and *Salmonella* these genes are grouped into three operons: gmhD, waaQ and kdtA

[107–109]. The first operon contains four genes: gmhD, waaF, waaC are involved in the biosynthesis and subsequent transfer of heptose residues, whereas the fourth one, waaL, is responsible for the ligation of the *O*-chain to the core-lipid A moieties [110]. The waaQ operon contains genes necessary for the biosynthesis of the outer core and for its decoration. Eventually, the operon kdtA presents the KdtA gene that is responsible for the addition of the two Kdo residues described above.

To complete the biosynthesis of a smooth LPS, the addition of an *O*-chain moiety to the nascent lipid A-core is required. Similar to the core oligosaccharide, the *O*-polysaccharide is synthesized on the cytoplasmic surface of the inner membrane, but the ligation to the core occurs at the periplasmic face of the membrane; thus, both the lipid A-core and the *O*-chain must be delivered to the periplasmic face by proteins MsbA and Wzx, respectively, to complete the LPS biosynthesis. In both *E. coli* and *Salmonella*, the enzymes involved in *O*-polysaccharide synthesis, assembly and export are encoded by genes composing the cluster *rfb* [12]. Unlike the core oligosaccharide, the repeating units of the *O*-chain are assembled on the undecaprenyl phosphate carrier, embedded in the plasma membrane, by using sugar nucleotides as donors. The elongation of the polysaccharide chain can occur in three different pathways termed Wzy-dependent, ATP-binding cassette (ABC)-transporter dependent and synthase-dependent [12]. The first process occurs in the case of heteropolymers or branched polysaccharides, in which the single repeating unit is synthesized in the cytoplasm by sequential glycosyltransferases and then is transferred to the periplasm; afterwards, the nascent chain is allocated on the translocated repeating unit by the Wzy *O*-chain polymerase enzyme [12]. The ABC-transporter pathway takes its name from the translocation of the chain, which is totally synthesized in the cytoplasm for homopolysaccharides, into the periplasm that requires the action of an ABC transporter [12]. Finally, the third pathway, the synthase-dependent, is rarely found in *O*-polysaccharide biosynthesis; indeed, it was observed only for some species such as *S. enterica* serovar *Borreze* O:54 [111]. The mechanism provides the presence of synthases that are integral membrane proteins with a double function: to polymerize the chain elongation acting as glycosyltransferases, and to extrude the nascent chain across the membrane playing as an exit pore [110–111]. Thus, the sugar's addition to the growing end of the chain extrudes the polymer from the cytoplasm to the periplasm. Independently from the pathway, once the undecaprenyl-linked *O*-chain is exported to the periplasm, the WaaL ligase links the *O*-chain to the preformed lipid A-core, thus concluding the LPS biosynthesis.

The nascent LPS, at this point, must be translocated from the periplasmic face of the inner membrane to the outer membrane. It has been reported that eight proteins, named lipopolysaccharide transporters (*lpts*), constitute the "Lpt machinery" that transports LPSs across the periplasm and inserts it in the outer membrane [112–117]. This machinery is composed of the cytosolic protein LptB, the periplasmic protein LptA, the inner membrane proteins LptC, LptF and LptG, and the outer membrane proteins LptD and LptE [112–117]. Briefly, the ABC transporter LptBFG translocates nascent LPSs to the inner leaflet of the outer membrane, functioning with LptC and LptA [113–114]. In particular, this latter appears as the key protein in the Lpt machinery, representing the protein bridge between the inner and the outer membrane as well as the binding protein to the LPS [118]. Once the endotoxin is in the inner leaflet of the outer membrane, LptD and LptE export it to the outer leaflet [118]. It is worth underlining that both of the detailed mechanisms of LPS translocation across the periplasm and its insertion in the outer membrane and the molecular requirements for LPS binding to LptA still remain unknown. In this scenario, a very interesting study demonstrated that the LptG protein, in the *Burkholderia* genus, is responsible for the recognition of Ara*p*N-modified LPSs, thus only exporting LPSs possessing the Ara*p*N-decoration to the outer membrane [31]. Since previous publications have reported that the presence of Ara*p*N on *Burkholderia* LPSs is essential for survival and viability of this bacterium, the abovementioned study has demonstrated that this essentiality is related to the proper export mechanism and the assembly of the *Burkholderia* LPS to the bacterial outer membrane [31].

3.7 LPSs in the Elicitation of Innate and Adaptive Immune Responses

In vertebrates, the early encounter with infective microorganisms contemplates the intervention of the innate immune system, which generates a rapid inflammatory process in order to arrest the growth and proliferation of the invading microbe. Following this, the adaptive immune response takes place, leading to the elimination of the infectious agent and to the establishment of a memory of this infection in order to respond more quickly and with higher efficiency in case of a possible new encounter with the same microbe. Despite the sophisticated adaptive immune response, the more crude innate immunity can count on diverse phagocytic cells, including monocytes and macrophages, which constitutively express on their surface pattern recognition receptors (PRRs) to recognize conserved microbial components, known as pathogen-associated molecular patterns (PAMPs) and also to recognize host-derived danger signals (damage-associated molecular patterns (DAMPs)) released in response to tissue damage. The recognition of these highly conserved microbial molecules allows the host to unequivocally signal the presence of bacterial infection. In the case of Gram-negative bacteria, LPSs, and in particular, the lipid A moiety, are recognized as PAMPs by a family of PRRs termed toll-like receptors (TLRs) [119–120]; moreover, the core oligosaccharide or, in smooth bacterial colonies, the *O*-chain, are the antigenic determinants recognized by the adaptive immune system. Among all TLRs, TLR4 is the receptor designated to recognize the LPS macromolecule, but its detection requires several accessory molecules. Indeed, since LPSs are present in oligomeric micelles, the first accessory protein, termed LPS binding protein (LBP), is responsible for the conversion to a monomeric LPS and for the delivery of the endotoxin to another protein, the glycosyl phosphatidylinositol-anchored cluster of differentiation 14 (CD14) [121]. This latter can chaperone LPSs from LBPs to TLR4s at the cell surface. The extracellular domain of TLR4 is linked to a small glycoprotein, MD-2, which turns out to be absolutely necessary for the proper functioning of TLR4s [122]. The binding of LPSs to the TLR4/MD-2 receptorial complex (Fig. 3.5) leads to the activation of two different immune response pathways, both resulting in the amplification of the transduction signal with the consequent massive production of inflammatory proteins, thus eliciting the inflammatory process [123]. As described above, the endotoxic properties of LPSs principally resides in the lipid A moiety and is strongly influenced by its primary structure; thus, each chemical variation that can occur in the general lipid A architecture can affect the interaction and the activation of the receptorial complex. A plethora of studies aimed at the elucidation of the LPS structure-activity relationship have highlighted that the hexa-acylated bisphosphorylated lipid A with an asymmetric 4:2 fatty acid distribution, found in the majority of enterobacteria as the *E. coli* lipid A, is considered to have the highest immunostimulatory capacity in mammalian cells. In contrast, the hypoacylated synthetic precursor of *E. coli* lipid A, the tetra-acylated lipid IV_A (Fig. 3.2), has been shown to have weak agonistic effects in some species of mammals and is well known to possess a strong antagonistic activity in human cells [37]. These different biological effects are correlated to two interconnected structural parameters: the molecular shape of the lipid A and the tilt angle between the diglucosamine backbone and fatty acid

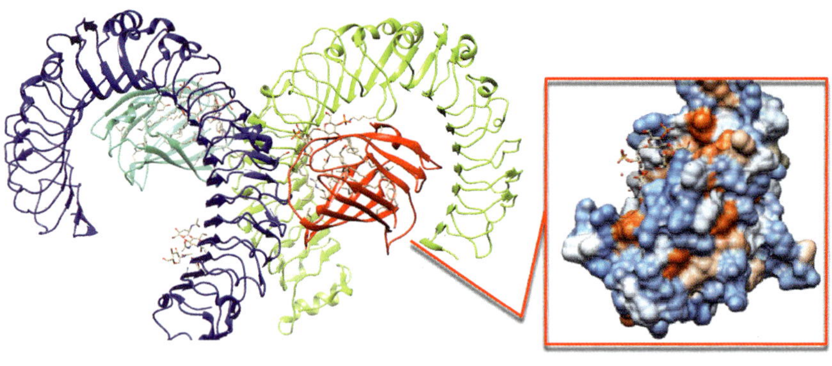

Figure 3.5 (On the left) 3D Model of the dimer of *E.coli* LPS core in complex with TLR4/MD-2. (On the right) Detail of the hydrophobic binding pocket of MD-2 protein.

chains, that is, the inclination of the lipid A hydrophilic moiety with respect to the hydrophobic portion [38]. In detail, the molecular shape possessed by lipid A influences its ability to be recognized by host immune system receptors; indeed, it has been shown that at 37°C, in aqueous solution and in physiological conditions, the most agonistic lipid A form has a truncated cone that drives to a hexagonal supra-structure, while lipid A moieties presenting an antagonistic activity assume a cylindrical shape leading to a lamellar structure [38]. Regarding the tilt angle, it has been found that the most active form has a conical structure with a tilt angle >50°, while the antagonist structures present smaller values of tilt angles [34]. Species with a tilt angle of <25°, such as lipid IV_A and the penta-acylated and symmetrically (3 + 3) hexa-acylated lipid A moieties, act as antagonists, while species with an angle between 25 and 50°, such as monophosphorylated lipid A, have a low bioactivity [34]. These physical parameters are reflected in the different binding-mode of the LPS molecule to the TLR4/MD-2 complex, since they regulate the lipid A affinity to the MD-2 hydrophobic pocket that is pivotal for the TLR4 activation. Briefly, the structure of *E. coli* lipid A provides that only five of its fatty acids are inside the binding task of the MD-2 protein, while the sixth one is extruded and interacts with TLR4, leading to its activation, thus promoting the activation of the immune response [123]. On the contrary, antagonist lipid A species, such as lipid IV_A (Fig. 3.2), binds completely into the MD-2 pocket with all four fatty acid chains, thus being unable to trigger any TLR4 activation [124]. In this scenario, a fundamental finding has been the discovery of monophosphoryl lipid A that is used as a vaccine adjuvant, since it has low toxicity and has been demonstrated to bc safe and effective [125].

An interesting issue regards the yet unclear mechanism of elicitation of innate immunity in plants. Indeed, despite some aspects displaying parallels with animals, very little is known about the mechanisms of detection of PAMPs by plants and about the associated signal transduction pathways that trigger plant immunity. LPSs seem to have diverse roles in the bacterial pathogenesis of plants. In particular, LPSs can act as other established elicitors, including flagellin and periplasmic oligosaccharides, being recognized by plants in order to elicit or potentiate plant defense-related responses [126]. It was previously assessed that LPSs are able to retard or block what is known as "hypersensitive responses" (HRs). HRs are programmed cell deaths that are induced by non-pathogenic bacteria and are considered resistant to plant hosts, since they are associated with a rapid decline of the number of viable microbes present in plant tissues followed by cell and tissue necrosis. Dow *et al.* [127] demonstrated that inoculation of leaves with heat-killed bacteria followed by inoculation with living bacteria prevents the HR effects. For instance, it was recently demonstrated that lipid A from *Xanthomonas campestris* pv. *campestris* is recognized as a PAMP by plant cells triggering the activation of immune response genes, as in mammals and insects [54]. Moreover, it was showed that core oligosaccharides from this bacterium also induce plant defense, but through a mechanism that is independent from that of lipid A [54]. The comprehension of the molecular mechanisms underlying the effects that LPSs have on plants has enormous importance in the control of bacterial plant diseases, as well as in the study of the beneficial bacteria–plants and bacteria–fungi interactions, which in some cases is demonstrated to occur as a consequence of recognition of LPSs [68,128]. Therefore, one of the major goals of research on LPSs is aimed at identifying the yet unknown LPS receptors involved in recognition and successive signal transduction in plant cells.

References

1. Lüderitz, O., Freudenberg, M. A., Calanos, C., Lehmann, V., Rietschel, E. Th., Shaw, D. H. (1982) *Current Topics in Membranes and Transport* **17**, Academic Press Inc. (New York). pp. 79–51.
2. Westphal, O., Lüderitz, O., Calanos, C., Mayer H., Rietschel, E. T. (1986) *Advances in Immunopharmacology* Chedid, L., Hadden, J. W., Spreafico, F. Eds. Pergamon Press (Oxford). pp. 13–34.
3. Wilkinson, S. C. (1977) *Surface Carbohydrates of the Procaryotic Cell.* Sutherland I. W. Ed. Academic Press Inc. (New York). pp. 97–105.
4. Pfeiffer, R. (1892) *Z Hygiene* **11**, 393–412.

5. Rietschel, E. T., Westphal, O. (1999) *Endotoxin in Health and Disease.* Brade, H., Opal, S. M., Vogel, S. N., Morrisson, D. C. Eds. Marcel Dekker (New York). pp. 1–30.
6. Westphal, O., Jann, K. (1965) *Methods Carbohydr. Chem.* **5**, 83–91.
7. Tanamoto, K., Zähringer, U., Mckenzie, G. R., Galanos, C., Rietschel, E. T., Lüderitz, O., Kusumoto, S., Shiba, T. (1984) *Infect. Immun.* **44**, 421–426.
8. Alexander, C., Rietschel, E. T. (2001) *J. Endoxin Res.* **7**, 167–202.
9. Cardoso, L. S., Araujo, M. I., Góes, A. M., Pacífico, L. G., Oliveira, R. R., Oliveira, S. C. (2007). *Microb. Cell Fact.* **6**, 1.
10. Silipo, A., De Castro, C., Lanzetta, R., Parrilli, M., Molinaro, M. (2010) *Prokaryotic Cell Wall Compounds Structure and Biochemistry.* Konig, H., Herald, C., Varma, A. Eds. Springer-Verlag (Berlin). pp. 133–154.
11. Raetz, C. R. (1990) *Annu. Rev. Biochem.* **59**, 129–170.
12. Raetz, C. R., Whitfield, C. (2002) *Annu. Rev. Biochem.* **71**, 635–700.
13. Holst O., Ulmer, A. J., Brade, H., Flad, H. D., Rietschel, E. T. (1996) *FEMS Immunol. Med. Microbial.* **16**, 83–104.
14. Lüderitz, O., Galanos, C., Risse, H. J., Ruschmann, E., Schlecht, S., Schmidt, G., Schulte-Holthausen, H., Wheat, R., Westphal, O., Schlosshardt, J. (1966) *Ann. NY Acad. Sci.* **133**, 347–349.
15. Kim, Y. B., Watson, D. W. (1967) *J. Bacteriol.* **94**, 1320–1326.
16. Kasay, N., Nowotny, A. (1967) *J. Bacteriol.* **94**, 1824–1836.
17. Galanos, C. (1975) *Z Immunol. Forsch* **149**, 214.
18. Zähringer, U., Lindner, B., Rietschel, E. T. (1999) *Endotoxin in Health and Disease.* Brade, H., Opal, S. M., Vogel, S. N., Morrison, D. C. Eds. Marcel Dekker (New York). pp. 93–114.
19. Hellerqvist, C. G., Lindberg, B., Samuelsson, K., Lindberg, A. A. (1971) *Acta Chem. Scand.* **25**, 955–961.
20. Hellerqvist, C. G., Lindberg, B., Svensson, S., Holme, T., Lindberg, A. A. (1969) *Carbohydr. Res.* **9**, 237–241.
21. Svenson, S. B., Lönngren, J., Carlin, N., Lindberg, A. A. (1979) *J. Virol.* **32**, 583–592.
22. Szafranek, J., Kumirska, J., Czerwicka, M., Kunikowska, D., Dziadziuszko, H., Glosnicka, R. (2006) *FEMS Immunol. Med. Microbiol.* **48**, 223–236.
23. Kaczynski, Z., Gajdus, J., Dziadziuszko, H., Stepnowski, P. J. (2009) *Pharm. Biomed. Anal.* **50**, 679–682.
24. Hellerqvist, C. G., Larm, O., Lindberg, B., Holme, T., Lindberg, A. A. (1969) *Acta Chem. Scand.* **23**, 2217–2222.

25. Di Fabio, J. L., Brisson, J. R., Perry, M. B. (1989) *Biochem. Cell Biol.* **67**, 278–280.
26. Lindberg, B., Leontein, K., Lindquist, U., Svenson, S. B., Wrangsell, G., Dell, A., Rogers, M. (1988) *Carbohydr. Res.* **174**, 313–322.
27. Schwudke, D., Linscheid, M., Strauch, E., Appel, B., Zähringer, U., Moll, H., Müller, M., Brecker, L., Gronow, S., Lindner, B. (2003) *J. Biol. Chem.* **278**, 27502–27512.
28. Ieranò, T., Silipo, A., Nazarenko, E. L., Gorshkova, R. P., Ivanova, E. P., Garozzo, D., Sturiale, L., Lanzetta, R., Parrilli, M., Molinaro, A. (2010) *Glycobiology* **20**, 586–593.
29. van Putten, J. P. (1993) *EMBO J.* **12**, 4043–4051.
30. Raetz, C. R., Reynolds, C. M., Trent, M. S., Bishop, R. E. (2007) *Annu. Rev. Biochem.* **76**, 295–329.
31. Hamad, M. A., Di Lorenzo, F., Molinaro, A., Valvano, M. A. (2012) *Mol. Microbiol.* **85**, 962–974.
32. Brandenburg, K., Mayer, H., Koch, M. H., Weckesser, J., Rietschel, E. T., Seydel, U. (1993) *Eur. J. Biochem.* **218**, 555–563.
33. Rietschel, E. T., Kirikae, T., Schade, F. U., Mamat, U., Schmidt, G., Loppnow, H., Ulmer, A. J., Zähringer, U., Seydel, U., Di Padova, F., Schreier, M., Brade, H. (1994) *FASEB J.* **8**, 217–225.
34. Seydel, U., Oikawa, M., Fukase, K., Kusumoto, S., Brandenburg, K. (2000) *Eur. J. Biochem.* **267**, 3032–3039.
35. Fukuoka, S., Brandenburg, K., Müller, M., Lindner, B., Koch, M. H., Seydel, U. (2001) *Biochim. Biophys. Acta* **1510**, 185–197.
36. Oikawa, M., Shintaku, T., Fukuda, N., Sekljic, H., Fukase, Y., Yoshizaki, H., Fukase, K., Kusumoto, S. (2004) *Org. Biomol. Chem.* **2**, 3557–3565.
37. Golenbock, D. T., Hampton, R. Y., Qureshi, N., Takayama, K., Raetz, C. R. (1991) *J. Biol. Chem.* **266**, 19490–19498.
38. Netea, M. G., van Deuren, M., Kullberg, B. J., Cavaillon, J. M., van der Meer, J. W. (2002) *Trends Immunol.* **23**, 135–139.
39. Que, N. L. S., Lin, S. H., Cotter, R. J., Raetz, C. R. (2000) *J. Biol. Chem.* **275**, 28006–28016.
40. Ieranò, T., Silipo, A., Sturiale, L., Garozzo, D., Brookes, H., Khan, C. M., Bryant, C., Gould, F. K., Corris, P. A., Lanzetta, R., Parrilli, M., De Soyza, A., Molinaro, A. (2008) *Glycobiology* **18**, 871–881.
41. Sforza, S., Silipo, A., Molinaro, A., Marchelli, R., Parrilli, M., Lanzetta, R. (2004) *J. Mass Spectrom.* **39**, 378–383.
42. Sturiale, L., Garozzo, D., Silipo, A., Lanzetta, R., Parrilli, M., Molinaro, A. (2005) *Rapid Communications in Mass Spectrometry* **19**, 1829–1834.

43. Hase, S., Rietschel, E. T. (1976) *Eur. J. Biochem.* **63**, 101–107.
44. Knirel, Y. A., Moll, H., Zähringer, U. (1996) *Carbohydr. Res.* **293**, 223–234.
45. Ribeiro, A. A., Zhou, Z. Raetz, C. R. H. (1999) *Magn. Reson. Chem.* **37**, 620–631.
46. Brecker, L. (2003) *Chem. Phys. Lipids* **125**, 27–39.
47. Silipo, A., De Castro, C., Lanzetta, R., Molinaro, A. Parrilli, M. (2004) *Glycobiology* **14**, 805–815.
48. Holst, O. (1999) *Endotoxin in Health and Disease.* Brade, H., Morrsion, D.C., Opal, S., Vogel, S. Eds. Marcel Dekker (New York). pp. 115–154.
49. Meredith, T. C., Aggarwal, P., Mamat, U., Lindner, B., Woodard, R. W. (2006) *ACS Chem. Biol.* **1**, 33–42.
50. Mamat, U., Meredith, T. C., Aggarwal, P., Kühl, A., Kirchhoff, P., Lindner, B., Hanuszkiewicz, A., Sun, J., Holst, O., Woodard, R. W. (2008) *Mol. Microbiol.* **67**, 633–648.
51. Vinogradov, E. V., Bock, K., Petersen, B. O., Holst, O., Brade, H. (1997) *Eur. J. Biochem.* **243**, 122–127.
52. Vinogradov, E. V., Korenevsky, A., Beveridge, T. J. (2003) *Carbohydr. Res.* **338**, 1991–1997.
53. De Soyza, A., Silipo, A., Lanzetta, R., Govan, J. R., Molinaro, A. (2008) *Innate Immun.* **14**, 127–144.
54. Silipo, A., Molinaro, A., Sturiale, L., Dow, J. M., Erbs, G., Lanzetta, R., Newman, M. A., Parrilli, M. (2005) *J. Biol. Chem.* **280**, 33660–33668.
55. De Castro, C., Molinaro, A., Lanzetta, R., Silipo, A., Parrilli, M. (2008) *Carbohydr. Res.* **343**, 1924–1933.
56. Tsai, C. M., Jankowska-Spephens, E., Mizamur, R. M., Cipollo, J. F. (2009) *J. Biol. Chem.* **284**, 4616–4625.
57. Preston, A., Mamdrell, R. E., Gibson, B. W., Apicella, M. A. (1996) *Crit. Rew. Microbiol.* **22**, 139–180.
58. Holst, O. (2000) In *Bacterial Toxins: Methods and Protocols. Methods in Molecular Biology.* Holst, O. Ed. Humana Press (Totowa). pp. 345–353.
59. Vinogradova, E. V., Bradea, L., Bradea, H., Holst, O. (2003) *Carbohyd. Res.* **338**, 2751–2756.
60. Zähringer, U., Knirel, Y. A., Lindner, B., Helbig, J. H., Sonesson, A., Marre, R., Rietschel, E. T. (1995) *Prog. Clin. Biol. Res.* **392**, 113–139.
61. Kittelberger, R., Hilbink, F. (1993) *J. Biochem. Biophys. Methods* **26**, 81–86.
62. Adinolfi, M., Corsaro, M. M., De Castro, C., Evidente, A., Lanzetta, R., Molinaro, A., Lavermicocca, P., Parrilli, M. (1996) *Carbohydr. Res.* **284**, 119–133.

63. Jansson, P. E., Brade, H., Morrison, D. C., Opal, S., Vogel, S. (1999) In *Endotoxin in Health and Disease.* Brade, H., Opal, S. M., Vogel, S. N., Morrison, D. C. Eds. Marcel Dekker (New York). pp. 155–178.
64. Whitfield, C. (2006) *Annu. Rev. Biochem.* **75**, 39–68.
65. Lerouge, I., Vanderleyden, J. (2001) *FEMS Microbiology Reviews* **26**, 17–47.
66. Adinolfi, A., Corsaro, M. M., De Castro, C., Evidente, A., Lanzetta, R., Mangoni, L., Parrilli, M. (1995) *Carbohydr. Res.* **274**, 223–232.
67. Adinolfi, A., Corsaro, M. M., De Castro, C., Evidente, A., Lanzetta, R., Molinaro, A., Parrilli, M. (1996) *Carbohydr. Res.* **284**, 111–118.
68. Silipo, A., Leone, M. R., Erbs, G., Lanzetta, R., Parrilli, M., Chang, W. S., Newman, M. A., Molinaro, A. (2011) *Angew. Chem. Int. Ed. Engl.* **50**, 12610–12612.
69. Di Lorenzo, F., Silipo, A., Costello, A., Sturiale, L., Garozzo, D., Callaghan, M., Lanzetta, R., Parrilli, M., McClean, S., Molinaro A. (2012) *Eur. J. Org. Chem.* **11**, 2243–2249.
70. Amano, Y., Kanda, T. (2002) *Trends in Glycoscience and Glycotechnology* **14**, 105–114.
71. Knirel, Y. A. (2011) *Bacterial Lipopolysaccharides. Structure, Chemical Synthesis, Biogenesis and Interaction with Host Cells.* Knirel, Y. A., Valvano, M. A. Eds. Springer Wien (New York). pp. 41–118.
72. Knirel, Y. A., Kocharova, N. A., Hynes, S. O., Widmalm, G., Andersen, L. P., Jansson, P. E., Moran, A. P. (1999) *Eur. J. Biochem.* **266**, 123–131.
73. Ciucanu, I., Kerek, F. (1984) *Carbohydr. Res.* **131**, 209–217.
74. Leontein, K., Lindberg, B., Lonngren, J. (1978) *Carbohydr. Res.* **62**, 359.
75. Doerrler, W. T. (2006) *Mol. Microbiol.* **60**, 542–552.
76. Wang, X., Quinn, P. J. (2010) *Progress in Lipid Research* **49**, 97–107.
77. Wyckoff, T. J., Lin, S., Cotter, R. J., Dotson, G. D., Raetz, C. R. (1998) *J. Biol. Chem.* **273**, 32369–32372.
78. Dotson, G. D., Kaltashov, I. A., Cotter, R. J., Raetz, C. R. H. (1998) *J. Bacteriol.* **180**, 330–337.
79. Anderson, M. S., Bull, H. G., Galloway, S. M., Kelly, T. M., Mohan, S., Radika, K., Raetz, C. R. (1993) *J. Biol. Chem.* **268**, 19858–19865.
80. Anderson, M. S., Robertson, A. D., Macher, I., Raetz, C. R. (1988) *Biochemistry* **27**, 1908–1917.
81. Young, K., Silver, L. L., Bramhill, D., Cameron, P., Eveland, S. S., Raetz, C. R., Hyland, S. A., Anderson, M. S. (1995) *J. Biol. Chem.* **270**, 30384–30391.

82. Stead C. M., Pride A. C., Trent M. S. (2011) In *Bacterial Lipopolysaccharides: Structure, Chemical Synthesis, Biogenesis and Interaction with Host Cells.* Knirel, Y. A., Valvano, M. A. Eds. Springer Wien (New York). pp. 163–194.
83. Onishi, H. R., Pelak, B. A., Gerckens, L. S., Silver, L. L., Kahan, F. M., Chen, M. H., Patchett, A. A., Galloway, S. M., Hyland, S. A., Anderson, M. S., Raetz, C. R. (1996) *Science* **274**, 980–982.
84. Jackman, J. E., Fierke, C. A., Tumey, L. N., Pirrung, M., Uchiyama, T., Tahir, S. H., Hindsgaul, O., Raetz, C. R. (2000) *J. Biol. Chem.* **275**, 11002–11009.
85. Clements, J. M., Coignard, F., Johnson, I., Chandler, S., Palan, S., Waller, A., Wijkmans, J., Hunter, M. G. (2002) *Antimicrob. Agents Chemother.* **46**, 1793–1799.
86. Kelly, T. M., Stachula, S. A., Raetz, C. R. H., Anderson, M. S. (1993) *J. Biol. Chem.* **268**, 19866–19874.
87. Babinski, K. J., Raetz, C. R. H. (1998). *FASEB J.* **12**, A1288.
88. Ray, B. L., Painter, G., Raetz, C. R. (1984) *J. Biol. Chem.* **259**, 4852–4859.
89. Radika, K., Raetz, C. R. (1988) *J. Biol. Chem.* **263**, 14859–14867.
90. Garrett, T. A., Que, N. L., Raetz, C. R. (1998) *J. Biol. Chem.* **273**, 12457–12465.
91. Garrett, T. A., Kadrmas, J. L., Raetz, C. R. (1997) *J. Biol. Chem.* **272**, 21855–21864.
92. Brozek, K. A., Hosaka, K., Robertson, A. D., Raetz, C. R. (1989) *J. Biol. Chem.* **264**, 6956–6966.
93. Belunis, C. J., Clementz, T., Carty, S. M., Raetz, C. R. (1995) *J. Biol. Chem.* **270**, 27646–27652.
94. Brabetz, W., Lindner, B., Brade, H. (2000) *Eur. J. Biochem.* **267**, 5458–5465.
95. Rund, S., Lindner, B., Brade, H., Holst, O. (2000) *Eur. J. Biochem.* **267**, 5717–5726.
96. Brozek, K. A., Raetz, C. R. H. (1990) *J. Biol.Chem.* **265**, 15410–15417.
97. White, K. A., Lin, S., Cotter, R. J., Raetz, C. R. (1999) *J. Biol. Chem.* **274**, 31391–31400.
98. Carty, S. M., Sreekumar, K. R., Raetz, C. R. (1999) *J. Biol. Chem.* **274**, 9677–9685.
99. Guo, L., Lim, K. B., Poduje, C. M., Daniel, M., Gunn, J. S., Hackett, M., Miller, S. I. (1998) *Cell* **95**, 189–198.
100. Bishop, R. E., Gibbons, H. S., Guina, T., Trent, M. S., Miller, S. I., Raetz, C. R. (2000) *EMBO J.* **19**, 5071–5080.

101. Reynolds, C. M., Ribeiro, A. A., McGrath, S. C., Cotter, R. J., Raetz, C. R., Trent, M. S. (2006) *J. Biol. Chem.* **281**, 21974–21987.
102. Trent, M. S., Pabich, W., Raetz, C. R., Miller, S. I. (2001) *J. Biol. Chem.* **276**, 9083–9092.
103. Gunn, J. S., Lim, K. B., Krueger, J., Kim, K., Guo, L., Hackett, M., Miller, S. I. (1998) *Mol. Microbiol.* **27**, 1171–1182.
104. Trent, M. S., Ribeiro, A. A., Doerrler, W. T., Lin, S., Cotter, R. J., Raetz, C. R. (2001) *J. Biol. Chem.* **276**, 43132–43144.
105. Doerrler, W. T., Gibbons, H. S., Raetz, C. R. (2004) *J. Biol. Chem.* **279**, 45102–45109.
106. Lee, H., Hsu, F. F., Turk, J., Groisman, E. A. (2004) *J. Bacteriol.* **186**, 4124–4133.
107. Heinrichs, D. E., Yethon, J. A., Whitfield, C. (1998) *Mol. Microbiol.* **30**, 221–232.
108. Regué, M., Climent, N., Abitiu, N., Coderch, N., Merino, S, Izquierdo, L., Altarriba, M., Tomás, J. M. (2001) *J. Bacteriol.* **183**, 3564–3573.
109. Roncero, C., Casadaban, M. J. (1992) *J. Bacteriol.* **174**, 3250–3260.
110. Whitfield, C., Amor, P. A., Köplin, R. (1997) *Mol. Microbiol.* 23, 629–638.
111. Keenleyside, W. J., Whitfield, C. (1996) *J. Biol. Chem.* **271**, 28581–28592.
112. Doerrler, W. T. (2006) *Mol. Microbiol.* **60**, 542–552.
113. Bos, M. P., Tefsen, B., Geurtsen, J., Tommassen, J. (2004) *Proc. Natl. Acad. Sci. USA* **101**, 9417–9422.
114. Wu, T., McCandlish, A. C., Gronenberg, L. S., Chng, S. S., Silhavy, T. J., Kahne, D. (2006) *Proc. Natl. Acad. Sci. USA* **103**, 11754–11759.
115. Sperandeo, P., Lau, F. K., Carpentieri, A., De Castro, C., Molinaro, A., Dehò, G., Silhavy, T. J., Polissi, A. (2008) *J. Bacteriol.* **190**, 4460–4469.
116. Ruiz, N., Gronenberg, L. S., Kahne, D., Silhavy, T. J. (2008) *Proc. Natl. Acad. Sci. USA* **105**, 5537–5542.
117. Chng, S. S., Ruiz, N., Chimalakonda, G., Silhavy, T. J., Kahne, D. (2010) *Proc. Natl. Acad. Sci. USA* **107**, 5363–5368.
118. Sperandeo P., Dehò G., Polissi A. *Bacterial Lipopolysaccharides. Structure, Chemical Synthesis, Biogenesis and Interaction with Host Cells.* Knirel, Y. A., Valvano, M. A. Eds. Springer Wien (New York). pp. 262–282.
119. Medzhitov, R., Janeway, C. Jr. (2000) *Immunol. Rev.* **173**, 89–97.
120. Akira, S., Takeda, K. (2004) *Nat. Rev. Immunol.* **4**, 499–511.
121. Medzhitov, R. (2001) *Nat. Rev. Immunol.* **1**, 135–145.

122. Ohto, U., Fukase, K., Miyake, K., Satow, Y. (2007) *Science* **316**, 1632–1634.
123. Park, B. S., Song, D. H., Kim, H. M., Choi, B. S., Lee, H., Lee, J. O. (2009) *Nature* **458**, 1191–1195.
124. Meng, J., Drolet J. R., Monks, B. G., Golenbock, D. T. (2010) *J. Biol. Chem.* **285**, 27935–27943.
125. Mata-Haro, V., Cekic, C., Martin, M., Chilton, P. M., Casella, C. R., Mitchell, T. C. (2007) *Science* **316**, 1628–1632.
126. Conrath, U., Thulke, O., Katz, V., Schwindling, S., Kohler, A. (2001) *Eur. J. Plant Pathol.* **107**, 113–119.
127. Dow, J. M., Newman, M. A., von Roepenack, E. (2000) *Annu. Rev. Phytopathol.* **38**, 241–261.
128. Leone, M. R., Lackner, G., Silipo, A., Lanzetta, R., Molinaro, A., Hertweck, C. (2010) *Angew. Chem. Int. Ed. Engl.* **49**, 7476–7480.

Part 2:

Towards Glycomics: Methodologies and Techniques for the Determination of Structure and Activity Relationship

Chapter 4

Mass Spectrometry: Updates in the Elucidation of Structure of Oligosaccharides

Maria Michela Corsaro

Dipartimento di Scienze Chimiche, Università degli Studi di Napoli "Federico II", Complesso Universitario Monte S. Angelo, Via Cinthia 4, 80126 Napoli, Italy

4.1 Introduction

Carbohydrates are the most abundant substances present on Earth. They can be organized either as polysaccharides, or oligosaccharides substituting proteins and lipids, giving more complex molecules called glycoconjugates. These can be classified as (1) glycolipids, when a lipophilic moiety is substituted by an oligo/polysaccharide; (2) glycoproteins, when one or more oligosaccharides are linked to a specific amino acid of the polypeptide; or (3) proteoglycans, when a protein is modified with linear anionic polysaccharides (glycosaminoglycan, GAG). In addition, monosaccharides can be a part of important biological molecules such as nucleic acids.

The oligomerization of carbohydrates can give rise to an almost boundless array of structures. For example, three hexoses could produce from 1,056 to 27,648 different and distinct trisaccharides. If the number of monosaccharides enhances in the structure, the possibility of different arrays rises exponentially. In addition, branched structures and/or the presence of non-carbohydrate substituents further increases the number of possible combinations [1]. In contrast, nucleic acids and proteins are linear polymers that can each have only one basic type of linkage. Moreover, unlike nucleic acids and proteins, the biosynthesis of carbohydrates is not template-driven; that is, there is no direct genetic code for this complex biosynthesis.

The structures of glycoconjugate glycans are regulated dynamically, as a result of biosynthetic reactions that are under complex regulation. Due to the fact that several of these reactions depend on many factors directly linked to both gene expression and metabolism, the result is that glycans are typically expressed as a mixture of glycoforms. This diversity multiplies the structural and functional complexity of proteins and lipids by orders of magnitude, and reflects both the types and the status of the cell in which the glycans are synthesized [2].

Both eukaryotic and prokaryotic organisms express abundant carbohydrates that are localized both in the cells and in the intercellular room. In particular, the glycoconjugates are concentrated on the surface of the cells, and therefore, are involved in the interactions between the cell and the environment. Recognition, adhesion, cell–cell communication, and cell–host interaction are the most important processes where the glycoconjugates are involved. Moreover, they participate in cell embryogenesis and change their structure according to the health or illness state of the cell [3]. Many pathologies can involve the alteration of the glycan moieties of glycoconjugates, for example in the case of cancer [4,5] and for congenital disorders of glycosylation, the latter due to the genetic alteration of enzymes involved in the carbohydrate biosynthesis. Therefore, they play a key role in both medical and pharmaceutical fields, with their structure being useful for both diagnostic and therapeutic purposes, as in the case of the development of carbohydrate-based vaccines [6].

Despite the central role played by carbohydrates in numerous important biological processes, their structural definition is still a challenge for analytical chemists [7]. Mass spectrometry (MS) represents one of the most powerful techniques to overcome separation problems and, with the introduction of tandem mass methodologies, provides additional structural information complementary to the previous analytical experiments.

The primary structure of an oligo/polysaccharide chain is defined by the following parameters: (1) qualitative and quantitative composition of monosaccharides; (2) absolute configurations (D or L); (3) anomeric configurations (α or β); (4) position of glycosidic linkages; (5) sequence of monosaccharides; and (6) type, number and position of non-carbohydrate substituents (phosphates, sulfates, acetates, etc).

Mass spectrometry is able to fulfill all of these items, with very low amounts of sample, except for the anomeric configuration determination, for which nuclear magnetic resonance (NMR) experiments are requested.

In this chapter, new methodological aspects about mass spectrometry of the glycan moiety of glycoconjugates will be reviewed. In particular, some mass spectrometry approaches will be described for the carbohydrate moieties of glycoproteins, glycolipids and glycosaminoglycans.

Many mass spectrometry instruments with different combinations of ion sources and analyzers that are able to investigate oligosaccharide structures are available today, the descriptions of which have been extensively reviewed in dedicated reviews [8–11]. Here, only a brief depiction of ion sources is reported.

4.2 Ion Sources

Mass spectrometry of glycoconjugates started in the eighties, when the field desorption mass spectrometry (FD-MS) of intact oligosaccharides derived from glycoproteins were obtained [12]. Until that moment, it was only possible to analyze derivatized monosaccharides with electron ionization (EI) as the source. Today, matrix-assisted laser desorption/ionization (MALDI) and electrospray ionization (ESI) soft ionizations are the methods used in the vast majority of published papers. It is important to underline that these two techniques are mainly used for the de-polymerization products obtained in a different way for each class of glycoconjugate.

MALDI entails a pulsed laser light that hits a metal target where an organic acidic substance (matrix) and the analyte are co-crystalized. This irradiation creates a molecular plume containing ionized analyte and matrix molecules that may be analyzed. The main advantage of this technique is the possibility of analyzing complex mixtures of oligosaccharides, even if some inconveniences may be encountered. One of these is the loss of important monosaccharides (such as fucose and sialic acid) or decoration of oligosaccharides, such as sulfates and phosphate groups, due to the matrix acidity. To bypass these problems, glycans derivatization has been introduced [13]. Another way is the analysis of the oligosaccharides, also by ESI MS, since the extent of fragmentation of acidic glycans and other fragile ions is much lower than the one observed using vacuum MALDI [14].

The ESI process occurs at atmospheric pressure by passing a solution-based sample through a small capillary (internal diameter < 250 μm) that is at a potential difference relative to a counter electrode. The analyte solution is thereby caused to form extremely fine droplets that readily evaporate to generate analyte ions. During the ESI process, multiple charged ions are created; thus, a high resolution analyzer is required in order to investigate oligosaccharide mixtures.

Electron ionization, one of the long-time ionization techniques, is very useful in carbohydrate analysis. This type of source, when connected with a gas chromatography-mass spectrometry (GC-MS), is a valuable tool to determine sugar composition and linkage analysis for unknown structures. Many methods for monosaccharide composition analysis have been developed since the seventies and all of them require sugar derivatization before GC-MS analysis. The most widely used are alditol acetates or *O*-methyl glycosides, both as peracetylated derivatives, and per-*O*-trimethylsilyl ethers, and all of them are obtained after cleaving the glycosidic linkages and subsequent derivatization. The identification occurs both by the retention time and the fragmentation pattern.

A smart approach for determining linkage position and the ring size of monosaccharides is the methylation analysis, which comprises the introduction on each hydroxyl group of the native oligo/polysaccharide a methyl group. After cleavage of the glycosidic linkages and reduction of anomeric carbons, alditols with free hydroxyl groups previously involved in a linkage and in the ring size, are produced. Acetylation of this alditol mixture affords partially methylated alditol acetates that are then analyzed by the GC-MS.

Another application of the GC-MS is the monosaccharide absolute configuration determination. The procedure consists of the differentiation of D and L monosaccharides through diastereoisomeric octylglycosides preparation, using (+)-2-octanol or (+)-2-butanol. After acetylation, the mixture is analyzed by GC-MS and the identification occurs of both the retention time and the fragmentation pattern [15].

4.3 Tandem Mass Spectrometry

As with other classes of biomolecules, tandem mass spectrometry is a very effective technique to obtain sequence information from oligosaccharide chains [16–18]. Tandem MS (MS^n) involves the isolation of specific ion species that are further investigated for structures after fragmentation. To gather information about fragment ions in MS/MS, ion activation and dissociation of the mass selected precursor is essential. Many methods have been developed; however, they can be classified into two groups based on the method of energy deposition. These two groups are vibrational and electronic [19]. The former includes low energy collision-induced dissociation (CID) and infrared multiphoton dissociation (IRMPD), whereas the latter includes high energy CID, electron capture dissociation (ECD), electron transfer dissociation (ETD), and ultraviolet (UV) photodissociation.

Recently, several novel fragmentation techniques have been applied to the analysis of carbohydrates, such as electron detachment dissociation (EDD) and electron excitation dissociation (EED). Unfortunately, these techniques are mostly limited to

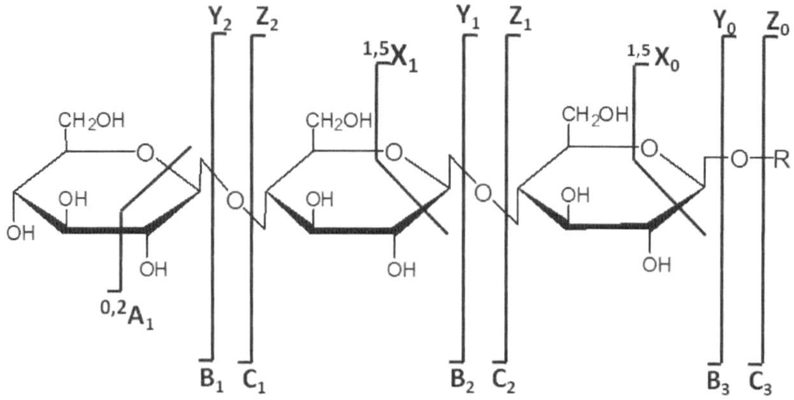

Figure 4.1 Nomenclature for glycans and glycoconjugates fragments observed in tandem mass spectra, according to Domon and Costello [21].

fourier transform ion cyclotron resonance (FT-ICR) MS, for which the cost and complexity of these instrument systems still limit their practice [20].

Two types of fragmentations occur in carbohydrates: the first concerns the glycosidic cleavage, while the second, a cross-ring cleavage. In each case, the charge retention can be either on the reducing or non-reducing end. In addition, a double fragmentation can sometimes be observed. The fragmentation of ions obtained from carbohydrates is assigned according to the nomenclature introduced by Domon and Costello (Fig. 4.1) [21].

4.4 Glycoproteins

Since the eighties, the incredible importance of glycoconjugates in biology and medicine has been realized [22], challenging the entire generation of chemists in solving the structural intricacies of carbohydrates. Many efforts have been focused on the development of new methodologies for studying the glycome, i.e. the repertoire of primary structures of the glycan moieties attached to proteins [23]. While glycomics is the study of the glycome, the central point of glycoproteomics is the study of the profile of glycosylated proteins (glycoproteome) in a biological system.

Initially identified in 1963 [24], glycosylation was thought of as a solely eukaryotic mechanism, although this view is no longer supported [25]. Glycoproteins are present in all cellular compartments of eukaryotic cells, and the glycosylation is highly specific at each site and generally stable for a given cell type and physiological state. However, the glycosylation of a protein can be altered significantly during a disease, such as inflammation, cancer and neurodegeneration. In particular, the alterations in glycoproteins can occur in one or both of two ways: (1) protein glycosylation sites are either hypo-, hyper-, or newly-glycosylated; and/or (2) the glycosylation form of the attached carbohydrate moiety is altered. Sensitive detection of such disease-associated glycosylation changes and abnormalities can provide a unique way to develop glycoprotein biomarkers for diagnosis and prognosis. In addition, intervention in the glycosylation and carbohydrate-dependent cellular pathways represents a potential new modality for cancer therapies [23].

In glycoproteins, the addition of sugar chains usually occurs at either asparagine (Asn) (termed *N*-glycosylation) residues within a defined consensus sequence (Asn-X-serine (Ser)/threonine (Thr)), where X is any amino acid residue except proline, or at hydroxylysine, hydroxyproline, serine, or threonine (termed *O*-glycosylation) [26]. Structurally, *N*- and *O*-glycans are also quite different. The *N*-linked glycans have a common core composed of $(GlcNAc)_2Man_3$ (*N*-acetylglucosamine, GlcNAc; mannose, Man) and other monosaccharides extend from the α(1–3)- and α(1–6)-branched core mannose units, including mannose, galactose, glucosamine, fucose, and sialic acids, usually *N*-acetyl neuraminic acid (NANA) or *N*-glycolyl neuraminic acid (NGNA), to complete the structure [27]. On the basis of their specific structural features, *N*-linked glycans are subcategorized into three main classes: high mannose, complex, and hybrid (Fig. 4.2). In contrast to the *N*-linked glycans, *O*-attached carbohydrates do not have a common core structure. Adding to the overall complexity of a given glycoprotein is the "microheterogeneity" for sites-of-glycosylation.

For a population of given glycoproteins, rarely there is a single site occupied only by a single carbohydrate structure; rather,

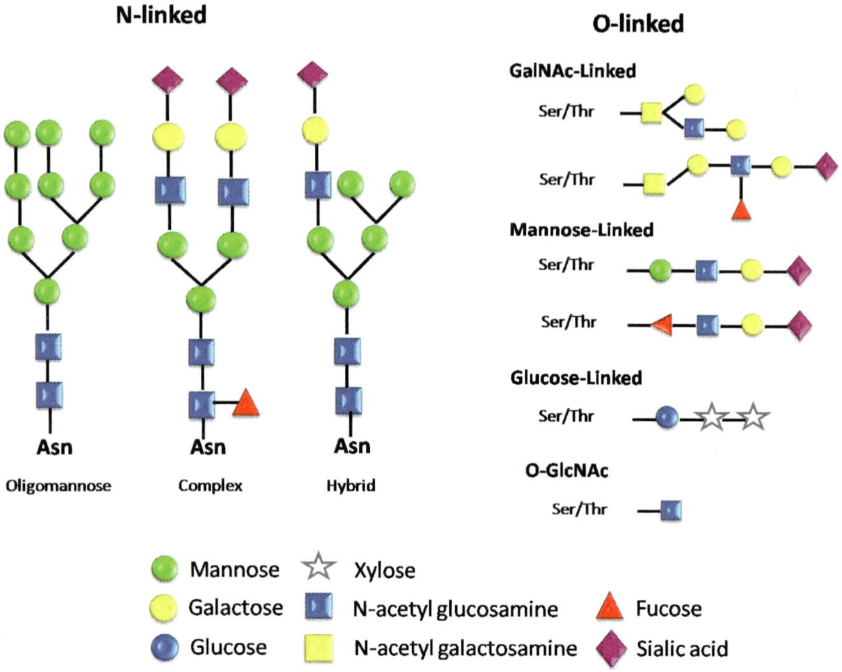

Figure 4.2 *N*-linked and *O*-linked glycans structure, according to Consortium for Functional Glycomics notation.

multiple glycans are frequently associated with glycosylation sites. Thus, a complete characterization of a glycoprotein is a truly daunting task when a thorough structural elucidation of the carbohydrates is required, along with locating the sites-of-glycosylation, their levels of occupancy, and their microheterogeneities.

Mass spectrometry plays a predominant role in the analysis of any glycoprotein, and MS-based approaches for identification, analysis, and characterization have been reviewed extensively [7,26,28,29].

Mass spectrometric analysis of glycans typically consists of the following steps: (1) glycan release by either peptide-*N*-glycosidase (PNGase) F treatment (for *N*-linked glycans) or β-elimination (for *O*-linked glycans); (2) glycan isolation; (3) glycan derivatization and ionization; and (4) annotation of the resulting spectra to list possible glycan structures [2,30,31].

Improvement of glycan release by PNGase F treatment can be obtained either by using ultra-high pressure during incubation or a microwave-assisted method [32]. Similarly, a more efficient approach to release *O*-glycans has been obtained by the use of a non-specific proteolysis in conjunction with a solid-phase permethylation [32]. All these methods provide an almost complete release of the oligosaccharides from glycoproteins, thus ensuring much more sensitive mass spectra.

Several methods have focused on liquid chromatography (LC)-coupled MS, using separation and enrichment of glycans by hydrophilic interaction liquid chromatography (HILIC), porous graphitized carbon (PGC), and reverse-phase (RP) liquid chromatography [27].

Derivatization of glycans prior to MS analysis enhances their detection. A number of modifications may be made to the glycan structure to aid in their overall analysis. Among these, permethylation is the most popular. It offers several advantages, such as an improved sensitivity of 10 to 20 times over the native glycans, the conversion of acidic structures to neutral analytes that permits the positive-ion mode mass spectra recording, and the enhancement of the cross-ring fragmentation during tandem MS procedures, thus enabling a more definitive structural characterization. Reductive amination using aniline, 2-aminopyridine, 2-aminobenzamide, 2- and 4-aminobenzoic acid, 2-aminoquinoline, and 1-phenyl-3-methyl-5-pyrazolone have been used to reduce the dynamic equilibrium between the closed-ring structure and an aldehyde and in order to improve MS sensitivity [27,33].

After acquisition of mass spectra, the glycome analysis can be completed by comparing acquired data with databanks. In addition, some programs exist to automate some aspects of data interpretation. Among these are Cartoonist, Glyco-Pep DB, SysBio Ware, GlycoWorkbench, GlycReSoft, and more recently, Multiglycan, even if software-based interpretation of glycan spectral data is still in its infancy [34–37].

For a glycoproteomic study, mass spectrometry has been applied to identify glycoproteins in sera, as well as biomarkers of healthy

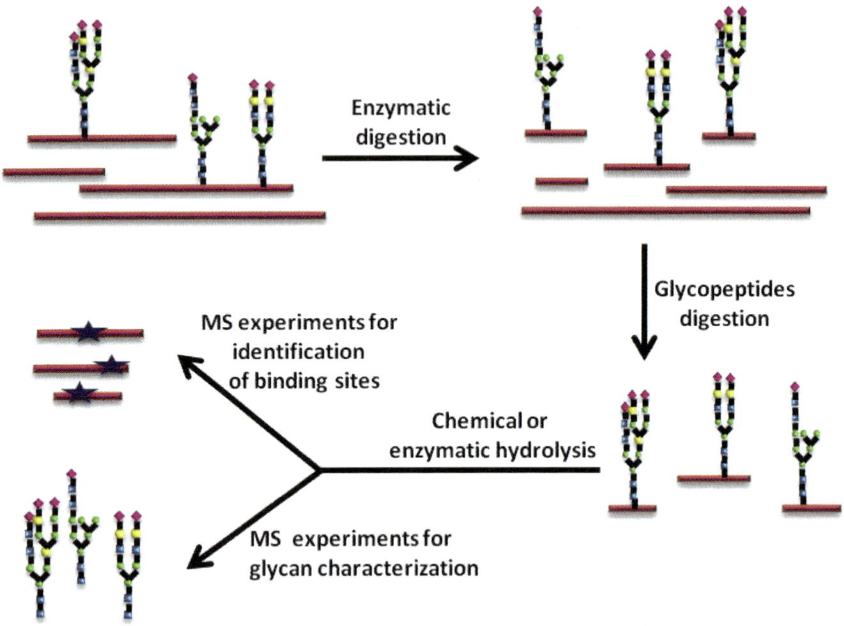

Figure 4.3 Experimental scheme for typical glycoproteomic or glycomic analysis by mass spectrometry.

and cancerous tissues [38,39]. In general, a typical glycoproteomic experiment involves the proteolytic digestion of a heterogeneous protein mixture, followed by glycopeptide enrichment, deglycosylation, and MS-based analysis of the corresponding peptide or proteome (Fig. 4.3). Glycopeptides can be enriched for analysis through the use of several methods; one of the most popular uses involves lectin affinity chromatography to subfractionate the glycoproteome [2,39]. Glycopeptides may also be isolated by hydrophilic interaction solid phase extraction or chromatography and graphitized carbon solid phase extraction. They may be enriched based on their high molecular mass using size exclusion chromatography or captured by binding to immobilized boronic acid derivatives [2].

As many of the observed signals at particular *m/z* values arise from a combination of several isomeric possibilities, tandem mass spectrometry are frequently mandatory to determine the fine details required for a more complete characterization of a certain oligosaccharide. Low- and high-energy CID, ECD, and ETD fragmentation techniques are widely used for a global characterization of complex glycoproteins enriched from biological samples [20,40]. However, for clinical applications, the quantification of glycoproteins, used as biomarkers, and their associated glycans using MS techniques is at a nascent stage. Quantitative determination of glycoproteins has been obtained by MALDI-TOF analysis of human serum from cancer patients and healthy individuals [41]. Most recently, great advances have been achieved with approaches based on immunoaffinity enrichment of peptides or proteins, followed by multiple reaction monitoring (MRM)/selected reaction monitoring (SRM) quantitative analysis [42]. With these methodologies, sensitivities at very low levels (ng/mL), where tumor biomarkers are found, were reached [41].

Other methods have recently been exploited [43,44], even if some difficulties, such as incomplete proteolytic digestion [45] and low efficiency ionization [46], are still present.

There are no doubts that MS-based approaches for biomarker quantification hold the greatest potential in the near future, even if significant technological progress has to be made before its application in a clinical context.

4.5 Lipopolysaccharides

Lipopolysaccharides (LPSs) (endotoxin) represent the main component of the outer leaflet of the external membrane of Gram-negative bacteria and play a key role during severe Gram-negative infection, sepsis, and septic shock [47,48]. Smooth-form lipopolysaccharides (S-LPSs) consist of three covalently linked regions: the lipid A, also known as the endotoxin for human pathogen microorganisms; the oligosaccharide region (core region); and the *O*-specific polysaccharide (*O*-chain). Chemically, a lipid A, which anchors the LPS to the outer membrane, consists of a glucosamine disaccharide β-GlcN-(1′-6)-α-GlcN, *N*- and *O*-acylated with fatty acids at the C-2, C-3, C-2′ and C-3′ positions, and phosphorylated at

O-1 and O-4′ positions. The core region is made up of several sugar units such as neutral hexoses, heptose and 3-deoxy-D-*manno*-oct-2-ulosonic acid (Kdo). Finally, the *O*-chain can be a homo- or a hetero-polysaccharide. The latter is usually built up with a variable number of an oligosaccharidic repeating unit. It is considered as an outer cell antigenic determinant (*O*-antigen). Rough-form lipopolysaccharides (R-LPSs), also named lipooligosaccharides (LOSs), lack the polysaccharidic portion [49].

Two reviews about the role of mass spectrometry in lipopolysaccharides structural elucidation have been recently published [50, 51]. Classical methodologies and MS-based application improvements will then be reported here.

The structural elucidation of LPSs and/or LOSs is a complex task, due to the lack of solubility of the intact molecules in any solvent. In order to override this problem, mild acid hydrolysis can be performed on smooth LPSs and the insoluble lipid A can be removed by centrifugation. The *O*-antigen polysaccharide can be separated from core oligosaccharides by gel filtration chromatography. Many functions can be ascribed to *O*-chain polysaccharides. These include the protection of bacterial cells from anti-biocide substances, the adhesion to mammalian tissues, and to be the first contact with the host during infections [47]. In addition, due to their antigenic properties, the *O*-chains are used for serotype classification. MALDI-TOF and ESI mass spectrometry of purified *O*-chain can be useful to confirm structures obtained by two-dimensional NMR and chemical methods [52–58], providing in some cases, the exact number and the size of repeating units [54]. Finally, the length of the *O*-antigen can influence the LPS activity [59].

The great majority of lipid A glycoforms has been identified and characterized by mass spectrometry [60–63]. These unique glycolipids are particularly suitable for being detected with both MALDI and ESI ion sources, because they are able to discriminate among the various species generated from the intrinsic heterogeneity, without the necessity to separate and purify them. This heterogeneity is due to the degree of phosphorylation of glucosamine units, the type of acyl residues, and the stoichiometry of acylation [47].

The different phosphorylation patterns of lipid A moieties greatly affects the endotoxic properties of LPSs [64]. Deletion of one phosphate group reduces the endotoxic activity of the resulting monophosphorylated LPS by approximately 100-fold [65]. For example, monophosphorylated lipid A is currently used as a vaccine adjuvant [66].

Lipid A modifications with aminoarabinose and/or phosphoethanolamine increase bacterial resistance to cationic antimicrobial peptides [67–69]. In addition, the deacylation and/or palmitoylation of lipid A helps bacteria evade immunosurveillance by decreasing the endotoxin activity of LPSs [70,71]. Therefore, the analysis of lipid A modifications helps us to examine the pathogenesis of bacteria, including clinical isolates. Detailed structural information, including the phosphorylation pattern of lipid A, can be obtained by tandem mass spectrometry. Several tandem mass spectrometry experiments have been used for lipid A characterization, all of them recently reviewed [50]. Tandem mass spectrometry experiments have been obtained both in negative and positive modes, by using fragmentation techniques ranging from CID to IRMPD.

A separate discussion is reserved for the core oligosaccharides, the structure of which can be generally deduced after an alkaline treatment. The most common method first entails a selective ester-linked fatty acid removal by hydrazine treatment, followed by aqueous potassium hydroxide (KOH) hydrolysis. Both samples can be analyzed by mass spectrometry, even if some biologically important functional group can be lost during these procedures. Then, the obtainment of the entire LPS/LOS mass spectrum is mandatory.

Many methods have been recently developed for obtaining the full mass spectra of LOSs. Both ESI and MALDI-MS have been employed towards this aim [72,73]. It is worth noting that the tandem mass spectrometry allows for the simultaneous characterization of both lipid A and core oligosaccharide moieties. In the MALDI spectra of LOSs, in-source fragmentation produces separated signals for lipid A and core oligosaccharide moieties, due to the well-known lability of the Kdo linkage. Then, by selecting the signals corresponding

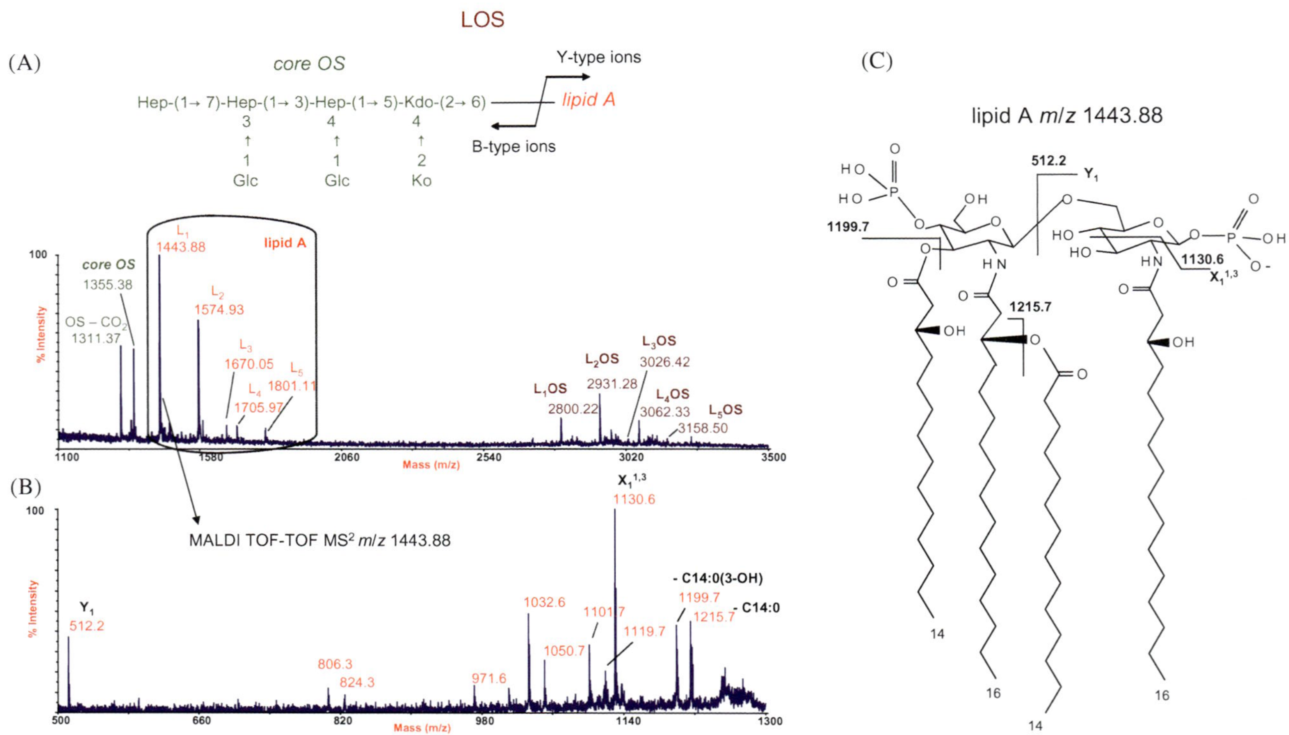

Figure 4.4 Structural investigation of LOS from *B. cenocepacia* strain K56–2 by reflectron MALDI TOF MS and MALDI TOF MS/MS. Reproduced with permission of [74].

to different acylated lipid A species, it is possible to obtain a fine characterization of the fatty acid distribution without the necessity of prior chemical degradation (Fig. 4.4) [74].

ESI FT-ICR mass spectra has revealed itself to be an excellent tool to identify both smooth and rough forms of lipopolysaccharides, due to the high mass resolution (more than 50,000) and accuracy (about 1 ppm) [73,75,76]. With this powerful technique, it is possible in some cases to reveal the linkage between the core oligosaccharide and the first *O*-chain repeating unit [75].

4.6 Glycosphingolipids

Glycosphingolipids (GSLs) are amphipathic lipid molecules consisting of a hydrophilic oligosaccharide chain and a hydrophobic moiety, named ceramide [77]. According to the most common way to classify them, neutral GSLs are glycolipids containing only uncharged sugars, such as glucose (Glc), galactose (Gal), GlcNAc, *N*-acetylgalactosamine (GalNAc) or fucose (Fuc) as substituents on the ceramide moiety, while acidic GSLs (gangliosides) are those that contain ionized functional groups attached to neutral sugars (phosphate or sulfate), or ionizable sugar residues, such as sialic acid.

GSLs are particularly abundant in the lipid domains of the plasma membrane, where they are inserted into the outer leaflet with the double-tailed lipid anchor and the oligosaccharide moiety facing the external environment [78,79]. Gangliosides are predominantly localized in the mammalian central nervous system, where they trigger a variety of biological events. Ganglioside composition may alter during brain development, growth, differentiation, neurodegeneration and oncogenesis. Therefore, gangliosides are considered to be diagnostic biomarkers and therapeutic targets [80–83].

From the beginning, mass spectrometry has revealed itself to be a powerful tool for the structural analysis of GSLs [84], and during the last three decades, a vast number of MS-based methods have been developed, and they were specifically tailored to obtain insight into the different aspects of GSL structures. Thin layer chromatography (TLC), the most commonly applied separation technique for lipid analysis, offers several advantages, such as simplicity, low cost, high resolution, high throughput, and easy visualization of samples. However, for *de novo* identification, it is necessary to couple TLC with a more informative device, such as mass spectrometry. One way is to run the TLC, then treat it with specific ligands (for example, antibodies, lectins, and bacterial toxins), and finally directly analyze the TLC by mass spectrometry [85].

Another very recent method consists of coupling TLC to ESI-MS either with a CAMAG TLC-MS interface or with a liquid extraction surface analysis (LESA, Advion Biosciences) [86]. This "direct sampling TLC-MS" overcomes the risk of sialic acid elimination in the MALDI source, due to the more gentle ionization method of electrospraying.

A procedure for transferring phospholipids and glycosphingolipids separated on a high performance (HP) TLC-plate to a poly vinylidene difluoride (PVDF) membrane has been developed since 1994 [87]. Then, the PVDF is moved onto a MALDI target plate and analyzed by TOF [87,88]. This elegant method solves instrument contamination problems related to direct TLC-MS. In addition, this method has been broadened to imaging mass spectrometry (IMS) [89], providing a powerful tool for sensitive identification of minor species and structural identification of novel structures (Fig. 4.5). The method has been applied to human brain gangliosides separated from the tissues of patients with Alzheimer's disease and control patients [90].

Recently, fully automated chip-based nanoelectrospray mass spectrometry approaches for glycosphingolipids analysis has been reviewed [92,93]. The MS platform has been coupled to a quadrupole time-of-flight (QTOF) [94], a high-capacity ion trap (HCT) [95], and more recently to an orbitrap instrument [96]. High resolved spectra and tandem mass spectrometry experiments of brain tissues showed the remarkable potential of this method in brain cancer research, for the determination of tumor-associated markers and for their structural determination.

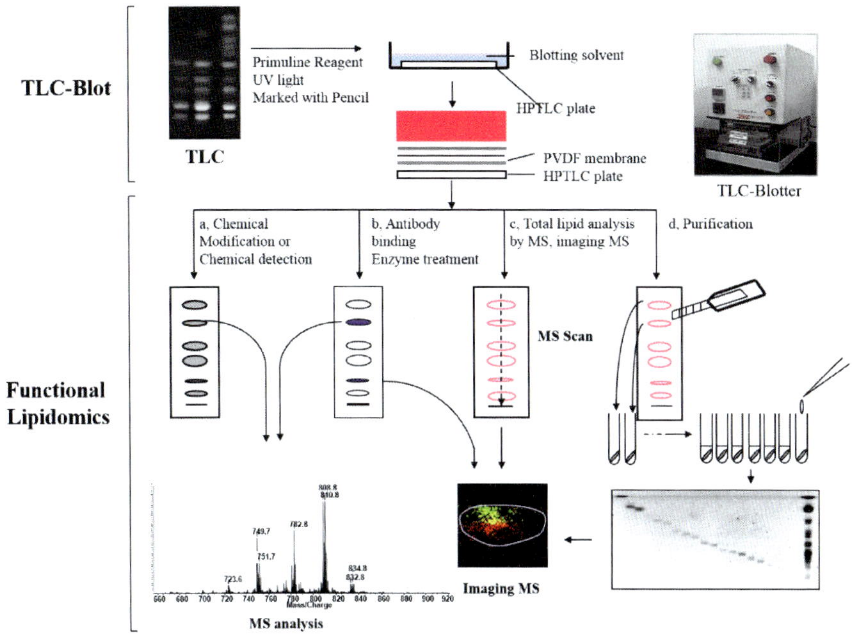

Figure 4.5 Scheme for functional lipidomics by TLC-Blot/MALDI TOF MS. Reproduced with permission of [91].

On-line capillary electrophoresis (CE)-MS methods have also been exploited for GSL analysis, but at the moment, the level of these approaches are functional only for simple, commercially available mixtures [97].

4.7 Glycosaminoglycans

Proteoglycans are composed of a core protein and one or more glycosaminoglycan (GAG) chains. GAGs can be divided into families: hyaluronic acid (HA), keratan sulfate (KS), heparan sulfate (HS), heparin (Hp), chondroitin sulfate (CS), and dermatan sulfate (DS) [98]. Each GAG is a polymer of repeating disaccharides comprising a hexosamine and a non-nitrogenous sugar, which is usually a uronic acid, except for KS, in which there is a hexose (galactose).

From a structural point of view, HA is the simplest GAG and is composed of alternating β-D-GlcNAc and β-D-glucuronic acid (GlcA) units, with glycosidic linkages at positions C-3 and C-4, respectively. KS has the basic disaccharide repeating unit of a D-galactose β(1→4) linked to the C-3 of a β-D-GlcNAc. The *O*-sulfonations can be found at C-6 position of either monosaccharide, that of uronic acid being the most sulfated one [99]. HS and Hp chains are synthesized as a disaccharide unit 4)-GlcA-β-(1–4)-GlcNAc-α-(1, and subsequently modified by an *N*-deacetylase/*N*-sulfotranferase enzyme that removes glucosamine (GlcN) acetyl groups and replaces them with sulfate groups. This enzyme produces domains of high *N*-sulfate content, interspersed with those containing high *N*-acetate content. The *N*-sulfated domains may be acted upon by glucuronic acid C-5 epimerase, resulting in the conversion of GlcA residues to iduronic acid (IdoA) and can be subsequently sulfated at the 3O- and/or 6O-positions of GlcN and/or the 2O-position of hexuronic acid (HexA). Chondroitin sulfate consists of repeating units of 4)-GlcA-β-(1–3)-GalNAc-β-(1 and may be sulfated on the 4- and/or 6-positions of GalNAc. DS is a CS variant in which a substantial fraction of GlcA residues are epimerized to IdoA, which may be sulfated at the 2-position.

Proteoglycans and glycosaminoglycans are components of extracellular matrices and cell surfaces. GAGs interact with a variety of ligands [1], most of them being proteins. At molecular and cellular level, these specific interactions are due to their great structural diversity [100–104], that is encompassed by GAG type, size, composition and charge density, degree of substitution, domain arrangement, and sequence.

Many functions can be ascribed to these complex molecules. CS and DS mediate physicochemical and adhesive properties of proteoglycans [105–107]. They mainly serve to modulate the activities of the fibroblast growth factor family and hepatocity growth factor. In addition, CS is the main component of cartilage and other connective tissues, and is responsible for the maintenance of tissue viscoelasticity [108]. Changes in its structure are found during development and progression of osteoarthritis.

Heparin and HS influence numerous physiological and pathophysiological processes, such as tumorigenesis, growth control, cell adhesion, inflammation, neural development, and regeneration. Cell-surface HS proteoglycans also work as adhesion receptors for many viruses and bacteria, augmenting the pathogen's ability to bind to secondary receptors responsible for internalization [109,110].

HA is abundant in skin, skeletal tissues, eye vitreous, umbilical cord tissue, and synovial fluid. It generally displays a high molecular mass, ranging from 10^5 to 10^7 Daltons. At high concentrations, HA confers viscoelasticity to tissues, which drops rapidly after depolimerization caused by stress, such as oxidative stress [111].

KS is mainly present in corneal tissue, where its abundance appears to be related to the maintenance of a level of tissue hydration critical for corneal transparency [112]. In addition, KS, together with CS, have been associated as inhibitors of restoring a correct neuronal network after spinal cord lesions [113], and it is involved in the development of other neurodegenerative diseases, such as amyothrophic lateral sclerosis (ALS) [114]. Thus, GAGs are macromolecules of great importance in the fields of biochemistry, pathology, and pharmacology.

Most GAGs are derived from animal sources by extraction processes to produce purified drugs and active macromolecules to use in multiple therapeutic applications [115]. Furthermore, these natural substances have been chemically and enzymatically modified, and synthetic analogs have been developed [115–117]. Isolation and purification of GAGs from biological samples can be achieved through many different approaches, but even this work presents many obstacles. In fact, the low amount that can be purified is not sufficient for many analytical approaches. However, the methods currently utilized include solubilizing the tissue with proteolytic enzymes, fractionating the mixture by either ion-exchange chromatography and/or gel-filtration columns and/or sequential precipitation by organic solvents [118]. Purified GAGs are then characterized mostly after degradation with highly specific lyases [119–121] into the building block disaccharides or oligosaccharides by a partial depolymerization. The mass spectrometry analysis of these molecules can give information about the sequence of oligosaccharides and/or the sulfation grade and positions that influence their interaction with specific proteins.

The topic of mass spectrometric analysis of the GAGs has been extensively reviewed recently [122,123], thus only some additional examples are reported here. The acidic nature of the oligomers of GAGs allows for the utilization of both MALDI and ESI as ion sources. GAGs obtained from tissues, biological fluids, or cells are treated with various enzymes to obtain disaccharides that are fluorescently labeled with 2-aminoacridone (AMAC) and are resolved by different LC systems for high-sensitivity detection by fluorescence, and they are then unambiguously characterized by MS [124]. On-line liquid chromatography combined with ESI mass spectrometry for the analysis of GAG oligomers comprises size-exclusion chromatography, hydrophilic interaction chromatography, graphitized carbon columns, and reverse phased ion pairing [122]. Besides them, capillary electrophoresis coupled with mass spectrometry is extensively used [115]. Multistage MS using collisionally-activated dissociation (CAD) [117] has been used for comparative studies of fine structure of unknown CS/DS samples relative to standard ones. Very recently, a new approach to distinguish isobaric mixtures of GAG oligomers has been developed [125]. A gas-phase ion separation method, namely the high-field asymmetric wave form ion mobility spectrometry (FAIMS), has been coupled with FT-ICR MS. In this paper, isobaric GAG gas phase ions, including that of stereoisomers or conformers with the same mass-to-charge ratio, have been successfully separated and analyzed.

Acknowledgments

The author is indebted to Prof. M. Parrilli, Prof. R. Lanzetta, Dr. A. Casillo and Dr. S. Carillo for their kind support.

References

1. Chrispeels, M., Manzi, A., Hindsgaul, O., Powell, L., Paulson, J. C., van Halbeek, H., Lowe, J. (1999) In *Essentials of Glycobiology*, Varki, A., Cummings, R. D., Esko, J. D., Freeze, H. H., Hart, G. W., Marth, J. Eds. Cold Spring Harbor, New York. pp. 1–15.
2. Zaia, J. (2010) *OMICS* **14**, 401–418.
3. Ohtsubo, K., Marth, J. D. (2006) *Cell* **126**, 855–867.
4. An, H. J., Kronewitter, S. R., de Leoz, M. L. A., Lebrilla, C. B. (2009) *Curr. Opin. Chem. Biol.* **13**, 601–607.
5. Adamczyk, B., Tharmalingam, T., Rudd, P. M. (2012) *Biochim. Biophys. Acta.* **1820**, 1347–1353.
6. Danishefsky, S. J., Allen, J. R. (2000) *Angew. Chem. Int. Edit.* **39**, 836–863.
7. Kailemia, M. J., Ruhaak, L. R., Lebrilla, C. B., Amster, I. (2014) *Anal. Chem.* **86**, 196–212.
8. Bhardwaj, C., Hanley, L. (2014) *Nat. Prod. Rep.* **31**, 756–767, doi: 10.1039/C3NP70094A.
9. Stedwell, C. N., Galindo, J. F., Roitberg, A. E., Polfer, N. C. (2013) *Annu. Rev. Anal. Chem.* **6**, 267–285.
10. Wu, C., Dill, A. L., Eberlin, L. S., Cooks, R. G., Ifa, D. R. (2013) *Mass Spectrom. Rev.* **32**, 218–243.
11. Himmelsbach, M. (2012) *J. Chromatogr. B.* **883–884**, 3–17.
12. Linscheid, M., D'angona, J., Burlingame A. L., Dell, A., Ballous C. E. (1981) *Proc. Natl. Acad. Sci.* **78**, 1471–1475.
13. Viseux, N., de Hoffmann, E., Domon, B. (1997) *Anal. Chem.* **69**, 3193–3198.
14. Zaia, J. (2007) *Encyclopedia of Mass Spectrometry* **6**. Gross, M. G., Caprioli, R. M. Eds. Elsevier (New York). pp. 889–903.
15. Leontein, K., Lindberg B., Lönngren, J. (1978) *Carbohydr. Res.* **62**, 359–362.
16. Harvey, D. J. (2005) *Expert Rev. Proteomic* **2**, 87–101.
17. Zhang, J. H., Schubothe, K., Li, B. S., Russell, S., Lebrilla, C. B. (2005) *Anal. Chem.* **77**, 208–214.
18. Hitchcock, A. M., Yates, K. E., Costello, C. E., Zaia, J. (2008) *Proteomics* **8**, 1384–1397.
19. Joo An, H., Lebrilla, C. B. (2011) *Mass. Spectrom. Rev.* **30**, 560–578.
20. Yu, X., Jiang, Y., Chen, Y., Huang, Y., Costello, C. E., Lin, C. (2013) *Anal. Chem.* **85**, 10017–10021.

21. Domon, B., Costello, C. E. (1988) *Glyconjugate J.* **5**, 397–409.
22. Hart, G. W., Copeland, R. J. (2010) *Cell* **143**, 672–676.
23. Pan, S., Chen, R., Aebersold, R., Brentnall, T. A. (2011) *Mol. Cell. Proteomics* **10**, R110.003251.
24. Marks, G. S., Marshall, R. D., Neuberger, A. (1963) *Biochem. J.* **87**, 274–281.
25. Iwashkiw, J. A., Vozza, N. F., Kinsella, R. L., Feldman, M. F. (2013) *Mol. Microbiol.* **89**, 14–28.
26. Tousi, F., Hancock, W. S., Hincapie, M. (2011) *Anal. Methods* **3**, 20–32.
27. Alley, W. R., Jr., Mann, B. F., Novotny M. V. (2013) *Chem. Rev.* **113**, 2668–2732.
28. Harvey, D. J. (2005) *Expert Rev. Proteomic* **2**, 87–101.
29. Haslam, S. M., North, S. J., Dell, A. (2006) *Curr. Opin. Struct. Biol.* **16**, 584–591.
30. North, S. J., Hitchen, P. G., Haslam, S. M., Dell, A. (2009) *Curr. Opin. Struct. Biol.* **19**, 498–506.
31. Morelle, E., Michalski. J. C. (2007) *Nat. Protoc.* **2**, 1585–1602.
32. Novotny, M. D., Alley Jr., W., Mann, B. F. (2013) *Glycoconj. J.* **30**, 89–117.
33. Dass, C. (2007) In *Fundamentals of Contemporary Mass Spectrometry.* Desiderio, D. M., Nibbering, N. M. Eds. Wiley (New York). pp. 397–421.
34. Vakhrushev, S. Y., Dadimov, D., Peter-Katalinić, J. (2009) *J. Anal. Chem.* **81**, 3252–3260.
35. Go, E. P., Rebecchi, K. R., Dalpathado, D. S., Bandu, M. L., Zhang, Y., Desaire, H. (2007) *Anal. Chem.* **79**, 1708–1713.
36. Goldberg, D., Sutton-Smith, M., Paulson, J., Dell, A. (2005) *Proteomics* **5**, 865–875.
37. Yu, C. Y., Mayampurath, A., Hu, Y., Zhou, S., Mechref, Y., Tang, H. (2013) *Bioinformatics* **29**, 1706–1707.
38. Mechref, Y., Hu, Y., Garcia, A., Hussein, A. (2012) *Electrophoresis* **33**, 1755–1767.
39. Rakus, J. F., Mahal, L. K. (2011) *Annu. Rev. Anal. Chem.* **4**, 367–392.
40. Nilsson, J., Halim, A., Grahn, A., Larson, G. (2013) *Glycoconj. J.* **30**, 119–136.
41. Kuzmanov, U., Kosanam, H., Diamandis, E. P. (2013) *BMC Medicine* **11**, 31.
42. Lemoine, J., Fortin, T., Salvador, A., Jaffuel, A., Charrie, J. P., Choquet-Kastylevsky, G. (2012) *Expert. Rev. Mol. Diagn.* **12**, 333–342.

43. Maresca, B., Cigliano, L., Spagnuolo, M. S., Dal Piaz, F., Corsaro, M. M., Balato, N., Nino, M., Ayala, F., Abrescia, P. (2012) *PLoS One* **7**, e52040.
44. Nakano, M., Nakagawa, T., Ito, T., Kitada, T., Hijioka, T., Kasahara, A., Tajiri, M., Wada, Y., Taniguchi, N., Miyoshi, E. (2008) *Int. J. Cancer* **122**, 2301–2309.
45. Song, E., Pyreddy, S., Mechref, Y. (2012) *Rapid. Commun. Mass. Spectrom.* **26**, 1941–1954.
46. Wada, Y., Azadi, P., Costello, C. E., Dell, A., Dwek, R. A., Geyer, H., Geyer, R., Kakehi, K., Karlsson, N. G., Kato, K., Kawasaki, N., Khoo, K. H., Kim, S., Kondo, A., Lattova, E., Mechref, Y., Miyoshi, E., Nakamura, K., Narimatsu, H., Novotny, M. V., Packer, N. H., Perreault, H., Peter-Katalinić, J., Pohlentz, G., Reinhold, V. N., Rudd, P. M., Suzuki, A., Taniguchi, N. (2007) *Glycobiology* **17**, 411–422.
47. Caroff, M., Karibian, D. (2003) *Carbohydr. Res.* **338**, 2431–2447.
48. Corsaro, M. M., Lanzetta, R., Parrilli, E., Parrilli, M., Tutino, M. L. (2001) *Eur. J. Biochem.* **268**, 5092–5097.
49. Corsaro, M. M., Dal Piaz, F., Lanzetta, R., Parrilli, M. (2002) *J. Mass Spectrom.* **37**, 481–488.
50. Kila, A., Dornyei, A., Kocsis, B. (2013) *Mass Spectrom. Rev.* **32**, 90–117.
51. Banoub, J. H., Aneed, A. E., Cohen, A. M., Joly, N. (2010) *Mass Spectrom. Rev.* **29**, 606–650.
52. Wang, Z., Larocque1, S., Vinogradov1, E., Brisson, J. R., Dacanay, A., Greenwell, M., Brown, L. L., Li, J., Altman, E. (2004) *Eur. J. Biochem.* **271**, 4507–4516.
53. Wang, Z., Li, J., Altman, E. (2008) *Carbohydr. Res.* **343**, 483–488.
54. Turska-Szewczuk, A., Guz, L., Lindner, B., Pietras, H., Russa, R., Holst, O. (2011) *Carbohydr. Res.* **346**, 815–821.
55. Czaja1, J., Jachymek, W., Niedziela, T., Lugowski, C., Aldova, E., Kenne, L. (2000) *Eur. J. Biochem.* **267**, 1672–1679.
56. Maciejewska, A., Lukasiewicz , J., Niedziela, T., Szewczuk, Z., Lugowski, C. (2009) *Carbohydr. Res.* **344**, 894–900.
57. Corsaro, M. M., Gambacorta, A., Iadonisi, A., Lanzetta, R., Naldi, T., Nicolaus, B., Romano, I., Ummarino, S., Parrilli, M. (2006) *Eur. J. Org. Chem.* **7**, 1801–1808.
58. Pieretti, G., Corsaro, M. M., Lanzetta, R., Parrilli, M., Canals, R., Merino, S., Tomas, J. M. (2008) *Eur. J. Org. Chem.* **18**, 3149–3155.
59. Capodici, C., Chen. S., Sidorczyk, Z., Elsbach, P., Weiss, J. (1994) *Infect. Immun.* **62**, 259–265.

60. Zhou, P., Altman, E., Perry M. B., Li, J. (2010) *Appl. Environ. Microb.* **76**, 3437–3443.
61. Corsaro, M. M., Dal Piaz, F., Lanzetta, R., Parrilli, M. (2002) *J. Mass Spectrom.* **37**, 481–488.
62. Corsaro, M. M., Dal Piaz, F., Lanzetta, R., Naldi, T., Parrilli, M. (2004) *Rapid Commun. Mass Spectrom.* **18**, 853–858.
63. Carillo, S., Pieretti, G., Lindner, B., Romano, I., Nicolaus, B., Lanzetta, R., Parilli, M., Corsaro, M. M. (2013) *Mar. Drugs* **11**, 184–193.
64. Rietschel, E. T., Kirikae, T., Schade, F. U., Mamat, U., Schmidt, G., Loppnow, H., Ulmer, A. J., Zahringer, U., Seydel, U., Di Padova, F., Schreier, M., Brade, H. (1994) *FASEB J.* **8**, 217–225.
65. Park, B. S., Song, D. H., Kim, H. M., Choi, B. S., Lee, H., Lee J. O. (2009) *Nature* **458**, 1191–1195.
66. Garçon, N., Chomez, P., Van Mechelen, M. (2007) *Expert Rev. Vaccines* **6**, 723–739.
67. Gunn, J. S., Lim, K. B., Krueger, J., Kim, K., Guo, L., Hackett, M., Miller, S. I. (1998). *Mol. Microbiol.* **27**, 1171–1182.
68. Lee, H., Hsu, F. F., Turk, J., Groisman, E. A. (2004) *J. Bacteriol.* **186**, 4124–4133.
69. Nummila, K., Kilpelainen, I., Zahringer, U., Vaara, M., Helander, I. M. (1995) *Mol. Microbiol.* **16**, 271–278.
70. Guo, L., Lim, K., Gunn, J. S., Bainbridge, B., Darveau, R., Hackett, M., Miller, S. I. (1997) *Science* **276**, 250–253.
71. Kawasaki, K., Ernst, R. K., Miller, S. I. (2004) *J. Endotoxin. Res.* **10**, 439–444.
72. Sturiale, L., Garozzo, D., Silipo, A., Lanzetta, R., Parrilli, M., Molinaro, A. (2005) *Rapid Commun. Mass Spectrom.* **19**, 1829–1834.
73. Kondakova, A. N., Lindner, B. (2005) *Eur. J. Mass Spectrom.* **11**, 535–546.
74. Sturiale, L., Palmigiano, A., Silipo, A., Knirel, Y. A., Anisimov, A. P., Lanzetta, R., Parrilli, M., Molinaro, A., Garozzo, D. (2011) *J. Mass Spectrom.* **46**, 1135–1142.
75. Turska-Szewczuk, A., Lindner, B., Komaniecka, I., Kozinska, A., Pekala, A., Choma, A., Holst, O. (2013) *Mar. Drugs* **11**, 1235–1255.
76. Klein, G., Lindner, B., Brabetz, W., Brade, H., Raina, S., (2009) *J. Biol. Chem.* **284**, 15369–15389.
77. Muthing, J., Distler, U. (2010) *Mass Spectrom. Rev.* **29**, 425–479.
78. Sonnino, S., Mauri, L., Chigorno, V., Prinetti, A. (2006) *Glycobiology* **17**, 1R–13R.

79. Sonnino, S., Prinetti, A., Mauri, L., Chigorno, V., Tettamanti, G. (2006b). *Chem Rev.* **106**, 2111–2125.
80. Potapenko, M., Shurin, G. V., de León, J. (2007) *Adv. Exp. Med. Biol.* **601**, 195–203.
81. Modak, S., Cheung, N. K. (2007) *Cancer Invest.* **25**, 67–77.
82. Gu, J., Tifft, C. J., Soldin, S. J. (2008) *Clin. Biochem.* **41**, 413–417.
83. Wuhrer, M. (2013) *Glycoconjugate J.* **30**, 11–22.
84. Iwamori, M., Nagai, Y. (1978) *J. Biochem.* **84**, 1601–1608.
85. Muthing, J., Distler, U. (2010) *Mass Spectrom. Rev.* **29**, 425–479.
86. Park, H., Zhou, Y., Costello, C. E. (2014) *J. Lipid Res.* **55**, 773–781.
87. Taki, T., Handa, S., Ishikawa, D. (1994) *Anal. Biochem.* **221**, 312–316.
88. Goto-Inouea, N., Hayasakaa, T., Takib, T., Gonzalezc, T. V., Setoua, M. (2009) *J. Chromatogr. A.* **1216**, 7096–7101.
89. Goto-Inouea, N., Hayasakaa, T., Sugiurab, Y., Taki, T., Li, Y., Matsumotoc, M., Setoua, M. (2008) *J. Chromatogr. B.* **870**, 74–83.
90. Valdes-Gonzales, T., Goto-Inoue, N., Hirano, W., Ishiyama, H., Hayasaka, T., Setou, M., Taki, T. (2011) *J. Neurochem.* **116**, 678–683.
91. Taki, T. (2013) *Proc. Jpn. Ac. B.* **89**, 302–319.
92. Flangea, C., Serb, A., Sisu, E., Zamfir, A. D. (2011) *Biochim. Biophys. Acta* **1811**, 513–535.
93. Bindila, L., Peter- Katalinić, J. (2009) *Mass Spectrom. Rev.* **28**, 223–253.
94. Zamfir, A. D., Lion, N., Vukelic, Z., Bindila, L., Rossier, J., Girault, H. H., Petre-Katalinic, J. (2005) *Lab. Chip* **5**, 298–307.
95. Mosoarca, C., Ghiulai, R. M., Novaconi, C. R., Vukelic, Z., Chiriac, A., Zamfir, A. D (2011) *Anal. Lett.* **44**, 1036–1047.
96. Sokol, E., Almeida, R., Hannibal-bach, H. K., Kotowoska, D., Vogt, J., Baumgart, J., Kristiansen, K., Nitsch, R., Knudsen, J., Ejsing, C. S. (2013) *Anal. Biochem.* **443**, 89–96.
97. Campa, C., Coslovi, A., Flamigni, A., Rossi, M. (2006) *Electrophoresis* **27**, 2027–2050.
98. Lindahl, U., Hook, M. (1978) *Ann. Rev. Biochem.* **47**, 385–417.
99. Pomin, V. H., Piquet, A. A., Pereira, M. S., Mourão, P. A. (2012) *Carbohyd. Polym.* **90**, 839–846.
100. Sasisekharan, R., Raman, R., Prabhakar, V. (2006) *Annu. Rev. Biomed. Eng.* **8**, 181–231.
101. Sasisekharan, R., Shriver, Z., Venkataraman, G., Narayanasami, U. (2002) *Nat. Rev. Cancer* **2**, 521–528.
102. Capila, I., Linhardt, R. J. (2002) *Angew. Chem. Int. Ed.* **41**, 391–412.
103. Linhardt, R. J. (2004) *Chem. Biol.* **11**, 420–422.

104. Mulloy, B., Linhardt, R. J. (2001) *Curr. Opin. Struct. Biol.* **11**, 623–628.
105. Lyon, M., Deakin, J. A., Gallagher, J. T. (2002) *J. Biol. Chem.* **277**, 1040–1046.
106. Penc, S. F., Pomahac, B., Winkler, T., Dorschner, R. A., Eriksson, E., Herndon, M., Gallo, R. L. (1998) *J. Biol. Chem.* **273**, 28116–28121.
107. Trowbridge, J. M., Rudisill, J. A., Ron, D. R., Gallo, L. (2002) *J. Biol. Chem.* **277**, 42815–42820.
108. Knudson, C. B., Knudson, W. (2001) *Semin. Cell. Dev. Biol.* **12**, 69–78.
109. Jones, C. J., Beni, S., Limtiaco, J. F. K., Langeslay, D., Larive, C. (2011) *Annu. Rev. Anal. Chem.* **4**, 439–465.
110. Tran, V. M., Nguyen, T. K. N., Raman, K., Kuberan, B. (2011) *Anal. Bioanal. Chem.* **399**, 559–570.
111. Corsaro, M. M., Pietraforte, D., Di Lorenzo, A. S., Minetti, M., Marino, G. (2004) *Free Radical. Res.* **38**, 343–353.
112. Funderburgh, J. L. (2000) *Glycobiology* **10**, 951–958.
113. Imagama, S., Sakamoto, K., Tauchi, R., Shinjo, R., Ohgomori, T., Ito, Z., Zhang, H., Nishida, Y., Asami, N., Takeshita, S., Sugiura, N., Watanabe, H., Yamashita, T., Ishiguro, N., Matsuyama, Y., Kadomatsu, K. (2011) *J. Neurosci.* **31**, 17091–17102.
114. Hirano, K., Ohgomori, T., Kobayashi, K., Tanaka, F., Matsumoto, T., Natori, T., Matsuyama, Y., Sakamoto, K., Takeuchi, H., Hirakawa, A., Suzumura, A., Sobue, G., Ischiguro, N., Imagama, S., Uchimura, K., Kadomatsu, K. (2013) *PLoS One* **8**, e66969.
115. Volpi, N., Maccari, F., Linhardt, R. (2008) *Electrophoresis* **29**, 3095–3106.
116. Bedini, E., De Castro, C., De Rosa, M., Di Nola, A., Restaino, O. F., Schiraldi, C., Parrilli, M. (2012) *Chem. Eur. J.* **18**, 2123–2130.
117. Bedini, E., De Castro, C., De Rosa, M., Di Nola, A., Iadonisi, A., Restaino, O. F., Schiraldi, C., Parrilli, M. (2011) *Angew. Chem. Int. Ed.* **50**, 6160–6163.
118. Yanagishita, M. (2001) In *Proteoglycan Protocols.* Iozzo, R. V. Ed. Humana Press (Totowa). pp. 3–8.
119. Gu, K., Liu, J., Pervin, A., Linhardt, R. J. (1993). *Carbohydr. Res.* **244**, 369–377.
120. Jandik, K. A., Gu, K., Linhardt, R. J. (1994) *Glycobiology* **4**, 289–296.
121. Ernst, S., Langer, R., Cooney, C. L., Sasisekharan, R. (1995) *Crit. Rev. Biochem. Mol. Biol.* **30**, 387–444.
122. Zaia, J. (2013) *Mol. Cell. Proteomics* **12**, 885–892.

123. Sisu, E., Flangea, C., Serb, A., Zamfir, A. D. (2011) *Amino Acids* **41**, 235–256.
124. Volpi, N., Galeotti, F., Yang, B., Linhardt, R. J. (2014) *Nat. Protocol.* **9**, 541–558.
125. Kailemia, M. J., Park, M., Kaplan, D. A., Venot, A., Boons, G. J., Li, L., Linhardt, R. J., Amster, J. (2013) *J. Am. Soc. Mass Spectrom.* **25**, 258–268, doi: 10.1007/s13361-013-0771-1.

Chapter 5

Advanced NMR Techniques: Defining Carbohydrate Structures and Ligand–Receptor Interactions

M. Álvaro Berbís,* Ángeles Canales,* Javier Sastre-Martinez,* Luca Unione,* M. C. Fernández-Alonso,* Pilar Blasco,* F. Javier Cañada* and Jesús Jiménez-Barbero*,†,‡

**Chemical and Physical Biology Department, Centro de Investigaciones Biológicas, CSIC, 28040 Madrid, Spain*
†CIC bioGUNE, 48160 Derido, Spain
‡Ikerbasque–Basque Foundation for Science, 48013 Bilbao, Spain

5.1 Introduction

Carbohydrates are among the most ubiquitous and complex types of biomolecules in nature. They are composed of smaller units, called monosaccharides, which are linked in the form of linear sequences, with or without sophisticated branches [1–3], allowing for a high potential for overall structural diversity. In addition, the introduction of substituents at the hydroxyl groups, such as phosphate, sulfate [4] and/or acetyl [5] groups, greatly increases carbohydrates' potential for complexity. This structural diversity allows carbohydrates to be specifically recognized by receptors with exquisite selectivity, making them very suitable for encoding biological information [6].

Carbohydrates, especially when conjugated as glycoproteins or glycolipids at the periphery of the cell, play a very important role in molecular recognition processes by establishing cell–cell and cell–matrix interactions, triggering a variety of intra- and extracellular physiological and pathological events. Subsequently, modifications of cell glycosylation may lead to or be involved in pathological phenomena, such as inflammation and cancer [7–9]. In this context, knowledge at the maximum resolution possible on the three-dimensional (3D) structure of carbohydrates, and characterization of their recognition by different entities, such as lectins, enzymes, viruses, and antibodies, is essential for a thorough understanding of many vital processes and to open the possibility of their modulation. Glycobiology thus a discipline lying at the interface between chemistry and biology, and with implications in biomedicine, is a challenging and active field of research today.

Over the last years, numerous investigations have looked deeply into the fine structural details of protein–carbohydrate interactions. Advances in different biophysical and spectroscopic techniques, such as X-ray crystallography and nuclear magnetic resonance (NMR) spectroscopy, complemented by quantitative data obtained through isothermal titration microcalorimetry (ITC) [10], have allowed for the gathering of information on protein–carbohydrate complexes [11] at atomic resolution and have permitted speculations on the forces involved in their interactions.

Among these, NMR is one of the most widely used techniques to characterize molecular recognition events [12], thanks to its versatility, and a spectacular progress that it has experienced in recent times, with advances in distinct areas to generate an improvement in the resolution and sensitivity of NMR experiments and the development of procedures for structural determination, as well as the study of the conformation and dynamics of complex oligosaccharides [13]. Also of note, NMR techniques provide the possibility to work in solution, i.e. in conditions similar to those of the medium physiological.

In addition to those suitable for the study of isolated carbohydrate molecules, a plethora of NMR techniques are also available for studying protein–ligand interactions. These methods, usually assisted by computational protocols, permit the elucidation of the structural features of the binding entities, as well as the characterization of the kinetic and thermodynamic parameters of the free-bound equilibrium.

5.2 Defining Carbohydrate Structures

The 3D structure of carbohydrates plays a pivotal role in their recognition by receptors. Thus, in order to fully comprehend the recognition process, the structure of both partners involved and the changes they may experience upon binding must be elucidated.

The high variability of structures displayed by carbohydrates is originated, at the monosaccharide level, by a number of sources, such as: the absolute configuration at each chiral center of the sugar backbone; the orientation of the substituent in the anomeric position, i.e. axial or equatorial in respect to the plane containing the sugar ring, which defines the α or β anomer; and the ring conformation [14] (Fig. 5.1). Furthermore, for each pair of monosaccharides linked by a glycosidic bond, several conformers around the Φ and ψ dihedral angles are possible [15].

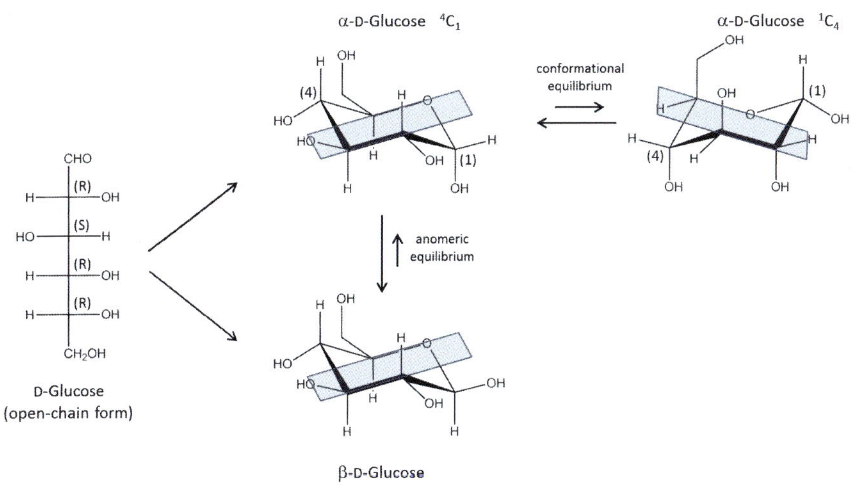

Figure 5.1 Structural variability of a monosaccharide (glucose). Schematic representation of D-glucose in its open-chain form and α or β anomeric cyclic forms. The absolute configuration is labeled with either the (*R*) or (*S*) descriptors for each stereocenter. The substituents at each carbon atom can be at axial or equatorial disposition with respect to the imaginary plane containing the C_2, C_3, C_5, and O_5 atoms. In β-D-glucose, the orientation of the hydroxyl group in C_1 is equatorial, while in the α anomer, it is axial. The chair conformation of the sugar ring is labeled as $^4C_1$ or $^1C_4$, depending on the position of carbons 1 and 4 (above or below the sugar ring plane, as noted by the position of the numeral).

Among several experimental restraints that can be gathered for the structural determination of carbohydrates, vicinal ^1H–^1H coupling constants ($^3J_{HH}$) are particularly useful, as they help determine the relative configuration of the sugar nuclei. For its part, nuclear overhauser enhancement (NOE) spectroscopy (NOESY) is employed for conformational analyses of the glycosidic linkage [12,16]. Finally, paramagnetic data are used to obtain longer-range restraints.

5.2.1 *Coupling constants*

In the late fifties, Nobel Laureate Martin Karplus described a relationship between the dihedral torsion angle established between vicinal hydrogens and the $^3J_{HH}$ coupling constant, by using a theoretical approach [17]:

$$^3J_{HH} = A + B\cos\Phi + C\cos^2\Phi \tag{5.1}$$

where Φ is the dihedral angle between the two protons.

According to Eq. 5.1, for dihedral angles of 0° or 180°, the observed coupling constant is large; while conversely, dihedral angles around 90° lead to $\cos^2\Phi$ values close to zero, and thus the observed coupling constant is small or null. For its part, coefficients *A*, *B* and *C* are parameterized for each particular kind of molecule, atom, and substituent. In this regard, and for instance, eight new Karplus relationships have recently been refined for use in conformational studies of saccharide *N*-acetyl side-chains [18].

A practical example of the usefulness of the Karplus equation to determine the relative configuration of the chiral centers is provided by the differentiation between two monosaccharides, such as β-glucose and β-mannose, which are epimers at position 2. The dihedral angle between H_1 and H_2 is around 180° in β-glucose and 60° in β-mannose. Subsequently, the $^3J_{HH}$ value is large in glucose, of around 10-15 Hz, and small in mannose, of around 3–5 Hz.

Also, the absolute configuration of the anomeric function in glycosides can be obtained with the same approach: in α-glucose, the $^3JH_1H_2$ is small due to the small angle established between these two protons, while their di-axial disposition in β-glucose entails a larger J coupling value.

Finally, saccharide units are characterized by a conformational equilibrium of the sugar ring. Among all different possible conformers, the two energy minima for β-glucose are dubbed as $^1C_4$ and $^4C_1$ (Fig. 5.1). In the case of the $^4C_1$ conformer, all the dihedral angles between protons of adjacent carbon atoms are large, being their respective vicinal coupling constants of around 10 Hz, while in the $^1C_4$ conformer, the trans di-equatorial disposition implies J coupling values smaller than 5 Hz. Of note, scalar coupling constants have the advantage that calculation of average values over an ensemble of conformations is simple; in fact, the scalar coupling values are simple linear averages of the ensemble of individual conformers [9].

Today, the Karplus equation is behind every conformational analysis studied by NMR. Variations of the Karplus equation are profusely used in elucidating the structure of organic compounds from NMR data, depending on the nature of the atoms or groups that compose the molecule or macromolecule.

Furthermore, not only $^3J_{HH}$ coupling constants, but also longer-range coupling ($^xJ_{HH}$, x > 3), as well as single and multiple-bond heteronuclear coupling ($^xJ_{HY}$, $1 \leq x \leq 4$, and Y = ^{13}C, ^{15}N, ^{19}F, etc.) and homonuclear $^xJ_{YY}$ coupling constants can provide a wealth of connectivity information on how molecular fragments are linked together. Specifically, ^1H–^{13}C correlations across carbon–carbon or carbon–heteroatom linkages give information on the molecular skeleton, providing one of the most powerful approaches to define the structure of organic molecules.

However, when they are added to each other, homo- and heteronuclear couplings can turn into a source of spectral complexity that hampers the obtention of detailed structural information. Fortunately, there are strategies that exist to circumvent this problem. For example, in a ^1H-coupled Y–NMR spectrum, where Y is, for instance, ^{13}C or ^{19}F, chemical shift and spin–spin coupling information may be obscured by overlapping multiplets. In the heteronuclear two-dimensional (2D) J-resolved experiment, however, these two parameters are separated and displayed on different axes of the 2D matrix. This way, H, Y multiplets can be observed with their natural line-width in one resolved dimension. Alternatively, a ^{1}H-decoupled Y–NMR spectrum can be acquired. In this experiment, the *J* constants arising from the coupling with the heteronucleus are deleted. The resulting spectrum is thus cleaner, permitting the measurement of J_{HH} constants, while, by comparing the coupled and the decoupled spectra, the heteronuclear *J* coupling constants are also obtained.

5.2.2 *The Nuclear Overhauser Effect (NOE). NOE Spectroscopy (NOESY)*

The 3D molecular geometry of carbohydrates becomes more intricate as the number of saccharide units increases. Knowledge on the spatial relationships among nuclei belonging to different units can be provided by both NOESY studies and paramagnetic restraints. These two sources of structural information are based on their dependency on the interproton distance through space.

The NOE is defined as the change in intensity of a nucleus' signal when the spin transitions of another nucleus cause a perturbation of its equilibrium populations. The two nuclei do not share a scalar (through bond) coupling; instead, they are sufficiently close in space to share a dipolar coupling. Thus, the NOE originates from dipolar cross-relaxation between pairs of protons, and depends on the proton–proton distance and on the molecular motion of the interproton vector:

$$I_{NOE} \approx \langle 1/r^6 \rangle f(\tau_c) \tag{5.2}$$

where I_{NOE} is the NOE intensity, *r* is the proton–proton distance, and *f* is a function that depends, among others factors, on the correlation time (τ_c) that describes the motion of the interproton vector.

The inverse-sixth relationship in Eq. 5.2 implies that the NOE only develops between nuclei that are close in space, usually within 5 Å. In 2002, Kurt Wüthrich was awarded the Nobel Prize in Chemistry for demonstrating that 2D NOE spectroscopy could be used to determine the 3D structures of biological macromolecules in solution [19–20]. In 2D NOESY experiments, a map of the NOE relationships of every hydrogen nucleus is obtained, thus allowing for a thorough structural characterization of the molecule.

Of course, not only homonuclear NOESY, but also heteronuclear studies are possible. For instance, ^{19}F-^{1}H heteronuclear NOE spectroscopy (HOESY) is used to estimate fluorine–proton distances [21].

5.2.3 *Paramagnetism-based restraints: Long range structural information*

Since NOEs and *J* couplings only provide local information and it is common for carbohydrates to be found in extended shapes, with no long range NOEs contacts, additional parameters are usually required to define their global shape and dynamics [22–24]. In this context, residual dipolar couplings (RDCs) [23–26], pseudo contact shifts (PCSs) [27–28], and paramagnetic relaxation enhancements (PREs) [29] emerge as key parameters in the structural characterization of complex carbohydrates.

An RDC is originated by the dipolar interaction occurring between a pair of nuclei (e.g. ^{1}H–^{13}C, ^{1}H–^{1}H), causing a splitting in the signal of each nucleus involved. The magnitude and the sign of the splitting depend on two geometrical parameters: the distance (r) between the two nuclei and the angle (θ) that the vector joining the two nuclei forms with the spectrometer's magnetic field. For a given pair of nuclei, for instance a ^{1}H–^{13}C pair in a carbohydrate, the dipolar interaction is given by Eq. 5.3:

$$D_{(^{1}\mathrm{H}-^{13}\mathrm{C})} = \frac{\mu_0 \gamma_H \gamma_C h}{\langle 2\pi\, r_{HC} \rangle^3} \left\langle \frac{3\cos^2\theta - 1}{2} \right\rangle \tag{5.3}$$

The terms enclosed within brackets in Eq. 5.3 correspond to the time average distance $\langle r_{HC}{}^{-3} \rangle$ and time angular averaging $\langle \cos^2\theta \rangle$. In a common solution, the solute tumbles isotropically; therefore, the time average of the angular term $\langle \cos^2\theta \rangle$ is zero, leading to a null dipolar splitting. A different situation occurs for a solute weakly oriented in an alignment medium. The weak alignment imposes certain anisotropy in the tumbling on the solute, making some orientations of the interatomic vector connecting the two coupled nuclei slightly more favorable than others. Because the solute still preserves a significant degree of rotational freedom, as in the isotropic case, $\langle \cos^2\theta \rangle$ is averaged out to a great extent, but the term is not necessarily null. Therefore, the observed splitting is reduced from the tenth of a kilohertz, as commonly found for strongly oriented molecules with a fix angle θ, to only a few tenths of a hertz; hence the term "residual" to refer to RDCs [26]. In practical terms, RDC measurements require the partial alignment of the molecule under study [30–31]. Several alignment media have been tested for carbohydrates using both external alignment systems [32–33] and paramagnetic tags [34].

The calculation of the dipolar interaction between a pair of dipolar-coupled nuclei requires the value of $\langle \cos^2\theta \rangle$ in Eq. 5.3, which can be determined using the alignment tensor methodology [35–36]. The alignment tensor is defined by five key parameters that represent the magnitude and rombicity of the alignment, and the three principal axes of the alignment of the molecule with respect to the direction of the magnetic field. The alignment tensor can be determined for a rigid molecule or fragment with a minimum of five RDCs pointing in nonparallel directions. In the so-called principal axis frame, in which the three axes of the alignment tensor are diagonal, Eq. 5.2 provides the theoretical value of the RDC for any ^{1}H–^{13}C vector of the carbohydrate, where A_{ax} and A_{rh} are the axial and rhombic components, respectively, θ is the angle that forms vector H–C with the z-axis in the molecular frame, and Φ is the azimuthal angle that describes the projection of vector H–C over the plane xy of the molecular frame (see Fig. 5.2).

$$D_{(^1\mathrm{H}-^{13}\mathrm{C})} = \frac{-\mu_0}{16\pi^2}\frac{\gamma_\mathrm{H}\gamma_C h}{r_{HC}^3}[A_{ax}(3\cos^2\theta - 1) + 3/2A_{rh}(\sin^2\theta\cos^2 2\phi)] \tag{5.4}$$

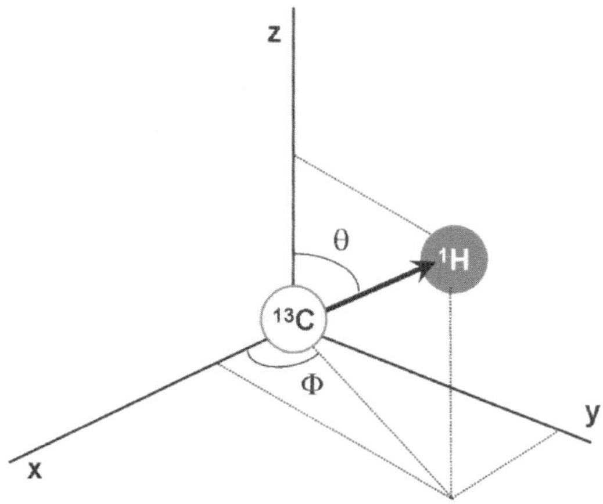

Figure 5.2 Definition of the angular parameters θ and Φ used in Eq. 5.4.

From the experimental viewpoint, RDCs simply add to the splitting due to the *J* scalar coupling. For a given signal in the spectrum, the magnitude of the RDC is determined from the difference in splitting between the sample oriented in the liquid crystal (RDC + *J*) and the isotropic non-aligned sample (J) [32,37].

The great advantage of RDCs in the carbohydrate field is that they help correlate the geometrical orientation between structural fragments (e.g. pyranose rings), no matter how far apart they are in the molecule [16]. In addition, RCSs are particularly useful to deduce the geometry of the glycosidic linkages, which have a major influence on the conformation of carbohydrates and their overall shape [38].

PCSs also arise from a dipolar interaction, but in this case, the effect is due to the dipolar interaction between the unpaired electrons of a paramagnetic entity (e.g. a lanthanide ion) and the nuclei in its vicinity. The effect is most easily described in terms of the magnetic susceptibility tensor χ, which governs the interaction of the paramagnetic dipole moment with the external magnetic field. PCSs are proportional to the anisotropy of the χ tensor ($\Delta\chi$) and independent of the isotropic component of the χ tensor. It is therefore sufficient to describe the PCS as a function of the axial and rhombic components of the $\Delta\chi$ tensor, $\Delta\chi_{ax}$ and $\Delta\chi_{rh}$ (Eq. 5.4), in close analogy to the common formulation of residual dipolar couplings [39]. The equation that governs PCS is given in Eq. 5.5, where $\Delta\delta_{PCS}$ is the difference in chemical shifts measured between diamagnetic and paramagnetic samples, r is the distance between the metal ion and the nuclear spin, and θ and Φ are the polar coordinates describing the position of the nuclear spin with respect to the principal axes of the $\Delta\chi$ tensor.

$$\Delta\delta_{PCS} = \frac{1}{12\pi r^3}[\Delta\chi_{ax}(3\cos^2\theta - 1) + 3/2\Delta\chi_{rh}\sin^2\theta\cos^2\Phi] \qquad \text{(Eq. 5.5)}$$

As seen in Eq. 5.5, PCSs give rise to large changes in chemical shifts that decrease with the distance between the metal ion and the nuclear spin with a $1/r^3$ dependence. Thus, PCSs provide long-range structural information.

Conformational studies of diamagnetic carbohydrates have been carried out by converting them to paramagnetic compounds through the attachment of an ethylenediaminetetraacetic acid (EDTA)-based paramagnetic tag, following the methodology already applied to proteins [40–41]. The use of this methodology has allowed for the characterization of disaccharides [34,41], trisaccharides [42], tetrasacharides [28], and very recently, a nonasaccharide [27].

5.3 Defining Ligand–Receptor Interactions: A 3D View by NMR

In addition to the structure of small molecules, such as carbohydrates, the interactions between them and their receptors can also be studied by using a plethora of NMR methods.

These techniques encompass both ligand- and receptor-detected experiments. Receptor-based methods are based on monitoring and comparing the receptor signals in the presence and absence of compounds of interest. By identifying perturbations of specific protein resonances, their binding sites can be delineated. These methods demand physicochemical properties of the protein target that are often challenging. First, they are usually applicable to relatively small receptors (<35 kDa) only. Also, milligram quantities of soluble, non-aggregated protein must be expressed and purified. Thus, suitable expression hosts must be found that permit isotope enrichment (e.g. ^{13}C, ^{15}N, and sometimes also ^{2}H) and are needed for resonance assignment of the macromolecules. In addition, it must be ensured that samples are stable for the time required for sequential resonance assignment (weeks).

For their part, ligand-detected methods are based on the monitoring of the NMR parameters of ligands in the presence and absence of the receptor. Saturation transfer difference (STD), WaterLOGSY, and transferred NOE spectroscopy (trNOESY) are used for detecting binding based on either changes in the motional properties of the saccharides upon binding to large macromolecular receptors, or the transfer of magnetization via their binding to the receptors. In contrast to protein-detected methods, the size of the protein is not an issue with these approaches. In fact, most ligand-based experiments become more sensitive when working with larger and multiple site receptors. Additionally, ligand observation bypasses the need to produce milligram quantities of isotopically labeled protein.

An obvious setback of ligand-based approaches is the impossibility to identify their binding site on the receptor. Also, ligand-based approaches rely on the exchange-mediated transfer of bound state information to the free state. This requirement biases ligand-detected methods toward the study of weakly binding ligands (i.e. in fast exchange) and the use of large ligand molar excesses.

In short, both receptor- and ligand-based approaches have distinct advantages and disadvantages, but they can be used together to provide a thorough vision on all the structural aspects that govern the sugar-receptor interaction.

5.3.1 *The ligand's point of view*

5.3.1.1 *Saturation transfer difference (STD)*

STD [43–44] is based on the transfer of magnetization between the receptor, which is selectively irradiated by a train of radiofrequency

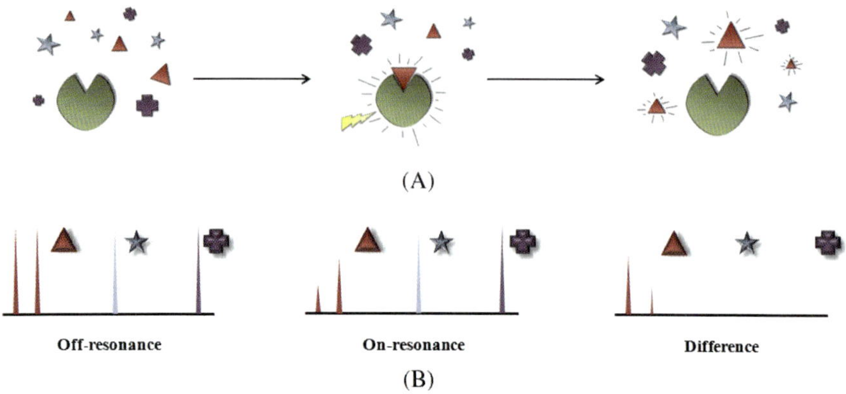

Figure 5.3 The STD experiment. A: Selective transfer of magnetization from the irradiated protein to a ligand that binds to it (red triangles). B: Simulated STD spectrum, in which saturation of the binding molecule leads to a decrease of its NMR signals in the on-resonance spectrum, yielding an STD (difference) spectrum revealing its NMR signals only.

pulses, and any ligands binding to the receptor. STD experiments (Fig. 5.3) involve the comparison of two one-dimensional (1D) ^{1}H spectra: the on-resonance spectrum, in which the saturation pulse is applied on the protein signals, and the off-resonance spectrum, in which this pulse is applied away from the protein signals. In the on-resonance spectrum, the irradiation pulse must be applied at a frequency devoid of ligand signals. Typically, irradiation is set around the aliphatic region (from −1 to 1 ppm) or alternatively around the aromatic region (usually 6–8 ppm) if the ligand has no aromatic signals [45]. The saturation then propagates along the macromolecule via ^{1}H–^{1}H cross relaxation mechanisms, and then from the macromolecule to the ligand's nuclei in close contact with the macromolecule ($r < 4$–5 Å), by a NOE effect. The dissociation of the ligand, according to binding exchange kinetics, allows saturated ligand molecules to accumulate, and when the NMR spectrum is recorded, saturated nuclei give rise to attenuated signals. To compare the on-resonance experiment, a spectrum called off-resonance is acquired. In the off-resonance spectrum, the saturation pulse is set at a frequency where no signals of either macromolecules or ligands appear (e.g. 100 ppm). Thus, the result of the off-resonance spectrum can be considered as a reference ^{1}H–NMR spectrum. The difference between the on-resonance and the off- resonance spectra gives a spectrum showing the signals of those nuclei that had received magnetization during their binding to the macromolecule, exclusively. Analysis of these signals not only permits the identification of binding ligands, but also provides information on their binding epitope, since nuclei of complex ligands, which are closer to the protein, become more saturated than those that do not participate directly in the binding [46]. STD experiments can also be used to study competition among ligands, and to calculate affinity constants [47].

In practical terms, STD experiments require small concentrations of receptor (low micromolar) in the presence of an excess of ligand (ratios from 1:10 up to 1:100), making this technique very convenient in terms of protein economy. However, STD also presents a number of limitations. Due to its intrinsic dependency on fast exchange, STD is only useful for systems with K_D in the 10^{-8} to 10^{-3} M range, thus hampering the detection and study of high affinity ligands.

5.3.1.2 *WaterLOGSY*

The WaterLOGSY experiment can be considered a variant of the STD, in which the saturated entity is the group of water molecules present at the ligand–receptor interface [48–49]. Water molecules bound to the receptor's binding pocket show characteristics similar to those of large molecules, whereas solvation water molecules of ligands behave like small molecules. As small molecules, solvation water molecules show positive NOEs, in contrast to large molecules, which show negative NOEs. The WaterLOGSY experiment takes profit of the magnetization of bulk water molecules to transfer it via different pathways. One pathway involves ^{1}H–^{1}H cross-relaxation mechanisms between the bound ligand and water molecules of the binding pocket. A second pathway occurs via cross-relaxation with exchangeable protons of the receptor (i.e. of amino (NH) or hydroxyl (OH) groups), which propagate inverted magnetization to bound ligand protons [50]. These pathways of magnetization transfer act constructively. In all cases, inverted magnetization is transferred to bound ligands, yielding an inversion of the NOE sign of its proton signals. Ligands that do not bind to the macromolecule maintain the positive sign of their NOEs, while ligands binding see the sign of their NOEs altered, adopting the opposite sign characteristic of large molecules. Experimental requirements are similar to those of STD, i.e. small amounts of protein and excesses of ligand. However, samples must be prepared in non-deuterated water.

Information obtained through WaterLOGSY experiments is comparable to that obtained through STD, allowing for the gathering of information on the ligand's orientation in the bound state. However, and in contrast with STD, WaterLOGSY is less dependent on spin diffusion, being the preferred method when studying receptors of low proton densities, such as nucleic acids [49].

5.3.1.3 *Transferred NOESY (trNOESY)*

Since its introduction by Bothner-By in the early seventies, trNOE spectroscopy [51–52] has been thoroughly used to study receptor–ligand interactions. This technique focuses on the change of sign in the ligand NOE peaks upon binding to the receptor. This is due to the dependence of the NOE sign on the correlation time of the molecule, which in turn, is related to its size. As a result, small molecules show positive NOEs, while macromolecules exhibit negative NOEs. In addition, transferred NOEs provide information on the conformation of the ligands in the bound state.

The trNOESY experiment is a regular 2D NOESY experiment in which the cross-peaks of the ligand change their NOE sign by virtue of the fact that, in the bound state, they adopt the motional properties of the macromolecule. As a result, non-binders show no NOE sign change upon the presence of receptor and their NOEs remain positive, while binders show negative NOE cross-peaks. Of note, the observed NOE is a contribution of both the free (positive NOE) and the bound ligand (negative NOE), but the cross-relaxation rate of the bound ligand is larger than that in the free state. Thus, the observed NOEs show a predominance of the contribution arisen by the bound form [45]. As a result, the trNOESY experiment not only helps detect interactions, but also provides information on the conformation of the ligand in the bound state.

5.3.2 *The protein's point of view*

The above mentioned techniques are useful to study the phenomenon of molecular recognition from the ligand's perspective, to detect binding events, and to provide insight into the structure of the bound saccharide. When it comes to studying the binding from the viewpoint of the other partner involved, i.e. the macromolecule, another set of methodologies focusing on monitoring the protein signals can provide information on the receptor binding epitopes. Given the high spectral overlap of proteins in the proton dimension, most protein-detected methods are 2D experiments (notably ^{1}H–^{15}N heteronuclear single-quantum coherence, HSQC), among which heteronuclear experiments require isotopic labeling, and thus protein expression in suitable systems. This poses a limitation to the size of the receptors studied, usually not exceeding a size of 35 kDa [45], although current advances are pushing this limit further. In addition, detailed study on receptor–ligand interactions focusing on the protein signals require a full or partial assignment of the protein ^{1}H, ^{13}C, and ^{15}N resonances, which is often very time- and material-consuming. However, these setbacks are compensated by the amount of information that these techniques provide.

In its simple version, the interaction between ligands and peptides, very small proteins or domains thereof can be studied from the receptor's perspective by monitoring their ^{1}H signals in mono- or two-dimensional (e.g. NOESY) experiments with unlabeled receptor. In such cases, the assignment of the receptor's ^{1}H–NMR signals can be achieved with the standard strategies used for spectral assignment of small ligands. While these instances constitute only an exiguous minority of all sugar/protein systems, these mono-nuclear strategies have long been applied, for example, to study the interaction between hevein, a 43-amino acid, 4.7 kDa lectin from the rubber tree, and β (1–4) linked *N*-acetylglucosamine glycans, a model system of protein–carbohydrate interactions [53–54].

For a majority of the sugar–protein systems, however, 2D heteronuclear experiments are required. By far, the most widely used protein-detected technique is ^{1}H–^{15}N HSQC. Heteronuclear coherence experiments are based on the transfer of magnetization between ^{1}H and a bonded heteronucleus via scalar couplings. The result of a ^{1}H–^{15}N HSQC experiment is a two-dimensional spectrum [55] showing a number of cross-peaks close to the number of residues. Prolines, which lack an amide proton, are not represented in the spectrum, while side-chain amines of nitrogen-containing residues provide additional cross-peaks. If the cross-peaks are assigned to their respective residues, ^{1}H–^{15}N HSQC experiments can provide valuable information on the protein epitopes for binding ligands, since cross-peaks belonging to the binding site residues become shifted upon the addition of the ligand (Fig. 5.4).

As already mentioned, obtaining ^{1}H–^{15}N HSQC data requires the use of ^{15}N isotopically-enriched protein and the spectral assignment of at least most ^{15}NH resonances, which is achieved by

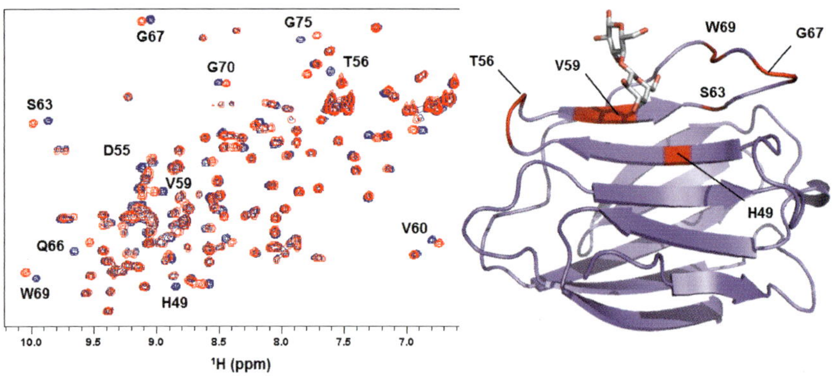

Figure 5.4 Left: ^{1}H–^{15}N HSQC spectrum of human galectin-7 in the absence (blue) and presence (red) of lactose, with labels indicating some of the most shifted peaks, according to its full spectral assignment [56]. Right: crystallographic structure of lactose-bound galectin-7 (PDB: 4GAL [57]), with labels indicating some of the residues with the most perturbed cross-peaks upon addition of the ligand in the ^{1}H–^{15}N HSQC experiment.

sequential assignment of the protein backbone in 3D triple $^{1}H/^{13}C/^{15}N$ resonance experiments. For this purpose, proteins are usually expressed in bacteria cultured in minimum media containing ^{13}C-glucose and/or ^{15}N-ammonium as sole carbon and nitrogen sources, respectively. Also, deuterated water can be included in culture media for proteins to also be labeled with ^{2}H [58]. While this usually leads to a decrease in bacterial growth and protein harvest efficiency, it is used for studying large proteins, as deuterium enrichment of the protein side-chains helps to minimize ^{1}H-mediated relaxation pathways that lead to signal broadening and loss [59].

In any case, fast relaxation is always an issue when dealing with high molecular weight macromolecules. However, technological advances in hardware and novel developments in pulse sequences, such as transverse relaxation optimized spectroscopy (TROSY) [60], are pushing this limit further. TROSY-edited experiments permit the obtaining of high-resolution spectra (e.g. ^{1}H–^{15}N TROSY–HSQC) of proteins up to 100 kDa [61] by selectively recording the most slowly relaxing signal component of the ^{15}N–^{1}H correlation.

Lastly, a variety of experiments that can be collectively referred to as selective labeling methods, are also utilized to gather insight into the ligand–protein binding from the perspective of the receptor. They all have in common the simplification of the protein NMR spectrum via selective spin-labeling of only a fraction of its nuclei, and are especially useful when they constitute an important part of the binding site. For instance, selective ^{13}C-labeling of valine (Val), isoleucine (Ile), and leucine (Leu) side-chain methyls can be achieved by bacterial culturing in the presence of labeled versions of their biochemical precursors [62], and information is obtained through ^{1}H–^{13}C–HSQC experiments. Also, the virtues of fluorine-19 NMR [63] can be exploited by ^{19}F-labeling of proteins using fluorine-containing analogs of natural amino acids, such as fluorophenylalanine [64], and information is obtained through clean, 1D ^{19}F-NMR spectra. Finally, other strategies for selective labeling that do not require manipulation at the pre-translational level have been reported. Such is the case of solvent exposed-amines (SEA)–HSQC [65] or SEA-TROSY [66], which allow for the selective detection of solvent-exposed NHs, and are useful for the screening of interactions, since binding mainly occurs on the surface of proteins.

5.3.2.1 *Combination with computational approaches*

In most of the cases, NMR-based conformational analysis of saccharides is assisted by molecular modeling protocols [67]. Considering that NMR parameters reflect the ensemble average of the different situations that take place in solution, the combination of the NMR experimental data with the parameters deduced from theoretical models is essential to provide the complete picture of the actual conformational behavior and provide an almost quantitative estimation of the populations of the different conformations which can coexist in solution [68].

During the past decades, the development of modeling techniques has allowed the successful prediction of the structure of ligands, as well as their interactions with different receptors. Indeed, the award of the Nobel Prize in Chemistry 2013 to M. Karplus, M. Levitt, and A. Warshel "for the development of multiscale models for complex chemical systems" shows the increasing importance that computer models have gained through the years.

There are several computational methods that can be applied for the study of structures and conformations of carbohydrates and their complexes [69]. The particular approach to be employed depends on the specific requirements and characteristics of the system under study. Molecular mechanics are based in laws of the classic mechanics and, as a model, will use a molecule composed of atoms, considered as point masses with charge, which are bound by bonds. Contributions to the potential energy due to stretching of bonds, distortion of the bond angle, distortion out of the plane, internal rotation around a bond (torsion), cross interactions between these movements, van der Waals attraction and repulsion forces between non-bonded atoms, and electrostatic interactions between atoms are considered. The explicit expression for each term of the equation defines what is known as a force field. Each kind of force field uses adjustable parameters (force constants, structural values from equilibrium geometries, etc.) to make the calculated the closest possible to the experimental. The two most used force fields in the field of carbohydrates are MM3 and AMBER [70–73]. Nevertheless, molecules are moving unless they stay at absolute zero temperature. In this context, molecular dynamics protocols solve the classical Newtonian movement equation at an atomic level. The availability of a precise potential (force field) that describes the interactions between atoms is essential. For carbohydrates, glycosylation-dependent cell adhesion molecule (GLYCAM) parameters [74] are the most commonly employed. Nevertheless, more sophisticated computational chemistry methods based on quantum chemistry may also be employed. *Ab initio* calculations solve the Schrodinger equation through a series of approximations, and have also been employed to model a variety of systems in the carbohydrate field [75].

As common protocols, the conformational behavior of sugars in their free state can be studied by characterizing the potential energy surface [76,77], which provides information about the geometries of local minima and the associated transition states. Each point of the surface will correspond to specific values of the considered freedom degrees, which will be associated with a certain molecular structure, correlated with its specific energy. The minimum is the low part of a valley in the surface, and any movement from there will lead to a higher energy conformer. Minima correspond to equilibrium geometries. Crests and peaks in the surfaces correspond to maximum energy points. The most interesting ones are the transition states that connect two structures in equilibrium. Using *ab initio* calculations, the different states can be characterized by the value of their imaginary frequencies (zero for minima and 1 for transition states) and they can be linked with intrinsic reaction coordinate (IRC) calculations [78,79]. Calculation of the reaction coordinates allow to assign the minima associated with each transition state. Characterization of the different conformations, minima, and transition states, as well as the calculation of the specific energy barriers between them, can provide very useful information about the relative stabilities, as well as the energies that have to be overcome for the selection of the preferred conformation. This information is also of high relevance when molecular recognition processes are evaluated. Examples of such calculations for six-membered rings as simple models have been presented [80,81]. Further analyses with real sugars have been performed using different computational approaches, relating the shape of the pyranose rings with the transition states and intermediates associated with carbohydrate-modifying enzymes [82].

A common procedure to evaluate the conformational behavior of a saccharide consists of the 1) definition of the torsion degrees of our ligand; 2) generation of a theoretical conformational map that defines the low energy regions (local minima) employing molecular mechanics calculations; and 3) characterization of the possible geometric restrictions for each of the conformations that can be related to NMR data (NOEs and/or coupling constants), such as distances and torsion angles. If these restrictions are unique for a specific conformer (i.e. exclusive NOEs [83]), the comparison with the experimental data will allow setting the theoretical conformations that are actually present in solution, as well as their populations [84]. In an analogous manner, the bound state geometry to a particular receptor can be derived, thus providing information on the nature of the molecular recognition process. The success of this approach has been shown in many examples [85].

The combination of NMR experiments with molecular dynamics (MD) protocols has allowed for the obtaining of the 3D structure of different carbohydrate–protein complexes [86–90]. In the best situation, experimental information from ligand-based and receptor-based experiments should be available, thus providing the full experimental information required to derive a precise 3D structure [91,92]. In other cases, only experimental information on the bound ligand conformation is accessible and the structure of the receptor has to be deduced from alternative methods [93]. In these cases, the combination of docking protocols and MD simulations is of paramount importance [94]. In this manner, information about the dynamics and flexibility of both ligand and protein during the complex formation, as well as about the interactions established between them and their evolution through time can be derived [95]. The MD simulations can be linked to other methods, such as molecular mechanic (MM)/Poisson–Boltzmann surface area (PBSA), which allows decomposing the binding energy in its components per residue giving the weight of its implication in the interaction [96].

For carbohydrate-processing enzymes, methods that combine quantum mechanics (QM) with MM have been developed [97]. They help define the region of the catalytic site (ligand and some other amino acids of interest) to be studied through *ab initio* calculations, leaving the rest of the protein subject to standard molecular mechanics. Catalytic processes or reaction pathways can be thus followed [98,99].

5.4 Conclusions

Carbohydrates are complex and diverse molecules playing key functions in all living organisms. NMR spectroscopy, assisted in many cases with molecular modeling protocols, provides scientists with the possibility to study, in solution and at atomic scale, the structure of carbohydrates and the mechanisms that govern their recognition by receptors. Advances in both instrumentation and novel methodologies are currently expanding the use and range of applicability of NMR for the study of carbohydrates and their interacting partners.

References

1. Gabius, H. J., André, S., Jiménez-Barbero, J., Romero, A., Solís, D. (2011) *Trends Biochem. Sci.* **36**, 298–313.
2. Silipo, A., Larsbrink, J., Marchetti, R., Lanzetta, R., Brumer, H., Molinaro, A. (2012) *Chem. Eur. J.* **18**, 13395–13404.
3. Silipo, A., Sturiale, L., De Castro, C., Lanzetta, R., Parrilli, M., Garozzo, D., Molinaro, A. (2012) *Carbohydr. Res.* **357**, 75–82.
4. Nieto, L., Canales, A., Gimenez-Gallego, G., Nieto, P. M., Jimenez-Barbero, J. (2011) *Chem. Eur. J.* **17**, 11204–11209.

5. Perepelov, A. V., Shekht, M. E., Liu, B., Shevelev, S. D., Ledov, V. A., Senchenkova, S. N., L'Vov V. L., Shashkov, A. S., Feng, L., Aparin, P. G., Wang, L., Knirel, Y. A. (2012) *FEMS Immunol. Med. Microbiol.* **66**, 201–210.
6. Gabius, H. J. Ed. (2009) *The Sugar Code: Fundamentals of Glycoscience.* Wiley-Blackwell.
7. Bewley, C. A., Shahzad-Ul-Hussan, S. (2013) *Biopolymers* **99**, 796–806.
8. Gabius, H. J., Siebert, H. C., Andre, S., Jimenez-Barbero, J., Rudiger, H. (2004) *Chembiochem.* **5**, 740–764.
9. Varki, A. (1993) *Glycobiology* **3**, 97–130.
10. Toone, E. (1994) *Curr. Opin. Struct. Biol.* **4**, 719–728.
11. Weis, W. I., Drickamer, K. (1996) *Annu. Rev. Biochem.* **65**, 441–473.
12. Roldós, V., Cañada, F. J., Jimenez-Barbero, J. (2011) *Chembiochem.* **12**, 1–17.
13. Weimar, T., Woods, R. J. (2002) In *NMR Spectroscopy of Glycoconjugates.* Jiménez-Barbero, J., Peters, T. Eds. Wiley-VCH (Berlin). pp. 109–144.
14. Angyal, S. J. (1969) *Angew. Chem. Int. Ed.* **8**, 157–166.
15. Nurisso, A., Blanchard, B., Audfray, A., Rydner, L., Oscarson, S., Varrot, A., Imberty, A. (2010) *J. Biol. Chem.* **285**, 20316–20327.
16. Calle, L. P., Canada, F. J., Jimenez-Barbero, J. (2011) *Nat. Prod. Rep.* **28**, 1118–1125.
17. Karplus, M. (1959) *J. Chem. Phys.* **30**, 11–15.
18. Hu, X., Carmichael, I., Serianni, A. S. (2010) *J. Org. Chem.* **75**, 4899–4910.
19. Williamson, M. P., Havel, T. F., Wuthrich, K. (1985) *J. Mol. Biol.* **182**, 295–315.
20. Riek, R., Hornemann, S., Wider, G., Glockshuber, R., Wuthrich, K. (1997) *FEBS Lett.* **413**, 282–288.
21. Combettes, L. E., Clausen-Thue, P., King, M. A., Odell, B., Thompson, A. L., Gouverneur, V., Claridge, T. D. (2012) *Chem. Eur. J.* **18**, 13133–13141.
22. Martin-Pastor, M., Bush, C. A. (2001) *J. Biomol. NMR* **19**, 125–139.
23. Azurmendi, H. F., Martin-Pastor, M., Bush, C. A. (2002) *Biopolymers* **63**, 89–98.
24. Martin-Pastor, M., Bush, C. A. (2000) *Carbohydr. Res.* **323**, 147–155.
25. Tian, F., Al-Hashimi, H. M., Craighead, J. L., Prestegard, J. H. (2000) *J. Am. Chem. Soc.* **123**, 485–492.
26. Canales, A., Jimenez-Barbero, J., Martin-Pastor, M. (2012) *Magn. Reson. Chem.* **50** Suppl 1, S80–85.
27. Canales, A., Mallagaray, A., Perez-Castells, J., Boos, I., Unverzagt, C., Andre, S., Gabius, H. J., Canada, F. J., Jimenez-Barbero, J. (2013) *Angew. Chem. Int. Ed.* **52**, 13789–13793.

28. Zhang, Y., Yamamoto, S., Yamaguchi, T., Kato, K. (2012) *Molecules* **17**, 6658–6671.
29. Jain, N. U., Venot, A., Umemoto, K., Leffler, H., Prestegard, J. H. (2001) *Protein Sci.* **10**, 2393–2400.
30. Kummerlowe, G., Kiran, M. U., Luy, B. (2009) *Chem. Eur. J.* **15**, 12192–12195.
31. Annila, A., Permi, P. (2004) *Concepts Magn. Reson.* **23A**, 22–37.
32. Tian, F., Bolon, P. J., Prestegard, J. H. (1999) *J. Am. Chem. Soc.* **121**, 7712–7713.
33. Martin-Pastor, M., Canales-Mayordomo, A., Jimenez-Barbero, J. (2003) *J. Biomol. NMR* **26**, 345–353.
34. Erdelyi, M., d'Auvergne, E., Navarro-Vazquez, A., Leonov, A., Griesinger, C. (2011) *Chem. Eur. J.* **17**, 9368–9376.
35. Saupe, A. (1968) *Angew. Chem. Int. Ed.* **7**, 97–112.
36. Losonczi, J. A., Andrec, M., Fischer, M. W., Prestegard, J. H. (1999) *J. Magn. Reson.* **138**, 334–342.
37. Pham, T. N., Hinchley, S. L., Rankin, D. W., Liptaj, T., Uhrin, D. (2004) *J. Am. Chem. Soc.* **126**, 13100–13110.
38. Martin-Pastor, M., Canales, A., Corzana, F., Asensio, J. L., Jimenez-Barbero, J. (2005) *J. Am. Chem. Soc.* **127**, 3589–3595.
39. Bertini, I., Luchinat, C., Parigi, G. (2002) *Prog. NMR Spec.* **40**, 249–273.
40. Leonov, A., Voigt, B., Rodriguez-Castaneda, F., Sakhaii, P., Griesinger, C. (2005) *Chem. Eur. J.* **11**, 3342–3348.
41. Mallagaray, A., Canales, A., Dominguez, G., Jimenez-Barbero, J., Perez-Castells, J. (2011) *Chem. Commun.* **47**, 7179–7181.
42. Yamamoto, S., Zhang, Y., Yamaguchi, T., Kameda, T., Kato, K. (2012) *Chem. Commun.* **48**, 4752–4754.
43. Mayer, M., Meyer, B. (1999) *Angew. Chem. Int. Ed.* **38**, 1784–1788.
44. Meyer, B., Peters, T. (2003) *Angew. Chem. Int. Ed.* **42**, 864–890.
45. Fernandez-Alonso, M. D. C., Berbís, M. A., Canales, A., Arda, A., Cañada, F. J., Jimenez-Barbero, J. (2013) In *New Applications of NMR in Drug Discovery and Development.* Garrido L., Beckmann, N. Eds. The Royal Society of Chemistry (Cambridge). pp 7–42.
46. Mayer, M., Meyer, B. (2001) *J. Am. Chem. Soc.* **123**, 6108–6117.
47. Angulo, J., Enríquez-Navas, P. M., Nieto, P. M. (2010) *Chem. Eur. J.* **16**, 7803–7812.
48. Dalvit, C., Fogliatto, G., Stewart, A., Veronesi, M., Stockman, B. (2001) *J. Biomol. NMR* **21**, 349–359.

49. Dalvit, C., Pevarello, P., Tatò, M., Veronesi, M., Vulpetti, A., Sundström, M. (2000) *J. Biomol. NMR* **18**, 65–68.
50. Otting, G., Liepinsh, E. (1995) *Acc. Chem. Res.* **28**, 171–177.
51. Balaram, P., Bothner-By, A. A., Breslow, E. (1972) *J. Am. Chem. Soc.* **94**, 4017–4018.
52. Balaram, P., Bothner-By, A. A., Dadok, J. (1972) *J. Am. Chem. Soc.* **94**, 4015–4017.
53. Fernández-Alonso, M. C., Diaz, D., Berbiş, M. A., Marcelo, F., Canada, J., Jimenez-Barbero, J. (2012) *Curr. Protein Pept. Sci.* **13**, 816–830.
54. Jimenez-Barbero, J., Javier Canada, F., Asensio, J. L., Aboitiz, N., Vidal, P., Canales, A., Groves, P., Gabius, H. J., Siebert, H. C. (2006) *Adv. Carbohydr. Chem. Biochem.* **60**, 303–354.
55. Bax, A., Lerner, L. (1986) *Science* **232**, 960–967.
56. Nesmelova, I. V., Berbis, M. A., Miller, M. C., Canada, F. J., Andre, S., Jimenez-Barbero, J., Gabius, H. J., Mayo, K. H. (2012) *Biomol. NMR Assign.* **6**, 127–129.
57. Leonidas, D. D., Vatzaki, E. H., Vorum, H., Celis, J. E., Madsen, P., Acharya, K. R. (1998) *Biochemistry* **37**, 13930–13940.
58. Gardner, K. H., Kay, L. E. (1998) *Annu. Rev. Biophys. Biomol. Struct.* **27**, 357–406.
59. LeMaster, D. M. (1990) *Annu. Rev. Biophys. Biophys. Chem.* **19**, 243–266.
60. Pervushin, K., Riek, R., Wider, G., Wuthrich, K. (1997) *Proc. Natl. Acad. Sci. USA* **94**, 12366–12371.
61. Vogtherr, M., Fiebig, K. (2003) *EXS* 183–202.
62. Sibille, N., Hanoulle, X., Bonachera, F., Verdegem, D., Landrieu, I., Wieruszeski, J. M., Lippens, G. (2009) *J. Biomol. NMR* **43**, 219–227.
63. Dalvit, C., Mongelli, N., Papeo, G., Giordano, P., Veronesi, M., Moskau, D., Kummerle, R. (2005) *J. Am. Chem. Soc.* **127**, 13380–13385.
64. Jackson, J. C., Hammill, J. T., Mehl, R. A. (2007) *J. Am. Chem. Soc.* **129**, 1160–1166.
65. Lin, D., Sze, K. H., Cui, Y., Zhu, G. (2002) *J. Biomol. NMR* **23**, 317–322.
66. Pellecchia, M., Meininger, D., Shen, A. L., Jack, R., Kasper, C. B., Sem, D. S. (2001) *J. Am. Chem. Soc.* **123**, 4633–4634.
67. Woods RJ, Tessier MB. (2010) *Curr. Opin. Struct. Biol.* **20**, 575–583.
68. Jiménez-Barbero, J., Asensio, J. L., Cañada, F. J., Poveda, A. (1999) *Curr. Opin. Struct. Biol.* **9**, 549–555.
69. Fadda, E., Woods, R. J. (2010) *Drug Discov. Today* **15**, 596–609.
70. Allinger, N. L., Yuh, Y. H., Lii, J. H. (1989) *J. Am. Chem. Soc.* **111**, 8551–9566.

71. Allinger, N. L., Yan, J. (1993) *J. Am. Chem. Soc.* **115**, 11918–11925.
72. Homans, S. W. (1990) *Biochemistry* **29**, 9110–9118.
73. Wang, J., Wolf, R. M., Caldwell, J. W., Kollman, P. A., Case, D. A. (2004) *J. Comput. Chem.* **25**, 1157–1174.
74. Kirschner, K. N., Yongye, A.B., Tschampel, S.M., González-Outeiriño, J., Daniels, C.R., Foley, B. L., Woods, R. J. (2008) *J. Comput. Chem.* **29**, 622–655.
75. French, A. D., Johnson, G. P. (2011) *Methods Mol. Biol.* **715**, 21–42.
76. Stortz, C. A., Johnson, G. P., French, A. D., Csonka, G. I. (2009) *Carbohydr. Res.* **344**, 2217–2228.
77. Jimenez-Barbero, J., Bouffar-Roupe, C., Rochas, C., Pérez, S. (1989) *Int. J. Biol. Macromol.* **11**, 265–272.
78. González, C. Schlegel, H. B. (1989) *J. Chem. Phys.* **90**, 2154–2161.
79. González, C. Schlegel, H. B. (1990) *J. Phys. Chem.* **94**, 5523–5527.
80. Fernández-Alonso Mdel, C., Asensio, J. L., Cañada, F. J., Jiménez-Barbero, J., Cuevas, G. (2003) *ChemPhysChem.* **4**, 754–757.
81. Fernández-Alonso Mdel, C., Cañada, F. J., Jiménez-Barbero, J., Cuevas, G. (2005) *ChemPhysChem.* **6**, 671–680.
82. Davies, G. J., Planas, A., Rovira, C. (2012) *Acc. Chem. Res.* **45**, 308–316.
83. Kozár, T., Nifant'ev, N. E., Grosskurth, H., Dabrowski, U., Dabrowski, J. (1998) *Biopolymers* **46**, 417–432.
84. Widmalm, G. (2013) *Carbohydr. Res.* **378**, 123–132.
85. Kogelberg, H., Solís, D., Jiménez-Barbero, J. (2003) *Curr. Opin. Struct. Biol.* **13**, 646–653.
86. Asensio, J. L., Ardá, A., Cañada, F. J., Jiménez-Barbero, J. (2013) *Acc. Chem. Res.* **46**, 946–954.
87. Wimmerová, M., Kozmon, S., Nečasová, I., Mishra, S. K., Komárek, J., Koča, J. (2012) *PLoS One* **7**, e46032.
88. Fernández-Alonso Mdel, C., Cañada, F. J., Jiménez-Barbero, J., Cuevas, G. (2005) *J. Am. Chem. Soc.* **127**, 7379–7386.
89. Vandenbussche, S., Díaz, D., Fernández-Alonso, M. C., Pan, W., Vincent, S. P., Cuevas, G., Cañada, F. J., Jiménez-Barbero, J., Bartik, K. (2008) *Chem. Eur. J.* **14**, 7570–7578.
90. Imberty, A., Pérez, S. (2000) *Chem. Rev.* **100**, 4567–4588.
91. Nieto, L., Canales, Á., Fernández, I. S., Santillana, E., González-Corrochano, R., Redondo-Horcajo, M., Cañada, F. J., Nieto, P., Martín-Lomas, M., Giménez-Gallego, G., Jiménez-Barbero, J. (2013) *Chembiochem.* **14**, 1732–1744.
92. Matei, E., Furey, W., Gronenborn, A. M. (2008) *Structure* **16**, 1183–1194

93. Martín-Santamaría, S., André, S., Buzamet, E., Caraballo, R., Fernández-Cureses, G., Morando, M., Ribeiro, J. P., Ramírez-Gualito, K., de Pascual-Teresa, B., Cañada, F. J., Menéndez, M., Ramström, O., Jiménez-Barbero, J., Solís, D., Gabius, H. J. (2011) *Org. Biomol. Chem.* **9**, 5445–5455.
94. Sandström, C., Hakkarainen, B., Matei, E., Glinchert, A., Lahmann, M., Oscarson, S., Kenne, L., Gronenborn, A. M. (2008) *Biochemistry* **47**, 3625–3635.
95. Nivedha, A. K., Makeneni, S., Foley, B. L., Tessier, M. B., Woods, R. J. (2014) *J. Comput. Chem.* **35**, 526–539.
96. Vidal, P., Roldós, V., Fernández-Alonso Mdel, C., Vauzeilles, B., Bleriot, Y., Cañada, F. J., André, S., Gabius, H. J., Jiménez-Barbero, J., Espinosa, J. F., Martín-Santamaría, S. (2013) *Chem. Eur. J.* **19**, 14581–14590.
97. Gómez, H., Polyak, I., Thiel, W., Lluch, J. M., Masgrau, L. (2012) *J. Am. Chem. Soc.* **134**, 4743–4752.
98. Lammerts van Bueren, A., Ardèvol, A., Fayers-Kerr, J., Luo, B., Zhang, Y., Sollogoub, M., Blériot, Y., Rovira, C., Davies, G. J. (2010) *J. Am. Chem. Soc.* **132**, 1804–1806.
99. Gómez, H., Lluch, J. M., Masgrau, L. (2013) *J. Am. Chem. Soc.* **135**, 7053–7063.

Chapter 6

Glycoarrays: An Invaluable Tool for Glycomics

Niels-Christian Reichardt, Sonia Serna and Juan Echevarria

Glycotechnology Laboratory, CIC biomaGUNO,
Paseo Miramon 182, 20009 San Sebastian, Spain

6.1 Carbohydrate Microarray Fabrication

6.1.1 *Introduction to microarray technology, application to glycomics*

A carbohydrate microarray is a high-throughput bioanalytical screening tool to evaluate thousands of individual glycan–receptor binding events in a single experiment and with minimum reagent use. The carbohydrate ligands are immobilized on a surface and put into contact with a solution containing a fluorescently tagged receptor of the receptor mixture. After a washing step, which removes unbound or only weakly bound material, the receptor binding profile can be determined by quantifying the fluorescence intensity of the bound material with a fluorescence scanner at micrometre resolution. Alternative methods for the detection of binding events, like the use of secondary antibodies, surface based mass spectrometry, or radioimaging will be discussed below.

Microarrays were initially developed for oligonucleotides [1], and subsequently applied to peptides, proteins [2] and cell [3] adhesion studies.

6.1.2 *First approaches*

Since more than ten years ago when the first proof of concept studies appeared for the use of microarrays to study carbohydrate–protein interactions, the technology has evolved rapidly into a major enabling technology for glycomics research [4].

Both bacterial polysaccharides [5] and neoglycolipids [6], which are oligosaccharides from natural sources chemically coupled to a lipid head group, were arrayed onto nitrocellulose-coated glass slides and interrogated with serum antibodies, lectins and other carbohydrate binding proteins (CBPs). Likewise, glycolipids formed by derivatization of oligosaccharides with saturated alkyl chains were non-covalently bound to a polystyrene microtitre well surface and used for lectin-binding assays, as well as on-chip enzymatic transformations [7]. Slides of plasma treated polystyrene were another material explored early on for the immobilization of polysaccharides, glycoproteins, proteoglycans or even cell extracts via hydrophobic interactions, which were subsequently probed with a panel of anti-glycan monoclonal antibodies [8]. The anomeric functionalization of free glycans with a linker in combination with chemoselective reactive surfaces opened the way to the covalent and oriented probe immobilization. Spacers, which increase the surface–probe distance, and diluters, which modify the inter-ligand distance, were introduced to improve the accessibility of CBPs to the immobilized carbohydrates and to increase assay sensitivity. These immobilization strategies have also helped to adapt carbohydrate microarrays later on to different read-out methods, like mass spectrometry [9].

6.1.3 *Carbohydrate sources*

A major challenge is the population of arrays with sufficient numbers of ligands to ideally cover all important functional epitopes of the glycome of a particular organism. Glycoproteins from tissue, serum, or food products (e.g. milk, eggs) are important sources for the isolation of a variety of even complex natural glycans by multi-dimensional chromatographic separation. After enzymatic release from the protein carrier, glycans are usually coupled by reductive amination to a bifunctional fluorescent tag, like aminobenzamide (AB), 2,6-diaminopyridine or 2-amino-*N*-(2-aminoethyl)benzamide (AEAB) for easy detection and subsequent printing [10]. Although separation technology is constantly improving, the separation of heterogenous mixtures of glycans with often only subtle differences between components is still very challenging, and often, mixtures cannot be completely resolved.

Glycan isolation from natural materials is an important source of glycans for microarray preparation, in particular when the interaction of glycan binding proteins (GBPs) with the entire glycome is of interest, or synthetic or commercial glycans are simply unavailable.

An alternative to glycans from natural sources is their chemo-enzymatic synthesis. This approach not only assures the purity and structure of each glycan synthesized, but can also provide a wide variety of similar or isomeric structures which, in combination with microarrays, is a valuable tool for the more precise definition of carbohydrate binding epitopess. However, even after the important advances in carbohydrate chemistry of the last decades and the development of automated chemical and chemo-enzymatic synthetic strategies, the targeted synthesis of glycans is still a time-consuming exercise even for the expert [11–13]. The combination of chemical and enzymatic methods is currently the most promising method to generate larger collections of well-defined glycans fairly rapidly. To this end, conserved glycan core structures incorporating a linker are synthesized as primers and elongated with a set of recombinant glycosyltransferases (GTs) to form a diverse collection of structures around a common glycan core. The high specificity of GTs allows for direct stereo- and regioselective glycosylation, omitting the need for any protecting groups. In spite of the limited commercial availability of GTs, the chemo-enzymatic approach has been very helpful in the preparation of glycan libraries by iterative elongation procedures [14].

A further method for the preparation of larger glycan collections is the parallel on-chips synthesis either by chemical or enzymatic methods. The Mrksich group explored the chemical synthesis of a small collection of disaccharide structures starting from a phenol-linked self-assembled alkylthiolate monolayer on a gold surface and subsequent deprotection and glycosylation steps on the chip [15]. The completion of every reaction step could be monitored by surface based matrix-assisted laser desorption/ ionization mass spectrometry (MALDI-MS). The use of GTs for the synthesis of oligosaccharides on solid supports implies less stringent requirements for the stability of the linkage than chemical synthesis and has been reviewed [16]. The precise robotic dispensers used for the fabrication of microarrays can also be used for the high-throughput parallel enzymatic diversification of glycan structures on the chip (Fig. 6.1). Selected immobilized glycans can be extended or core-modified by overlay deposition with nanodroplets containing GTs and cofactors, giving rise to high density arrays with spot diameters of 100–200 micrometres [17].

6.1.4 *Immobilization strategies*

The non-covalent attachment of free glycans to hydrophobic surfaces like nitrocellulose [5] or plasma treated polystyrene [8] usually only works well for large polysaccharides (Fig. 6.2). By tagging with a fluorous or lipid head group prior to printing, smaller glycans can be non-covalently immobilized to per-fluorinated [18] or hydrophobic [6] surfaces. Usually, the stability of these setups is sufficient to resist the washing procedures required for the microarray experiments and can be further improved by choice of increasing the linker length or branching.

Both non-covalent attachment methods (via hydrophobic and fluorous interactions) have been exploited for the analysis of enzyme activity and specificity by on-chip mass spectrometry on functionalized aluminium oxide slides and [19] on indium tin oxide (ITO) slides [20]. Alternatively, biotinylated glycans can be efficiently immobilized on streptavidine coated slides [21] via very strong non-covalent interactions. Finally, mixtures of deoxyribonucleic acid (DNA) or peptide nucleic acid (PNA) functionalized glycans can be incubated in solution with potential binding partners

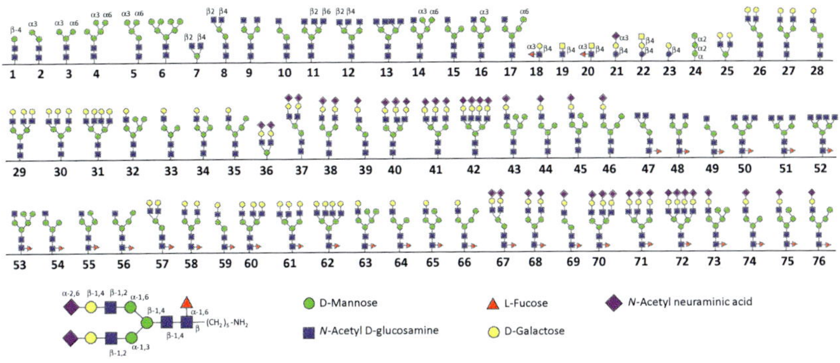

(A)

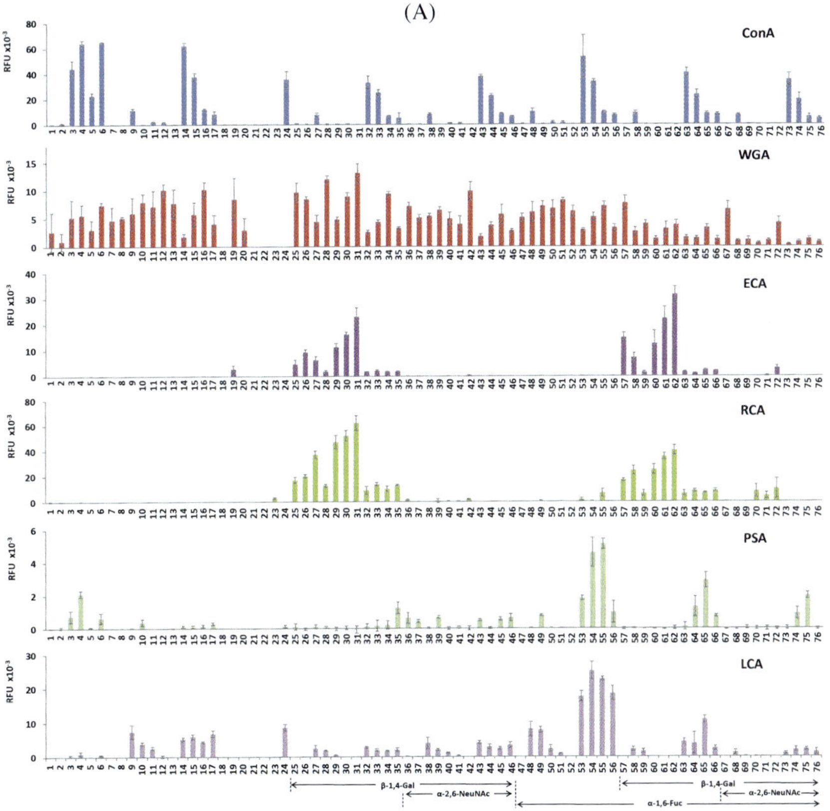

(B)

Figure 6.1. (A) On-chip enzymatic diversification of a panel of 25 N-glycan structures immobilized on NHS activated hydrogel coated glass slides. (B) Histograms showing the interaction of the immobilized N-glycans in the microarray with a set of fluorescently labeled lectins [32].

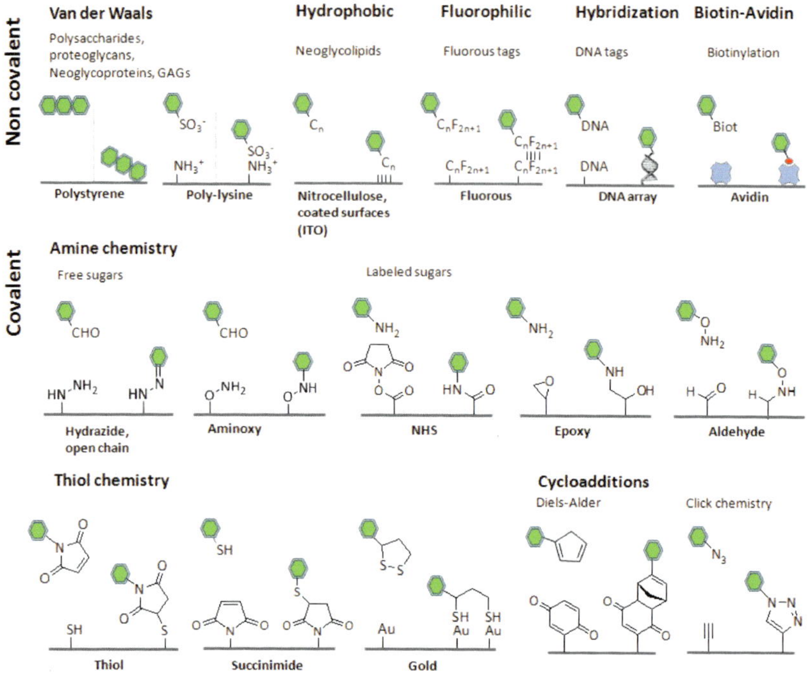

Figure 6.2 Immobilization strategies for the fabrication of glycan microarrays. Figure adapted from [4].

and then sorted by hybridization to a DNA array carrying the complement sequences [22]. Every immobilization method will require a specific passivation protocol to avoid non-specific interactions between analytes and the functionalized surface.

The covalent attachment of glycans, usually via an anomeric spacer to the surface, is the most common method for the immobilization of oligosaccharides and leads to glycoconjugates with uniform orientation on the surface. A large variety of coupling chemistries have been proposed, but most arrays today are made by spotting alkylaminoglycosides onto *N*-hydroxysuccinimide (NHS) or epoxide activated chips, partly due to the high coupling efficiency and partly to make ligand libraries interchangeable [23,24]. Thiol functionalized ligands can be immobilized directly to gold surfaces, form thioethers with maleimide and alkene linkers or form disulfide bonds with thiol functionalized surfaces. The self assembled monolayer formation of thiols on gold surfaces gives rise to a fairly stable linkage, which is compatible with mass spectrometry and surface plasmon resonance (SPR) analysis [25]. Free reducing oligosaccharides can also be immobilized by direct printing onto surfaces functionalized with alkyloxyamines, hydrazides or amines to form oximes, hyraxzones or instable Schiff bases, which are reduced to secondary amines with sodium cyanoborohydride. Direct immobilization avoids tagging of the carbohydrates prior to printing, which is very useful for the immobilization of purified natural material after enzymatic or chemical cleavage [26]. A major drawback is the formation of open ring forms for the reducing end sugar residues, which has been shown to alter the recognition with biological receptors, in particular, for core fucose residues [27]. Free reducing sugars can also be immobilized on surfaces functionalized with photo-active groups, like phthalimide chromophores, tetrafluorarylazides or aryl-trifluoromethyldiazirines by robotic printing of a collection of carbohydrates and subsequent illumination with ultraviolet (UV) light [28].

As carbohydrate–protein interactions are generally weak with binding constants in the micro- to millimolar range, the use of microarrays for their study does not seem to be obvious in the first instance. In nature, a multivalent glycan presentation on the cell surface is responsible for the strong binding enhancement observed, compared to the monovalent interaction in solution. This concept is also applied in glycan arrays, which are a close mimic of the dense glycan representation found on a eukaryotic cell. The high number of potential ligands allows the rapid rebinding upon rupture and the binding of multiple ligands in case of multimeric receptors, like many lectins or antibodies. Both mechanisms can strengthen the interaction by up to three orders of magnitude. Other means to further improve assay sensitivity are to increase the ligand density per spot (e.g. by the immobilization of ligands within a three-dimensional (3D) hydrogel mesh) or to omit the washing step to maintain the weaker interactions and subsequent analysis of the arrays by evanescent field fluorescence detection. With evanescent field fluorescent detection, only those fluorescently tagged analytes that are within the evanescent field of around 100 nm from the surface are excited and contribute to the fluorescence signal [29], making a washing step unnecessary.

The way a ligand is presented on an array can have a profound effect on the strength of the carbohydrate interaction. In addition to the above mentioned 3D hydrogels with multiple functionalities [30,31], neoglycopolymers and monovalent polymeric glycosylated spacers [32] have been introduced primarily to override size or charge dependent differences in immobilization yields for different ligands, which can compromise the comparability of binding data. Synergistic effects of different ligands spotted together have been investigated by several groups and provide further insights into the complex carbohydrate–protein recognition found in natural systems [33]. In addition, chemo-informatic tools have been applied for the rational design of carbohydrate microarrays with minimal ligand coverage and maximizing epitope coverage [34].

6.1.5 *Microarray quality*

Despite huge progress in the field over the last 12 years, there is still considerable room for improving the quality, sensitivity and reproducibility of glycan microarrays. Platform comparison studies have highlighted the effects of immobilization chemistry, linker ligand purity and incubation conditions on binding data, the need for a more stringent control and standardization of procedures and reporting guidelines, similar to minimum information about a microarray experiment (MIAME) standards, which have been put into practice for other microarray platforms [35,36].

6.2 General Glycomics Applications

6.2.1 *Glycan binding proteins*

Many of the specific biological roles of carbohydrates are mediated through recognition via GBPs or lectins. This family of proteins is involved in immune responses [37], pathogen adhesion and recognition [38], cell–cell interactions and fertility, among others [39].

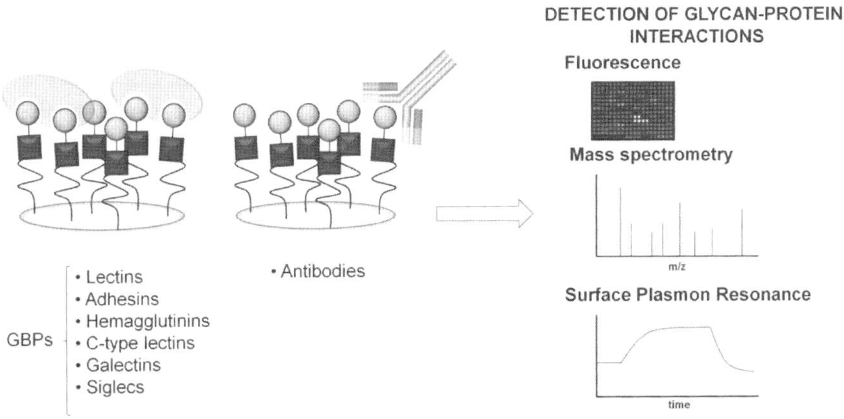

Figure 6.3 Schematic representation of protein and antibody binding to glycan array.

GBPs have been found in viruses, bacteria, plants, fungi, invertebrates and mammals. The study of the structure and function of GBPs is a main topic in glycobiology research.

In this context, glycan arrays have become powerful tools for the high-throughput study of GBPs (Fig. 6.3). Carbohydrate libraries printed in microarray format have helped in the discovery of novel GBPs, as well as in the determination of the exact binding specificities of known GBPs. Lectins were first isolated from plants in 1888 by Peter Hermann Stillmark. Stillmark found that extracts of castor bean seeds contained a protein that could agglutinate animal red blood cells. Therefore, some plant lectins also receive the name "agglutinin". Since then, many other lectins able to recognize different carbohydrates have been isolated from plants, seeds and fungi. Glycan arrays have been used for the characterization of exact binding epitopes of these family of proteins [40–42], and also for the discovery and characterization of carbohydrates presented on glycoproteins. GBPs have found application as diagnostic tools for the detection of tumor associated carbohydrate antigens, and some lectins are under investigation as therapeutic agents. Lectin microarrays have become useful tools for comparative sample glycoprofiling [43].

Carbohydrate–protein interactions are also implicated in the regulation of immunity. Dendritic cells express a large array of receptors, by which they scan their surroundings for recognition and uptake of pathogens. One of the receptor families is the C-type lectin (calcium (Ca^{+2}) dependent type lectin) family, which binds carbohydrate structures and internalizes antigens upon recognition. The exact glycan binding motifs of two receptors from this family, dendritic cell-specific intercellular adhesion molecule-3-grabbing non-integrin (DC-SIGN) and Langerin, were studied by using glycan arrays [44]. Both receptors are implicated in the human immunodeficiency virus-1 (HIV-1) infection, but they show very different binding properties. While DC-SIGN interacts with the glycan-containing shield of the virus to facilitate virus infection and transmission, Langerin promotes viral uptake and degradation. Both C-type lectins share a structural tripeptide Glu-Pro-Asn (EPN) motif that predicts binding towards mannose and fucose, but after glycan microarray incubation, more specific information about the binding epitopes of each protein could be obtained, showing important differences between the two structurally related proteins. This information could be applied for the rational design of microbicides against HIV-1 that block the interaction of the virus with DC-SIGN, but do not block the protective action of Langerin. Another C-type lectin that was characterized with the help of glycan arrays is human C-type lectin domain family 4, member C (CLEC4C) [45]. This receptor is presented in plasmacytoid dendritic cells and it is implicated in the production of type I interferon (type I IFN). The secretion of type I IFN regulates antiviral and antitumor responses, as well as autoimmune disorders. The interrogation of recombinant CLEC4C using the printed Glycan Array (version 3.0) by the Consortium for Functional Glycomics (CFG), the protein showed important binding towards complex *N*-glycan structures containing both β1-3-galactose and β1-4-galactose as terminal sugars in the non-reducing end.

Another important family of animal lectins are galectins, which share a common specificity to β-galactopyranosides and possess a conserved homologous carbohydrate recognition domain (CRD).

Despite their homology, galectins are multifunctional proteins that participate in several cellular mechanisms, including cell–cell interactions, apoptosis, cell growth, inflammation and tumorigenesis. Although this family of lectins could bind in a similar way to certain glycoconjugates, each member should display more specific binding towards discrete structures and, in such a way, galectins display such a broad range of biological activities. Systematic study of galectin binding specificity has been performed using glycan arrays. The different binding of human galectin-1, -2 and -3 versus sialylated glycans and blood group antigens was investigated, and data obtained from microarray experiments were further corroborated in cellular experiments [46]. A systematic discrimination of binding affinities of major human galectins (hGal-1, hGal-2, hGal-3, hGal-4, hGal-7, hGal-8, hGal-9) was studied using a small family of chemically synthesized β-galactoside glycoconjugates comprising the most common non-reducing terminal sequences presented in natural *N*- and *O*-glycans [47]. High affinity ligands that are presumably involved in the biological function of galectins were determined by this method. Galectin-1 and galectin-3 binding profiles were studied against isolated natural glycans. Natural glycans were labelled with a bifunctional fluorescence linker that facilitates chromatographic isolation and enables the immobilization onto NHS activated glass slides. Galectins were tested against a panel of putative natural ligands giving two different binding profiles [48].

Siglecs are a family of sialic acid binding proteins that are expressed in various subsets of white blood cells. Siglecs constitute interesting therapeutic targets due to their ability to internalize cargo into cells by endocytosis. For cell-specific delivery, high affinity ligands for the Siglec family that naturally recognize Neu5Acα2-6Gal and/or Neu5Acα2-3Gal need to be determined. Glycan microarrays have been used as a platform for the chemical synthesis of analogs through copper (Cu)(I)-catalyzed azide–alkyne cycloadditions (CuAAC). Modified sialic acids with alkyne groups in positions five and nine were immobilized in NHS modified glass slides, and subjected to CuAAC reactions with 47 different azido compounds. The modified sialic acids generated on chip were screened towards Siglec-7 and Siglec-10, permitting characterization of a high affinity ligand for Siglec-7 [49].

Glycan microarray profiling of whole cells could be beneficial, as GBPs are displayed in their natural environment with their native spacing and orientation. However, the interpretation of the results could be more complicated because of the presence of more than one GBP on the cell surface. A glycan microarray of 120 glycans immobilized on epoxy modified glass slides was used to determine the nature of GBPs presented in five different human cancer cell lines (such as breast cancer MDA-MB435 cells, cervical cancer HeLa (Henrietta Lacks) cells and colon cancer (Caco)-2, HCT-116 and HCT-116-FM6 cells). Significant binding was observed towards many glycan structures presented in the microarray and a different binding profile was observed for the five cell lines [50].

Although fluorescence is the most commonly employed method of detection of carbohydrate–protein interactions in microarrays, alternative methods have been described. SPR imaging has been employed for the real time, label-free detection of carbohydrate–protein interactions. As proof of concept, a panel of biotinylated monosaccharides were immobilized onto commercially available neutravidin-coated gold sensor chips and interrogated with the corresponding plant lectin binders. Additionally, an array of 40 sialylated and unsialylated glycans were employed for determining the preferential binding of human Siglec-7, with α2-8 and α2-6 sialic acid being better ligands than α2-3 sialyl *N*-acetyllactosamines (LacNAcs) [51]. A similar approach using SPR imaging as a detection system was employed to screen natural and unnatural sugars that bind *Ricinus communis agglutinin* (RCA_{120}), which could be used as model systems for the development of SPR-based sensors to detect the lethal protein ricin [52]. Cantilever array biosensors have been described as a label-free method for detecting interactions in real time with high sensitivity and selectivity. A series of eight individual cantilevers were modified by thiol ending carbohydrate moieties and were used to detect the interaction between different oligomannosides with Cyanovirin-*N* in the picomolar range [53].

Mass spectrometry analysis of peptides after trypsin digestion is a classical method for the identification of proteins. A glycan array prepared on indium tin oxide allowing for MALDI-TOF detection was employed for the protein capture, *in situ* trypsin reaction, and detection of the peptidic fragments. This approach employed model protein *Concanavalin A* capture with high mannose *N*-glycan structures, but could provide a novel strategy towards the discovery and characterization of CBPs in complex samples [20].

6.2.2 *Glycan processing enzymes*

Glycosyltransferases and glycosidases are the enzymes involved in the *in vivo* synthesis of glycoconjugates, being responsible for the complexity of the glycome. Glycosyltransferases comprise one of the most abundant protein families, being highly conserved between different organisms. Of the putative glycosyltransferase sequences deposited in databases, such as the Carbohydrate-Active Enzymes (CAZy) database (http://www.cazy.org/), only a small part of them are fully characterized. Along with the development of glycan microarray platforms in recent years, different formats have been developed for the direct study of glycan modifying enzymes employing immobilized carbohydrate substrates. The use of glycan arrays allows for the parallel interrogation of various substrates with different enzymes employing only minimum amounts of immobilized substrates, nucleotide donors and precious glycosyltransferases. Tedious purification steps are unnecessary, as the immobilized products could be separated from the glycosyltransferase and excess of glycosidic donor and buffers/salts by simple washings. For the success of the enzymatic assay, highly sensitive detection techniques to assess product formation need to be available.

One simple strategy for the detection of newly introduced sugar moieties by the action of glycosyltransferases is the incubation with fluorescently labelled specific CBPs or specific antibodies [17,54,55]. As quantification is often compromised by the specific lectin affinity, this approach only provides qualitative data. Nevertheless, important information about the substrate specificity of the glycosyltransferases could be obtained [56]. Glycosyltransferases exhibit high specificity for their donor and acceptor substrates, giving rise to complete regio- and stereoselective transformations, and therefore, they are powerful tools in the synthesis of complex carbohydrate structures without the need for protection and deprotection steps required in classical organic synthesis. A glycan array focussed on chemically synthesized *N*-glycan structures was used for the screening of the substrate specificity of recombinant core α1,6 and core α1,3-fucosyltransferases. The detection of fucose moieties introduced by the enzyme was performed by incubation with four fluorescently labelled lectins and one antibody. After the activity screening, some selected positive substrates were fully converted in solution into fucosylated *N*-glycans in high yields and with complete stereoselectivity. This provided an efficient approach for the assembly of new core fucosylated *N*-glycan structures that could be employed for studying the biological relevance of core fucosylation [57].

Biotin modified cytidine-5′-monophospho-*N*-acetylneuraminic acid (CMP-Neu5Ac) was applied to glycan microarrays for high-throughput acceptor specificity screening of various recombinant sialyltransferases. The transferred biotinylated sialic acid moieties were detected by incubation with fluorescein conjugated streptavidin. The known specificity of the recombinant enzymes was confirmed and additional specificity information was revealed [58].

One of the most common solution assays for determining glycosyltransferase activity is based on the employment of radioactive labelled nucleotide donors. This could also be applied to immobilized acceptors on glycan arrays, after incubation with a solution containing the radiolabelled nucleotide donor and the glycosyltransferase. The introduced radioactivity could be measured using autoradiography [59]. The radioactive readout on immobilized substrates on microarray slides is compatible with fluorescent lectin or antibody incubation and could offer complementary information about the stereo- and/or regiochemistry of the newly introduced sugar moieties. This approach has been applied to the glycosyltransferase profiling of crude samples [60].

Carbohydrate arrays generated by the formation of modified self-assembled monolayers (SAMs) on gold surfaces constitutes a powerful label-free technique for the detection of enzymatic activity by matrix-assisted laser desorption-ionization mass spectrometry [61,62]. When the monolayer is irradiated with the laser, the alkanethiol layer is desorbed from the gold surface providing a spectra of the alkanethiol modified ligands. This strategy has recently been applied to the discovery of new glycosyltransferases. As a high-throughput platform for mass spectrometry, the authors employed metal plates containing an array of 384 gold coated islands. Thousands of combinations of putative glycosyltransferases with different acceptor substrates and donors were screened, leading to the discovery of four new glycosyltransferases that were further biochemically characterized [63].

The immobilization of carbohydrates modified with lipid chains through hydrophobic interactions in MALDI-TOF compatible platforms has also allowed the study of both glycosyltransferases and glycosidases. Initial experiments were performed in gold coated MALDI plates. After the generation of a SAM of alkane thiols, lipid tagged carbohydrate structures were immobilized by hydrophobic interactions and the activity of bovine milk β-1,4-galactosyltransferase and *Aspergillus oryzae* β-galactosidase was tested following the formation of reaction products by mass spectrometry [64]. This platform was further applied to the study of glycan hydrolases. Various lipid-tagged polysaccharide structures were used in the activity screening and characterization of purified enzymes and enzymes isolated from environmental samples [65]. Mass spectrometry based readout could be done using ITO as a platform, enabling the study of interactions by fluorescence. A bilayer constituting of NHS-activated bidentate linker immobilized onto a SAM of alkane ligands was generated in ITO microscope type slides. *N*-glycans modified by an amino linker were robotically printed on top and the enzymatic activity of eight recombinant glycosyltransferases was studied [20].

6.3 Biomedical Research Applications

6.3.1 *General considerations*

Glycans conjugated to lipids or proteins cover the extracellular side of most cell membranes and are involved in the majority of biological processes at the cellular level, including host defense, cell development and differentiation. Membrane glycan composition is species- and tissue-specific, and used as a cellular postcode in cell–cell and cell–pathogen recognition. Glycans interact with a species-specific repertoire of lectins found at the cell surface of host or pathogen cells, which recognize monosaccharides or larger epitopes with high specificity. Individual glycan–protein interactions are usually weak with dissociation constants in the milli- to micromolar range. Through the multivalent presentation of glycans on the cell surface, which is effectively mimicked by glycan arrays, this interaction can increase up to three orders of magnitude. Disease-related changes in glycosylation or unique pathogenic glycoconjugates are best detected from serum or urine samples via specific anti-glycan antibodies, or by glycan profiling employing mass spectrometry or lectin arrays. Glycan arrays on the other hand, can be used to evaluate the immune response towards aberrant or pathogenic glycans through the binding of serum anti-carbohydrate antibodies or by detecting specific viral or bacterial glycan receptors. Other examples of applications of glycan arrays in biomedical research and clinical diagnostics include the screening of inhibitors for glycosyltransferases and glycosidases in competitive assays on glycan arrays, the evaluation of the immune response after immunization trials in vaccine development [66], or the search for new antagonists targeting specific lectin receptors.

6.3.2 *Detection of serum anti-carbohydrate antibodies anticancer vaccines*

Aberrant glycosylation is a general hallmark of many cancers and, consequently, a promising parameter for the development of diagnostic cancer markers. Changes in glycosylation linked to cancer can be related to changes in the expression levels of enzymes intervening in glycan biosynthesis, such as glycosyltransferases, glycosidases or sugar nucleotide transporters, as well as metalloproteases, the latter increasing the liberation of glycoproteins from tumor tissue into the blood stream. Enzyme expression can be up- or down-regulated, giving rise to specific glycosylation patterns that can vary strongly between different cancer types. These tumor associated carbohydrate antigens (TACAs), often truncated and oversialylated forms of the structures found in healthy tissue, can elicit a weak immune response, giving rise to antibodies, which tend to be non-protective, but show diagnostic potential. Here, glycan arrays have been used to screen for specific serum antibody binding profiles to cancer related glycan structures. More importantly, TACAs have been developed into carbohydrate conjugate vaccines to specifically boost the immune response against altered surface glycans. Binding strength and specificity of antibodies generated against the synthetic vaccines can be rapidly evaluated on glycan arrays. This is an important step in conjugate vaccine development to detect potential antibody cross reactivity with other non-tumor related glycan structures potentially leading to autoimmune reactions. The glycoprotein mucine MUC1, for example, is overexpressed in epithelial tumors and has been an attractive target for immune targeting and diagnostics alike. Microarrays of glycopeptides with variations in the amino acid sequence, number of glycosylation sites and type of glycan have been employed by various groups to evaluate the binding specificity of serum antibodies after vaccination with synthetic therapeutic vaccines targeting MUC1. Likewise, arrays of selected MUC1 glycopeptides are being developed into a rapid cancer diagnostic tool to detect serum antibodies against tumor-associated aberrant glycans [67]. In a similar fashion, serum antibodies against the surface glycolipids, globohexaosylceramide (Globo-H) and stage-specific embryonic antigen (SSEA)-4, which are over expressed in breast cancer [68,69], can be detected on glycan arrays and guide the development of vaccine candidates based on these glycan antigens.

Serum antibodies produced against bacterial, viral or parasite glycans using glycan arrays can be detected with the help of glycan arrays and are valuable tools for seroepidemiological studies [70], or in the development of pathogen glycan targeted vaccines, while their use in the diagnosis of infection is probably reduced to the detection of short-lived antibodies, which accompany the disease, but disappear in the absence of the pathogen. The screening of pooled serum samples has shown broad recognition of glycan epitopes, many of them reminiscent of past infections and general exposure to bacterial polysaccharides [71]. In spite of the presence of a large pool of anti-glycan serum antibodies with no apparent diagnostic value, glycan arrays have found their application in autoimmune disease diagnostics for the detection of Crohn's disease [72], inflammatory bowel disease or colitis [73,74].

Glycan arrays have also been evaluated for the detection of pathogens via interaction with bacterial or viral surface receptors. As many bacteria have lectins with similar glycan binding specificities, a detection method based on the binding to host glycans is prone to lack the selectivity required for a detection method. Work from the Seeberger group suggests that assaying bacterial binding to glycan arrays could be used in the screening of lectin inhibitor libraries as anti-adhesive therapeutics [75].

Glycan arrays have also found an important application in determining differences in binding specificities of influenza virus receptors of pandemic human, avian or porcine strains. Influenza viruses attach to the sialylated surface glycans via hemagglutinin, and their binding patterns on glycan arrays including α-2,3 and α-2,6 sialosides have helped to understand new appearing transmissibility between species [76,77]. Arrays comprising a confined set of sialoside ligands could prove useful for influenza strain subtyping and for the detection of newly evolving human pathogenic strains [78]. While human influenza viruses recognize α-2,6-sialic acids moieties in the respiratory tract, avian influenza viruses recognize α-2,3-sialic acids in intestinal epithelial cells. The transmission of avian influenza to humans seems to be related to changes in the binding specificity of viral hemagglutinins presented on different strains [76,78–80].

Pathogens display surface GBPs that selectively recognize carbohydrate epitopes presented in host cells. This recognition is a fundamental interaction for the initial infection and posterior growth of the pathogen. Several pathogen lectins, known as adhesins for their ability to adhere to their host organisms, have been characterized using glycan microarrays. Some examples are the *N*-acetylglucosamine (GlcNAc) binding protein A from *Vibrio cholerae*, a gram positive bacterium that causes cholera [81]; the galactose binding *Pseudomonas aeruginosa* lectin I (PA-IL) isolated from the opportunistic bacterium *Pseudomonas aeruginosa* that infects the lungs of immune-compromized patients [82]; and the fucose-binding agglutinin-like protein Als1 from *Candida albicans*, a diploid fungus that produces candidiasis [83].

A second example where glycan arrays have helped to determine viral binding specificities is HIV. The viral envelope protein glycoprotein (gp120) is decorated with *N*-glycans that facilitate viral entry via interactions with the lectins DC-SIGN or Langerin [84,85]. As some broadly neutralizing antibodies against HIV have shown to bind strongly to high-mannose and complex *N*-glycan structures present on the viral surface, glycan arrays comprising these and similar structures have also been important tools in the development of an HIV vaccine based on glycan or glycopeptide antigens.

Most pathogens additionally express specific carbohydrates in their cell surfaces that the host organism recognizes as non-self and could elicit the production of specific antibodies in infected humans. In this context, glycan microarrays have served for the characterization of specific broadly neutralizing antibodies against HIV. The binding specificity of human monoclonal 2G12 antibody against oligomannose structures has been extensively studied using glycan arrays [86,87]. In contrast, PGT121 type broadly neutralising antibodies showed detectable binding towards complex *N*-glycan structures [88] and recently, a glycan array of synthetic sialylated *N*-glycan structures was used to identify potential natural and unnatural high binders towards antibody PG16 [89]. All this could find application in the rational design of carbohydrate based vaccines against HIV.

Rather than being a diagnostic device themselves, glycan arrays can also be a valuable aid in finding antigens of diagnostic value for the subsequent production of monoclonal diagnostic antibodies used for enzyme-linked immunosorbent assay (ELISA) tests [90].

While the previous examples have concentrated on the diagnostic approaches measuring the binding of lectins and antibodies to specific glycan structures, we close this chapter with a few examples of the use of glycan arrays in carbohydrate-based drug development.

The importance of glycosyltransferases and glycosidases for the processing of carbohydrate structures and, subsequently, their indirect implication in many biological processes, makes them interesting targets for the development of therapies against viral infections, genetic disorders or cancer [91]. If coupled with a mass spectrometric readout technique, glycan arrays can be employed as high-throughput enzyme assays for the rapid evaluation of large compound libraries as potential enzyme inhibitors [63, 92]. The potency of a potential inhibitor can be measured as the ratio of the peak intensities of substrate and reaction product(s) after incubation with the target enzyme.

Finally, the on-chip synthesis of inhibitor libraries is an attractive strategy for the future rapid creation of potent and selective lectin antagonists or enzyme inhibitors in a highly parallel fashion, taking privileged glycan structures as starting scaffolds [93–95]. First examples in this direction are the enzymatic on-chip diversification of glycan scaffolds in nano-droplets [17], or the on-chip synthesis of potential Siglec antagonists employing click reactions for the introduction of a second library of structural substituents [49].

References

1. Schena, M., Shalon, D., Davis, R. W., Brown, P. O. (1995) *Science* **270**, 467–470.
2. Zhu, H., Bilgin, M., Bangham, R., Hall, D., Casamayor, A., Bertone, P., Lan, N., Jansen, R., Bidlingmaier, S., Houfek, T., Mitchell, T., Miller, P., Dean, R. A., Gerstein, M., Snyder, M. (2001) *Science* **293**, 2101–2105.
3. Ziauddin, J., Sabatini, D. M. (2001) *Nature* **411**, 107–110.

4. Rillahan, C. D., Paulson, J. C. (2011) *Annu. Rev. Biochem.* **80**, 797–823.
5. Wang, D., Liu, S., Trummer, B. J., Deng, C., Wang, A. (2002) *Nat. Biotechnol.* **20**, 275–281.
6. Fukui, S., Feizi, T., Galustian, C., Lawson, A. M., Chai, W. (2002) *Nat. Biotechnol.* **20**, 1011–1017.
7. Fazio, F., Bryan, M. C., Blixt, O., Paulson, J. C., Wong, C. H. (2002) *J. Am. Chem. Soc.* **124**, 14397–14402.
8. Kristensen, T., Mikkelsen, J. D., Knox, J. P. (2002) *Proteomics* **2**, 1666–1671.
9. Zhi, Z.-L., Laurent, N., Powell, A. K., Karamanska, R., Fais, M., Voglmeir, J., Wright, A., Blackburn, J. M., Crocker, P. R. D., Russell, A., Flitsch, S., Field, R. A., Turnbull, J. E. (2008) *ChemBioChem,* **9**, 1568–1575.
10. Song, X., Xia, B., Stowell, S. R., Lasanajak, Y., Smith, D. F., Cummings, R. D. (2009) *Chem. Biol.* **16**, 38–47.
11. Zhang, Z., Ollmann, I. R., Ye, X. S., Wischnat, R., Baasov, T., Wong, C. H. (1999) *J. Am. Chem. Soc.* **121**, 734–753.
12. Plante, O. J., Palmacci, E. R., Seeberger, P. H. (2001) *Science* **291**, 1523.
13. Hsu, C. H., Hung, S. C., Wu, C. Y., Wong, C. H. (2011) *Angew. Chem. Int. Ed.* **50**, 11872–11923.
14. Homann, A., Seibel, J. (2009) *Nat. Prod. Rep.* **26**, 1555–1571.
15. Ban, L., Mrksich, M. (2008) *Angew. Chem. Int. Ed.* **47**, 3396–3399.
16. Voglmeir, J., Sardzik, R., Weissenborn, M. J., Flitsch, S. L. (2010) *OMICS* **14**, 437–444.
17. Serna, S., Etxebarria, J., Ruiz, N., Martin-Lomas, M., Reichardt, N. C. (2010) *Chem. Eur. J.* **16**, 13163–13175.
18. Ko, K. S., Jaipuri, F. A., Pohl, N. L. (2005) *J. Am. Chem. Soc.* **127**, 13162–13163.
19. Chang, S. H., Han, J. L., Tseng, S. Y., Lee, H. Y., Lin, C. W., Lin, Y. C., Jeng, W. Y., Wang, A. H. J., Wu, C. Y., Wong, C. H. (2010) *J. Am. Chem. Soc.* **132**, 13371–13380.
20. Beloqui, A., Calvo, J., Serna, S., Yan, S., Wilson, I. B. H., Martin-Lomas, M., Reichardt, N. C. (2013) *Angew. Chem. Int. Ed.* **52**, 7477–7481.
21. Galanina, O., Mecklenburg, M., Nifantiev, N., Pazynina, G., Bovin, N. (2003) *Lab. Chip* **3**, 260–265.
22. Chevolot, Y., Bouillon, C., Vidal, S., Morvan, F., Meyer, A., Cloarec, J. P., Jochum, A., Praly, J. P., Vasseur, J. J., Souteyrand, E. (2007) *Angew. Chem. Int. Ed.* **46**, 2398–2402.

23. Blixt, O., Head, S., Mondala, T., Scanlan, C., Huflejt, M. E., Alvarez, R., Bryan, M. C., Fazio, F., Calarese, D., Stevens, J., Stevens, D. J., Skehel, J. J., van Die, I., Burton, D. R., Wilson, I. A., Cummings, R., Bovin, N., Wong, C. H., Paulson, J. C. (2004) *Proc. Natl. Acad. Sci. USA* **101**, 17033–17038.
24. Schwarz, M., Spector, L., Gargir, A., Shtevi, A., Gortler, M., Altstock, R. T., Dukler, A. A., Dotan, N. (2003) *Glycobiology* **13**, 749–754.
25. Zhi, Z.-L., Laurent, N., Powell, A. K., Karamanska, R., Fais, M., Voglmeir, J., Wright, A., Blackburn, J. M., Crocker, P. R., Russell, D. A., Flitsch, S., Field, R. A., Turnbull, J. E. (2008) *ChemBioChem* **9**, 1568–1575.
26. Lee, M. R., Shin, I. (2005) *Org. Lett.* **7**, 4269–4272.
27. de Boer, A. R., Hokke, C. H., Deelder, A. M., Wuhrer, M. (2007) *Anal. Chem.* **79**, 8107–8113.
28. Carroll, G. T., Wang, D., Turro, N. J., Koberstein, J. T. (2006) *Langmuir* **22**, 2899–2905.
29. Tateno, H., Mori, A., Uchiyama, N., Yabe, R., Iwaki, J., Shikanai, T., Angata, T., Narimatsu, H., Hirabayashi, J. (2008) *Glycobiology* **18**, 789–798.
30. Zhou, X., Turchi, C., Wang, D. (2009) *J. Proteome Res.* **8**, 5031–5040.
31. Zhang, J., Zhou, X. (2011) *Biosens. Biolectron.* **28**, 355–361.
32. Etxebarria, J., Serna, S., Beloqui, A., Martin-Lomas, M., Reichardt, N. C. (2013) *Chem. Eur. J.* **19**, 4776–4785.
33. Liang, C. H., Wang, S. K., Lin, C. W., Wang, C. C., Wong, C. H., Wu, C. Y. *Angew. Chem. Int. Ed.* **2011**, 50, 1608–1612.
34. Rademacher, C., Paulson, J. C. (2012) *ACS Chemical Biology* **7**, 829–834.
35. Wang, L., Cummings, R. D., Smith, D. F., Huflejt, M., Campbell, C. T., Gildersleeve, J. C., Gerlach, J. Q., Kilcoyne, M., Joshi, L., Serna, S., Reichardt, N.-C., Parera Pera, N., Pieters, R., Eng, W., Mahal, L. K. (2014) *Glycobiology* **24**, 507–517.
36. Brazma, A., Hingamp, P., Quackenbush, J., Sherlock, G., Spellman, P., Stoeckert, C., Aach, J., Ansorge, W., Ball, C. A., Causton, H. C., Gaasterland, T., Glenisson, P., Holstege, F. C. P., Kim, I. F., Markowitz, V., Matese, J. C., Parkinson, H., Robinson, A., Sarkans, U., Schulze-Kremer, S., Stewart, J., Taylor, R., Vilo, J., Vingron, M. (2001) *Nat. Gen.* **29**, 365–372.
37. Johnson, J. L., Jones, M. B., Ryan, S. O., Cobb, B. A. (2013) *Trends Immunol.* **34**, 290–298.
38. Longwell, S.A., Dube, D. H. (2013) *Curr. Opin. Chem. Biol.* **17**, 41–48.

39. Varki, A., Cummings, R. D., Esko, J. D., Freeze, H. H., Stanley, P., Bertozzi, C. R., Hart, G. W., Etzler, M. E. Eds. (2009) *Essentials in Glycobiology*, 2nd ed. Cold Spring Harbor Laboratory Press, Cold Spring Harbor, NY.
40. Smith, D. F., Song, X., Cummings, R. D., (2010) *Meth. Enzymol.* **480**, 417–444.
41. Manimala, J. C., Roach, T. A., Li, Z., Gildersleeve, J. C. (2006) *Angew. Chem. Int. Ed.* **45**, 3607–3610.
42. Porter, A., Yue, T., Heeringa, L., Day, S., Suh, E., Haab, B. B. (2010) *Glycobiology* **20**, 369–380.
43. Ribeiro, J. P., Mahal, L. K. (2013) *Curr. Opin. Chem. Biol.* **17**, 827–831.
44. Holla, A., Skerra, A. (2011) *Protein Eng. Des. Sel.* **24**, 659–669.
45. Riboldi, E., Daniele, R., Parola, C., Inforzato, A., Arnold, P. L., Bossisio, D., Fremont, D. H., Bastone, A., Colonna, M., Sozzani, S. (2011) *J. Biol. Chem.* **286**, 35329–35333.
46. Stowell, S. R., Arthur, C. M., Mehta, P., Slanina, K. A., Blixt, O., Leffler, H., Smith, D. F., Cummings, R. D. (2008) *J. Biol. Chem.* **283**, 10109–10123.
47. Horlacher, T., Oberli, M. A., Werz, D. B., Kröck, L., Bufali, S., Mishra, R., Sobek, J., Simons, K., Hirashima, M., Niki, T., Seeberger, P. H. (2010) *ChemBioChem.* **11**, 1563–1573.
48. Song, X., Xia, B., Stowell, S. R., Lasanajak, Y., Smith, D. F., Cummings, R. D. (2009) *Chem. Biol.* **16**, 36–47.
49. Rillahan, C. D., Schwartz, E., Rademacher, C., McBride, R., Rangarajan, J., Fokin, V. V., Paulson, J. C. (2013) *ACS Chem. Biol.* **8**, 1417–1422.
50. Arndt, N. X., Tiralongo, J., Madge, P. D., von Itzstein, M., Day, C. J. (2011) *J. Cell. Biochem.* **112**, 2230–2240.
51. Karamanska, R., Clarke, J., Blixt, O., MacRae, J. I., Zhang, J. Q., Crocker, P. R., Laurent, N., Wright, A., Flitsch, S. L., Russell, D. A., Field, R. A. (2008) *Glycoconj. J.* **25**, 69–74.
52. Fais, M., Karamanska, R., Allman, S., Fairhurst, S. A., Innocenti, P., Fairbanks, A. J., Donohoe, T. J., Davis, B. G., Russell, D. A., Field, R. A. (2011) *Chem. Sci.* **2**, 1952–1959.
53. Gruber, K., Horlacher, T., Castelli, R., Mader, A., Seeberger, P. H., Hermann, B. A. (2011) *ACS Nano* **5**, 3670–3678.
54. Park, S., Myung-Ryul, L., Shin, I. (2007) *Nat. Protoc.* **2**, 2747–2758.
55. Park, S., Lee, M., Pyo, S. J., Shin, I. (2004) *J. Am. Chem. Soc.* **126**, 4812–4819.
56. Yan, S., Serna, S., Reichardt, N. C., Paschinger, K., Wilson, I. B. H. (2013) *J. Biol. Chem.* **288**, 21015–21028.

57. Serna, S., Yan, S., Martin-Lomas, M., Wilson, I., Reichardt, N. (2011) *J. Am. Chem. Soc.* **133**, 16495–16502.
58. Blixt, O., Allin, K., Bohorov, O., Liu, X., Andersson-Sand, H., Hoffmann, J., Razi, N. (2008) *Glycoconjugate J.* **25**, 59–68.
59. Shipp, M., Nadella, R., Gao, H., Farkas, V., Sigrist, H., Faik, A. (2008) *Glycoconjugate J.* **25**, 49–58.
60. Serna, S., Hokke, C. H., Weissenborn, M., Flitsch, S., Martin-Lomas, M., Reichardt, N. C. (2013) *ChemBioChem* **14**, 862–869.
61. Houseman, B. T., Mrksich, M. (2002) *Chem. Biol.* **9**, 443–454.
62. Z Zhi, Z. L., Laurent, N., Powell, A. K., Karamanska, R., Fais, M., Voglmeir, J., Wright, A., Blackburn, J. M., Crocker, P. R., Russell, D. A., Field, R. A., Turnbull, J. E. (2008) *ChemBioChem* **9**, 1568–1575.
63. Ban, L., Pettit, N., Li, L., Stuparu, A. D., Cai, L., Chen, W., Guan, W., Han, W., Wang, P. G., Mrksich, M. (2012) *Nat. Chem. Biol.* **8**, 769–773.
64. Sanchez-Ruiz, A., Serna, S., Ruiz, N., Martin-Lomas, M., Reichardt, N. C. (2011) *Angew. Chem. Int. Ed.* **50**, 1801–1804.
65. Beloqui, A., Sanchez-Ruiz, A., Martin-Lomas, M., Reichardt, N. A. (2012) *Chem. Commun.* **48**, 1701–1703.
66. Wang, C. H., Li, S. T., Lin, T. L., Cheng, Y. Y., Sun, T. H., Wang, J. T., Cheng, T. J. R., Mong, K. K. T., Wong, C. H., Wu, C. Y. (2013) *Angew. Chem. Int. Ed.* **52**, 9157–9161.
67. Matsushita, T., Takada, W., Igarashi, K., Naruchi, K., Miyoshi, R., Garcia-Martin, F., Amano, M., Hinou, H., Nishimura, S. I. (2013) *Biochim. Biophys. Acta* **1840**, 1105–1116.
68. Wang, C. C., Huang, Y. L., Ren, C. T., Lin, C. W., Hung, J. T., Yu, J. C., Yu, A. L., Wu, C. Y., Wong, C. H. (2008) *Proc. Natl. Acad. Sci. USA* **105**, 11661–11666.
69. Hung, T. C., Lin, C. W., Hsu, T. L., Wu, C. Y., Wong, C. H. (2013) *J. Am. Chem. Soc.* **135**, 5934–5937.
70. Thirumalapura, N., Morton, R., Ramachandran, A., Malayer, J. (2005) *J. Immunol. Methods* **298**, 73–81.
71. Von Gunten, S., Smith, D. F., Cummings, R. D., Riedel, S., Miescher, S., Schaub, A., Hamilton, R. G., Bochner, B. S. (2009) *J. Allergy Clin. Immunol.***123**, 1268–1276.
72. Chevalier, R., Esnault, J., Vandewalle, P., Sendid, B., Colombel, J. F., Poulain, D., Mallet, J. M. (2005) *Tetrahedron* **61**, 7669–7677.
73. Conrad, K., Roggenbuck, D., Reinhold, D., Sack, U. (2012) *Autoimmun. Rev.* **11**, 207–211.

74. Dotan, N., Altstock, R., Schwarz, M., Dukler, A. (2006) *Lupus* **15**, 442–450.
75. Disney, M. D., Seeberger, P. H. (2004) *Chem. Biol.* **11**, 1701–1707.
76. Childs, R. A., Palma, A. S., Wharton, S., Matrosovich, T., Liu, Y., Chai, W., Campanero-Rhodes, M. A., Zhang, Y., Eickmann, M., Kiso, M., Hay, A., Matrosovich, M., Feizi, T. (2009) *Nature Biotechnol.* **27**, 797–799.
77. Stevens, J., Blixt, O., Glaser, L., Taubenberger, J. K., Palese, P., Paulson, J. C., Wilson, I. A. (2006) *J. Mol. Biol.* **355**, 1143–1155.
78. Liao, H. Y., Hsu, C. H., Wang, S. C., Liang, C. H., Yen, H. Y., Su, C. Y., Chen, C. H., Jan, J. T., Ren, C. T., Chen, C. H., Cheng, T. J., Wu, C. Y., Wong, C. H. (2010) *J. Am. Chem. Soc.* **132**, 14849–14856.
79. Crusat, M., Liu, J., Palma, A. S., Childs, R. A., Liu, Y., Wharton, S. A., Lin, Y. P., Coombs, P. J., Martin, S. R., Matrosovich, M., Chen, Z., Stevens, D. J., Hien, V. M., Thanh, T. T., Nhu, L. N. T., Nguyet, L. A., Ha, D. Q., van Doorn, H. R., Hien, T. T., Conradt, H. S., Kiso, M., Gamblin, S. J., Chai, W., Skehel, J. J., Hay, A. J., Farrar, J., de Jong, M. D., Feizi, T. (2013) *Virology* **447**, 326–337.
80. Walther, T., Karamanska, R., Chan, R. W., Chan, M. C., Jia, N., Air, G., Hopton, C., Wong, M. P., Dell, A., Peiris, J. M., Haslam, S. M., Nicholls, J. M. (2013) *PLoS Pathog.* **9**, e1003223.
81. Wong, E., Vaaje-Kolstad, G., Ghosh, A., Hurtado-Guerrero, R., Konarev, P. V., Ibrahim, A. F. M., Svergun, D. I., Eijsink, V. G. H., Chatterjee, N. S., van Aalten, D. M. F. (2012) *PLoS Pathog.* **8**, e1002373.
82. Blanchard, B., Nurisso, A., Hallville, E., Tétaut, C., Wiels, J., Pokorná, M., Wimmerová, M., Varrot, A., Imberty, A. (2008) *J. Mol. Biol.* **383**, 837–853.
83. Donohue, D. S., Ielasi, F. S., Goosens, K. V. Y., Willaert, R.G. (2011) *Mol. Microbiol.* **80**, 1667–1679.
84. Pejchal, R., Doores, K. J., Walker, L. M., Khayat, R., Huang, P. S., Wang, S. K., Stanfield, R. L., Julien, J. P., Ramos, A., Crispin, M., Depetris, R., Katpally, U., Marozsan, A., Cupo, A., Maloveste, S., Liu, Y., McBride, R., Ito, Y., Sanders, R. W., Ogohara, C., Paulson, J. C., Feizi, T., Scanlan, C. N., Wong, C. H., Moore, J. P., Olson, W. C., Ward, A. B., Poignard, P., Schief, W. R., Burton, D. R., Wilson, I. A. (2011) *Science* **334**, 1097–1103.
85. Shivatare, S. S., Chang, S. H., Tsai, T. I., Ren, C. T., Chuang, H. Y., Hsu, L., Lin, C. W., Li, S. T., Wu, C. Y., Wong, C. H. (2013) *J. Am. Chem. Soc.* **135**, 15382–15391.

86. Wang, S.-K., Liang, P.-H., Astronomo, R. D., Hsu, T.-L., Hsieh, S.-L., Burton, D. R., Wong, C.-H. (2008) *Proc. Natl. Acad. Sci. USA* **105**, 3690–3695.
87. Adams, E. W., Ratner, D. M., Bokesch, H. R., McMahon, J. B., O'Keefe, B. R., Seeberger, P. H. (2004) *Chemistry & biology* **11**, 875–881.
88. Mouquet, H., Scharf, L., Euler, Z., Liu, Y., Eden, C., Scheid, J. F., Halper-Stromberg, A., Gnanapragasam, P. N., Spencer, D. I. R., Seaman, M. S., Schuitemaker, H., Feizi, T., Nussenzweig, M. C. Bjorkman, P. J. (2012) *Proc. Natl. Acad. Sci. USA* **109**, E3268–E3277.
89. Shivatare, S. S., Chang, S. H., Tsai, T. I., Ren, C. T., Chuang, H. Y., Hsu, L., Lin, C. W., Li, S. T., Wu, C. Y., Wong, C. H. (2013) *J. Am. Chem. Soc.* **135**, 15382–15391.
90. Anish, C., Martin, C. E., Wahlbrink, A., Bogdan, C., Ntais, P., Antoniou, M., Seeberger, P. H. (2013) *ACS Chem. Biol.* **8**, 2412–2422.
91. Asano, N. (2003) *Glycobiology* **13**, 93R.
92. Blixt, O., Han, S., Liao, L., Zeng, Y., Hoffmann, J., Futakawa, S., Paulson, J. C. (2008) *J. Am. Chem. Soc.* **130**, 6680–6681.
93. Rillahan, C. D., Schwartz, E., McBride, R., Fokin, V. V., Paulson, J. C. (2012) *Angew. Chem. Int. Ed.* **51**, 11014–11018.
94. Gurard-Levin, Z. A., Scholle, M. D., Eisenberg, A. H., Mrksich, M. (2011) *ACS. Comb. Sci.* **13**, 347–350.
95. Steiner, A. J., Stütz, A. E., Wrodnigg, T. M., Tarling, C. A., Withers, S. G., Hermetter, A., Schmidinger, H. (2008) *Bioorg. Med. Chem. Lett.* **18**, 1922–1925.

Chapter 7

Informatics and Analytical Tools for Glycan Analysis and the Development of Biotherapeutics

Catherine A. Hayes,* Giorgio Carta,† Niclas G. Karlsson*
Fergal Duffy† and Pauline M. Rudd†

**Department of Medical Biochemistry and Cell Biology, University of Gothenburg, 40530 Gothenburg, Sweden*
†National Institute for Bioprocessing Research and Training, Fosters Avenue, Mount Menion, Co. Dublin, Ireland

7.1 Introduction

Over the past decades, high resolving, powerful analytical techniques, such as High Pressure Liquid Chromatography (HPLC), capillary electrophoresis (CE) and mass spectrometry (MS) have been optimized to reveal the diverse structures of *N*- and *O*-glycans present on glycoproteins [1–3]. It is evident that the structures of these oligosaccharides significantly influence a wide range of biological functions and recognition processes, demonstrating the importance of glycomic information for understanding the complex mechanisms of living organisms [4,5].

In comparison to proteomic, and especially genomic research, the rate of data production in glycomics has been rather limited due to long sample preparation, separation times, and the difficulty of analyzing data. The advent of Ultra Performance Liquid Chroma-tography (UPLC) and improved chromatographic phases [6–8], together with automated liquid handling systems and advanced MS techniques [9,10] are ensuring that these bottlenecks will eventually be overcome [11].

The increasingly large amounts of glycomic data that are being generated requires powerful and sophisticated database software, as well as tools for storage, manipulation, and easy access by scientists around the world (Fig. 7.1) [12]. The provision of these experimental and literature databases not only allow analysis of the data that is already present, but provides a source of material from which chemical, molecular, and biochemical rules can be deduced. This will allow the development of methodologies, such as glycosylation-site prediction [13–18], site occupancy [19–22], and binding partners [23–25], among others.

This review summarizes the current software and databases that have been developed to date (Table 7.1).

7.2 Glycan Databases — Current

7.2.1 *BCSDB*

The Bacterial Carbohydrate Structure DataBase (BCSDB, http://csdb.glycoscience.ru/bacterial/) [48] is based in Russia and was first presented in 2005. The database is a compilation of structural, bibliographic, taxonomic, and related information on bacterial carbohydrate structures from over 4,000 publications dating from 1941 to the present day.

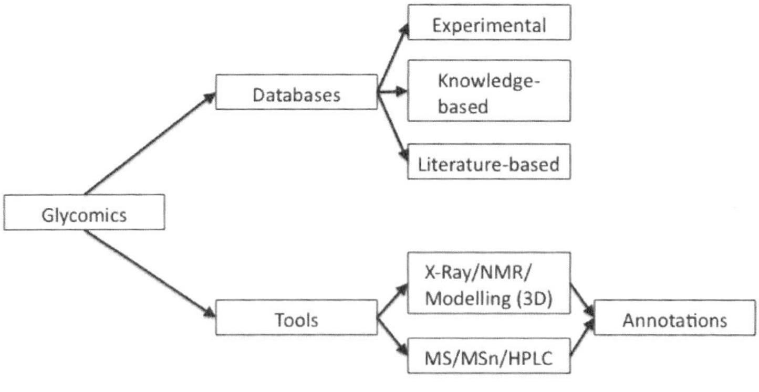

Figure 7.1 The current field of glycomics [26–28].

Table 7.1 Glycan associated databases & tools (both past & current).

Name	Description	Link	Licensing	Technology
UniCarb-DB [29]	LC-MS/MS spectra for oligosaccharides	www.unicarb-db.org	open-source	Database
UniCarbKB [30]	Online knowledgebase resource for glycomics/glycobiology	www.unicarbdb.org	open-source	Database (Knowledgebase)
GlycoBase (NIBRT) [31]	HPLC/UPLC/CE experimental database for released oligosaccharides	glycobase.nibrt.ie	free for academics	Database & tools
GlycomeDB [32]	Unique references to the existing databases	http://www.glycome-db.org/	free	Databse & tools
Bacterial Carbohydrate Structure Database [33]	Structural, bibliographic, taxonomic & related info on bacterial carbohydrate structures	http://csdb.glycoscience.ru/bacterial/	free	Database
CCDS-CarbBank [34]		http://www.glycome-db.org/getDownloadPage.action?page=carbbank	free	Database
Glycosciences.de [35]	Carbohydrate 3D structure tools	http://www.glycosciences.de/tools/	free	Database & tools
KEGG Glycan [36]		http://www.genome.jp/kegg/glycan/	free	Database, tools & API
Glycan Fragment DB [37]	Oligosaccharide classification system based on primary sequence & glycosidic linkage	http://www.glycanstructure.org/	free	Database & tools

(Continued)

Table 7.1 (*Continued*)

Name	Description	Link	Licensing	Technology
Glycoconjugate Data Bank [38]	Database of glycan structures in the PDB	http://www.glycostructures.jp/	free	Database
GlycoMapsDB [39]	Management of conformational maps & profiles	http://www.glycosciences.de/modeling/glycomapsdb/	free	Database & tools
GlycoSuiteDB [40]	Glycan Structure Database	http://glycosuitedb.expasy.org/glycosuite/glycodb	free	Database
Monosaccharide Browser [41]	View Fischer projections of glycans	http://www.terravivida.com/vivida/monosaccharide/	free	Tool
SuperSweet [42]	Sweetening Agents Database	http://bioinf-applied.charite.de/sweet/index.php?site=home	free	Database
PhosphoSitePlus [43]	PTM database including *O*-glcnacylation	http://www.phosphosite.org/	free	Database & tools
ProGlycProt [44]	Online collection of experimentally verified glycosites & glycoproteins from prokaryotes	http://www.proglycprot.org/	free	Database
SugaBase [45]		http://glycomics.ccrc.uga.edu/GlycomicsPortal/showEntry.action?id=48	free	Database
Phosida [46]	Database of PTMs including N-linked oligosaccharides	http://www.phosida.com/	free	Database
GlycoPattern [47]	Database & tools for glycan array	https://glycopattern.emory.edu/	free	Database

7.2.2 *CAZy*

The Carbohydrate-Active Enzymes (CAZy) database (www.cazy.org) [49] is focused on the genomic, structural and biochemical information on CAZys. These enzymes are involved in some way in the metabolism of carbohydrates, and in the degradation, modifications, and creation of glycosidic bonds. CAZy was started in 1998 and is based in the Centre national de la recherche scientifique (CNRS) in Marseille, France.

7.2.3 *CFG*

The Consortium for Functional Glycomics [50] is a research initiative based in the United States that provides glycomics resources to the scientific community. As part of this effort, they have created an experimental-based database that includes glycan array screening data, glycan profiling data, gene microarray data, and mouse phenotyping data. This is supplemented with molecular information about glycan structures, glycan-binding proteins and glycosyltransferases. This impressive array of resources allows for elegant searches allowing for the comparison of gene expressions, for example, or for investigating the effect of glycosyl transferase activity on certain cell types.

7.2.4 *GlycoBase (Lille)*

GlycoBase [51] is an experimental database of nuclear magnetic resonance (NMR) data for glycans. It is possible to search the database on glycan sequences from a number of different animal species, as well as download the NMR files and annotated spectra in joint photographic experts group (jpg) format. It is possible to search the site by sugar, molecule, experiment, trivial name, or species.

7.2.5 *GlycoBase (NIBRT)*

GlycoBase [31] is an experimental database for the storage, classification, and reporting of glycan structures and their associated experimental values obtained using chromatographic techniques, such as hydrophilic interaction liquid chromatography (HILIC), reverse-phase (RP) HPLC, UPLC, and CE. This was a part of the original EuroCarbDB [52] project, and shares a code–base and basic database structure with both UniCarbKB and UniCarb-DB. However, it has evolved to suit the needs of the chromatography community. The GlycoBase glycan collections contain retention data (expressed as Glucose Units or GU values) for more than 300 2AB-labeled *N*-linked glycan structures derived from a diverse set of glycoproteins of biotherapeutic and diagnostic interest. These values were obtained by strict and systematic analysis of 2AB labeled released glycans, analyzed using Waters leading HPLC/UPLC technologies as part of the National Institute for Bioprocessing Research and Training (NIBRT) glycan analytical platform. Hydrophilic interaction liquid chromatography combined with fluorescence detection (HILIC-fluorescence), supplemented by exoglycosidase sequencing and mass spectrometric confirmation, was used to generated this high confidence labeled glycan library. The resulting database has been made accessible through a customized Web-application containing a simple and intuitive interface for assignment and confirmation of glycan structures. The combined efficiencies of high resolution glycan analysis and simplified data processing will enable the biopharmaceutical industry to perform released glycan analysis with greater confidence and speed than has not been possible using existing analysis workflows.

7.2.6 *GlycomeDB*

GlycomeDB [32] is a depository for glycan structures and their associated meta-data. It is an effort to allow for the standardization of various structural databases and cross-talking between them. As part of the effort, the structural data was analyzed and any inconsistencies were highlighted to the relevant curators and corrected. GlycomeDB has also been incorporated into Brukers Proteinscape software. It is used as a tool to search for known glycan m/z values. GlycomeDB contains references to the existing public databases.

Glycans are indexed and defined using sequence format based on extensible markup language (XML) (GlycoCT [53]), capable of storing all structural information of carbohydrate sequences. With GlycomeDB, it is possible to get an overview of all carbohydrate structures in the different databases and to crosslink common structures in the different databases. Scientists are now able to search for a particular structure in the meta database and get information about the occurrence of this structure in the five carbohydrate structure databases. GlycomeDB currently maps glycan structural entries obtained from a variety of databases.

7.2.7 *GlycoO*

The Glycomics ontology (GlycoO), [54] is an ontology focused on providing a schema for the classification of concepts commonly found in the field of glycobiology. The resource is seen as a support mechanism for the development of automated processes for analytical interrogations.

7.2.8 *Glycosciences.de*

Glycosciences.de [35] is a collection of databases and tools for the glyco-analytical field, with a specific focus on three-dimensional (3D) applications. The databases are divided into the following types: bibliographic, structural (based on the Protein Data Bank (PDB)), and experimental (NMR and MS). This resource also contains a number of software tools to aid in 3D modeling.

7.2.9 *GlycoSuiteDB*

GlycoSuiteDB [40] is a curated database of glycan structure information that was collated from literature ranging from 1990 to 2005. The database is searchable through a number of different filters, including glycan composition, actual structure, associated protein, taxonomy, etc. This database has recently been incorporated into the UniCarbKB suite and is also being updated.

7.2.10 *JCGGDB*

The Japanese Consortium for Glycobiology and Glycotechnology Database (JCGGDB) [55] is a collection of glycan-related databases, covering glycogens, glycan-related diseases, glycan function, as well as analysis and detection. The databases can be queried using keywords, glycan structures, by disease, experimental protocol, glycogene, and so on.

7.2.11 *KEGG GLYCAN*

The Kyoto Encyclopedia of Genes and Genomes (KEGG) [56] is a reference knowledge base for high-throughput analytical techniques in the biological sphere. KEGG Glycan is an informatics resource integrating genomics and chemistry. This subset of the resource includes Pathway Maps for glycans, information about glycosyltransferases, as well as other glycan-binding proteins and structural information.

7.2.12 *PDB*

The PDB [57] is described as an information portal to biological macromolecular structures. It contains 3D X-Ray/NMR structures of proteins. As part of some glycan-binding proteins, it is possible to find 3D coordinates for carbohydrate structures. However, as carbohydrates are quite flexible, the resolution of these ligands is usually not quite as high as the protein backbone. The 3D coordinates are still quite an important research in the modeling of protein–sugar interactions and binding.

7.2.13 *SugarBind*

SugarBind (http://sugarbind.expasy.org/sugarbind) [58] is a database of pathogen and biotoxin lectin interactions along with their carbohydrate ligands. The database was originally published by the MITRE Corporation in 2005, but just recently has found a new home on the ExPASy server, as part of the UniCarbKB family of glycan resources.

7.2.14 *UniCarb-DB*

UniCarb-DB [29] is a glycan structural database with associated liquid chromatography (LC)-MS peak lists. The database currently holds both *N*- and *O*-linked oligosaccharide data from porous graphitized carbon (PGC)-LC-MS/MS run on an ion-trap with collision-induced dissociation (CID). The structural data is linked back to UniCarbKB and, therefore, to all the inherent protein/literature, etc. type data that is held there.

7.2.15 *UniCarbKB*

UniCarbKB [30] is a knowledge-based data repository aiming to promote the creation of an online information storage and search platform for glycomics and glycobiology research. The knowledge base will offer a freely accessible and information-rich resource supported by querying interfaces, annotation technologies and the adoption of common standards to integrate structural, experimental, and functional data. Currently, UniCarbKB creates a unique access point to three public databases: UniCarb-DB [29], GlycoSuiteDB [40], and GlycoBase [31]. The aim is to allow the users to navigate between the information retained in the three databases from a single web interface.

7.3 Glycan Databases — Historical

7.3.1 *Complex carbohydrate structure database*

The Complex Carbohydrate Structure Database (CCSD) [59] was a database started in the eighties to fill the growing needs of the glycobiology community for a curated resource of glycan structures and related information. Again, even though this resource has long gone un-funded, it has been incorporated into a number of meta-database projects, including EuroCarb-DB, GlycomeDB, and UniCarbKB.

7.3.2 *EuroCarbDB*

EuroCarbDB [52] was a European Union (EU)-funded initiative that created the technical framework to support glycobiology research. Even though the project itself is discontinued, the framework lives on as the underlying code in UniCarb-DB [29], UniCarbKB [30], GlycoBase (Dublin) [31], and the continuation of sub-projects, such as GlycanBuilder and GlycoWorkbench [60].

7.4 Glycan Tools (Computational Applications)

7.4.1 *ExPASy*

The Expert Protein Analysis System (ExPASy) [61] is the Swiss Institute of Bioinformatics (SIB) portal that hosts various databases and tools for the life sciences. Within the Protein Informatics Group (PIG), whose main focus up to now has been proteomics, they have amassed a number of glycan related tools and databases (mentioned above).

7.4.1.1 *GlycanMass*

GlycanMass [62] is an online tool for calculating the mass of a glycan. It is possible to input the composition by specifying the number of different residues, whether they are derivitized or not, and whether you would like the monoisotopic or average mass.

7.4.1.2 *GlycoMod*

GlycoMod [63] is a tool to find all possible glycan compositions from a precursor m/z. The tool can be used with any kind of MS data, and allows for parameters, such as derivitization of the reducing end, minimum and maximum number of residues, and labeling.

7.4.2 *KEGG*

The KEGG [56] has already been described in the context of available databases. This resource also contains some tools to support glycomics analysis.

7.4.2.1 *KegArray*

KegArray is a Java application for microarray data analysis that includes both gene expression and metabolome data. This software allows for the mapping of microarray data to other resources in the KEGG database family.

7.4.2.2 *KegDraw*

KegDraw is a simple editor to edit and/or draw glycans and other small molecules at atomic level. It supports mol-type file formats and is also capable of performing searches within the KEGG database.

7.4.3 *Glycam*

Glycam [64] is a forcefield for 3D prediction of glycans and glycoprotein complexes. There are a number of tools available for use on the site, including generation of input files in various 3D formats, generation of parameter files (for example, for input into AutoDock [65]), and an oligosacchride library.

7.4.4 *Glycosciences.de*

As mentioned previously, glycosciences.de [35] is a resource for the glyco-analytical field, with a specific focus on 3D applications. There are a whole suite of tools available for use with the 3D-structure database, the PDB [57], including tools for the detection and validation of carbohydrates, as well as for the statistical analysis of properties derived from the PDB. There is also a modeling section which includes tools for *in silico* modeling of glycosylated proteins, as well as portals for access to molecular dynamics simulations and tutorials.

7.4.5 *GlycoWorkBench*

The GlycoWorkBench [60] suite of tools has been developed to aid the interpretation of mass spectrometry of glycans. It includes the GlycanBuilder for easy visualization of glycans and the Glyco-PeakFinder [66] for the semi-automatic annotation of ms/ms fragment data.

7.4.6 *NetNGlyc/NetOGlyc*

The NetNGlyc [67] and NetOGlyc [16] servers are hosted at the Centre of Biological Sequence Analysis in the Technical University of Denmark. They use neural networks to predict the sites of *N*- and *O*-linked glycosylation on protein sequences.

7.4.7 *RINGS*

The RINGS [68] website is a resource providing algorithmic and data-mining tools for the glycobiology community. A lot of the tools are available for download and use as web services freely for academic research.

7.5 MIRAGE Initiative and the Definition of an Exchange Format between Glycan Databases

Of particular relevance in the field of glyco-bioinformatic and glycomic databases is the Minimum Information Required for a Glycomics Experiment (MIRAGE) initiative [69]. This project began in November 2012 after the glycomics community agreed that there is a common need for a standardized reporting effort. MIRAGE is run by a working group, which consists of glyco-analytical and bioinformatics experts from around the world. These panels are responsible for drawing up recommendations and guidelines in communication with the scientific community. The main MIRAGE aims are to: (i) develop a set of guidelines required for the publication of glycomics experiments in scientific journals; and (ii) propose standards for data storage and exchange formats. In this regard, the MIRAGE will define specific namespaces or dictionaries for different analytical techniques. The aim is thus to standardize (normalize) the majority of glycan related databases currently available. This will make raw data available in a commonly readable format so it can be reprocessed and re-interpreted in light of new knowledge (iii) to create a software framework for data integration and mining. As per June 2013, MS has been defined and LC guidelines are in the last draft. Glycan Array guidelines are also been developed.

7.6 Glycan Data Formats

7.6.1 *Glycan file formats*

7.6.1.1 *GLYDE — II*

Glyde [70] is an XML-based standard for the representation of glycans as chemical structures. This format is based on connection tables at the level of atoms, and so, this format is particularly useful for chemistry-based applications and structural calculations.

7.6.1.2 *GlycoCT (XML — Condensed)*

GlycoCT [53] is a format for the representation of glycan structures at the monosaccharide level. This format treats the glycan as a graph, and has three levels of description: a formal XML description (GlycoCT), a human-readable format (GlycoCT-condensed), and a short version that uses a third less space (GlycoCT-compressed). This format has been adopted as the standard for storage of glycan sequences in a number of related databases, including UniCarbKB, UniCarb-DB, GlycoBase, etc.

7.6.1.3 *IUPAC*

The International Union of Pure and Applied Chemistry (IUPAC) abbreviated format [71] for glycan sequences is a linear format where the smallest single unit is a monosaccharide. It was developed to combat the "unwieldy" nature of portraying large poly/oligosaccharides. It stems from the extended IUPAC format, which allows the presentation of different branches on different lines; in contrast, the condensed format is all contained in one line.

7.6.1.4 *KCF*

The KEGG Chemical Function (KCF) [72] format is a graph-like format for glycan structures used in the RINGS [68] resource. The resource also contains converters to and from many of the commonly used glycan formats.

7.6.1.5 *LINUCS*

Linear notation for unique description of carbohydrate sequences (LINUCS) [73] is another linear format for glycan structure that fulfills a number of criteria. For example, it is a linear code type, non-graphical, and is approved by IUPAC. It also allows for un-assigned linkages.

7.6.1.6 *SequenceGWS*

The sequenceGWS format is a linear format used in the GlycoWorkbench [60] suite of tools. GlycoWorkbench can be used to save structures and the underlying file format is XML-based. The sequenceGWS linear format is used in this XML schema. However, there are a number of converters within the tool allowing for the output of structures in a number of other formats, including GlycoCT-condensed.

7.6.2 *Cartoon glycan notation*

7.6.2.1 *Oxford glycan notation*

The Oxford Glycobiology group have proposed the University of Oxford Format (UOXF) [74,75], a rational cartoon format for oligosaccharides that enables the monosaccharide residue and the molecular weight to be deduced directly. The notation can be used in black and white or in color. The anomericity and linkage information is embedded into the representations and is designed for the easy working of data.

7.6.2.2 *Text*

The text format for displaying glycans in a graphical way was first used in the GlycoSuite [40] database. It uses abbreviations for the residues and writes the anomericity, anomeric carbon, and linked carbon along the bond between the residues.

7.6.2.3 *CFG*

The CFG [76] nomenclature for glycan representation is based on shape and color. The linkage information is written along the bonds.

7.7 Summary/Conclusion

The first studies linking genomics with glycomics have now been published [77–80]. These have involved the analysis of tens of thousands of glycomes and transcriptomes. The tools are now available to obtain data from different — omics experiments, highlighting the need for bioinformatics software that can encompass many kinds of data, including glycosylation.

If genomics is perceived as a starting point, then glycomics is the end. It represents the sum total of all the influences on the genome. The expectation is that an overall view of the data will provide insight into the critical features and pathways that influence glycosylation. One aim is to identify the tipping points where the flexibility of the processing pathways can no longer compensate for disease related changes, including in the epigenetics, genetics, metabolomics, lipidomics, and proteomics.

Acknowledgments

This work has been supported under the European Union's Seventh Framework Programme Glyco BioM project (FP7-HEALTH-259869).

References

1. Wuhrer, M., Deelder, A. M., Hokke, C. H. (2005) *J. Chromatogr. B Analyt. Technol. Biomed. Life Sci.* **825**, 124–133.
2. Chen, G., Pramanik, B. N. (2008) *Expert Rev. Proteomics* **5**, 435–444.
3. Mariño, K., Bones, J., Kattla, J. J., Rudd, P. M. (2010) *Nat. Chem. Biol.* **6**, 713–723.
4. Haltiwanger, R. S., Lowe, J. B. (2004) *Annu. Rev. Biochem.* **73**, 491–537.
5. Rudd, P. M., Elliott, T., Cresswell, P., Wilson, I. A., Dwek, R. A. (2001) *Science* **291**, 2370–2376.
6. Ahn, J., Bones, J., Yu, Y. Q., Rudd, P. M., Gilar, M. (2010) *J. Chromatogr. B Analyt. Technol. Biomed. Life Sci.* **878**, 403–408.
7. Knežević, A., Bones, J., Kračun, S. K., Gornik, O., Rudd, P. M., Lauc, G. (2011) *Analyst* **136**, 4670–4673.
8. Zauner, G., Deelder, A. M., Wuhrer, M. (2011) *Electrophoresis* **32**, 3456–3466.
9. Williams, J. P., Grabenauer, M., Holland, R. J., Carpenter, C. J., Wormald, M. R., Giles, K., Harvey, D. J., Bateman, R. H., Scrivens, J. H., Bowers, M. T. (2010) *Int. J. Mass Spectrom.* **298**, 119–127.
10. Harvey, D. J., Scarff, C. A., Crispin, M., Scanlan, C. N., Bonomelli, C., Scrivens, J. H. (2012) *J. Am. Soc. Mass Spectrom.* **23**, 1955–1966.
11. Pompach, P., Chandler, K. B., Lan, R., Edwards, N., Goldman, R. (2012) *J. Proteome Res.* **11**, 1728–1740.
12. National Research Council (US) Committee on Assessing the Importance and Impact of Glycomics and Glycosciences (2012) *Transforming Glycoscience: A Roadmap for the Future.* National Academic Press (Washington DC).
13. Julenius, K., Mølgaard, A., Gupta, R., Brunak, S. (2005) *Glycobiology* **15**, 153–164.
14. Blom, N., Sicheritz-Ponten, T., Gupta, R., Gammeltoft, S., Brunak, S. (2004) *Proteomics* **4**, 1633–1649.
15. Gupta, R., Brunak, S. (2002) *Pac. Symp. Biocomput.* **2002**, 310–322.

16. Hansen, J. E., Lund, O., Tolstrup, N., Gooley, A. A., Williams, K. L., *et al.* (1998) *Glycoconj. J.* **15**, 115–130.
17. Hansen, J. E., Lund, O., Engelbrecht, J., Bohr, H., Nielsen, J. O. (1995) *Biochem. J.* **308**, 801–813.
18. Caragea, C., Sinapov, J., Silvescu, A., Dobbs, D., Honavar, V. (2007) *BMC Bioinformatics* **8**, 438.
19. Senger, R. S., Karim, M. N. (2005) *Biotechnol. Prog.* **21**, 1653–1662.
20. Petrescu, A. J., Milac, A. L., Petrescu, S. M., Dwek, R. A., Wormald, M. R. (2004) *Glycobiology* **14**, 103–114.
21. Lu, D., Yang, C., Liu, Z. (2012) *J. Phys. Chem. B* **116**, 390–400.
22. Shelikoff, M., Sinskey, A. J., Stephanopoulos, G. (1996) *Biotechnol. Bioeng.* **50**, 73–90.
23. Grant, O. C., Woods, R. J. (2014) *Curr. Opin. Struct. Biol.* **28C**, 47–55.
24. Kletter, D., Cao, Z., Bern, M., Haab, B. (2013) *Curr. Protoc. Chem. Biol.* **5**, 157–169.
25. Porter, A., Yue, T., Heeringa, L., Day. S., Suh, E., Haab, B. B. (2010) *Glycobiology* **20**, 369–380.
26. Aoki-Kinoshita, K. F. (2013) *Mol. Cell Proteomics* **12**, 1036–1045.
27. Li, F., Glinskii, O. V., Glinsky, V. V. (2013) *Proteomics* **13**, 341–354.
28. Artemenko, N. V., McDonald, A. G., Davey, G. P., Rudd, P. M. (2012) *Methods Mol. Biol.* **899**, 325–350.
29. Hayes, C. A., Karlsson, N. G., Struwe, W. B., Lisacek, F., Rudd, P. M., Packer, N. H., Campbell, M. P. (2011) *Bioinformatics* **27**, 1343–1344.
30. Campbell, M. P., Peterson, R., Mariethoz, J., Gasteiger, E., Akune, Y., Aoki-Kinoshita, K. F., Lisacek, F., Packer, N. H. (2014) *Nucleic Acids Res.* **42**, D215–221.
31. Campbell, M. P., Royle, L., Radcliffe, C. M., Dwek, R. A, Rudd, P. M. (2008) *Bioinformatics* **24**, 1214–1216.
32. Ranzinger, R., Herget, S., von der Lieth, C. W., Frank, M. (2011) *Nucleic Acids Res.* **39**, D373–376.
33. Toukach, P. V. (2011) *J. Chem. Inf. Model* **51**, 159–170.
34. Doubet, S., Albersheim, P. (1992) *Glycobiology* **2**, 505.
35. Lutteke, T., Bohne-Lang, A., Loss, A., Goetz, T., Frank, M., von der Lieth, C. W. (2006) *Glycobiology* **16**, 71R–81R.
36. Hashimoto, K., Goto, S., Kawano, S., Aoki-Kinoshita, K. F., Ueda, N., Hamajima, M., Kawasaki, T., Kanehisa, M. (2006) *Glycobiology* **16**, 63R–70R.
37. Jo, S., Im, W. (2013) *Nucleic Acids Res* **41**, D470–D474.

38. Nakahara, T., Hashimoto, R., Nakagawa, H., Monde, K., Miura, N., Nishimura, S. (2008) *Nucleic Acids Res.* **36**, D368–D371.
39. Frank, M., Lutteke, T., von der Lieth, C. W. (2007) *Nucleic Acids Res.* **35**, 287–290.
40. Cooper, C. A., Joshi, H. J., Harrison, M. J., Wilkins, M. R., Packer, N. H. (2003) *Nucleic Acids Res.* **31**, 511–513.
41. Terra Vivida (1999) Monosaccharide Browser. Available: http://www.oxfordjournals.org/our_journals/nar/database/summary/535. Accessed 4 November 2014.
42. Ahmed, J., Preissner, S., Dunkel, M., Worth, C. L., Eckert, A. (2011) *Nucleic Acids Res* **39**, D377–382.
43. Hornbeck, P. V., Kornhauser, J. M., Tkachev, S., Zhang, B., Skrzypek, E., Murray, B., Latham, V., Sullivan, M. (2012). *Nucleic Acids Res.* **40**, D261–D270.
44. Bhat, A. H., Mondal, H., Chauhan, J. S., Raghava, G. P., Methi, A., Rao, A. (2012) *Nucleic Acids Res.* **40**, D388–D393.
45. Van Kuik, J. A., Hård, K., Vliegenthart, J. F. (1992) *Carbohydr. Res.* **235**, 53–68.
46. Gnad, F., Gunawardena, J., Mann, M. (2011) *Nucleic Acids Res.* **39**, D253–260.
47. Agravat, S. B., Saltz, J. H., Cummings, R. D., Smith, D. F. (2014) *Bioinformatics* **30**, 3417–3418.
48. Toukach. P, Joshi, H. J., Ranzinger, R., Knirel, Y., von der Lieth, C.-W. *Nucleic Acids Res.* **35**, D280–D286.
49. Cantarel, B. L., Coutinho, P. M., Rancurel, C., Bernard, T., Lombard, V., Henrissat, B. (2009) *Nucleic Acids Res.* **37**, D233–D238.
50. Consortium for Functional Glycomics (2010) *Functional Glycomics Gateway*. Available at: www.functionalglycomics.org. Accessed 4 November 2014.
51. Unité de Glycobiologie Structurale et Fonctionnelle (2007) *GlycoBase*. Available at: http://glycobase.univ-lille1.fr/base/. Accessed 4 November 2014.
52. von der Lieth, C.-W., Freire, A. A., Blank, D., Campbell, M. P., Ceroni, A., Damerell, D. R, Dell, A., Dwek, R. A., Ernst, B., Fogh, R., Frank, M., Geyer, H., Geyer, R., Harrison, M. J., Henrick, K., Herget, S., Hull, W. E., Ionides, J., Joshi, H. J., Kamerling, J. P., Leeflang, B. R., Lutteke, T., Lundborg, M., Maass, K., Merry, A., Ranzinger, R., Rosen, J., Royle, L., Rudd, P. M., Schloissnig, S., Stenutz, R., Vranken, W. F., Widmalm, G., Haslam, S. M. (2011) *Glycobiology* **21**, 493–502.

53. Herget, S., Ranzinger, R., Maass, K., Lieth, C.-W. (2008) *Carbohydr. Res.* **343**(12), 2162–2171.
54. Sahoo, S. S., Sheth, A., York, W. (2005) *GlycoO and ProPreO: A case study of two deep domain life science ontologies.* University of Georgia Large Scale Distributed Information Systems Laboratory Technical Report. University of Georgia: Athens.
55. Japan Consortium for Glycobiology and Glycotechnology (2007) *Japan Consortium for Glycobiology and Glycotechnology DataBase.* Available at: http://jcggdb.jp/database_en.html. Accessed 4 November 2014.
56. Kanehisa, M., Goto, S., Sato, Y., Kawashima, M., Furumichi, M., Tanabe, M. (2014) *Nucleic Acids Res.* **42**, 199–205.
57. Berman, H. M., Westbrook, J., Feng, Z., Gilliland, G., Bhat, T. N., Weissig, H., Shindyalov, I. N., Bourne, P. E. (2000) *Nucleic Acids Res.* **28**(1), 235–242.
58. Shakhsheer, B., Anderson M., Khatib K., Tadoori L., Joshi L., Lisacek, F., Hirschman, L., Mullen, E. (2013) *J. Mol. Recognit.* **26**(9), 426–431.
59. Doubet, S., Bock, K., Smith, D., Darvill, A., Albersheim, P. (1989) *Trends Biochem. Sci.* **14**(12), 475–477.
60. Damerell, D., Ceroni, A., Maass, K., Ranzinger, R., Dell, A., Haslam, S. M. (2012) *Biol. Chem.* **393**(11), 1357–1362.
61. Gasteiger, E., Gattiker, A., Hoogland, C., Ivanyi, I., Appel, R. D., Bairoch, A. (2003) *Nucleic Acids Res.* **31**, 3784–3788.
62. Swiss Institute of Bioinformatics (2011) *GlycanMass.* Available at: http://web.expasy.org/glycanmass/. Accessed 4 November 2014.
63. Cooper, C. A., Gasteiger, E., Packer, N. H. (2001) *Proteomics* **1**(2), 340–349.
64. Kirschner, K. N., Yongye, A. B., Tschampel, S. M., Gonzalez-Outeirino, J., Daniels, C. R., Foley, B. L., Woods, R. J. (2008) *J. Comput. Chem.* **29**(4), 622–655.
65. Morris, G. M., Huey, R., Lindstrom, W., Sanner, M. F., Belew, R. K., Goodsell, D. S., Olson, A. J. (2009) *J. Comput. Chem.* **30**(16), 2785–2791.
66. Maass, K., Ranzinger, R., Geyer, H., von der Lieth, C.-W., Geyer, R. (2007) *Proteomics* **7**(24), 4435–4444.
67. Gupta, R., Jung, E., Brunak, S. (2004) Prediction of N-glycosylation sites in human proteins. Available at: http://www.cbs.dtu.dk/services/NetNGlyc/. Accessed 4 November 2014.
68. Akune, Y., Hosoda, M., Kaiya, S., Shinmachi, D., Aoki-Kinoshita, K. F. (2010) *OMICS* **14**(4), 475–486.

69. Kolarich, D., Rapp, E., Struwe, W. B., Haslam, S. M., Zaia, J., McBride, R., Agravat, S., Campbell, M. P., Kato, M., Ranzinger, E., Kettner, C., Yorka, W. S. (2013) *Mol. Cell Proteomics* **12**(4), 991–995.
70. Sahoo, S. S., Thomas, C., Sheth, A., Henson, C., York, W. S. (2005) *Carbohydr. Res.* **340**(18), 2802–2807.
71. (1983). *Arch. Biochem. Biophys.* **220**(1), 325–329.
72. Hattori, M., Okuno, Y., Goto, S., Kanehisa, M. (2003) *J. Am. Chem. Soc.* **125**, 11853–11865.
73. Bohne-Lang, A., Lang, E., Forster, T., von der Lieth, C.-W. (2001) *Carbohydr. Res.* **336**(1), 1–11.
74. Harvey, D. J., Merry, A. H., Royle, L., Campbell, M. P., Rudd, P. M. (2009) *Proteomics* **11**(22), 4291–4295.
75. Harvey, D. J., Merry, A. H., Royle, L, Campbell, M. P., Dwek, R. A., Rudd, P. M. (2011) *Proteomics* **9**(15), 3796–3801.
76. Consortium for Functional Glycomics (2010) Symbol and Text Nomenclature for Representation of Glycan Structure. Available at: http://www.functionalglycomics.org/static/consortium/Nomenclature.shtml. Accessed 4 November 2014.
77. Lauc, G., Huffman, J. E., Pučić, M., Zgaga, L., Adamczyk, B., Ana Mužinić, *et al.* (2013) *PLoS Genet.* **9**, e1003225.
78. Lauc, G., Essafi, A., Huffman, J. E., Hayward, C., Knežević, A., *et al.* (2010) *PLoS Genet.* **6**, e1001256.
79. Huffman, J. E., Knezevic, A., Vitart, V., Kattla, J., Adamczyk, B., *et al.* (2011) *Hum. Mol. Genet.* **20**, 5000–5011.
80. Kutalik, Z., Benyamin, B., Bergmann, S., Mooser, V., Waeber, G., *et al.* (2011) *Hum. Mol. Genet.* **20**, 3710–3717.

Part 3:

Synthetic Challenges: Towards Drug Development

Chapter 8
Classical Tools for Oligosaccharide Synthesis

Daniel B. Werz

Institut für Organische Chemie,
Technische Universität Braunschweig, Germany

8.1 Introduction

The chemical synthesis of complex carbohydrates has been a challenging and difficult task for more than a century [1]. The big difference with the other two major classes of biopolymers, oligonucleotides and proteins, is that several points for further attachment (hydroxyls and amines) are commonly available at one monosaccharidic building block. Thus, even with two subunits, a variety of different dimers can be obtained. With more than two subunits, branched structures might arise as well. The situation becomes even more difficult, since a glycosidic linkage between two subunits creates a new stereogenic center. A recent statistical analysis regarding the diversity of mammalian carbohydrate structures, based on the "10 mammalian monosaccharides" without considering any further attachments, reveals astonishing numbers that dwarf the complexity of both peptides and oligonucleotides by orders of magnitude [2]. Although the number of structural combinations encountered in nature is much smaller than the theoretical ones, it is evident that glycochemists have to face a variety of challenges during the preparation of an oligosaccharide by chemical means.

8.2 From the Target Structure to the Final Glycan

In contrast to syntheses of alkaloids and many other secondary metabolites, the synthesis of glycans is much easier to plan. Nevertheless, a careful synthesis planning is also a prerequisite. A retrosynthetic analysis should include the glycosylation strategy (type of anomeric leaving groups) and some kind of temporary and permanent protecting group for the different hydroxyls [3–5]. Afterwards, mono- or disaccharide building blocks need to be prepared. With these compounds in hand, the assembly process consisting of repeating glycosylation and deprotection steps can be started (Fig. 8.1). Depending on the fate of the carbohydrate assembled, one might think about a linker functionalization. To access the native structure, a global deprotection is required. Afterwards, several purification steps might be necessary. The methods of choice are dialysis or the use of high-performance liquid chromatography (HPLC). Finally, a quality control including mass spectrometric and

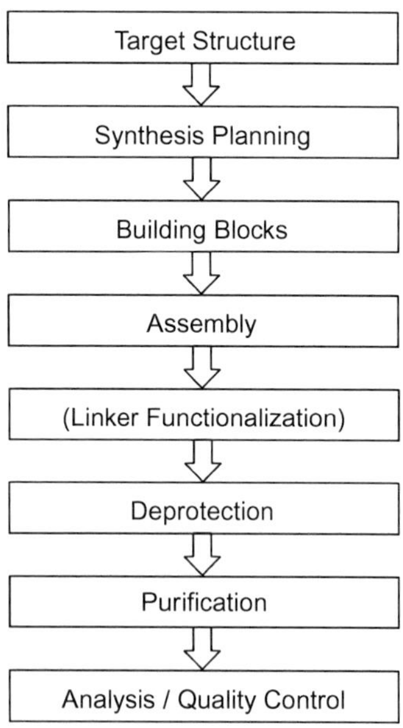

Figure 8.1 From the target structure to the final carbohydrate [5].

proton (^{1}H) and carbon (^{13}C) nuclear magnetic resonance (NMR) analyses should take place.

8.3 Building Blocks

A typical monosaccharidic building block used in oligosaccharide assembly is equipped with different protecting groups that mask the various hydroxyl and amine groups. Permanent protecting groups have to be distinguished from temporary ones. The former mask hydroxyl moieties to be unveiled at the end of the synthesis, whereas the latter mark sites of further glycosylation. Permanent protecting groups are commonly ethers (mostly benzyl (Bn) ethers), which can be cleaved by hydrogenolysis at the end of the assembly, but rather stable esters (such as Pivaloyl (Piv) or even benzoate (Bz) esters) are also used. Temporary protecting groups are functional groups that are easy to cleave, but should be stable during the glycosylation reaction, such as trityl ethers, fluorenylmethoxycarbonyl (Fmoc) groups, and several silyl groups, though acetates (Acs) are often employed as temporary protecting groups as well. However, whether a protecting group is used as permanent or temporary often depends strongly on the glycosylation strategy [3,5]. Besides controlling regioselectivity by orthogonal protecting groups to account for possible branching of the carbohydrate chain, the stereochemistry at the anomeric carbon must be controlled during the formation of the glycosidic bond. Placement of participating protective groups at the C-2 hydroxyl or amine groups ensures the formation of *trans*-glycosidic linkages (via an acyloxonium ion, cp. Fig. 8.6). In contrast, non-participating groups and the use of low temperatures during the glycosylation often lead to the preferential installation of *cis*-glycosides. However, also remote protecting groups might have an influence on the stereochemical outcome of the glycosylation via remote participation. Axial ester groups at position 4 might form bridged bicyclic intermediates.

Of course, the linchpin of every glycosylation is the anomeric leaving group, which is needed and should be easily activated to induce the formation of the glycosidic linkage [3,6]. Numerous

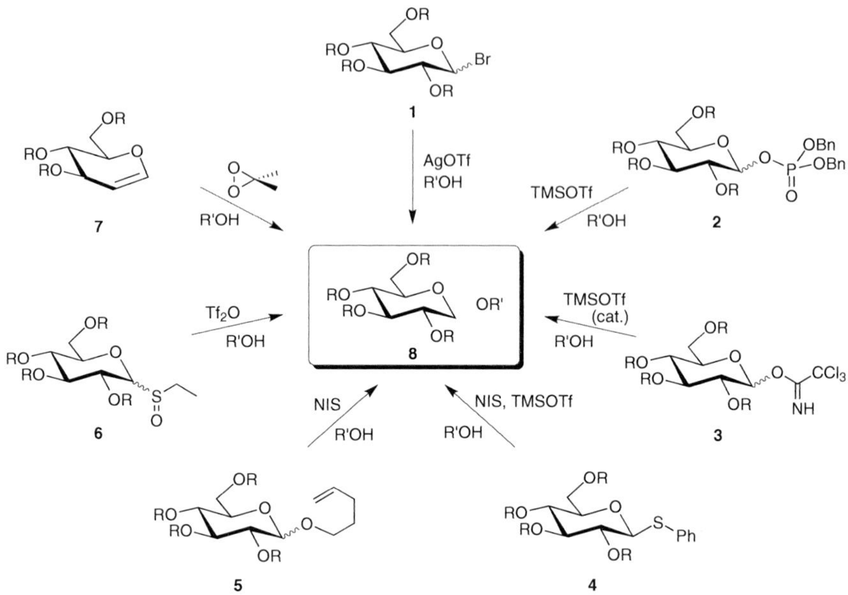

Figure 8.2 Examples of glycosyl donors differing in their anomeric leaving group. The corresponding mode of activation is given as well [5].

anomeric leaving groups have been examined and several of them are widely utitized in the synthesis of complex carbohydrates. Figure 8.2 depicts a selection of such anomeric leaving groups (1–7) with their corresponding activation protocols to perform a glycosylation reaction [3]. The oldest anomeric leaving groups already developed by the pioneer of carbohydrate chemistry, Emil Fischer, are glycosyl halides (such as 1) that are easily activated by silver salts. The driving force of this process is the precipitation of hardly soluble silver halides. Today, the most commonly used anomeric leaving groups are phosphates (2), trichloroacetimidates (3), and thioethers (4) [3,5]. Phosphates commonly need a stoichiometric amount of an activating agent, such as the Lewis acid trimethylsilyl trifluoromethanesulfonate (TMSOTf), whereas for the activation of trichloroacetimidates, a catalytic amount of this reagent is sufficient. Thioethers are shelf-stable compounds and can be stored for years at room temperature. For their activation, a thiophilic reagent such as

N-iodosuccinimide (NIS) is necessary. Pentenyl glycosides (5) are activated by NIS as well; first, an iodonium ion is formed, which leads to a tetrahydrofuran (THF) derivative as a leaving group. Sulfoxides are easily obtained from thioethers by oxidation and are activated by triflic anhydride. Glycals of type (7) are a special case because the glycosyl donor, an acetalic epoxide, is only formed *in situ* by epoxidation with dimethyldioxirane. This reagent forms epoxides under neutral conditions, releasing acetone; the use of *meta*-chloroperoxybenzoic acid (*m*CPBA) would decompose the system. Numerous other glycosyl donors are known and for special solutions, they might come into focus.

The preparation of building blocks is still the most time-consuming process of oligosaccharide synthesis. Rather than synthesizing each building block separately, integrated synthetic paths that grant access to several building blocks from a common precursor are desirable (cp. Fig. 8.5). Commonly, the differently protected and functionalized monosaccharides are accessed from commercially available sugars (e.g. glucose, mannose, or galactose) through a series of protection–deprotection steps. Such a process establishes the desired protecting group pattern and typically requires five to twenty steps or more, depending on the type of monosaccharide, the various protecting groups, and the type of anomeric leaving group.

It is of utmost importance to find ways to differentiate between the different hydroxyl groups of a sugar. The reactivity of the anomeric hydroxyl group being part of a hemiacetal differs significantly from that of the other hydroxyls. The selective removal of esters, such as acetate in the anomeric position, and the readily occurring substitution of esters by alcohols, such as *n*-pentenol or *p*-methoxybenzyl alcohol in the presence of Lewis acids, are well established [7,8]. The primary hydroxyl at position 6 also exhibits special reactivity, since it is the only primary hydroxyl of the common monosaccharide and is sterically exposed [8].

An alternative for the selective protection of four or five hydroxyls relies on rendering two of them silent. Different methods exist. *cis*-Hydroxyls easily form with acetone dimethylacetal in the

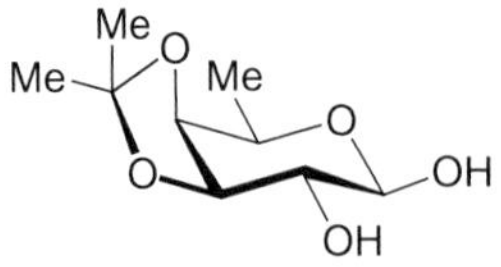

9
(Isopropylidene)

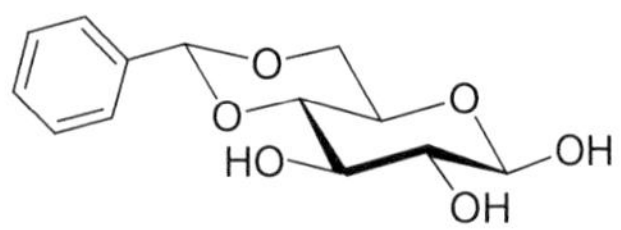

10
(Benzylidene)

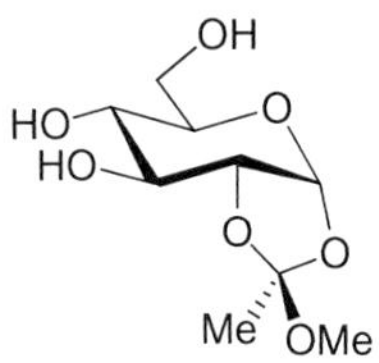

11
(Orthoester)

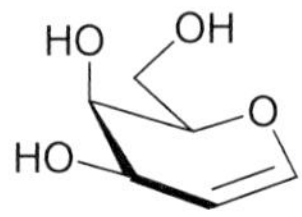

12
(1,2-Glycal)

Figure 8.3 Methods used in building block synthesis to render two hydroxyls "silent" [5].

presence of Lewis acid 5-membered rings, so-called isopropylidenes. Benzaldehyde and benzaldehyde dimethylacetal are ideal reagents to generate 6-membered rings, benzylidene acetals (10) involving the *C*-4 and the *C*-6 hydroxyl. A very useful method to lock the *C*-1 and *C*-2 hydroxyls is the formation of an orthoester (11) (Fig. 8.3). Usually, orthoesters are synthesized by nucleophilic attack of an alcohol at the oxocarbenium ion that can be generated by a glycosyl bromide with the neighboring ester group in the presence of a non-nucleophilic base [8]. This method commonly yields two stereoisomeric products. Even the installation of a double bond, instead of an adjacent diol, might be a way to render two hydroxyls silent. The remaining hydroxyls of these 1,2-glycals (12) are more easily distinguished. Finally, a selective oxidation, mostly an epoxidation by dimethyldioxirane, allows the re-installation of the eliminated hydroxyls at a later stage [5,8].

Besides the knowledge of general principles, the presentation of some case studies for building block syntheses might facilitate ease of understanding. Figure 8.4 depicts the preparation of a glucosyl phosphate with Fmoc as a temporary protecting group in position 4.

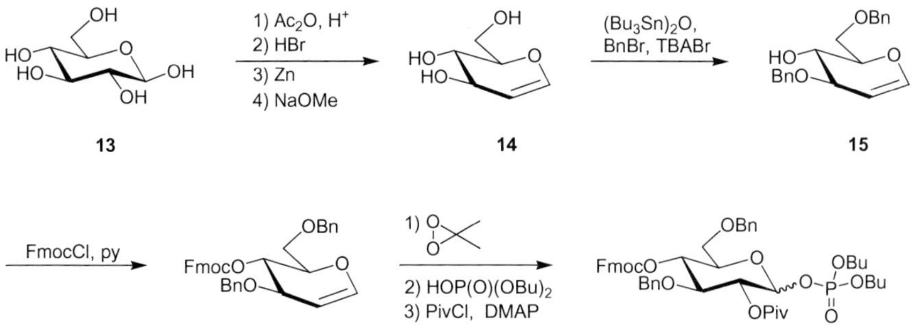

Figure 8.4 Typical procedure to prepare a glucosyl phosphate (17) with permanent and temporary protecting groups from glucose [10].

D-Glucose serves as starting material. Peracetylation and anomeric substitution of acetate by bromide is followed by zinc (Zn)-mediated reduction, eliminating two hydroxyl groups of the glucose (13) to afford glucal (14). Protection with permanent protecting groups (Bn) in positions 4 and 6 mediated via a stannyl ketal, and protection of the free hydroxyl with Fmoc-chloride (Cl) as temporary protecting group generates (16). Epoxidation with dimethyl dioxirane (DMDO) forms an intermediary epoxide being opened with dibutyl phosphate. Protection of the ensuing C2 hydroxyl group with a participating group (e.g. pivaloyl chloride (PivCl)) produced the desired glucosyl phosphate (17) in excellent yield [9,10].

To provide the reader with a feeling of which ways the different hydroxyls of a glucosamine might be addressed, Fig. 8.5 demonstrates three pathways to reach regioisomeric building blocks (22), (28), and (29) with a temporary protecting group either in position 3, 4, or 6. Common precursor (18) is a glucosamine with a silyl group (*tert*-butyldimethylsilyl (TBS)) as an anomeric protecting group, and a benzylidene to protect the 4- and 6-hydroxyls. To access the first glucosamine building block, an acetate is attached in position 3 to afford (19). Afterwards, the benzylidene protecting group is opened in such a way that a free 6-hydroxyl is formed. Due to the lability of acetate under basic conditions, a non-basic method for benzylation of the 6-hydroxyl is crucial, leading to (21).

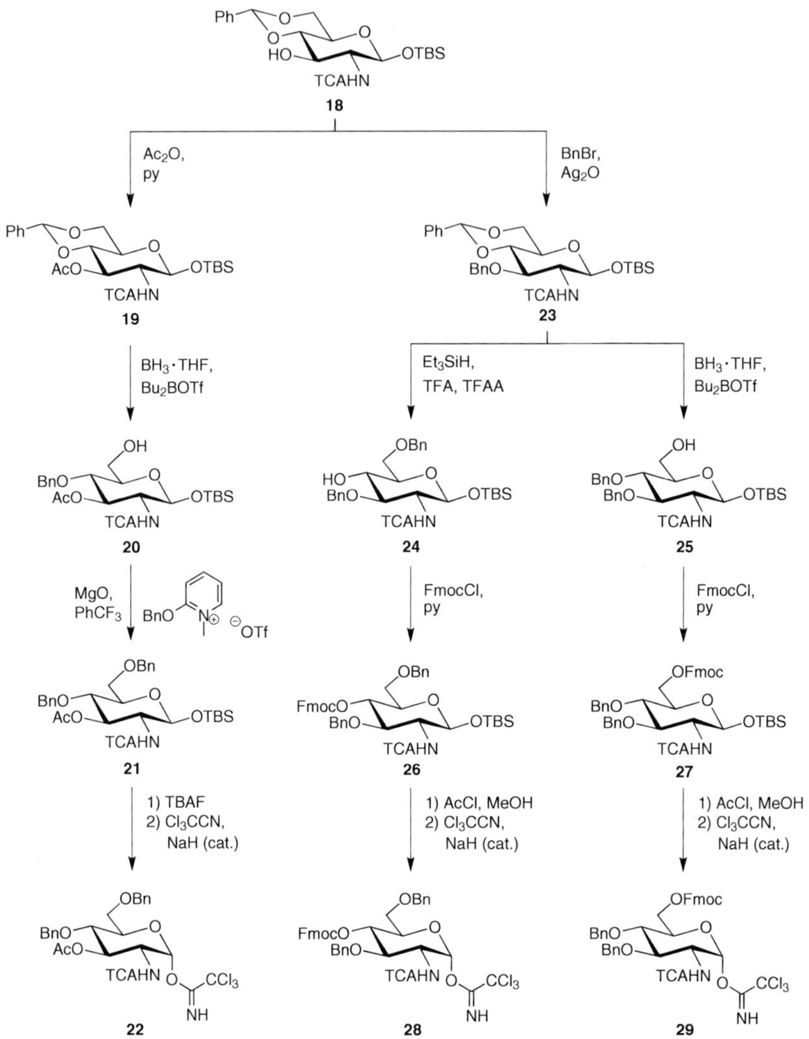

Figure 8.5 Integrated synthetic pathways from one common precursor leading to three regioisomeric glucosaminyl trichloroacetimidates.

Deprotection of the silyl group followed by attack of the ensuing hydroxyl to trichloroacetonitrile generates building block (22). The other two regioisomeric building blocks are obtained by 3-*O*-benzylation to furnish (23). Depending on the reagent mixture, a selective benzylidene opening to either (24) (triethylsilane (Et_3SiH), trifluoroacetic acid (TFA), trifluoroacetic anhydride (TFAA)) or (25) (borane (BH_3), THF, dibutylborinic acid trifluoromethanesulfonate (Bu_2BOTf)) is possible. Protection with FmocCl in pyridine installs Fmoc as a temporary protecting group. Due to the basic nature of fluoride anions, a deprotection with tetrabutylammonium fluoride (TBAF) would lead to a removal of the Fmoc group. To finalize the building block syntheses, an acidic cleavage of the TBS ether is chosen, followed by formation of the trichloroacetimidate.

8.4 Glycosylation and Deprotection Steps

It definitely would go beyond the scope of this brief overview to elucidate in detail all the parameters that affect the outcome of a glycosylation reaction [3,4,11]. However, a short summary of the mechanistic ideas, which have been developed for this important reaction, seems to be appropriate. It is important to note that detailed mechanistic information about glycosylation reactions is still often fragmented. Commonly, the glycosylation reaction involves nucleophilic displacement at the anomeric carbon. Due to the endocyclic oxygen next to the anomeric center, a positive charge is strongly stabilized. Thus, after activation of (30), one may consider a glycosyl cation (31) (oxocarbenium ion) as an intermediate if no further stabilizing substituents are in reach (Fig. 8.6, Case A). A nucleophile, such as a hydroxyl group, can attack this species either from the top or from the bottom face; a mixture of *trans*- and *cis*-glycosides (32) and (33) results. In the case of a participating group in position 2, such a carbocation might also be stabilized by neighboring-group participation, leading to an acyloxonium ion as a major intermediate (Fig. 8.6, Case B). Such an acyloxonium ion (35) shields either the top or the bottom face (depending on the stereochemistry at C-2. As a result the attack of the nucleophilic hydroxyl group can only take place from the non-shielded side, leading to *trans*-glycoside (36) as a major isomer. Using this model, it is easy to understand why participating groups must not be used for

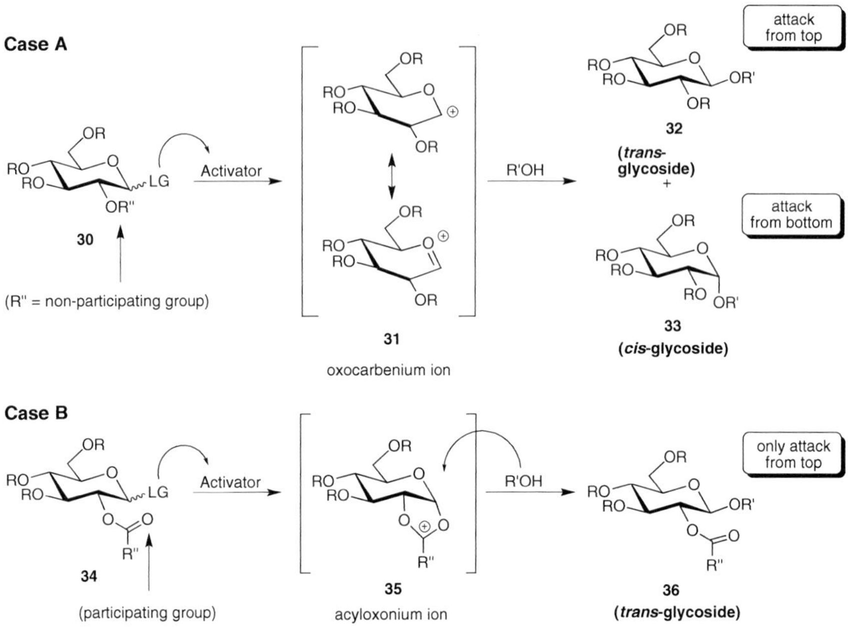

Figure 8.6 Simplified mechanistic picture of a glycosylation reaction with either a non-participating group (Case A) or with a participating group (Case B) in position 2 [5, 8].

the formation of *cis*-glycosides. For the creation of *cis*-glycosides, one has to rely, in most cases, on thermodynamic considerations. In the case of glucose, galactose, and fucose, as well as their corresponding amines and acids, the α product is thermodynamically preferred due to the anomeric effect. Low reaction temperatures favor their formation. However, a complete α selectivity is often difficult to achieve. In the case of mannose, the *cis*-mannosides are the β product. Therefore, the creation of β-mannosides requires other approaches that have been described in the literature [11,12].

In addition to the glycosylation reaction, the second most important reaction in oligosaccharide assembly is the highly selective removal of protecting groups [5,8]. For branched oligosaccharides, careful synthetic planning with respect to the temporary protecting group ensemble is required. Permanent protecting groups stay unaffected until the very final steps of the synthesis in order to procure the completely deprotected carbohydrate. The cleavage of temporary protecting groups after each glycosylation reaction is essential to liberate hydroxyls that are needed as nucleophiles for the next glycosylation step. Silyl protecting groups (e.g. in (37)) are cleaved by fluoride sources or strong acids, whereas Fmoc groups (e.g. in (39)) are cleaved by weak bases, such as piperidine. Stronger bases, such as sodium methoxide, are required for the cleavage of Ac groups (as shown for (41)), whereas levulinoate (Lev) esters in (43) only need hydrazine to be cleaved. An overview of four deprotection methods is provided in Fig. 8.7. Caution has to be taken in employing two temporary protecting groups next to each other. In some instances, a migration of the remaining protecting group might occur when one of the two is removed, especially in the presence of bases [7].

8.5 Global Deprotection

Once the oligosaccharide is assembled all permanent protecting groups have to be removed. This process is called global deprotection. A first possibility consists of a single-step procedure using elementary sodium, carried out in liquid ammonia. Glycochemists call it Birch reduction, since the same conditions are used for the reduction of benzene to 1,4-cyclohexadiene [7].

Benzyl ether groups, as well as esters and carbonates, are removed. The fact that olefinic moieties (except allyl) stay unaffected is the major advantage, since terminal carbon–carbon double bonds are often used as a chemical handle, as a future point of attachment to the surface of a microarray or to a carrier protein [5,7,13]. Thus, it should not be attacked during the deprotection sequence. Besides olefinic moieties, the reducing end hemiacetal of an oligosaccharide is also maintainable during Birch debenzylation [14]. Sometimes, the commonly employed trichloroacetyl (TCA) amino protecting group causes problems, resulting in a tremendous reduction of yield. To circumvent these difficulties, the transformation of the TCA group into an acetyl group by a radical initiated reduction with Bu_3SnH is advantageous. However, the Birch

Silyl Groups (*e.g.* TBS)

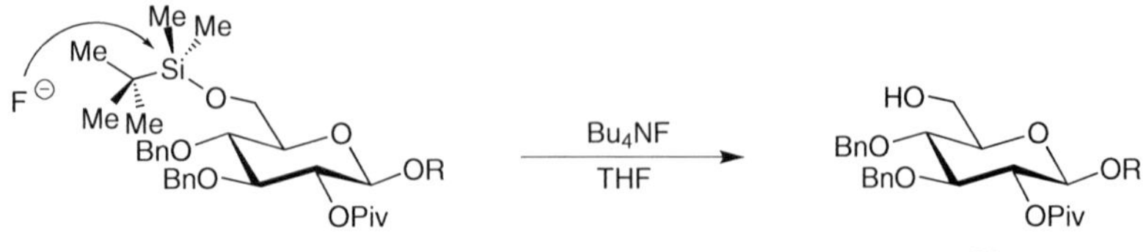

37 **38**

Fluorenylmethoxycarbonyl (Fmoc) Groups

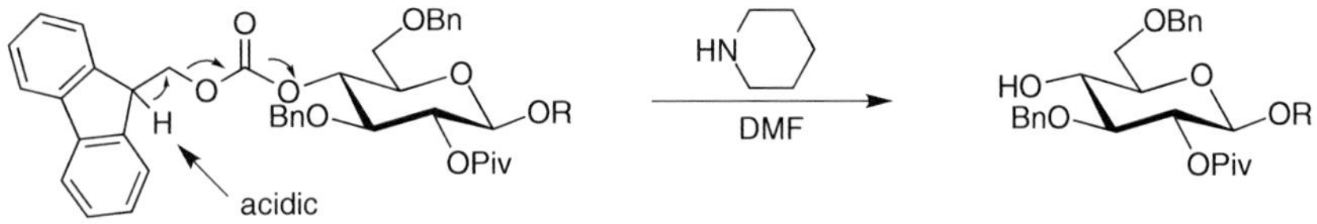

39 **40**

Acetate (Ac) Groups

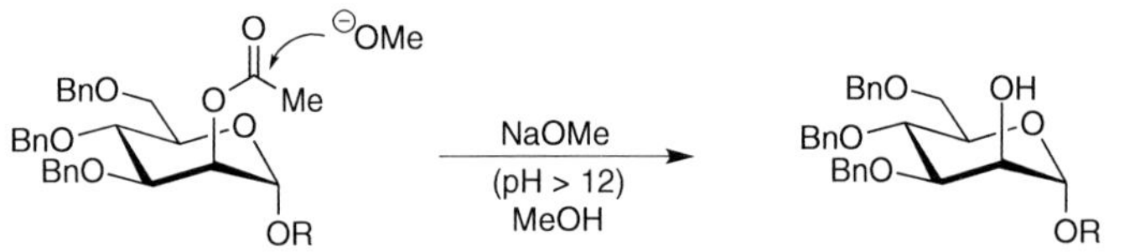

41 **42**

Levulinoyl (Lev) Groups

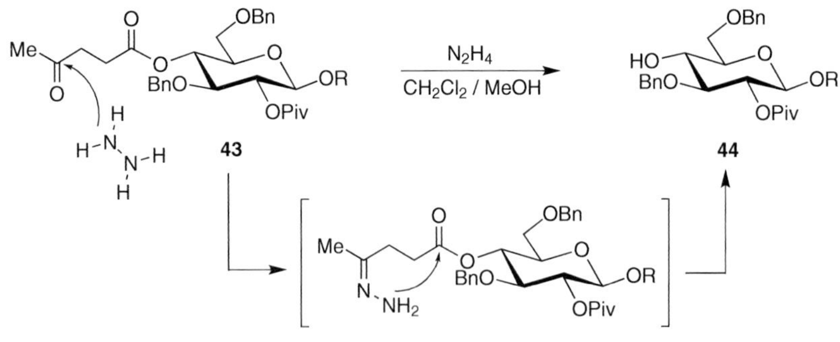

45

Figure 8.7 Four examples of temporary protecting groups (TBS, Fmoc, Ac, and Lev) and methods for their selective removal [5].

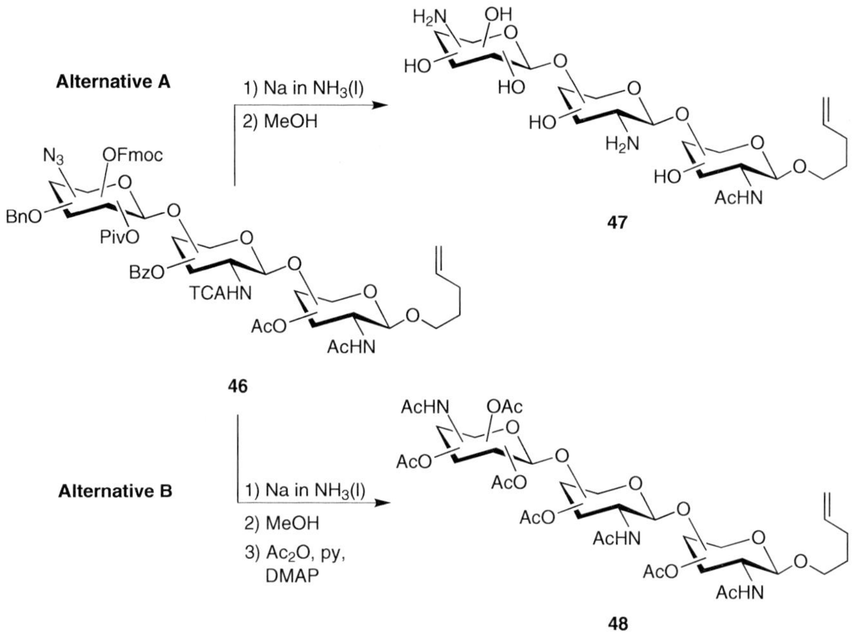

Figure 8.8 Birch reduction for global deprotection, either without (Alternative A) or with peracetylation (Alternative B) [5].

reduction might also work well without this preceding step. In order to facilitate the further purification process, the unprotected sugar, obtained by Birch reduction, is often peracetylated using acetic anhydride, pyridine, and dimethylamino pyridine (DMAP) (Fig. 8.8, Alternative B) [8,9].

The experimental set-up for the Birch reduction using sodium in liquid ammonia is relatively high. Minor impurities in the starting material may cause a serious decrease in yield. A major advantage compared to hydrogenolysis procedures is avoidance of heavy metals. Traces of such metals may cause problems in biological experiments employing the synthetic carbohydrates [5].

The second possibility to cleave the permanent protecting groups consists of a two-step procedure. In its first step, ester groups are removed by a base, before hydrogenolysis cleaves the Bn ethers [8]. In the latter case, palladium on charcoal or $Pd(OH)_2$ on charcoal (Pearlman's catalyst) are widely used as catalysts. The reaction itself takes place under an atmosphere of hydrogen. Commonly, the reaction proceeds well and is finished over night. If problems with the deprotection of several benzyl ethers (e.g. sterically hindered Bn groups) occur, hydrogen pressure up to 100 bar may be applied. It is evident that the conditions employed for hydrogenolysis also reduce carbon–carbon double bonds. If a carbon–carbon double bond is required as a handle for further attachment, the Birch reduction is the preferred method for global deprotection.

8.6 Case Study: Assembly of Blood Group A Antigen

To provide the reader with a feeling of the complete process of oligosaccharide synthesis, the following pages will demonstrate a prime example of glycan assembly. Figure 8.9 shows the blood group A antigen (49) with a thiol linker [5,15]. This carbohydrate is a branched tetrasaccharide consisting of galactosamine, fucose, galactose, and glucosamine. For the installation of a chemical handle (such as the thiol (–SH) group), which is necessary for surface immobilization or for attachment to proteins, a pentenyl group was attached to the reducing end of the carbohydrate. Such an olefinic moiety can be transformed easily at the end of the total synthesis into a thiol unit. The retrosynthetic analysis showing 5 building blocks (51–55) is depicted in Fig. 8.9, as permanent protecting groups for hydroxyls, Bn ethers, and Piv and Ac esters are employed. The amine moieties are protected in the case of galactosamine (51) as azide (N_3), and in the case of glucosamine (55), as trichloroacetamide (TCANH). In the latter case, a participating group, namely TCA, is necessary to achieve β selectivity, whereas a non-participating nitrogen functionality, namely the azide, is necessary to achieve α selectivity. For the branching galactose (52), two orthogonal temporary protecting groups are required. Fmoc in position 2 ensures β selectivity with its ability for neighboring group-participation, and can be cleaved later on by the weak base, piperidine, without affecting the levulinoate in position 3. In the case of the fucose building block (54) it is again important to have

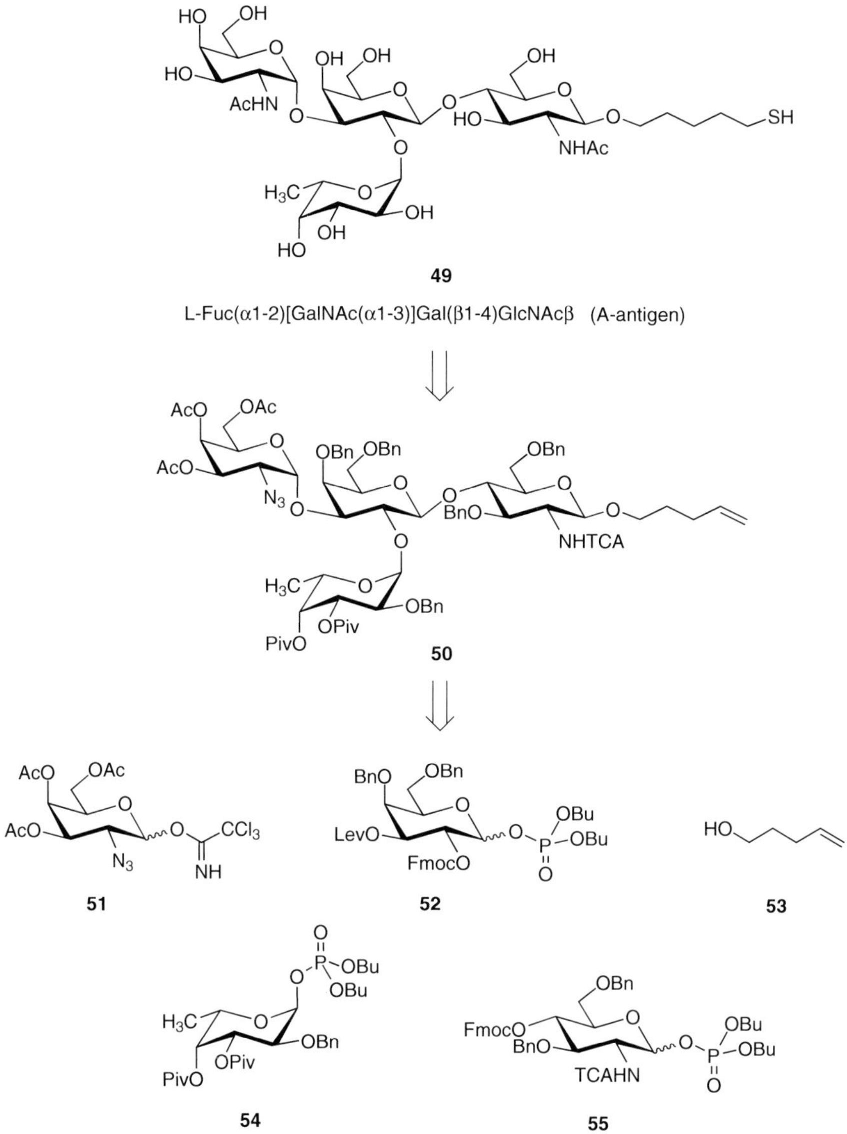

Figure 8.9 Retrosynthesis of the A antigen (49) with a thiol linker, building blocks (51–55) [5, 15].

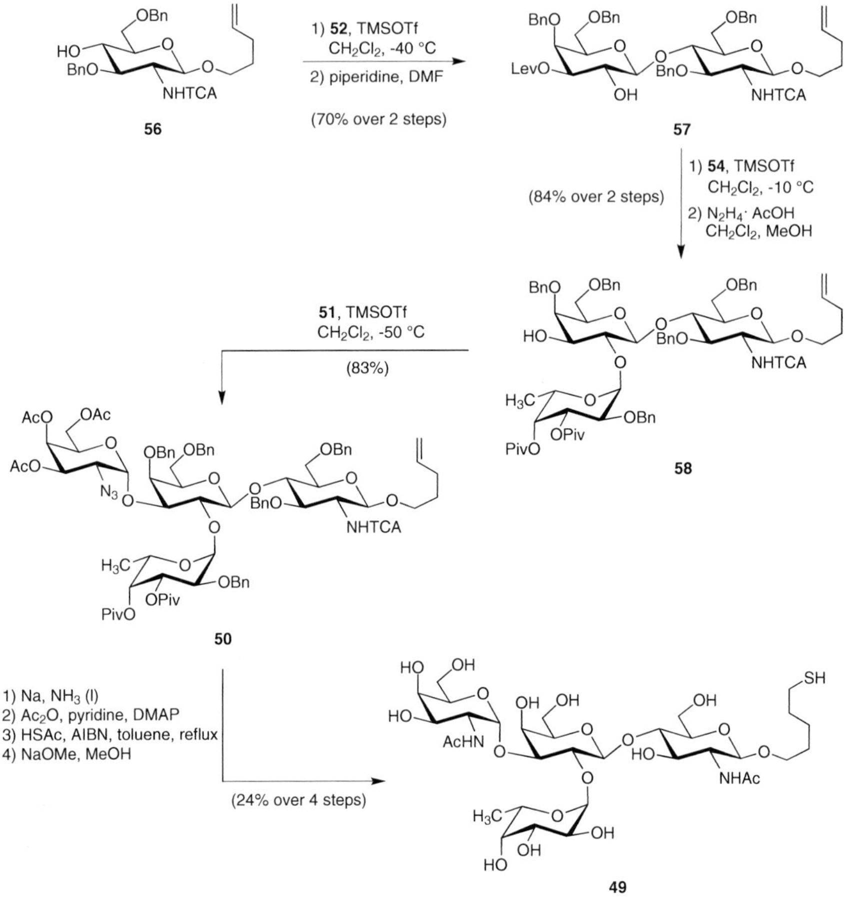

Figure 8.10 Synthesis of the blood group A antigen (49) with a thiol linker for further attachment [5, 15].

a non-participating group in position 2 in order to afford mainly the α product. For the anomeric leaving groups, a mixed strategy consisting of one trichloroacetimidate and three phosphates was chosen [5,15].

Figure 8.10 depicts all the steps from the first monosaccharide building block until the final target molecule [5,15]. The pentenyl functionalized monosaccharide (56) and galactosyl phosphate (52) were reacted in the presence of the Lewis acid, TMSOTf, as an activating agent, to produce a disaccharide whose Fmoc group was immediately deprotected to afford (57). Further fucosylation and removal of levulinoate in the presence of hydrazine yielded trisaccharide (58). In order to achieve high α selectivity, the last glycosylation was carried out at low temperatures. The completely protected tetrasaccharide (50) was subjected to Birch reduction. Sodium in liquid ammonia removed all permanent protecting groups and transformed the azide moiety, as well as the trichloroacetamide into "naked" amines, thus achieving global deprotection. The reaction with acetic anhydride and pyridine converted all hydroxyls to the corresponding acetates and all amines to the respective acetamido groups. This product is much less polar than the compound directly obtained after Birch reduction and can be more easily purified. A radical-initiated addition of thioacetic acid introduced the sulfur moiety [15]. In the final step, all acetates are cleaved to generate the native structure (49). Dialysis might be a good choice to remove salts, which might have accumulated during the last steps of the synthesis. As the very last step, lyophilization is commonly performed in order to remove all the water from the sample. The tetrasaccharide with the thiol linker can then further be used to attach it either to activated proteins (via Michael addition), to glass sides for microarray fabrication, or to gold surfaces.

8.7 Conclusion

The very first step in the assembly of a complex oligosaccharide is its retrosynthetic analysis and, afterwards, the choice of appropriate building blocks. These building blocks, commonly monosaccharide units with temporary and permanent protecting groups, and the good anomeric leaving group (e.g. trichloroacetimidates, phosphates, or thioethers), have to be built up before glycosylation reactions to connect the different monosaccharidic subunits that come into play. The creation of the corresponding building blocks is the most time-consuming part of the complete synthesis and, often, numerous steps have to be performed to generate the desired protecting group pattern. Therefore, integrated pathways with common intermediates to reach different building blocks are highly welcome. The choice of the protecting groups, especially the protecting group in position 2 of the monosaccharide, is highly influential in order to adjust the stereochemistry of the newly created stereogenic centre in position 1, the so-called anomeric centre. Neighboring participating groups, such as esters, form bridged intermediates of the intermediary oxonium ion and force the hydroxyl in the glycosylation reaction to attack from the other side. Thus, a 1,2-*trans*-stereochemistry is easily obtained. The assembly itself consists of a repetitive series of glycosylation and deprotection steps. During the glycosylation, the anomeric leaving group is activated, whereas in the next step, a temporary protecting group (often Ac or Fmoc) is cleaved to free the next hydroxyl needed for the subsequent glycosylation step. At the end of the synthesis, all permanent protecting groups, often benzyl groups, have to be removed (global deprotection) to afford the naked oligosaccharide. For this purpose, one uses either hydrogenation or Birch reduction (sodium in liquid ammonia).

References

1. Seeberger, P. H., Werz, D. B. (2007) *Nature* **446**, 1046–1051.
2. Werz, D. B., Ranzinger, R., Herget, S., Adibekian, A., von der Lieth, C. W., Seeberger, P. H. (2007) *ACS Chem. Biol.* **2**, 685–691.
3. Demchenko, A. V. Ed. (2008) *Handbook of Chemical Glycosylation.* Wiley-VCH (Weinheim).
4. Zhu, X., Schmidt, R. R. (2009) *Angew. Chem. Int. Ed.* **48**, 1900–1934.
5. Werz, D. B. (2012) In *Carbohydrate Microarrays: Methods and Protocols.* Chevolot, Y. Ed. Humana Press (New York). pp. 13–29.
6. Schmidt, R. R., Castro-Palomino, J. C., Retz, O. (1999) *Pure Appl. Chem.* **71**, 729–744.
7. Werz, D. B., Seeberger, P. H. (2005) *Angew. Chem. Int. Ed.* **44**, 6315–6318.
8. Lindhorst, T. K. (2007) *Essentials of Carbohydrate Chemistry and Biochemistry,* 3rd ed. Wiley-VCH (Weinheim).
9. Werz, D. B., Castagner, B., Seeberger, P. H. (2007) *J. Am. Chem. Soc.* **129**, 2770–2771.
10. Love, K. R., Seeberger, P. H. (2004) *Angew. Chem. Int. Ed.* **43**, 602–605.

11. Crich, D. (2010) *Acc. Chem. Res.* **43**, 1144–1153.
12. Crich, D. (2007) *ACS Symposium Series* **960**, 60–72.
13. Tamborrini, M., Werz, D. B., Frey, J., Pluschke, G., Seeberger, P. H. (2006) *Angew. Chem. Int. Ed.* **45**, 6581–6582.
14. Dudkin, V. Y., Miller, J. S., Danishefsky, S. J. (2004) *J. Am. Chem. Soc.* **126**, 736–738.
15. Horlacher, T., Oberli, M. A., Werz, D. B., Kröck, L., Bufali, S., Mishra, R., Sobek, J., Simons, K., Hirashima, M., Niki, T., Seeberger, P. H. (2010) *ChemBioChem* **11**, 1563–1573.

Chapter 9

When Enzymes Do It Better: Enzymatic Glycosylation Methods

Antoni Planas, Magda Faijes and Victoria Codera

Laboratory of Biochemistry, Institut Químic de Sarrià, Universitat Ramon Llull, 08017 Barcelona, Spain

9.1 Glycoside Bond Formation by CAZymes

Carbohydrates, in the forms of oligosaccharides, polysaccharides, and glycoconjugates (glycoproteins, glycolipids, glycosylated metabolites, etc) are key biomolecules in all living organisms. They play roles in a variety of biological functions as structural components of cell walls, reserves of stored energy, and mediators of cell–cell interactions, being central components in many important biological recognition processes, such as cell adhesion and signaling, cancer progression, host–pathogen interactions, and immune responses [1–5].

Synthetic tools to access natural and non-natural glycosides and glycoconjugates have become a central issue in glycochemistry and glycobiology. Both for industrial applications in diverse fields, such as medicine, pharmacology, food, cosmetics, etc., as well as for biological research in functional glycomic studies, there is the need for efficient synthetic approaches for structurally defined oligosaccharides and glycoconjugates. In contrast to conventional chemical synthesis that requires complex and tedious protection/deprotection manipulations and activation strategies to achieve regio- and stereo-selectivity (for a comprehensive review of chemical glycosylation methods, see [6]), enzymatic glycosylation offers many advantages, since enzymes provide selectivity, acting with perfect control of the anomeric configuration and with high regioselectivity, without the need of protecting groups.

The natural repertoire of carbohydrate acting enzymes (CAZymes, www.cazy.org, [7]) involved in the formation and breakdown of glycosydic bonds are glycosyltransferases (GTs), which catalyze glycoside bond formation by the transfer of a sugar unit from an activated glycosyl donor (nucleoside phosphate, phosphate, or lipid phosphate group) to an acceptor, and glycosidases (GH, glycoside hydrolases), which hydrolyze glycosidic bonds. The latter can also catalyze glycoside bond formation by reversal of the hydrolytic reaction or by transglycosylation reactions. In comparison with glycosyltransferase-catalyzed reactions that use complex sugar nucleotide donors and often stringent substrate specificities [8,9], additionally complicated by the limited availability or stability of the GT enzymes, glycosidase-catalyzed transglycosylations have many advantages, including the easy access to enzymes, their often high stability, the use of readily available donor substrates, and the relaxed substrate specificity for acceptors.

This chapter is intended to give a comprehensive overview of enzymatic glycosylation, mainly focused on the strategies to turn glycosidases, which naturally hydrolyze glycosidic bonds, into efficient synthetic tools for catalyzing glycoside bond formation. This will be developed in the following sections: mechanistic background of glycosidases in hydrolytic reactions, the use of glycosidases as synthetic tools, and glycosynthases as engineered glycosidases for glycoside bond formation.

9.2 Glycosidases in Hydrolytic Reactions: Mechanistic Background

Glycosidases (or glycoside hydrolases) fall into two mechanistic categories, according to the stereochemical outcome of the bond cleavage: inverting enzymes, which hydrolyze the glycosidic bond with inversion of configuration, and retaining enzymes, which do so with net retention of the anomeric configuration. Depending on the location of the glycoside bond to be hydrolyzed within the oligosaccharide chain, glycosidases are grouped either as *exo-* or *endo*-enzymes. *Exo*-glycosidases remove a monosaccharide (or disaccharide) from one of the ends of the chain (most commonly from the non-reducing end), whereas *endo*-glycosidases act on internal glycoside bonds within the oligosaccharide chain (either randomly or processively). This regiospecificity for bond cleavage is a consequence of the active site topology. In spite of the diversity of three dimensional folds of over 120 glycosyl hydrolase families [7], the carbohydrate binding site structure is reduced to three basic topologies: pocket, encountered in *exo*-glycosidases and optimal for recognition of a saccharide on the chain end; cleft, where an open grove structure allows binding of several sugar units in polymeric substrates and is commonly found in *endo*-acting enzymes; and tunnel, derived from the cleft topology with long loops that cover part of the cleft, as those found in cellobiohydrolases, where the resulting tunnel enables a polysaccharide chain to be threaded through it [10].

Both inverting and retaining glycosidases operate by general acid–base catalysis involving asparagines (Asp), glutamate (Glu), or tyrosine (Tyr) as catalytic residues, but differ in their mechanisms as a consequence of their active site topology [11,12]. Inverting glycosidases operate by a single-step mechanism in which a water molecule (with general base catalysis) affects a direct displacement at the anomeric center with protonic assistance by the general acid residue on the departing glycosidic oxygen (Fig. 9.1A). The catalytic residues are located approximately 10 Å apart from each other, allowing for the binding of the substrate and a water molecule in a ternary productive complex.

By contrast, retaining glycosidases follow a double displacement reaction via formation and hydrolysis of a glycosyl–enzyme intermediate, and have the catalytic residues located closer to each other at *c.a.* 5.5 Å (Fig. 9.1B). In the first step (glycosylation), the amino acid residue acts as the general acid and protonates the glycosidic

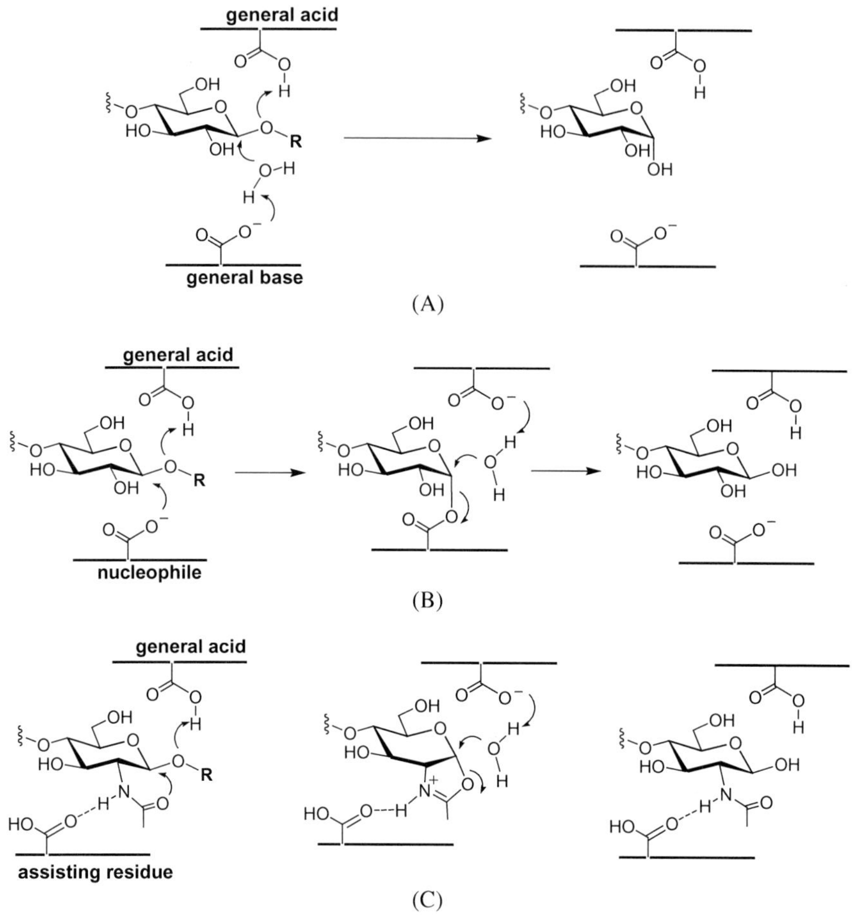

Figure 9.1 Glycosidase mechanism. A: Inverting glycosidase. B: Retaining glycosidase with enzyme nucleophile, via a glycosyl–enzyme intermediate in the two-step displacement mechanism. C: Retaining glycosidase by substrate-assisted catalysis, via an oxazolinium ion intermediate.

oxygen, while the deprotonated carboxylate functions as the nucleophile and attacks the anomeric center with concomitant carbon–oxygen breaking of the scissile glycosidic bond, leading to a covalent glycosyl–enzyme intermediate. The second deglycosylation step involves the attack by a molecule of water assisted by the conjugate base of the general acid residue, which renders the free sugar with overall retention of configuration, and the enzyme returns to its initial protonation state.

Some retaining glycosidases catalyzing glycoside bond hydrolysis in 2-acetamido sugars (i.e. GH18 chitinases, GH20 hexosaminidases, GH52 hyaluronidases, and GH85 *endo*-β-acetylglucosaminidases) follow a variation of the retaining mechanism (Fig. 9.1C); they lack the enzyme's catalytic nucleophile and operate by substrate-assisted catalysis, where the *N*-acetyl group of the substrate acts as an internal nucleophile, forming an oxazolinium intermediate, which is then attacked by a water molecule assisted by the conjugate base of the general acid residue to yield the product with net retention of the anomeric configuration [13]. In this case, an auxiliary residue that hydrogen bonds with the –NH of the acetamido group assists the formation of the oxazolinium intermediate.

The last two decades have been highly productive in biochemical, structural, and computational studies, which have provided many details that delineate the sequence of events in glycosidase catalysis: substrate binding, ring distortion, oxocarbenium ion-like transition states, glycosyl–enzyme intermediates, logarithmic acid dissociation constant (pKa) cycling of the general acid/base during catalysis, etc. [12,14–19].

9.3 Glycosidases as Synthetic Tools

9.3.1 *Retaining glycosidases with enzyme nucleophile*

Retaining glycosidases have been long used as biocatalysts for the synthesis of oligosaccharides and glycoconjugates. Classical approaches with wild type retaining glycosidases are based on reversal of their hydrolytic action. This is achieved either by displacing the equilibrium towards glycoside bond formation (thermodynamically controlled synthesis) or by using activated glycosyl donors (kinetically controlled transglycosylation).

The hydrolysis–synthesis equilibrium is balanced by approximately 4 kcal·mol^{-1} towards bond cleavage under ambient aqueous conditions. The thermodynamically controlled synthesis approach involves the shift of the equilibrium towards products by altering the reaction conditions, such as using high substrate concentrations, adding water-miscible organic co-solvents to reduce the activity of water, or performing the reaction at higher temperatures. Reverse hydrolysis in non-aqueous solutions are generally not useful, since the sugar substrates become insoluble. Most of the reported examples by the thermodynamic approach are for the preparation of simple glycosides (mainly disaccharides) or glycosides of hydrophilic alcohols using *exo*-glycosidases [20,21]. Novel approaches using lipid-coated enzymes in non-aqueous systems have improved the efficiency of reverse hydrolysis reactions, as it will be discussed in Section 9.4.

The most common kinetically controlled transglycosylation approach by retaining GHs with catalytic nucleophiles depends on the more rapid trapping of the reactive glycosyl–enzyme intermediate by the glycosyl acceptor than by water. Activated glycosyl donors (glycosyl fluorides, aryl glycosides, or oligosaccharides) are used to increase the concentration of the glycosyl–enzyme intermediate after the first step of the normal retaining glycosidase mechanism (Fig. 9.2A). Under appropriate conditions, a nucleophilic acceptor other than water can intercept the glycosyl–enzyme intermediate and form a new glycosidic bond (transglycosylation product) with the same anomeric configuration in the product than in the glycosyl donor (overall retaining reaction). Although glycoside bond formation may be favored kinetically, hydrolysis always remains favored thermodynamically, because the product itself is a substrate for the enzyme, and the equilibrium is slowly shifted towards hydrolysis. The competing hydrolysis over transglycosylation is often referred to as two processes: primary hydrolysis, meaning the hydrolysis of the glycosyl–enzyme intermediate (water attack in the deglycosylation step), and secondary hydrolysis, meaning hydrolysis of the transglycosylation product that is a substrate for the hydrolase activity of the enzyme (Fig. 9.2A).

Different strategies aimed to modify the reaction conditions and favor transglycosylation over hydrolysis have been developed with variable success, which is highly dependent on the particular enzyme and donor/acceptor pair.

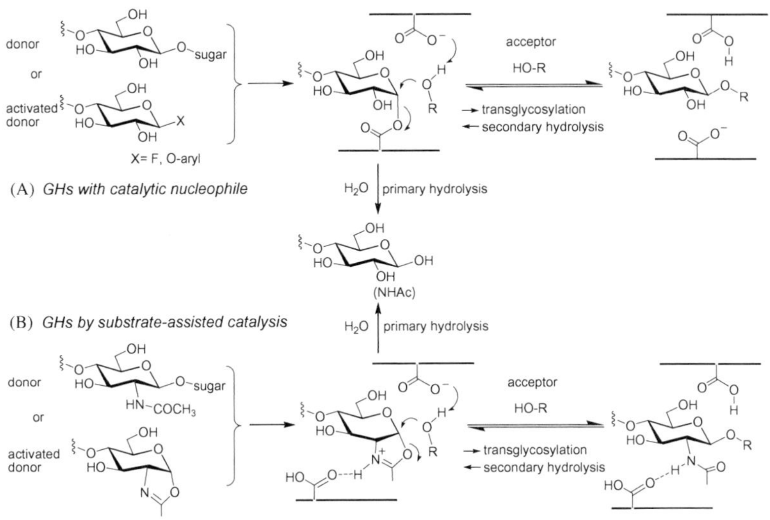

Figure 9.2 Transglycosylation reactions catalyzed by retaining glycosidases. (A) Glycosidases with catalytic nucleophile. (B) Glycosidases operating by substrate-assisted catalysis.

9.3.1.1 *The use of high acceptor concentration*

A higher acceptor concentration above its dissociation constant (K_d) for the acceptor site of the enzyme can be used to favor *trans*-glycosylation over the competing hydrolysis of the glycosyl–enzyme intermediate.

9.3.1.2 *The use of highly reactive glycosyl donors*

In many cases, disaccharides have been used as donors with *exo*-glycosidases (the overall reaction thus being a monosaccharyl transfer from the disaccharide to the acceptor), for example, in the production of functional food oligosaccharides [22,23]. However, the saccharide released as a leaving group can itself compete as an acceptor. Activated glycosides, most commonly aryl glycosides and glycosyl fluorides, are a better choice. First, the departed aglycone (fluoride ion or a phenol) is a poor acceptor and will not compete with the actual acceptor molecule. Secondly, the activated glycosyl donor is often a better substrate of the enzyme (in terms of the specificity constant (k_{cat}/K_M)) than the saccharide product formed, thus reducing the competing hydrolysis reaction of the product. In other words, a fast-reacting donor keeps the reaction times short, hence decreasing the extent of product hydrolysis, and tight binding of the donor (low Michaelis constant (K_M)) also inhibits product hydrolysis. In addition to glycosyl fluorides and aryl glycosides, other less common glycosyl donors, such as vinyl glycosides, nitropyridyl glycosides, or glycols, have also been used.

9.3.1.3 *The use of organic co-solvents or ionic liquids to reduce the activity of water and decrease hydrolysis*

Using organic co-solvents or ionic liquids reduces the activity of water and decreases hydrolysis. This is a general strategy when using water-miscible organic co-solvents, but the amount of co-solvent is often limited (typically 10–20% dimethyl sulfoxide (DMSO), dimethylformamide (DMF), acetonitrile, etc.) due to protein instability (denaturation at higher organic solvent concentrations) and reduced solubility of the carbohydrate substrates in organic solvents [24,25].

Recently, ionic liquids have been introduced as novel solvents in green chemistry, and shown to be a new class of organic co-solvents for enzyme-catalyzed reactions [26]. Ionic liquids are organic salts that are liquid at low temperature, characterized by their ionic character, practically have no vapor pressure, and are solvents that, compared to normal organic solvents, lead to novel and unusual chemical reactivity [27–29]. Glycosidase-catalyzed transglycosylation in ionic liquids have shown improved yields, not only because of hydrolysis reduction, but also because they decrease donor self-condensation reactions [30,31].

9.3.1.4 *Removal of the transglycosylation product from the reaction mixture*

Ideally, if the products formed were continuously and selectively removed from the reaction without affecting the enzyme, high *trans*glycosylation yields would be possible. Although limited to specific cases, successful examples include: crystallization or insolubilization of the product *in situ* in the presence of a high concentration of ammonium sulfate [32], absorption of the product on an activated charcoal column [33], or coupling the reaction to another enzymatic process, which will selectively transform the transglycosylation product into another compound that is no longer a hydrolyzable substrate of the glycosidase [34].

9.3.1.5 *Enzyme immobilization*

It is a widespread methodology in chemoenzymatic synthesis [35], also used with retaining glycosidases, that provides many advantages compared with free enzymes in solution. The different reaction conditions, such as higher enzyme stability in organic solvents and higher temperature, continuous operation in column reactors, and modified pH range of operation, has allowed to improve transglycosylation yields in a number of cases [36,37].

A particularly simple and effective immobilization method without a carrier support is the crosslinking of enzyme aggregates (CLEAs), where the precipitated enzyme (either crystalline or aggregate) is reacted with a cross-linking reagent (typically glutaraldehyde) to afford a catalyst with improved storage and operational stability towards denaturation by heat, organic solvents, and autoproteolysis [38]. A β-galactosidase-CLEA has been successfully used for the conversion of lactose to lactulose or the synthesis of galacto-oligosaccharides for use as prebiotic food additives.

9.3.1.6 *Lipid-coated glycosidases*

Enzyme–lipid complexes, in which the enzyme is coated with an amphiphilic lipid, are water insoluble systems, but soluble, stable, and active in organic media. They are prepared by mixing an aqueous solution of the enzyme and an aqueous dispersion of the lipid, followed by lyophilization of the precipitate. The resulting complex is insoluble in buffer solutions, but is freely soluble in most organic solvents. Lipid-coated glycosidases have shown to efficiently catalyze transglycosylation reactions in monophasic organic solvents or supercritical fluids with hydrophobic donors (aryl glycosides) and acceptors (hydrophobic alcohols), or in biphasic aqueous–organic systems when the glycosyl donor is a water-soluble saccharide. Lipid-coated β-mannosidase catalyzes the transglycosylation of p-nitrophenyl β-mannoside as a donor to 5-phenyl-1-pentanol in isopropyl ether at 30°C with yields close to 80% in condensation product [39]. With β-galactosidase from *Escherichia coli* in a two-phase aqueous–organic system, the lipid-coated enzyme and hydrophobic alcohol acceptors are in the organic phase, whereas lactose, as the donor, is mainly in the aqueous phase. The transglycosylation product was obtained in 66% yield and no galactose (from donor or product hydrolysis) was detected [40]. In supercritical carbon dioxide or fluoroform, lipid-coated β-galactosidase gave transglycosylation products in 80–90% yields, with rates faster than reactions in isopropyl ether as the organic solvent [41].

Among the glycoside hydrolase (GH) enzymes, a small number are natural transglycosylases (TGs). They essentially or exclusively catalyze transglycosylation in a water environment, despite the fact that they are closely related to their homologous hydrolases in the same GH families. Best studied naturally-occurring TGs include xyloglucan *endo*-transglycosylase (XET in GH16), sucrase-type enzymes using sucrose as a glycosyl donor (glucansucrases in GH13 and 7, and fructansucrases in GH68 and 32), cyclodextrin glucanotransferase (CGTase in GH13), and *trans*-sialidases (in GH33). How these enzymes modulate the transglycosylation/hydrolysis balance (T/H) in favor of transglycosylation is a key aspect of research with no general answer [42,43], and a subject which is guiding protein engineering approaches in the improvement of transglycosylation activity in GH enzymes [44,45]. Mutations on the acceptor subsites of the enzyme intended to increase hydrophobicity and improve acceptor binding while reducing water access, or mutations on the donor subsite to accumulate the glycosyl–enzyme intermediate to favor nucleophilic attack by the acceptor, are novel approaches under study addressed to engineer efficient transglycosylases from hydrolases [45].

9.3.2 *Stereo- and regio-selectivity of the reaction*

Another central issue in glycosidase-catalyzed transglycosylation is the stereo- and regio-selectivity of the reaction. Transglycosylation is a stereoselective process where the α or β linkage formed is dictated by the enzyme, but regioselectivity is not necessarily absolute or predictable. *Endo*-glycosidases (such as cellulases, β-amylases, 1,3-1,4-β-glucanases, xylanases, etc.) are highly regioselective in transglycosylation reactions, maintaining the same specificity as that of their corresponding hydrolytic reaction on oligo- and polysaccharides. For *exo*-acting glycosidases, however, mixtures of products with different glycosidic bonds are often observed. Selectivity can be controlled to some extent by the methods outlined below.

9.3.2.1 *Selecting an appropriate donor/acceptor combination*

The configuration of the anomeric center of the acceptor, or the aglycone group of the acceptor, influences the regioselectivity of glycosylation. α-Galactosidase, with p-nitrophenyl α-galatoside as a donor, gives mainly α-1,6-linked disaccharide with a methyl β-galactoside acceptor, whereas an α-1,3 disaccharide was obtained as the major product with methyl α-galactoside [46]. The effect of the aglycon is illustrated by the *E. coli* β-galactosidase-catalyzed *trans*-glycosylation of *p*-nitrophenyl β-galactoside with two acceptors, methyl β-xyloside, which exclusively gives the β-1,4 disaccharide, and benzyl β-xyloside, which provides the β-1,3 regioisomer as the major product [47]. Likewise, a 6-*O*-acetyl protected monosaccharide acceptor prevents 1,6-glycosidic bond formation, redirecting specificity towards the β-1,3-glycosyl linkage [48].

9.3.2.2 *The preference for different hydroxyl groups from the same acceptor*

Glycosidases from different species often show altered regioselectivities with preference for different hydroxyl groups of the same acceptor. For instance, galactosylation of D-xylose produces different mixtures of regioisomers depending on the source of the β-galactosidase. For the *Aspergillus oryzae* enzyme, 64% yield of the total disaccharide products was the β-1,4 isomer and only 7% was the β-1,2, whereas 58% of β-1,2 and 10% of β-1,4 disaccharides were obtained with the *Salix fragilis* enzyme [49]. Likewise, the type and amount of organic co-solvent added to the buffered aqueous solution may modify the proportions of different regioisomers [50].

9.3.3 *Retaining glycosidases by substrate-assisted catalysis*

In the case of retaining GH acting by substrate-assisted catalysis, the glycosyl donor for transglycosylation is an oligosaccharide substrate (*N*-acetylglucosamine (GlcNAc) sugar), or preferably, an activated sugar oxazoline, which binds the enzyme to directly form the reaction intermediate (Fig. 9.2B). Like the former GH enzymes with catalytic nucleophiles, transglycosylation efficiency depends on how the enzyme could accommodate a sugar acceptor to compete with water, and the same methodologies described above also apply in this case. Enzymes of GH18 (chitinases), GH20 (hexosaminidases), GH56 (hyaluronidases), and GH85 (*endo*-β-*N*-acetylglucosaminidases) families have been studied for oligosaccharide and glycoconjugate synthesis.

In recent years, methodologies for the construction of homogeneous *N*-glycopeptides and *N*-glycoproteins have attracted much attention. The biological roles of glycans in the context of these glycoconjugates is of fundamental importance, but functional glycomic studies are often hampered by the lack of homogeneous glycoconjugates, which are difficult to obtain in pure forms from natural sources. *Endo*-β-*N*-acetylglucosaminidases hydrolyze any *N,N′*-diacetylchitobiose moiety present in oligosaccharides that are *N*-linked to the asparagine residues of glycoproteins. Few of them possess transglycosylation activity in addition to their hydrolytic activity. They can transfer the intact released *N*-glycan to a suitable acceptor, such as an *N*-acetylglucosaminyl-peptide, yielding bioactive glycopeptides [51]. The major limitation of this approach is the rapid enzymatic hydrolysis of the products formed because of the inherent hydrolytic activity of the enzyme and the requirement for natural *N*-glycans to be used as donor substrates. A significant improvement was the use of synthetic sugar oxazolines as activated donor substrates for *endo*-β-*N*-acetylglucosaminidase (ENGase)-catalyzed transglycosylations [52]. Relevant examples of glycosylation remodeling of glycoproteins have been reported with Endo-M, Endo-A, and Endo-D of the GH85 ENGases family [53–57]. A further step forward in *N*-glycoprotein synthesis has been the ENGase-based glycosynthase approach, described in Section 9.5.

9.4 Artificial Polysaccharides by Glycosidase-Catalyzed Polymerization

Of particular relevance is the *in vitro* synthesis of polysaccharides by *endo*-glycosidases, where the regioselectivity is almost absolute as opposed to *exo*-glycosidases, providing a valuable methodology for the preparation of biopolymers. Four main different approaches have been developed.

9.4.1 *Polymerization by glycosidase-catalyzed transglycosylation of glycosyl fluoride donors in hydro-organic media*

Artificial cellulose was the first example produced by cellulase-catalyzed polycondensation of β-cellobiosyl fluoride in an acetonitrile/acetate buffer [58]. Other unnatural cellulose derivatives have been prepared by enzymatic polymerization of modified disaccharide donors with cellulases (i.e. alternating 6-*O*-methylated cellulose, thiocello-oligosaccharides, and alternating cellulose-chitin hybrids), and a hybrid cellulose–xylan polysaccharide with xylanases [59]. The methodology has been extended to other polysaccharides, such as amylose, xylan, and mixed-linked β-glucans with amylases, xylanases, and 1,3-1-4-β-glucanases, respectively [60]. In general, the polymers obtained have a low degree of polymerization (DP) (up to 30, molecular weight (Mw) 3–5 kDa) due to the product insolubility and hydrolase activity of the enzymes.

9.4.2 *Polymerization by reverse hydrolysis in non-aqueous media*

Using the above methodology with the activated β-cellobiosyl fluoride donor, pure cellulose was prepared successfully, but with a low DP, due to major problems that inevitably stop the chain-elongation reaction: the cellulase is gradually inactivated by acetonitrile, and the cellulose produced is not at all soluble in this reaction medium. A recent novel strategy is based on reverse hydrolysis catalyzed by a cellulase using unactivated cellobiose as the substrate, the utilization of a nonaqueous solvent system comprising lithium chloride in dimethylacetamide (DMAc), which dissolves cellulose to give a homogeneous condensation reaction, and functional preservation of cellulase in an aprotic solvent by a specific nonionic surfactant (lipid-coated enzyme) [61]. Further improvement was the addition of a protic acid in the aprotic organic solvent (protic acid-assisted lipid-coated-mediated reaction), which afforded cellulose polymers with a DP of up to 120 glucosyl units [62].

9.4.3 *Polymerization (oligomerization) by transglycosylation with unmodified sugar donors*

Besides activated glycosyl donors, unmodified sugars in which the leaving group is a saccharide may also self-condense in transglycosylation reactions. In general, these donors are less activated and hydrolysis predominates, requiring engineering of the reaction conditions (enzyme immobilization, organic solvents, etc.) and/or enzyme engineering (site-specific mutagenesis and directed evolution approaches) to make transglycosylation practical. Although these approaches have been extensively studied for the preparation of oligosaccharides and glycoconjugates (see Section 9.3.1), there are a few examples of application to polysaccharide synthesis. Relevant examples are the production of food-grade oligosaccharides [63] and poly-*N*-acetyl-lactosamines [64].

In contrast to hydrolytic glycosidases, there are natural transglycosylases, enzymes with the same mechanism as retaining glycosidases and classified in the same enzyme families, but whose predominant activity is transglycosylation. A number of them are involved in the biosynthesis of natural polysaccharides and, therefore, have found diverse applications in *in vitro* enzymatic syntheses. The most explored are glucansucrases and fructosyltransferases, which use sucrose as the donor substrate. Glucansucrases transfer the glucosyl moiety of sucrose to afford different β-glucans. Depending on the new linkages, amylose, dextran, mutan, alternan, and reuteran are synthesized. By contrast, fructosyltransferases transfer the fructosyl moiety, resulting in the synthesis of fructooligosaccharides and/or fructan polymers, such as levan or inulin [65–67]. Current efforts for enzyme development focus on side-directed mutagenesis, directed evolution, and chimeric enzyme construction to modulate the different activities and improve product properties and yields [68].

9.4.4 *Polymerization of oxazoline donors by retaining glycosidases acting by substrate-assisted catalysis*

Family GH18 (chitinases) and GH56 (hyaluronidases) are able to catalyze the ring-opening polyaddition of oxazoline donors and proceed consecutively. Polymeric products were obtained in yields between 20% and 75% with DPs from 6 to 88 depending on the enzyme and the reaction conditions. With this methodology, homopolysaccharides such as chitin, chitosan, and derivatives [69,70], and heteropolysaccharides, such as hyaluronic and chondroitin derivatives [71, 72], have been successfully produced.

9.5. Glycosynthases: Engineered Glycosidases

The major drawback of glycosidase-catalyzed transglycosylation is product hydrolysis due to the inherent hydrolase activity of the enzymes. Engineering the active site has led to a different strategy aimed at abolishing the hydrolase activity. The glycosynthase concept was introduced in 1998 on both an *exo*-glycosidase [73] and an *endo*-glycosidase [74]. These enzymes have their catalytic nucleophile substituted by a non-nucleophilic residue (typically a mutation of the Glu or Asp nucleophile to alanine (Ala), serine (Ser), or glycine (Gly)) and are able to catalyze transglycosylation from an appropriate activated glycosyl donor. Mutation of the catalytic nucleophile disables the enzyme as a hydrolase because no glycosyl–enzyme intermediate can be formed. However, an activated glycosyl donor with an anomeric configuration opposed to that of the donor substrate in the wild-type reaction (i.e. an α-glycosyl fluoride for a β-glycosidase) mimics the glycosyl–enzyme intermediate, and is therefore able to react with an acceptor (Fig. 9.3A). The cavity created in the active site by mutation of the carboxylate nucleophile in the wild-type enzyme by a smaller residue allows binding of the "inverted" glycosyl fluoride. Glycoside transfer to an acceptor is kinetically favored over hydrolysis, and the transglycosylation product is no longer hydrolyzed because of the lack of the catalytic nucleophile, and accumulates to give high transglycosylation yields.

Currently, a wide repertoire of glycosynthases has been developed, covering 16 different retaining GH families (for reviews see [60,75–79]). The initial concept was based on glycosyl fluoride donors with mutated retaining glycosidases at the catalytic nucleophile (Fig. 9.3A) and extended to other approaches including the *in situ* generation of an activated glycosyl donor with an external nucleophile, glycosynthases derived from glycosidases acting by substrate-assisted catalysis, and glycosynthases derived from inverting glycosidases (Fig. 9.3B–D).

9.5.1 *Glycosynthases from retaining GH with enzyme nucleophile and glycosyl fluoride donors*

Since the initial discovery, a number of enzymes were converted to glycosynthases upon mutation of the catalytic nucleophile to Ala, Gly, or Ser and using glycosyl fluoride donors, the efficiency of the different mutations being dependent on the particular enzyme. *Exo*-glycosynthases will add a monosaccharide on the non-reducing end of an acceptor, whereas *endo*-glycosynthases will be able to condense oligosaccharides for the synthesis of complex glycoconjugates

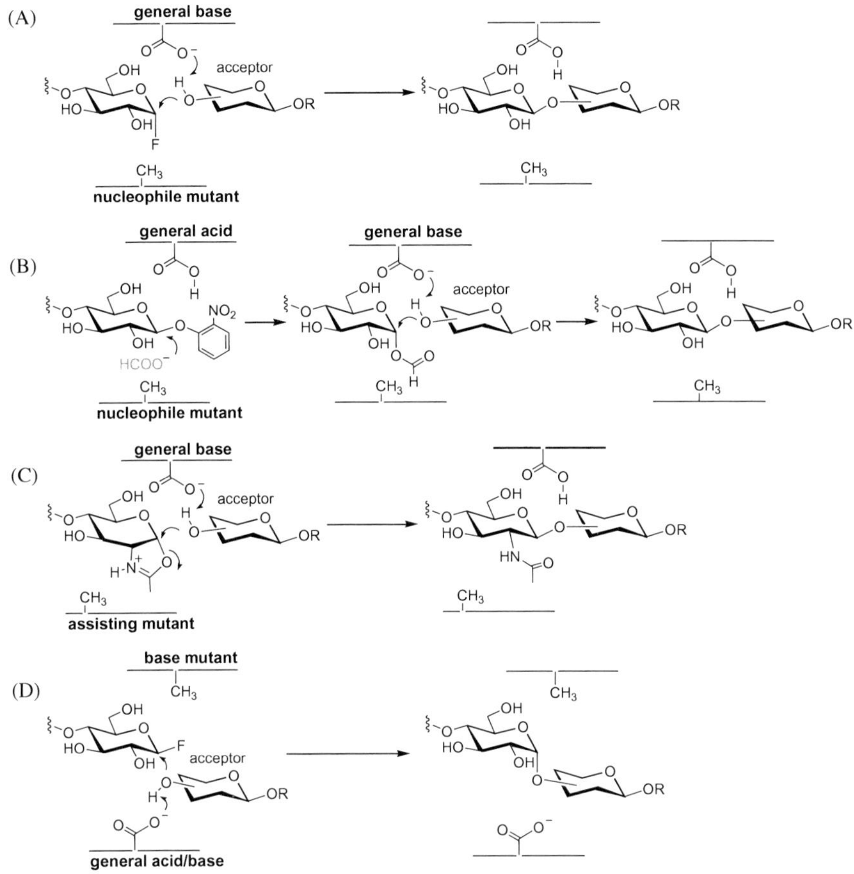

Figure 9.3 Glycosynthase enzymes. A: Nucleophile mutant of a retaining glycosidase with glycosyl fluoride donor. B: Nucleophile mutant of a retaining glycosidase (thermophilic) with *in situ* generation of the glycosyl donor by and exogenous nucleophile. C: Mutant at the assisting residue of a retaining glycosidase acting by substrate-assisted catalysis and a sugar oxazoline donor. D: Glycosynthase-like mutant derived from an inverting glycosidase (general base mutant of an inverting β-glycosidase in the example).

as they accept oligosaccharide donors. As observed for most glycosynthases, mixtures of transglycosylation products are often obtained as a consequence of polycondensation reactions. The glycosyl fluoride donor and the transglycosylation product may also act as acceptors, provided that the configuration of the condensable hydroxyl group on the non-reducing end of the donor has the same stereochemistry as the normal acceptor, then leading to self-condensation of the donor or elongation of the transglycosylation product, respectively. Kinetic analysis evidence reveals that whereas donor hydrolysis is negligible under normal reaction conditions, donor–acceptor condensation and donor self-condensation are competing processes with comparable rate constants [80]. It is therefore important to modulate the different reactions for an efficient use of the glycosynthase methodology in preparative synthesis of target oligosaccharides. As best illustrated with the *endo*-glycosynthases derived from the *Bacillus licheniformis* 1,3-1,4-β-glucanase and the *Humicola insolens* cellulase, different strategies to control the undesired reactions have been designed, based on the use of a temporary protecting group on the donor substrate to prevent self-condensation and elongation reactions. This may be achieved by chemical protection, such as the introduction of a tetrahydropyranyl group in the non-reducing end of the donor, which will be later removed from the glycosynthase product by acid treatment [81], or using a sugar as a protecting group, which is introduced in the non-reducing end of the donor by a glycosynthase prior to donor activation as glycosyl fluoride, and removed after the desired donor–acceptor condensation by treatment with the corresponding hydrolase. This is nicely illustrated in the synthesis of a library of xyloglucooligosaccharides that were designed for biochemical studies of plant xyloglucan endotransglycosylases [82,83]: (i) the oligosaccharide to be activated as the donor is first protected on the non-reducing end by a glycosynthase-catalyzed addition of a saccharide having a non-reactive non-reducing end (i.e. a lactosyl group, where the axial 4-OH will not act as the acceptor in the glycosynthase reactions); (ii) the resulting "protected" oligosaccharide is chemically activated as a glycosyl fluoride; (iii) the glycosynthase reaction is then performed for the transglycosylation of the donor to the oligosaccharide acceptor, leading to a single product since donor self-condenstion is prevented; and (iv) the saccharyl moiety on the non-reducing end (the lactosyl group in this example) is finally removed by a specific hydrolase to give the desired product.

Glycoconjugates have also been synthesized by the glycosynthase methodology. Glycosphingolipids, difficult to obtain from natural sources because of their structural heterogeneity, have been produced by an engineered endoglycosylceramidase. Mutation of the catalytic nucleophile renders a glycosynthase able to use sialydated lactosyl fluoride donors to glycosylate sphingosine derivatives [84]. Further directed evolution of this enzyme has expanded the synthetic scope to different sphingoid acceptors [85,86]. Likewise, glycosylated small-molecule natural products are accessible by the glycosynthase technology. The broad acceptor specificity of the glycosynthase mutant E197S from cellulase Cel7B has shown that it can directly glycosylate flavonoids with high efficiency [87].

9.5.2 *Glycosynthases from retaining GH with enzyme nucleophile and in situ generation of the glycosyl donor*

The glycosyl fluoride donors used in the "classical" glycosynthase approach are unstable at temperatures higher than 35–40°C, and have found limited applicability with enzymes from thermophilic organisms, since reaction conditions at low temperature are far from the expected high activity of thermophilic enzymes at high temperature. An alternative strategy was developed, consisting in the *in situ* generation of the glycosyl donor in the active site of the glycosynthase mutant. The nucleophile mutant is devoid of hydrolase activity, but activity on an aryl glycoside substrate can be rescued by an exogenous nucleophile. With sodium formate as an exogenous nucleophile, a transient glycosyl formate adduct is formed [88], which may act as a glycosyl donor for a glycosynthase-type reaction (Fig. 9.3B). The strategy has been applied to thermophilic β-glycosidases [89,90] and α-fucosidases with either formate or azide as an exogenous nucleophile [91,92]. However, other thermophilic glycosidases seem to operate better by the classical glycosynthase methodology with glycosyl fluoride donors at low temperatures than being rescued by exogenous formate at higher temperatures [93].

9.5.3 *Glycosynthases from retaining GH acting by substrate-assisted catalysis with oxazoline donors*

The conventional approach to generate a glycosynthase by knocking out the nucleophile residue could not be applied to glycosidases that operate by a substrate-assisted mechanism. The first study aiming to generate a glycosynthase from GH85 endo-β-N-acetylglucosaminidase (Endo-M) identified that a mutation of the assisting residue promoted transglycosylation with a sugar oxazoline donor, but did not hydrolyze the resulting glycopeptide product [94]. In the substrate-assisted mechanism, a conserved Asn residue in family GH85 enzymes plays a role in orientating the 2-acetamido group, and ultimately promoting the nucleophilic attack of the 2-acetamido group at the anomeric position to form the oxazolinium ion intermediate in the hydrolase mechanism (Fig. 9.3C). Mutation to alanine rendered an enzyme incompetent of oxazolinium ion formation with concomitant lost of hydrolase activity, but the mutant was still able to take a sugar oxazoline substrate for transglycosylation, where the newly formed glycoside bond was no longer hydrolyzed. The methodology has been extended to other GH85 enzymes (i.e. Endo-A, Endo-M, Endo-D), where mutations at the assisting residue in combination with activated sugar oxazoline donors may act as glycosynthases for the chemoenzymatic synthesis of *N*-glycoproteins [77,95]. Endo A has also shown to transfer *N*-glycans to natural products, such as glucosylated steroids and flavonoids [96]. Family GH18 enzymes, which comprise chitinases and endo-β-N-acetylglucosaminidases, also operate by substrate-assisted catalysis and are currently being engineered as glycosynthases [97,98].

9.5.4 *Glycosynthase-like activity from inverting glycosidases*

The glycosynthase concept is genuine to retaining glycosidases, acting via a double displacement mechanism: a mutant at an essential residue that renders an inactive hydrolase (catalytic nucleophile or assisting residue, depending on the retaining GH family), in combination with an activated glycosyl donor (with anomeric configuration inverted in relation to the normal substrate of the wt hydrolyse activity) that mimics the reaction intermediate (glycosyl–enzyme or oxazolinium ion intermediate). Inverting glycosidases operate via a single displacement mechanism (Fig. 9.3D) and enzyme engineering for glycosynthesis is based on a similar, but different concept. The first glycosynthase-like enzyme from an inverting glycosidase was that from a GH8 reducing-end-xylose releasing exo-oligoxylanase (Rex) [99]. Mutation of the catalytic base residue and use of an activated glycosyl fluoride donor with the same anomeric configuration as the normal hydrolysis product render a transglycosylating enzyme where the normal hydrolytic process is substantially slowed, but still significant. An alternative approach was the mutation of a neighboring residue that holds the nucleophilic water molecule that interacts with the general base, while keeping the general base residue intact, which gave higher transglycosylation yield because of the reduced hydrolysis [100]. Few other glycosynthase-like inverting glycosidases have been reported including a β-fucosidase [101] and a GH19 chitinase [102].

9.6 Artificial Polysaccharides by Glycosynthases

Glycosynthases expand the access to artificial polysaccharides (Section 9.4, endo-glycosidases by kinetically-controlled transglycosylation) with many advantages due to the lack of hydrolytic activity, which, in principle, may lead to higher molecular weight biopolymers in higher yields. When the glycosyl donor (glycosyl fluoride or oxazoline derivative) serves as a donor and an acceptor in the glycosynthase-catalyzed reaction, self-condensation proceeds consecutively leading to polymers. Only a few examples have yet been reported, but this is an emerging field for the preparation of new biomaterials with defined structures that either mimic natural polysaccharides or have unnatural structures and functionalities.

The E197A mutant of cellulase Cel7B from *H. insolens* was shown to polymerize α-cellobiosyl fluoride to give an insoluble β-1,4-glucan that is consistent with low molecular mass cellulose II [103]. This enzyme was also able to polymerize cellobiosyl fluoride donors modified at C-6^{II} with amino-, bromo-, or thioglycosyl groups to generate functionalized polysaccharides. These chemical functionalities open up possibilities to later modify and graft the polysaccharides with other biomolecules. The degree of polymerization of the generated polysaccharides are rather low (DP≈30) due to product insolubility.

Glycosynthase mutants of β-1,3-glucanase from barley was able to synthesize oligomers of β-1,3-linked glycosyl units [104]. The E231G mutant catalyzes the self-condensation of α-laminaribiosyl fluoride to yield an insoluble β-1,3-glucan with a DP of about 30, which is structurally similar to recrystallized curdlan. This mutant was also capable of self-condensing 3-thio-α-laminaribiosyl fluoride and producing oligosaccharides with alternating S- and O-linkages. The rate of formation and the degree of polymerization were lower than those of the homologous homo-polysaccharide, but the novel hetero-polysaccharide was more resistant to hydrolysis by β-1,3-glucanases.

Endo-xylanases from the GH10 family have been converted to glycosynthases that polymerize α-xylobiosyl fluoride to afford polymeric β-1,4-xylan with DP≈30 [105], as well as from GH52 β-xylosidase [106].

The glycosynthase derived from *B. licheniformis* 1,3-1,4-β-glucanase was able to polymerize glycosyl fluoride donors $(Glc\beta4)_n$Glcβ3GlcβF (n = 0–2), leading to artificial mixed-linked β-glucans with regular sequences and variable β-1,3 to β-1,4 linkage ratios [107,108]. With the E134A glycosynthase mutant, polymers had average Mw of 10–15 kDa, and the morphology of the water-insoluble polysaccharides was dependent on the repeating unit (highly porous crystalline spherulites for polymers from di- and tetrasaccharide donors, n = 0,2, whereas amorphous precipitates for polymers from trisaccharide donor, n = 1). With the more active E134S glycosynthase mutant, polymerization led to high molecular mass polysaccharides, where Mw was linearly dependent on enzyme concentration. Remarkably, a homo-polysaccharide $[4Glc\beta4Glc\beta4Glc\beta3Glc]_n$ with Mw as high as 30 kDa (n≈42) was obtained, which contained a small fraction of products up to 70 kDa, a value that is in the range of the molecular masses of low viscosity cereal 1,3-1,4-β-glucans, and among the largest products produced by a glycosynthase [108].

Xyloglucan polymers have also been obtained by different glycosynthases. The *H. insolens* GH7 glycosynthase, HiCel7B E197S, is capable of synthesizing non-galactosylated, XXXG-based homoxyloglucan up to Mw of 60,000 Da [G=Glcβ(1→4); X=Xylα(1→6) Glcβ(1→4); L=Galα(1→2)Xylβ(1→6)Glcβ(1→4)]. More complex xyloglucans have been achieved by the glycosynthase based on the GH16 xyloglucan hydrolase from *Tropaeolum majus* (TmNXG1), which is capable of synthesizing XLLG-based xyloglucan oligosaccharides, but not higher polymers [109]. Interestingly, the glycosynthase derived from GH5 *Paenibacillus pabuli* endo-xyloglucanase is a broad-specificity xyloglucan glycosynthase able to polymerize the fluoride donors XXXGαF and XLLGαF to afford high molecular mass xyloglucan polymers [110]. The product range could be further extended by combination with an α(1→2)-fucosyltransferase to achieve the *in vitro* synthesis of fucosylated xyloglucans typical of dicot primary cell walls [110]. Altogether, these glycosynthases provide a versatile method for the preparative synthesis of homogeneous xyloglucans with regular substitution patterns not available in nature.

The glycosynthase derived from rice β-glucosidase is an *exo*-glycosynthase that can produce long chain cellooligosaccharides. In contrast to the donor self-condensation ability of *endo*-glycosynthases, this enzyme exploits its long acceptor binding subsite and elongates condensation products to insoluble polymers with DP ≈11 with yields up to 80% [111].

GH18 chitinases, acting by substrate-assisted catalysis, are also amenable to behave as glycosynthases for polymerization reactions with oxazoline donors, but yet few studies have analyzed the usage of catalytically deficient GH18 chitinases as glycosynthases. *Bacillus circulans* chitinase (BcChiA1) mutants at Asp202 (which is assumed to be the stabilizer residue in the hydrolase mechanism) has shown to yield chito-oligomers of low molecular weight [97], but higher polymers (artificial chitin or chitosans) have not yet been produced by this technology.

9.7 Directed Evolution of Glycosynthases, Screening Methods

Improving the performance of glycosynthases with regard to increasing reaction rates and introducing novel substrate specificities is being addressed by enzyme-directed evolution approaches [112]. The success of these approaches not only depends on the strategy to create mutant libraries (random or focused libraries), but also depends on the selection or screening methodology for the property of interest. Few methods have been reported for the screening of directed evolution libraries of glycosynthases. Most of them rely on the detection of the product formed in the glycosynthase reaction: (i) coupled enzyme assay, in which the product formed by the glycosynthase is a substrate of another glycosidase that releases an easily detectable chromophoric aglycone (e.g. β-glucosidase from *Agrobacterium* [113]; (ii) enzyme-linked immunosorbent assay (ELISA)-based assay, where the acceptor is immobilized on a plate and the glycosynthase product is detected by a specific antibody (e.g. *Rhodococcus* sp. endoglycoceraminidase [85,114]); and (iii) chemical complementation, in which a yeast three-hybrid system is used to link glycosynthase activity to the transcription of a reporter gene making cell growth dependent on glycosynthase product formation (e.g. *Humicola insolens* cellulase 7B [115,116]. These methods have shown good screening capacity, yet at the price of a narrow applicability since they are specific for each enzyme. The first universal method to screen glycosynthase activity was a pH-based assay. It takes advantage of the hydrofluoric acid released as a by-product of the glycosynthase reaction, which is detected by color change in a pH indicator (e.g. β-xylosidase from *Geobacillus stearothermophilus*) [117]. Recently, a sensitive and robust general assay, independent of the enzyme specificity, has been developed. It is based on the quantification of the fluoride ion released as a by-product in all glycosynthase reactions with glycosyl fluoride donors by means of a specific chemical sensor to transduce fluoride concentration into a fluorescence signal, which enables the quantification of glycosynthase activity in high-throughput screening (HTS) format [118].

9.8 Conclusion

Glycosidases are versatile enzymatic tools for glycoside bond formation, where the enzyme provides stereo- and regioselectivity on free sugar substrates as opposed to chemical synthesis, which requires tedious protection/deprotection steps. With wild-type glycosidases, reversal of the hydrolase activity towards glycosynthesis is realized mainly under kinetically controlled conditions using activated glycosyl donors and engineering the reaction conditions to favor transglycosylation over hydrolysis. Current efforts are ongoing to improve the transglycosylation/hydrolysis ratio by enzyme engineering. A different strategy came with the introduction of glycosynthases, engineered glycosidases mutated at an essential catalytic residue that abolishes the hydrolase activity, but retains transglycosylation activity with specifically activated glycosyl donors (glycosyl fluorides or oxazoline derivatives). Both enzymatic strategies are an open field of research, with significant applications in glycosciences and industrial biotechnology, addressing the need for simple and efficient methodologies to access structurally defined oligosaccharides, polysaccharides, and glycoconjugates. Screening for new enzymes and protein engineering for novel specificities and improved performance are fast developing topics in the glycoscience and biocatalysis fields.

References

1. Dwek, R. A. (1996) *Chem. Rev.* **96**, 683–720.
2. Helenius, A., Aebi, M. (2001) *Science* **291**, 2364–2369.
3. Schwarz, F., Aebi, M. (2011) *Curr. Opin. Struct. Biol.* **21**, 576–582.
4. Dube, D. H., Bertozzi, C. R. (2005) *Nat. Rev. Drug Discov.* **4**, 477–488.
5. Adamczyk, B., Tharmalingam, T., Rudd, S. P. M. (2012) *Biochim. Biophys. Acta* **1820**, 1347–1353.
6. Demchenko, A. V. (2008) In *Handbook of Chemical Glycosylation: Advances in Stereoselectivity and Therapeutic Relevance.* Demchenko, A. V. Ed. Wiley-Vch Verlag GmbH & Co. (Weinheim). pp. 1–27.
7. Cantarel, B. L., Coutinho, P. M., Rancurel, C., Bernard, T., Lombard, V., Henrissat, B. (2008) *Nucl. Acids Res.* **37**, D233–238.

8. Luzhetskyy, A., Mendez, C., Salas, J. A., Bechthold, A. (2008) *Curr. Top. Med. Chem.* **8**, 680–709.
9. Harle, J., Bechthold, A. (2009) *Methods Enzymol.* **458**, 309–333.
10. Davies, G. J., Henrissat, B. (1995) *Structure.* **3**, 853–859.
11. Sinnott, M. L. (1990) *Chem. Rev.* **90**, 1171–1202.
12. McCarter, J. D., Withers, S. G. (1994) *Curr. Opin. Struct. Biol.* **4**, 885–892.
13. Tews, I., Vanscheltinga, A. C., Perrakis, A., Wilson, K. S., Dijkstra, B. W. (1997) *J. Am. Chem. Soc.* **119**, 7954–7959.
14. McIntosh, L. P., Hand, G., Johnson, P. E., Joshi, M. D., Korner, M., Plesniak, L. A., Ziser, L., Wakarchuk, W. W., Withers, S. G. (1996) *Biochemistry* **35**, 9958–9966.
15. Davies, G. J., Mackenzie, L., Varrot, A., Dauter, M., Brzozowski, A. M., Schülein, M., Withers, S. G. (1998) *Biochemistry* **37**, 11707–11713.
16. Davies, G., Sinnott, M. L., Withers, S. G. (1998) In *Comprehensive Biological Catalysis.* Sinnott, M. L. Ed. Academic Press Ltd. (London). pp. 119–209.
17. Withers, S. G. (2001) *Carbohyd. Polym.* **44**, 325–337.
18. Vasella, A., Davies, G. J., Böhm, M. (2002) *Curr. Opin. Chem. Biol.* **6**, 619–629.
19. Davies, G. J., Planas, A., Rovira, C. (2012) *Acc. Chem. Res.* **45**, 308–316.
20. Fernández-Mayoralas, A. (1997) *Top. Curr. Chem.* **186**, 1–20.
21. van Rantwijk, F., Oosterom, M. W., Sheldon, R. A. (1999) *J. Mol. Catal. B. Enzym.* **6**, 511–532.
22. Park, A. R., Oh, D. K. (2010) *Appl. Microbiol. Biotechnol.* **85**, 1279–1286.
23. Torres, D. P. M., Gonçalves, M. D. P. F., Teixeira, J. A., Rodrigues, L. R. (2010) *Compr. Rev. Food Sci F* **9**, 438–454.
24. Klibanov, A. M. (2001) *Nature* **409**, 241–246.
25. Carrea, G., Riva, S. (2000) *Angew. Chem. Int. Ed Engl.* **39**, 2226–2254.
26. Hernaiz, M. J., Alcantara, A. R., Garcia, J. I., Sinisterra, J. V. (2010) *Chemistry* **16**, 9422–9437.
27. Wasserscheid, P., Keim, W. (2000) *Angew. Chem. Int. Ed. Engl.* **39**, 3772–3789.
28. Roosen, C., Muller, P., Greiner, L. (2008) *Appl. Microbiol. Biotechnol.* **81**, 607–614.
29. Goldfeder, M., Fishman, A. (2014) *Appl. Microbiol. Biotechnol.* **98**, 545–554.
30. Sandoval, M., Cortes, A., Civera, C., Trevino, J., Ferreras, E., Vaultier, M., Berenguer, J., Lozano, P., Hernaiz, M. J. (2012) *RSC Adv.* **2**, 6303–6314.

31. Bayón, C., Cortés, A., Berenguer, J., Hernaiz, M. J. (2013) *Tetrahedron* **69**, 4973–4978.
32. Usui, T., Matsui, H., Isobe, K. (1990) *Carbohydr. Res.* **203**, 65–77.
33. Fujimoto, H., Hiroko, N., Ajisaka, K. (1988) *Agric. Biol. Chem.* **52**, 1345–1351.
34. Herrmann, G. F., Ichikawa, Y., Wandrey, C., Gaeta, F. C. A., Paulson, J. C., Wong, C. H. (1993) *Tetrahedron Lett.* **34**, 3091–3094.
35. Hanefeld, U., Gardossi, L., Magner, E. (2009) *Chem. Soc. Rev.* **38**, 453–468.
36. Hernaiz, M. J., Crout, D. H. (2000) *Enzyme Microb. Technol.* **27**, 26–32.
37. Stepankova, V., Bidmanova, S., Koudelakova, T., Prokop, Z., Chaloupkova, R., Damborsky, J. (2013) *ACS Catal.* **3**, 2823–2836.
38. Sheldon, R. A. (2011) *Appl. Microbiol. Biotechnol.* **92**, 467–477.
39. Mori, T., Okahata, Y. (1997) *Tetrahedron Lett.* **38**, 1971–1974.
40. Okahata, Y., Mori, T. (1998) *J. Mol. Catal. B. Enzym.* **5**, 119–123.
41. Mori, T., Li, M., Kobayashi, A., Okahata, Y. (2002) *J. Am. Chem. Soc.* **124**, 1188–1189.
42. Eklof, J. M., Shojania, S., Okon, M., McIntosh, L. P., Brumer, H. (2013) *J. Biol. Chem.* **288**, 15786–15799.
43. Saburi, W., Kobayashi, M., Mori, H., Okuyama, M., Kimura, A. (2013) *J. Biol. Chem.* **288**, 31670–31677.
44. Bissaro, B., Saurel, O., Arab-Jaziri, F., Saulnier, L., Milon, A., Tenkanen, M., Monsan, P., O'Donohue, M. J., Faure, R. (2014) *Biochim. Biophys. Acta* **1840**, 626–636.
45. Arab-Jaziri, F., Bissaro, B., Dion, M., Saurel, O., Harrison, D., Ferreira, F., Milon, A., Tellier, C., Faure, R., O'Donohue, M. J. (2013) *N. Biotechnol.* **30**, 536–544.
46. Nilsson, K. G. I. (1987) *Carbohydr. Res.* **167**, 95–103.
47. Lopez, R., Fernandez-Mayoralas, A. (1994) *J. Org. Chem.* **59**, 737–745.
48. MacManus, D. A., Vulfson, E. N. (2000) *Biotechnol. Bioeng.* **69**, 585–590.
49. Montero, E., Alonso, J., Canada, F. J., Fernandez-Mayoralas, A., Martin-Lomas, M. (1997) *Carbohydr. Res.* **305**, 383–391.
50. Murata, T., Usui, T. (2000) *Trends Glycosci. Glycotech.* **12**, 161–174.
51. Yamamoto, K. (2001) *J. Biosci. Bioeng.* **92**, 493–501.
52. Li, B., Zeng, Y., Hauser, S., Song, H., Wang, L. X. (2005) *J. Am. Chem. Soc.* **127**, 9692–9693.
53. Zeng, Y., Wang, J., Li, B., Hauser, S., Li, H., Wang, L. X. (2006) *Chemistry* **12**, 3355–3364.

54. Li, H., Li, B., Song, H., Breydo, L., Baskakov, I. V., Wang, L. X. (2005) *J. Org. Chem.* **70**, 9990–9996.
55. Rising, T. W., Claridge, T. D., Moir, J. W., Fairbanks, A. J. (2006) *ChemBioChem.* **7**, 1177–1180.
56. Rising, T. W., Heidecke, C. D., Moir, J. W., Ling, Z., Fairbanks, A. J. (2008) *Chemistry* **14**, 6444–6464.
57. Wang, L. (2008) *Carbohydrate Research* **343 (10–11)**, 1509–1522.
58. Kobayashi, S., Kashiwa, K., Kawasaki, T., Shoda, S. (1991) *J. Am. Chem. Soc.* **113**, 3079–3084.
59. Kobayashi, S., Sakamoto, J., Kimura, S. (2001) *Prog. Polym. Sci.* **26**, 1525–1560.
60. Faijes, M., Planas, A. (2007) *Carbohydr. Res.* **342**, 1581–1594.
61. Egusa, S., Kitaoka, T., Goto, M., Wariishi, H. (2007) *Angew. Chem. Int. Ed. Engl.* **46**, 2063–2065.
62. Egusa, S., Goto, M., Kitaoka, T. (2012) *Biomacromolecules* **13**, 2716–2722.
63. Chockchaisawasdee, S., Athanasopoulos, V. I., Niranjan, K., Rastall, R. A. (2005) *Biotechnol. Bioeng.* **89**, 434–443.
64. Murata, T., Honda, H., Hattori, T., Usui, T. (2005) *Biochim. Biophys. Acta* **1722**, 60–68.
65. Leemhuis, H., Pijning, T., Dobruchowska, J. M., van Leeuwen, S. S., Kralj, S., Dijkstra, B. W., Dijkhuizen, L. (2013) *J. Biotechnol.* **163**, 250–272.
66. Monsan, P., Remaud-Simeon, M., Andre, I. (2010) *Curr. Opin. Microbiol.* **13**, 293–300.
67. van Hijum, S. A., Kralj, S., Ozimek, L. K., Dijkhuizen, L., van Geel-Schutten, I. G. (2006) *Microbiol. Mol. Biol. Rev.* **70**, 157–176.
68. Andre, I., Potocki-Veronese, G., Morel, S., Monsan, P., Remaud-Simeon, M. (2010) *Top. Curr. Chem.* **294**, 25–48.
69. Ochiai, H., Ohmae, M., Kobayashi, S. (2004) *Carbohydr. Res.* **339**, 2769–2788.
70. Makino, A., Kurosaki, K., Ohmae, M., Kobayashi, S. (2006) *Biomacromolecules* **7**, 950–957.
71. Kobayashi, S., Fujikawa, S., Ohmae, M. (2003) *J. Am. Chem. Soc.* **125**, 14357–14369.
72. Kobayashi, S., Ohmae, M., Ochiai, H., Fujikawa, S. (2006) *Chemistry* **12**, 5962–5971.
73. Mackenzie, L. F., Wang, Q., Warren, R. A. J., Withers, S. G. (1998) *J. Am. Chem. Soc.* **120**, 5583–5584.

74. Malet, C., Planas, A. (1998) *FEBS Lett.* **440**, 208–212.
75. Perugino, G., Trincone, A., Rossi, M., Moracci, M. (2004) *Trends Biotechnol.* **22**, 31–37.
76. Shaikh, F. A., Withers, S. G. (2008) *Biochem. Cell Biol.* **86**, 169–177.
77. Wang, L., Huang, W. (2009) *Curr. Opin. Chem. Biol.* **13**, 592–600.
78. Cobucci-Ponzano, B., Moracci, M. (2012) *Nat. Prod. Rep.* **29**, 697–709.
79. Armstrong, Z., Withers, S. G. (2013) *Biopolymers* **99**, 666–674.
80. Faijes, M., Perez, X., Perez, O., Planas, A. (2003) *Biochemistry* **42**, 13304–13318.
81. Faijes, M., Fairweather, J. K., Driguez, H., Planas, A. (2001) *Chem. Eur. J.* **7**, 4651–4655.
82. Faure, R., Saura-Valls, M., Brumer, H., Royal Institute of Technology, Planas, A., Cottaz, S., Driguez, H. (2006) *J. Org. Chem.* **71**, 5151–5161.
83. Saura-Valls, M., Faure, R., Ragas, S., Piens, K., Brumer, H., Teeri, T. T., Cottaz, S., Driguez, H., Planas, A. (2006) *Biochem. J.* **395**, 99–106.
84. Vaughan, M. D., Johnson, K., deFrees, S., Tang, X., Warren, A. J., Withers, S. G. (2006) *J. Am. Chem. Soc.* **128**, 6300–6301.
85. Hancock, S. M., Rich, J. R., Caines, M. E., Strynadka, N. C., Withers, S. G. (2009) *Nat. Chem. Biol.* **5**, 508–514.
86. Rich, J. R., Withers, S. G. (2012) *Angew. Chem. Int. Ed. Engl.* **51**, 8640–8643.
87. Yang, M., Davies, G. J., Davis, B. G. (2007) *Angew. Chem. Int. Ed. Engl.* **46**, 3885–3888.
88. Viladot, J. L., Canals, F., Batllori, X., Planas, A. (2001) *Biochem. J.* **355**, 79–86.
89. Moracci, M., Trincone, A., Perugino, G., Ciaramella, M., Rossi, M. (1998) *Biochemistry* **37**, 17262–17270.
90. Cobucci-Ponzano, B., Perugino, G., Strazzulli, A., Rossi, M., Moracci, M. (2012) *Methods Enzymol.* **510**, 273–300.
91. Cobucci-Ponzano, B., Conte, F., Mazzone, M., Bedini, E., Corsaro, M. M., Rossi, M., Moracci, M. (2008) *Biocatal. Biotransform.* **26 (1–2)**, 18–24.
92. Cobucci-Ponzano, B., Conte, F., Bedini, E., Corsaro, M. M., Parrilli, M., Sulzenbacher, G., Lipski, A., Dal Piaz, F., Lepore, L., Rossi, M., Moracci, M. (2009) *Chem. Biol.* **16**, 1097–1108.
93. Pozzo, T., Plaza, M., Romero-GarcÃa, J., Faijes, M., Karlsson, E. N., Planas, A. (2014) *J. Mol. Catal. Enzym.* **107**, 132–139.
94. Umekawa, M., Huang, W., Li, B., Fujita, K., Ashida, H., Wang, L., Yamamoto, K. (2008) *J. Biol. Chem.* **283**, 4469–4479.

95. Yamamoto, K. (2013) *Biotechnol. Lett.* **35**, 1733–1743.
96. Huang, W., Ochiai, H., Zhang, X., Wang, L. X. (2008) *Carbohydr. Res.* **343**, 2903–2913.
97. Martinez, E. A., Boer, H., Koivula, A., Samain, E., Driguez, H., Armand, S., Cottaz, S. (2012) *J. Mol. Catal. B. Enzym.* **74**, 89–96.
98. Huang, W., Giddens, J., Fan, S. Q., Toonstra, C., Wang, L. X. (2012) *J. Am. Chem. Soc.* **134**, 12308–12318.
99. Honda, Y., Kitaoka, M. (2006) *Biol. Chem.* **281**, 1426–1431.
100. Honda, Y., Fushinobu, S., Hidaka, M., Wakagi, T., Shoun, H., Taniguchi, H., Kitaoka, M. (2008) *Glycobiology* **18**, 325–330.
101. Wada, J., Honda, Y., Nagae, M., Kato, R., Wakatsuki, S., Katayama, T., Taniguchi, H., Kumagai, H., Kitaoka, M., Yamamoto, K. (2008) *FEBS Lett.* **582**, 3739–3743.
102. Ohnuma, T., Fukuda, T., Dozen, S., Honda, Y., Kitaoka, M., Fukamizo, T. (2012) *Biochem. J.* **444**, 437–443.
103. Fort, S., Boyer, V., Greffe, L., Davies, G.J., Moroz, O., Christiansen, L., Schuelein, M., Cottaz, S., Driguez, H. (2000) *J. Am. Chem. Soc.* **122**, 5429–5437.
104. Hrmova, M., Imai, T., Rutten, S. J., Fairweather, J. K., Pelosi, L., Bulone, V., Driguez, H. (2003) *J. Biol. Chem.* **277(33)**, 30102–30111.
105. Sugimura, M., Nishimoto, M., Kitaoka, M. (2006) *Biosci. Biotech. Biochem.* **70**, 1210–1217.
106. Ben-David, A., Bravman, T., Balazs, Y. S., Czjzek, M., Schomburg, D., Shoham, G., Shoham, Y. (2007) *ChemBioChem.* **8**, 2145–2151.
107. Faijes, M., Imai, T., Bulone, V., Planas, A. (2004) *Biochem. J.* **380**, 635–641.
108. Perez, X., Faijes, M., Planas, A. (2011) *Biomacromolecules* **12**, 494–501.
109. Gullfot, F., Ibatullin, F. M., Sundqvist, G., Davies, G. J., Brumer, H. (2009) *Biomacromolecules* **10**, 1782–1788.
110. Spadiut, O., Ibatullin, F. M., Peart, J., Gullfot, F., Martinez-Fleites, C., Ruda, M., Xu, C., Sundqvist, G., Davies, G. J., Brumer, H. (2011) *J. Am. Chem. Soc.* **133**, 10892–10900.
111. Pengthaisong, S., Withers, S. G., Kuaprasert, B., Svasti, J., Cairns, J. R. (2012) *Protein Sci.* **21**, 362–372.
112. Kittl, R., Withers, S. G. (2010) *Carbohydr. Res.* **345**, 1272–1279.
113. Kim, Y. W., Lee, S. S., Warren, R. A., Withers, S. G. (2004) *J. Biol. Chem.* **279**, 42787–42793.
114. Hancock, S. M., Tarling, C. A., Withers, S. G. (2008) *Anal. Biochem.* **382**, 48–54.

115. Lin, H., Tao, H., Cornish, V. W. (2004) *J. Am. Chem. Soc.* **126**, 15051–15059.
116. Tao, H., Peralta-Yahya, P., Decatur, J., Cornish, V. W. (2008) *ChemBioChem* **9**, 681–684.
117. Ben-David, A., Shoham, G., Shoham, Y. (2008) *Chem. Biol.* **15**, 546–551.
118. Andres, E., Aragunde, H., Planas, A. (2014) *Biochem. J.* **458**, 355–363.

Chapter 10

Automated Chemical Synthesis of Oligosaccharides and Glycoconjugates

Salvatore G. Pistorio, Keith J. Stine and Alexei V. Demchenko

Department of Chemistry and Biochemistry,
University of Missouri–St. Louis, St. Louis, Missouri 63121, USA

10.1 General Introduction

It has long been known that carbohydrates are involved in a wide range of fundamental biological processes and are often called "essential molecules of life" [1]. Indeed, our life begins with fertilization, which takes place via a selective carbohydrate–protein recognition [2]. In addition, carbohydrates help us maintain a healthy lifestyle via their active involvement in anti-inflammation, immune responses, joint lubrication, cell growth, antigenic determination, etc. [3] The explosive growth of glycosciences in the recent years also brought the understanding of the roles of sugars as "molecules of death", due to their sizable contribution to harmful processes. Bacterial, parasitic, and viral infections; development and growth of tumors; metastasis; tissue rejection; septic shock; and congenital disorders are only a few to mention [4]. Elucidating the roles of carbohydrates in the pathogenesis of cancer, acquired immunodeficiency syndrome (AIDS), pneumonia, septicemia, diabetes, hepatitis, and malaria has been particularly stimulating for major efforts in the field of modern glycosciences [5–7].

It is already appreciated that both chemical and enzymatic syntheses [8] could lead to natural oligosaccharides of glycoconjugates. These compounds are typically needed for the study of their composition [9], conformation [10], interaction with other molecules [11–12], and biological roles [13]. Isolation from natural sources offers another viable means to obtain sugars. Only the synthetic approach, however, can provide unnatural glycomimetics, which are often of interest due to their therapeutic [4,14] and/or diagnostic potential [15–17]. Manufacturers have shown an interest in producing carbohydrate-based pharmaceuticals; however, even with significant progress, controlled chemical synthesis of complex carbohydrates remains difficult. Resultantly, the entire area of glycosciences remains somewhat underdeveloped, whereas other major classes of natural biopolymers, peptides, and polynucleotides can be studied by advanced methods and their synthesis can be automated using commercially available synthesizers. Low accessibility and high cost of complex carbohydrates hinders their large-scale development, and examples such as heparin analogs [18–19], oligosaccharide antibiotics [20–21], oligosaccharide- and glycoconjugate-based vaccines [22–26], etc. [6–7,27] are still rare.

Practically all complex carbohydrates have an oligomeric sequence wherein monosaccharide residues are linked via *O*-glycosidic linkages. This linkage is obtained by a glycosylation reaction that typically involves the nucleophilic displacement of a leaving group (LG; Fig. 10.1) on the glycosyl donor by a hydroxyl group of the acceptor (ROH) in the presence of an electrophilic promoter or activator [28]. The remaining functional groups of both components are temporarily masked with protecting (P) groups. A detailed mechanism has not yet been elucidated, although certain conventions and the involvement of the key reaction intermediates (A–C; Fig. 10.1) have been established [29–52]. In spite of significant recent progress, chemical glycosylation remains challenging due to the requirement of achieving complete stereocontrol and suppressing side reactions [28].

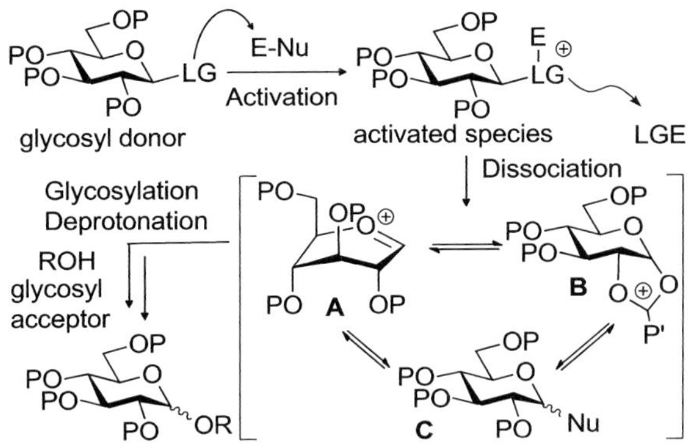

Figure 10.1 Chemical glycosylation.

The single-step glycosylation is only one challenge faced by researchers working on oligosaccharide synthesis: often, additional protecting and/or leaving group modifications between each glycosylation step are required. Because these manipulations become increasingly inefficient with larger oligosaccharides, many advanced strategies focus on minimizing the intermediate reaction steps [53]. A few relevant examples include Fraser–Reid's armed–disarmed approach [54–55], Nicolaou's selective activation [56], Danishefsky's glycal-based assembly [57], Kahne's one-pot synthesis [58], Roy's [59] and Boons' [60] active–latent concept, Ogawa's orthogonal strategy [61–62], Huang's preactivation concept [63–65], etc. [66–68].

In light of recent progress made in the areas of glycobiology and glycomics the discovery of efficient methods and technologies for the synthesis of complex carbohydrates is critical because reliable access to these compounds is becoming truly essential for research, development, and application in therapeutics and diagnostics. This chapter will discuss recent development of novel technologies for the automation of oligosaccharide synthesis, both on solid phase and in solution. Following numerous attempts to streamline oligosaccharide synthesis using predictive or computational reactivity assessment performed by the Fraser–Reid, Ley, and Wong groups in the nineties, the first example of the truly automated synthesis is credited to Seeberger and co-workers, who reported the first automated synthesizer in 2001. In this application, they used a peptide synthesizer ABI 433 that, after a few modifications, was optimized for the synthesis of oligosaccharides [69–71]. From that moment, an automated approach to oligosaccharide synthesis became possible, and a new era in carbohydrate research had been plotted. Another important milestone is the commercialization of this automated platform, Glyconeer 2.1, that is now available via GlycoUniverse. It should be mentioned that Activotec has also developed a peptide synthesizer-based oligosaccharide synthesizer that is now commercially available.

10.2 Automation of the Polymer-Supported Synthesis

Solid phase synthesis using insoluble polymer supports [72–73] has been widely used in the preparation of many organic molecules [72–75], including oligopeptides [76] and oligonucleotides [77]. The early example of solid phase peptide synthesis by Merrifield [78] was followed by oligosaccharide synthesis by Fréchet and Schuerch that emerged shortly thereafter [79]. Since the early attempts, the polymer-supported oligosaccharide synthesis has become a viable means for the rapid synthesis of oligosaccharide sequences without the necessity of purifying (and characterizing) the intermediates [80–83]. Another important advantage of oligosaccharide synthesis on solid support is the ease of excess reagent removal (usually can be achieved by filtration and rinsing).

There are two main strategies for solid phase saccharide synthesis that differ in the type of the attachment (Fig. 10.2). In strategy A, the glycosyl acceptor unit is bound to the solid support either at the anomeric position or at another suitable hydroxyl. In this case, an excess of the glycosyl donor and promoter are in the solution. In approach B, the glycosyl donor unit, linked to the solid support via a suitable hydroxyl group, is reacted with the solution phase acceptor. Two-directional techniques, combining approaches

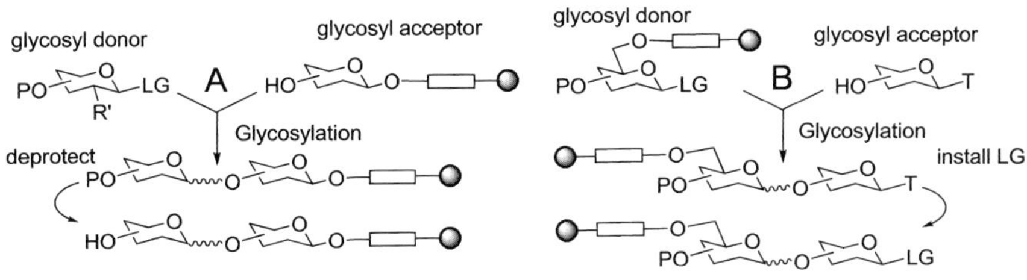

Figure 10.2 Glycosylation on polymer support.

A and B, are also known [84–85]. To improve the operational simplicity of the oligosaccharide assembly, Seeberger developed an automated approach, which is based on strategy A: Using a solid phase acceptor and a liquid phase donor [69]. Since this approach is already well documented in the literature, including a number of review articles [27,70,86–89], herein only a brief historical perspective with a few representative examples and milestones will be discussed.

10.2.1 *Peptide synthesizer-based automation of polymer-supported synthesis*

The automation was initially accomplished by using a peptide synthesizer that had been modified to perform oligosaccharide assembly at low temperature [69]. Merrifield resin was chosen as the solid support, for its good swelling properties in organic solvents commonly used in glycosylation. The design of the linker between the sugar and the solid support is of key importance. In this application, the polymer support was equipped with an olefin-type linker due to its high stability in both basic and acidic media. Thus, the octene-diol-functionalized resin (1) was glycosylated with trichloroacetimidate donor (2) (10-fold excess) in the presence of trimethylsilyl trifluoromethanesulfonate (TMSOTf; Fig. 10.3). The use of an acetyl group at C-2 of the glycosyl donor (2) was chosen to control the anomeric stereoselectivity for the formation of oligosaccharides (3), and to be selectively deprotected with sodium methoxide (NaOMe) in methanol/dichloromethane to obtain the next generation glycosyl acceptor (4). To reduce the formation of side products and

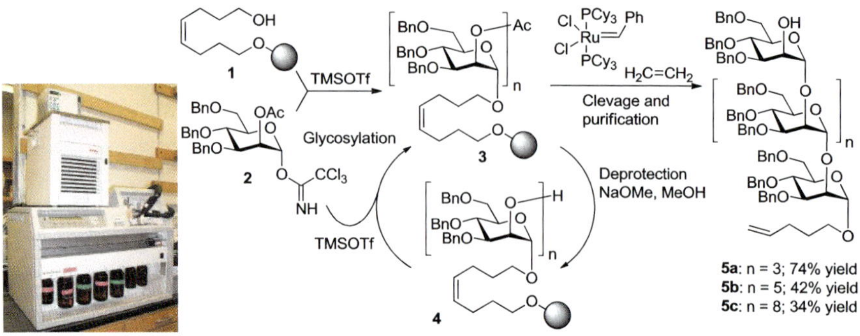

Figure 10.3 First automated oligosaccharide synthesis.

truncated sequences, both glycosylation and deprotection steps were performed twice. This cycle of glycosylation–washing–deprotection was repeated to obtain the desired linear oligosaccharide sequence. After that, the linker was removed using Grubbs' catalyst to afford penta-, hepta- and decasaccharides (5a–c) as pentenyl glycosides. For instance, using the automated approach, heptasaccharides (5b) were obtained in 24 hours in 42% overall yield, whereas the manual synthesis was much more laborious (14 days) and less efficient (9% overall yield) [90].

Working on expanding this promising technology, Seeberger and co-workers approached the synthesis of oligosaccharides containing other important residues and linkages, including aminosugars [91–94], sialic acids [95], furanosides [96], 1,2-*cis* glycosides [97], glycopeptides [98], and branched oligosaccharide sequences [69]. Codee *et al.* reported the synthesis of β-mannosides [99–100] using essentially the same automation platform. The acquired knowledge was applied to a very effective synthesis of Globo-H hexasaccharide. This oligosaccharide is expressed on surfaces of different types of cancer cells, and this represents a key target for the development of anti cancer vaccines and therapeutics [24,27,101–105]. The appreciation of the biological significance of Globo-H stimulated the interest in the synthetic community and the antigen has been synthesized by different methods [65,85,97, 106–108].

This effort culminated in the automated synthesis by Seeberger, which was the first attempt to conquer the challenge on 1,2-*cis* glycosidic bond formation using the automated approach [97]. Careful refinement of reaction conditions allowed for 1,2-*cis* galactosylation in dichloromethane–ether and the Globo-H sequence was assembled as depicted in Fig. 10.4. First, the glycosyl phosphate donor (6) was linked to the resin (1) via glycosylation using TMSOTf (repeated once) as the promoter, followed by deprotection of the fluorenylmethoxycarbonyl (Fmoc) substituent with piperidine (repeated twice) to provide a polymer-bound acceptor. In this synthesis, Fmoc group was chosen as a protecting group, for its high stability in acid and easy cleavage in the presence of mildly basic amines. In addition, after cleavage, Fmoc provides dibenzofluorene adduct that can be used to monitor the efficiency of deprotection via quantitative colorimetric assay [109]. The general synthetic protocol consists of repetitive cycles of glycosylation using

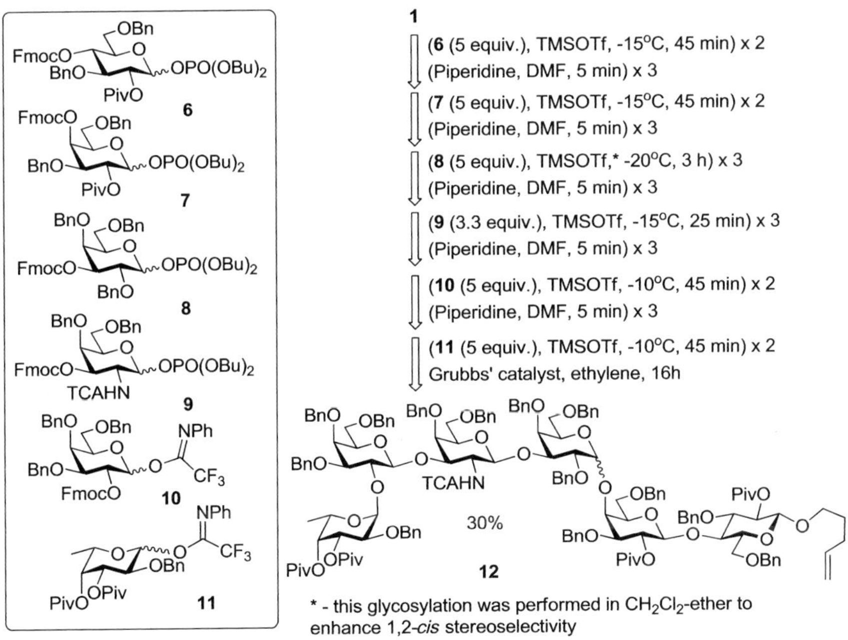

Figure 10.4 Automated synthesis of Globo-H hexasaccharide.

either glycosyl phosphate (6–9) or glycosyl *N*-phenyl trifluoroacetimidate donors (10,11) followed by the deprotection with piperidine. The final product (12) was obtained under an atmosphere of ethylene in the presence of Grubbs' catalyst [110] with an overall yield of 30%.

10.2.2 *Dedicated oligosaccharide synthesizer for the automation of polymer-supported synthesis*

More recently, Seeberger and co-workers introduced "the first fully automated solid phase oligosaccharide (OS) synthesizer" [111]. The new automated synthesizer was developed by combining the following equipment: syringe pumps, solenoid valves, a cryostat with the temperature controlling system capable of maintaining the reaction temperature between –50°C and 90°C, and a fraction collector (Fig. 10.5). The reaction vessel is equipped with the porous glass filter that allows for delivering the inert gas to maintain anhydrous reaction conditions. The gas is also used for rapid removal of the solution phase reagents and solvents. The glycosylation and deprotection protocol can be set using a computer program. This allows for the full automation of reactions, temperature control, cleavage, and the collection of the released oligosaccharide product.

To illustrate the versatility of this new synthesizer, Seeberger and co-workers performed the synthesis of a few oligosaccharides, including a high mannose type branched glucan (18), as depicted in Fig. 10.5. To ensure the feasibility of the complete automation, a new type of linker (13) stable under the most common glycosylation and protecting group removal conditions was introduced. The synthesis of pentasaccharide (18) began with the assembly of the chitobiose portion using two glycosylation–deprotection cycles with the glycosyl donor (14). A challenging β-mannosidic linkage was then obtained with a donor (15), equipped with a 2-(hydroxycarbonyl)benzyl leaving group developed by Kim [112]. Subsequently, the treatment with tetrabutylammonium fluoride (TBAF) and the selective opening of benzylidene acetal gave 3,6-diol. The latter was then subjected to bis-mannosylation using a donor (16) to

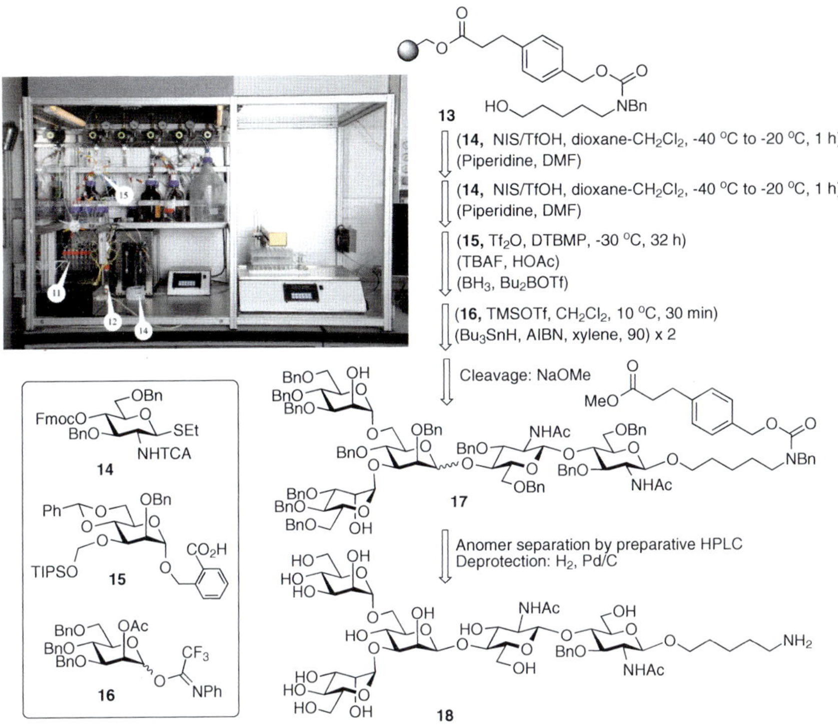

Figure 10.5 Automated synthesis of branched oligomannan (18) using second generation synthesizer.

obtain the final branched sequence. Cleavage from the solid support was affected in the presence of NaOMe to afford a precursor (17) ($\alpha/\beta = 1/3$). The undesired α-linked anomer formed as the by-product during the β-mannosylation step was separated off by preparative high-performance liquid chromatography (HPLC), and the resulting diastereomerically pure compound was deprotected by hydrogenation to afford the target pentasaccharide (18) in 3.5% yield overall.

Very recently, making use of essentially the same technology, Seeberger and co-workers obtained an α-(1→6)-linked linear oligosaccharide sequence containing 30 mannose residues [113]. To undergo this impressive synthesis, a modified Merrifield resin (19) equipped with the photocleavable *p*-nitrophenyl linker was repeatedly glycosylated with a phosphate donor (20) in the presence of TMSOTf (Fig. 10.6). The capping of the unreacted hydroxyls (21) was affected by acetylation with acetic anhydride (Ac_2O) in the presence of pyridine. The deprotection of the 6-*O*-Fmoc substituent was then affected with piperidine to afford a glycosyl acceptor of the next generation (22). Benzoyl protecting groups were used at other positions to ensure high stability, stereoselectivity, and yields throughout the synthesis. Upon execution of 29 glycosylation cycles with the interim capping–deprotection steps, the resulting 29-mer was then glycosylated with the donor (23). The latter was equipped with the spacer moiety to allow for a very

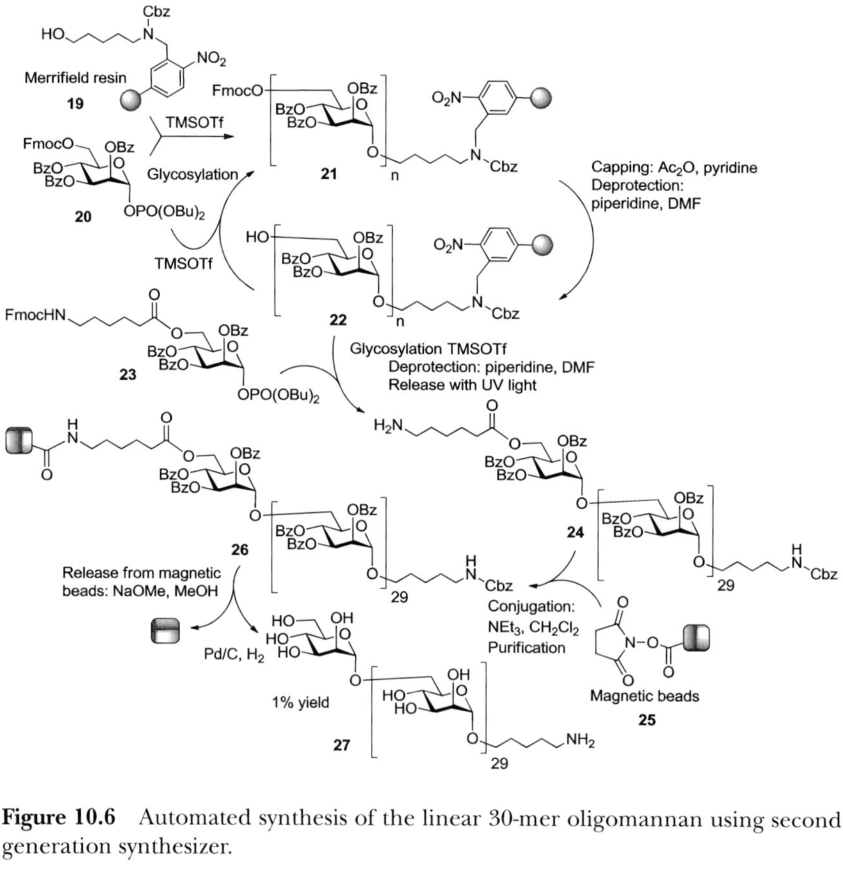

Figure 10.6 Automated synthesis of the linear 30-mer oligomannan using second generation synthesizer.

effective cap-and-tag purification technique [114–117]. Thus, upon release from the polymer support using ultraviolet (UV) light, the resulting 30-mer (24) was conjugated to magnetic beads via the amino group of the terminal mannose unit equipped with an ε-aminocaproic ester spacer. This approach allowed for magnetic separation of the tagged oligosaccharide (26), followed by simple washing with dichloromethane and methanol. Subsequent release from the magnetic beads and concomitant debenzoylation with sodium methoxide in methanol followed by hydrogenation to remove a carboxybenzyl (Cbz) group led to the fully uprotected 30-mer manno-oligosaccharide (27). Resultantly, the longest oligosaccharide sequence ever made by chemical synthesis was obtained in a total yield of 1%, which accounts for 96% average yield per synthetic step.

10.2.3 *HPLC-based automation of polymer-supported synthesis*

Demchenko, Stine, and their co-workers developed a new experimental set-up based on an unmodified HPLC instrument. In brief, an Omnifit chromatography column was packed with the preswelled polymer resin, TentaGel-NH_2. The column was then connected to the HPLC system, consisting of a reciprocating pump containing three chambers, a variable UV range detector, and a computer with standard HPLC-operating software installed (Fig. 10.7) [118]. The column was packed with the glycosyl acceptor loaded on resin, purged with the solvent, and then two separate solutions containing the glycosyl donor and promoter were delivered concomitantly. After a relatively short reaction time, typically 30–60 minutes, the system was purged (washed) with solvent. At this time, the resin is loaded with the disaccharide derivative, and the oligosaccharide elongation can be continued via alternating deprotection–glycosylation steps. Although the versatility of the HPLC-assisted method is still to be demonstrated, the synthesis of linear pentasaccharide (30) has already been accomplished (Fig. 10.7). For the synthesis of this molecule, the TentaGel-NH_2

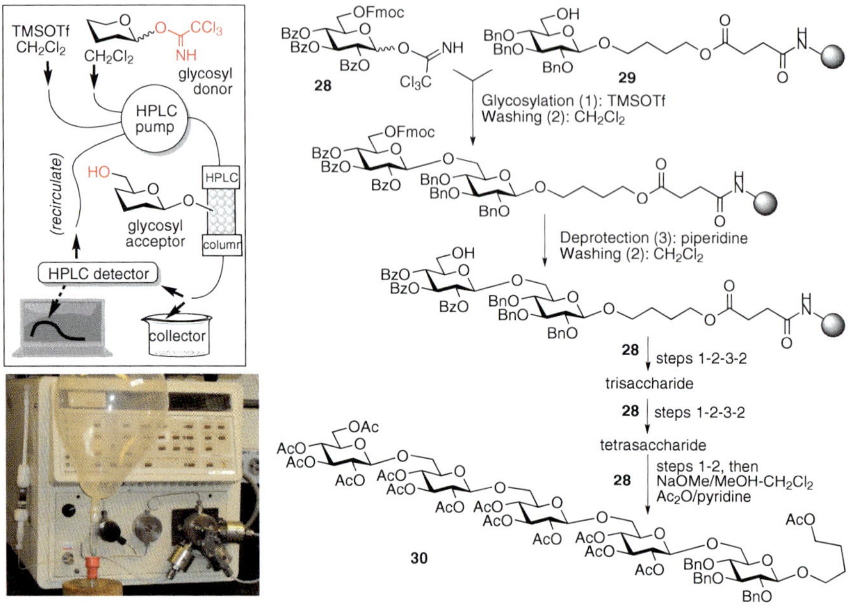

Figure 10.7 HPLC-assisted automated oligosaccharide synthesis.

resin loaded glycosyl acceptor (29) was packed into the Omnifit column (alternatively, the loading could be performed using the HPLC-based set-up). The elongation of the glucan sequence was then performed using a glycosyl donor (28) equipped with a 2-*O*-benzoyl group to ensure the stereoselectivity control and 6-*O*-Fmoc as the selectively removable temporary substituent. Two reagent bottles, one with a solution of donor (28) and one with a solution of promoter (TMSOTf), were then used to deliver the solutions to the column via dedicated pumps. After one hour, the two pumps were stopped and the system was purged with dichloromethane using a separate pump. Subsequently, a solution of piperidine in dimethylformamide (DMF) was delivered to perform Fmoc-deprotection, with a typical reaction time of five minutes, after which the system was purged with dichloromethane. The cycle (glycosylation–washing–deprotection–washing) was repeated until the desired oligosaccharide was assembled. The latter was then cleaved off from the polymer support by using a recirculating solution of NaOMe in methanol–dichloromethane. Resultantly, pentasaccharide (5) was isolated in 62% yield with an important reduction in time with respect to the same pentasaccharide obtained using a manual approach (seven hours vs. seven days).

It was also demonstrated that all steps of the HPLC-assisted synthesis could be monitored using a standard HPLC detection system set to record changes in the UV absorbance of the solution eluting off the column [118]. A solution of reagents can be recirculated to reduce the amount of reagents and to ensure complete conversion at each reaction step. This experimental set up offers the following advantages in comparison to that of conventional (manual) oligosaccharide synthesis on polymer supports: faster reaction times, real-time reaction monitoring using an HPLC detection system, and all steps and sequences can be automated using the standard HPLC-managing computer software. Novel *O*-benzoxazolyl (OBox) imidates were found to be promising glycosyl donors for HPLC-based applications [119].

10.3 Other Platforms for the Automated Synthesis

In spite of remarkable progress, the solid phase technique still suffers from significant limitations: large reagent excess, limited use of molecular sieves, large volume of waste solvent, cumbersome analysis of intermediates, lower stereoselectivity, loss and poisoning of resin, reagent trapping, etc. In addition, the necessity to carefully select (match) reaction components, linkers, resins, activators, solvents, temperature, etc. have also become apparent and require thorough refinement. This section will discuss major efforts to automate the synthesis of oligosaccharides, either in solution with or without using soluble tags or with the use of alternative solid supports.

10.3.1 *Automation of the one-pot oligosaccharide synthesis in solution*

Oligosaccharide synthesis, using the principle of chemoselectivity introduced by Fraser–Reid, has been used in a variety of ways and has found broad application in synthesis [54–55,120]. With the discovery of multiple reactivity levels ranging from the superdisarmed to the superarmed building blocks and systems [68, 121–124], the versatility of the chemoselective approach to oligosaccharide synthesis was enhanced. Wong and co-workers devised a mathematical approach, assigning relative reactivity values (RRVs) to a wide library of building blocks that were then used for oligosaccharide assembly in one-pot [125]. The determination of RRVs was made in standardized reaction conditions: tolyl thioglycoside donors in the presence of an *N*-iodosuccinimide/trifluoromethanesulfonic acid (NIS/TfOH) promoter system. The cumulative reactivity data was then compiled into a predictive computer program called Optimer [125]. Following these studies, a well-rounded technology for one-pot oligosaccharide synthesis based on RRVs emerged. A relevant example is shown in Fig. 10.8 (synthesis of (35)), wherein sequential activation of building blocks (31), (32), and (33) was based on their relative reactivity, which was found to be 17,000, 162.8, and 13.1, respectively [125]. In this context, the reactivity difference between similarly protected sugars of different series also has to be taken into consideration. For example, the reactivity ratio between per-benzylated S-(*p*-methylphenyl) glycosides of L-fucose, D-galactose, and D-glucose series was found to be 27.1, 6.4, and 1, respectively [125]. This approach has been used to obtain various oligosaccharides and glycoconjugates of biological significance and medicinal relevance [126–128]. Similar reactivity scales have also been devised by Fraser–Reid [129], Ley [130–132], and others [120].

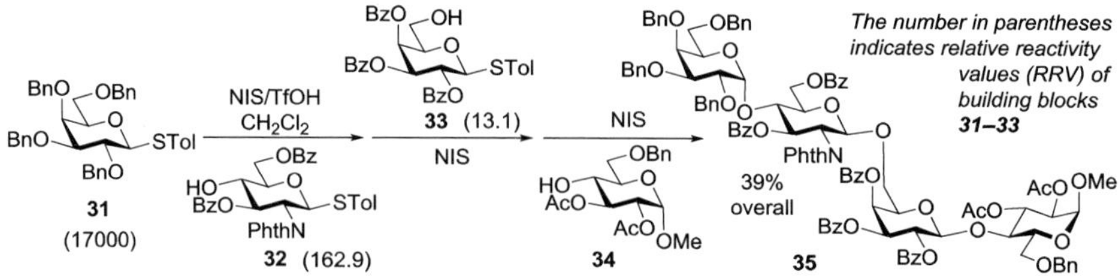

Figure 10.8 Programmable oligosaccharide synthesis assisted by the predictive computer program Optimer.

Takahashi and co-workers probed a number of automation platforms for the solution-based one-pot oligosaccharide synthesis [133]. In a majority of applications, selective activation of different leaving groups, another common approach to expeditious oligosaccharide synthesis [53,67], was executed. For instance, a parallel synthesis instrument, Quest 210 by Argonaut Technologies, was applied to the automated one-pot synthesis of linear and branched oligosaccharides shown in Fig. 10.9A and 10.9B [134]. Thus, for the synthesis of a trisaccharide (39), a glycosyl bromide donor (36) was selectively activated for reaction with a thioglycoside acceptor (37) in the presence of silver trifluoromethanesulfonate (AgOTf) as the promoter. The anomeric thiophenyl

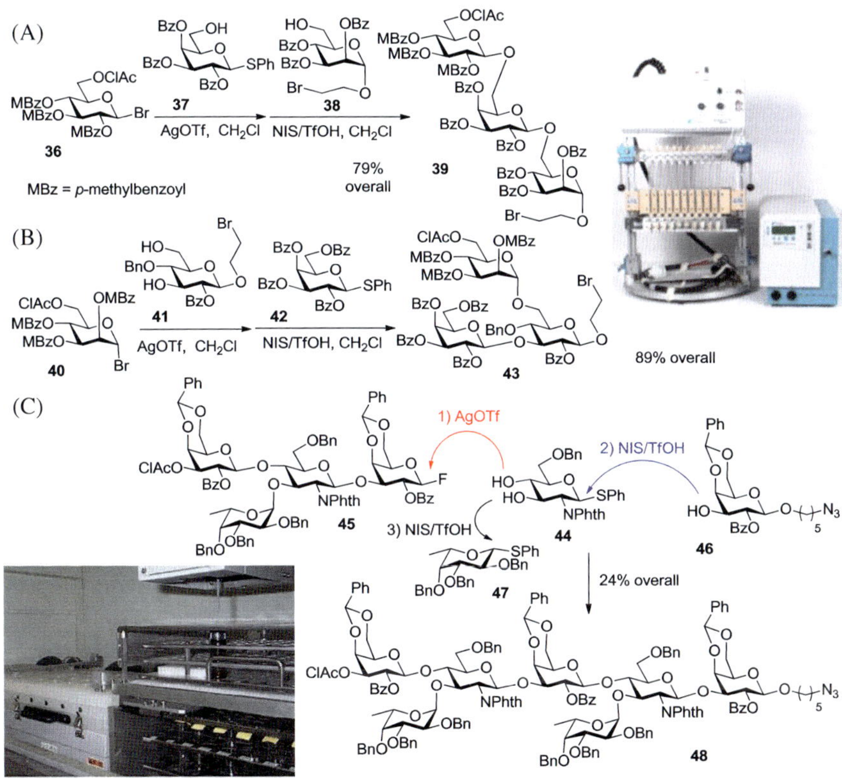

Figure 10.9 Automation of one-pot synthesis using parallel synthesizers Quest-210 (A and B) and Moritex L-cos (C).

leaving group of the disaccharide intermediate was then directly activated by the addition of NIS/TfOH and an acceptor (38) to provide a linear trisaccharide (39) in 79% yield over two steps. Fig. 10.9B illustrates the formation of the branched trisaccharide (43), which was obtained by two-step sequential glycosylation of a diol (41). First, the primary hydroxyl of the acceptor (41) was glycosylated with a bromide (40) in the presence of AgOTf. Subsequent addition of a thiophenyl donor (42) and the NIS/TfOH promoter system resulted in the glycosylation of the remaining secondary hydroxyl to afford a trisaccharide (43) in 89% yield over two steps.

In an effort to obtain more complex oligosaccharide sequences, Takahashi adapted another instrument, Liquid Crystal on Silicon (L-COS) by Moritex, which was used to make a dimeric Le^x (48) and its analogs. Using this instrument, Takahashi *et al.* achieved automation of stirring, temperature control, and rate of reagent addition for each glycosylation and deprotection step shown in Fig. 10.9C [133, 135]. The synthesis began by the coupling of the fluoride donor (45) with the diol acceptor (44) in the presence of AgOTf as the promoter. This reactions proceeded regioselectively at the C-4 position, and the resulting pentasaccharide equipped with the anomeric thiophenyl leaving group was used as the glycosyl donor for the coupling with the glycosyl acceptor (46), which was added along with NIS/TfOH to affect the activation. Finally, the remaining 3′–OH of the resulting hexasaccharide intermediate was fucosylated by the addition of a donor (47) and NIS/TfOH. Resultantly, the desired heptasaccharide (48) was obtained in eight hours in 24% yield (average 79% yield per step). The automated synthesizer could be combined with Combi Flash automated column chromatograph, which was used for purification of the resulting oligosaccharides.

10.3.2 *Automation of the fluorous tag-supported synthesis*

Fluorous tag-supported synthesis has emerged as a new and attractive strategy for obtaining oligosaccharides with good prospects for automation. Apparently, fluorous protecting groups (groups incorporating a fluorinated alkyl chain) allow for the separation of fluorinated species (typically glycosyl acceptors) from non-fluorinated species (glycosyl donors) by simple partitioning between perfluorohexanes and methanol (or toluene) [136].

Solution-based microreactors, developed in the late nineties, have been adapted to the fluorous tag-supported synthesis of oligosaccharides by Seeberger and co-workers [137]. A five-port silicon microfluidic reactor was utilized for the synthesis of the homotetramer (52). For this purpose, a glycosyl phosphate donor (48) was first glycosidated with a fluorous linker (49) in the presence of TMSOTf [137]. To accomplish the glycosylation reaction in a microfluidic device, three separate solutions in dichloromethane were prepared: glycosyl acceptor, phosphate donor, and the activator (TMSOTf). Using three syringe pumps, the solutions of acceptor, donor, and promoter were connected via inlets 1, 2, and 3, respectively (Fig. 10.10). By diffusion-controlled sampling, each solution was then delivered into the mixing zone and the concentration of each component was controlled by the concentration inside the syringe and by the flow rate of each stream. After mixing, the reaction then took place inside the reaction zone. The reaction was monitored using different types of instruments connected at the outlet of the microreactor, including UV-vis, infrared (IR), and mass spectrometers. The glycosylation step was then followed by the removal of the Fmoc group with piperidine and TBAF (used to remove 6-*O*-TMS formed as a result of competing silylation) to afford the fluorous glycosyl acceptor (50). The latter was then glycosylated with a glycosyl donor (48) and the deprotection–glycosylation sequence was repeated until the desired tetrasaccharide (51) had been assembled.

The tetrasaccharide was then cleaved from the fluorous support by the treatment with second-generation Grubbs' catalyst to provide a tetrasaccharide (52). The reaction times for glycosylations were 20 seconds for the formation of the disaccharide, and 60 seconds each for the tri- and tetrasaccharides. The yields for the

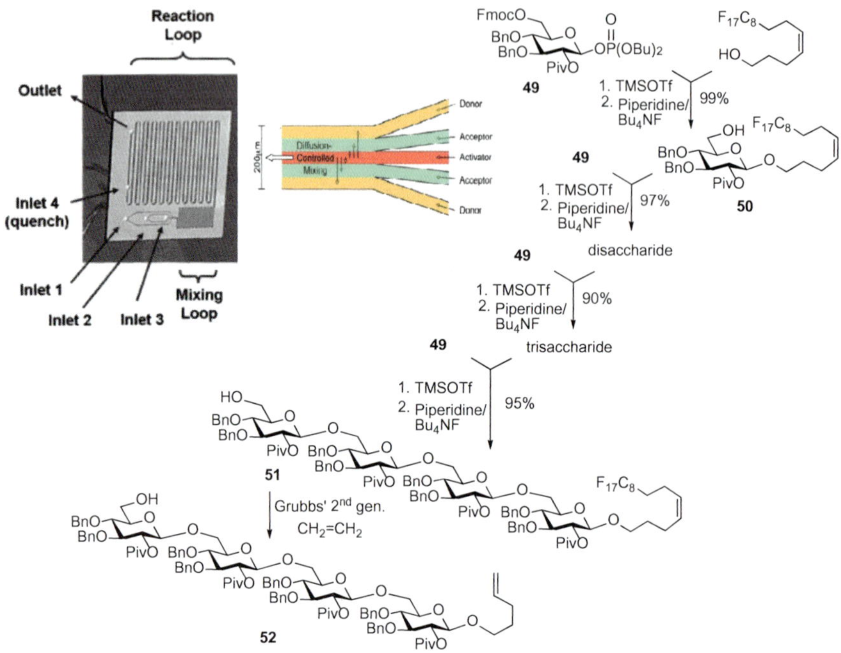

Figure 10.10 Fluorous tag-supported synthesis in the microreactor.

reactions after purification were 97, 90, and 95% for di-, tri-, and tetrasaccharides, respectively. The benefits of using microfluidic reactors in oligosaccharide synthesis include ease of scale up, the ability to readily optimize reaction conditions and control the reaction outcome by varying the flow rate and amounts of reagents, improved safety, a greater control of the reaction temperature, and the compatibility with various analytical techniques and automation platforms. Inherent microfluidic system drawbacks include possibly longer times if a low flow-rate is required and the incompatibility with solids.

Jaipuri and Pohl also used the fluorous tag-assisted glycosylation strategy for developing their own automation platform using the fluorous solid phase extraction protocol (Fig. 10.11) [138]. Their approach involved the use of a fluorous tag attached

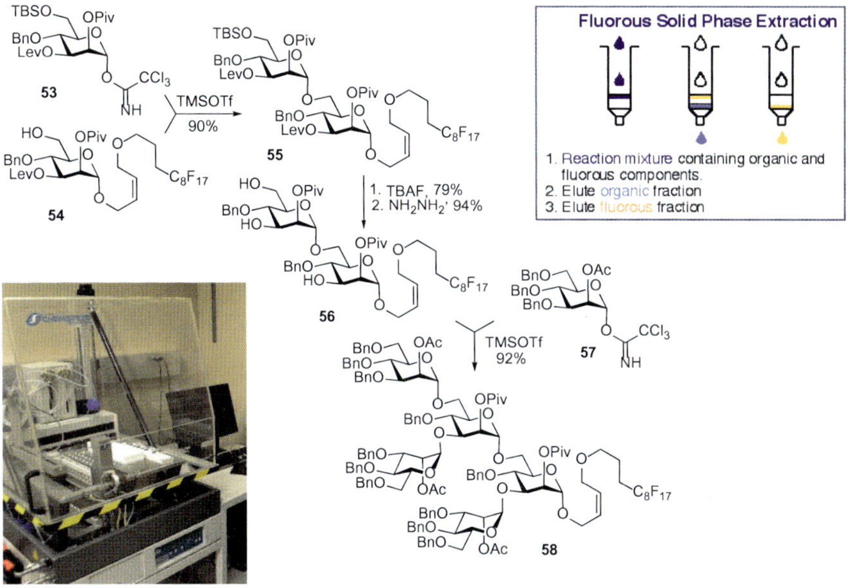

Figure 10.11 Pohl automated synthesis using fluorous tag approach.

glycosidically to a glycosyl acceptor (54), which was glycosylated using excess of the trichloroacetimidate donor (53) in the presence of TMSOTf. The resultant disaccharide (55) was separated using an automated three-step fluorous solid phase extraction protocol. First, the mixture containing organic and fluorous components were loaded on a separation column. All fluorine-free molecules eluted off the column using 20% solution of water in methanol. After that, fluorous molecules that were retained on the column were released using a fluorophilic wash, such as methanol or tetrahydrofuran (THF). This procedure can be automated using commercially available equipment that is able to create a positive pressure or vacuum. Subsequently, a disaccharide (55) was subjected to the two-step deprotection of TBS (with TBAF) and Levulinoate (Lev) groups (with hydrazine) to afford a triol (56). The latter was then glycosylated with excess trichloroacetimidate donor (57) to afford a pentasaccharide (58) in a high yield of 92%.

10.3.3 *STICS: Surface-Tethered Iterative Carbohydrate Synthesis*

In the attempt to address major drawbacks of the polymer-supported synthesis, predominantly swelling, reagent trapping, and poisoning, Demchenko and Stine introduced the Surface-Tethered Iterative Carbohydrate Synthesis (STICS), a novel approach to oligosaccharide synthesis using nanoporous gold (NPG) as a solid support [139]. NPG is a high-surface area, sponge-like nanomaterial. It can be utilized as a set of NPG plates (8 × 8 × 0.2 mm) assembled in a stacked Teflon mini reactor, as shown in Fig. 10.12. The glycosyl acceptor can be anchored to NPG using thiolated linkers and the oligosaccharide assembly is accomplished via alternating glycosylation, washing, deprotection, and drying steps, as depicted in Fig. 10.12. Thus, the 6-*O*-tert-butyldiphenylsilyl (TBDPS)-protected S-benzoxazolyl (SBox) glycosyl donor (60) was coupled to the lipoic acid anchored acceptor (59) in the presence of methyl trifluoromethanesulfonate (MeOTf). Then, after a quick rinse, the tethered disaccharide intermediate was treated with tetrabutylammonium fluoride (Bu_4NF) to deprotect the silyl group to afford the second-generation glycosyl acceptor. The latter was dried and reacted with the SBox donor (61), and the resulting trisaccharide

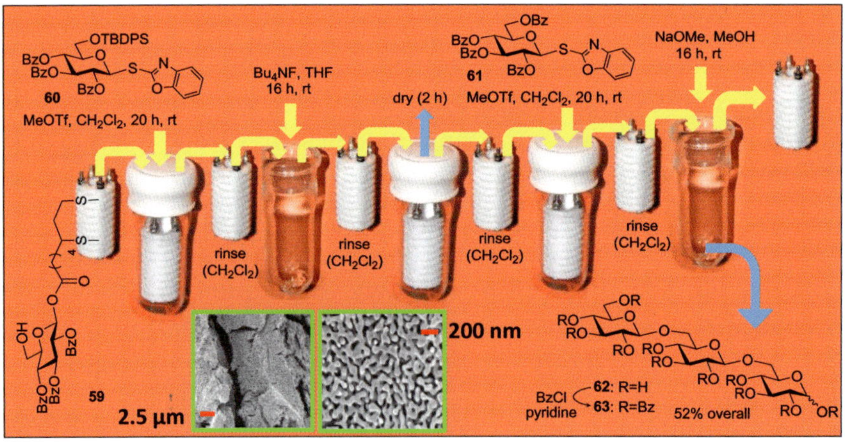

Figure 10.12 STICS: Surface-Tethered Iterative Carbohydrate Synthesis.

was released by the treatment with NaOMe in methanol (MeOH). Finally, for the purification and characterization purpose, the trisaccharide was benzoylated to afford compound (63) in 52% yield overall. The pictures in Fig. 10.12 are scanning electron microscope (SEM) images of NPG plates with two different bar scales (2.5 μm and 200 nm).

Overall, the STICS approach offers a promising new platform for oligosaccharide synthesis. The NPG surface only requires wetting; upon reaction, the access reagent is simply rinsed off. Reagent trapping/resin poisoning is not an issue with NPG because it does not trap any reagent and can be regenerated to bare surface by electrochemical methods if necessary, and gold is fully recoverable. The limitations of the STICS approach are similar to other standard drawbacks of the polymer-supported synthesis and include: difficulty of monitoring the reaction and analyzing intermediates, longer reaction times than those for reactions in solution, and the requirement to use large (5–10 fold) excess of reagents. To address some of these drawbacks, the subsequent study made use of the HPLC-assisted automated technology for oligosaccharide synthesis [140]. The HPLC pump was used to force the circulation of the donor inside the column and the NPG chips placed in the Omnifit column (Fig. 10.13). Also investigated was the role of the spacer in the lipoic

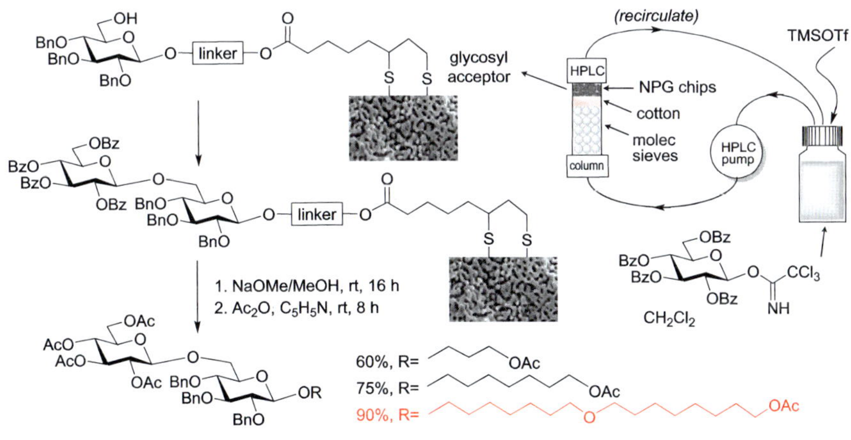

Figure 10.13 HPLC-assisted surface-tethered iterative carbohydrate synthesis.

acid anchoring systems. Using this strategy in combination with a long chain spacer between the acceptor and lipoic acid anchor, it was possible to increase the disaccharide yield from 60% to 90% [140].

10.3.4 *Solution phase-based automation using electrochemical activation of thioglycosides*

Electrochemical activation of chalcogenoglycosides has been known for years [141–146]. Very recently, Nokami and co-workers introduced an automated oligosaccharide synthesis based on the electrochemical activation of thioglycosides via the corresponding glycosyl triflates [147]. The automated synthesizer was specifically developed for this application, and the synthesis of a series of β-(1–6)-linked glycans consisting of multiple *N*-acetylglucosamine units has been accomplished. As depicted in Fig. 10.14, the aryl thioglycoside donor (64) was preactivated via the corresponding glycosyl triflate (65) by electrochemical oxidation at –80°C and 1.0 F/mol at 1.73 Volt for 40 min. The temperature was then increased to –60–60°C and the glycosyl acceptor (66) equipped with the S-aryl leaving group was added, resulting in the formation of a disaccharide (67) in 30 min. Essentially, the same two-step preactivation–glycosylation sequence could be repeated multiple times to afford the desired

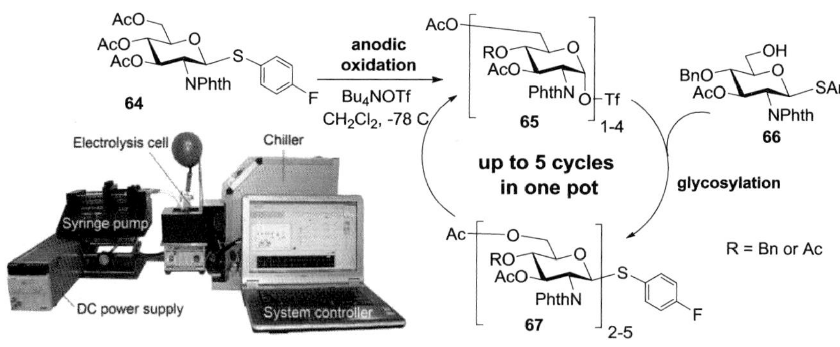

Figure 10.14 Automation of the electrochemical activation of thioglycosides in solution.

pentasaccharide (68) in 9–15% overall yield (62–68%, average yield per cycle) in ten hours [147].

10.4 Conclusions and Outlook

To keep pace with the exploding area of glycobiology, it is critical to make complex carbohydrates more accessible to the general chemical, biochemical, and industrial audience. This lofty goal can be achieved only by the development of reliable, stereoselective, and expeditious methods for glycoside synthesis and oligosaccharide assembly that are applicable to both laboratory- and industrial-scale preparations. Traditional chemical assembly of oligosaccharides in solution involves multiple protecting group and/or leaving group manipulations and requires tedious purification between each glycosylation step. Expeditious strategies for oligosaccharide assembly in solution require specialist knowledge of all aspects of carbohydrate chemistry and fine-tuning of the reaction conditions and reactivity levels. Manual polymer-supported synthesis helps to streamline the synthesis and purification, and the automated platform developed by Seeberger introduces an idea of operational simplicity. Yet, it requires a sophisticated and expensive synthesizer.

Recent developments of automation with the use of inexpensive laboratory equipment including syringe pumps, HPLC components, microreactors, and parallel synthesizers offer a promise of simple automation using commonly available and relatively inexpensive equipment. The modular character of many new wave synthesizers allows for endless opportunities to implement attachments, reagent delivery modes, detecting systems, accessories, and software packages. Some automation platforms allow for real-time reaction monitoring, which, in turn, helps reduce the reaction time and the amount of reagents needed. Further development of existing and new platforms for automated synthesis remains a significant area of research. While most automated platforms remain underdeveloped, solution-based manual (standard) synthesis will still remain an important (major) tool to obtain complex oligosaccharides or particularly challenging sequences.

Acknowledgments

The authors are indebted to the NIGMS for support of their work on the development of new strategies and automation platforms for oligosaccharide synthesis (GM090254 and GM077170).

References

1. Stick, R. V., Williams, S. J. (2009) *Carbohydrates: The Essential Molecules of Life*, 2nd ed. Elsevier (Amsterdam — Boston — Heidelberg).
2. Benoff, S. (1997) *Mol. Hum. Reprod.* **3**, 599–637.
3. Varki, A., Cummings, R. D., Esko, J. D., Freeze, H. H., Bertozzi, C. R., Stanley, P., Hart, G. W., Etzler, M. E. (2009) *Essentials of Glycobiology*, 2nd ed. CSH Laboratory Press (New York).
4. Cipolla, L., Arajo, A. C. , Bini, D., Gabrielli, L., Russo, L., Shaikh, N. (2010) *Expert Opin. Drug Discovery* **5**, 721–737.
5. Witczak, Z. J. (1997) In *Carbohydrates in Drug Design*. Witczak, Z. J., Nieforth, K. A. Eds. Marcel Dekker, Inc. (New York). pp. 1–37.
6. Wong, C. H. (2003) *Carbohydrate-Based Drug Discovery*. Wiley-VCH (Weinheim).
7. Klyosov, A. A., Witczak, Z. J., Platt, D. (2006) In *ACS Symp. Ser., Vol. 932*. ACS (Washington).
8. Muthana, S., Cao, H., Chen, X. (2009) *Curr. Opin. Chem. Biol.* **13**, 573–581.
9. Duus, J. O., Gotfredsen, C. H., Bock, K. (2000) *Chem. Rev.* **100**, 4589.
10. Wormald, M. R., Petrescu, A. J., Pao, Y. L., Glithero, A., Elliott, T., Dwek, R. A. (2002) *Chem. Rev.* **102**, 371–386.
11. Spillmann, D., Burger, M. M. (2000) In *Carbohydrates in Chemistry and Biology*. Ernst, B., Hart, G. W., Sinay, P. Eds. Wiley-VCH (Weinheim, New York). pp. 1061–1088.
12. Jin, S., Cheng, Y., Reid, S., Li, M., Wang, B. (2010) *Med. Res. Rev.* **30**, 171–257.
13. Varki, A. (1993) *Glycobiology* **3**, 97–130.
14. Cheng, Y., Li, M., Wang, S., Peng, H., Reid, S., Ni, N., Fang, H., Xu, W., Wang, B. (2010) *Sci. China Chem.* **53**, 3–20.
15. Dube, D. H., Bertozzi, C. R. (2005) *Nature Rev.* **4**, 477–488.
16. Murrey, H. E., Hsieh-Wilson, L. C. (2008) *Chem. Rev.* **108**, 1708–1731.
17. Chaubard, J. L., Krishnamurthy, C., Yi, W., Smith, D. F., Hsieh-Wilson, L. C. (2012) *J. Am. Chem. Soc.* **134**, 4489–4492.

18. Poletti, L., Lay, L. (2003) *Eur. J. Org. Chem.* 2999–3024.
19. Linhardt, R. J., Toida, T. (2004) *Acc. Chem. Res.* **37**, 431–438.
20. Kotra, L. P., Mobashery, S. (2001) *Curr. Org. Chem.* **5**, 193–205.
21. Ito, Y., Manabe, S. (2001) In *Glycoscience: Chemistry and Chemical Biology,* Vol. 3. Fraser-Reid, B., Tatsuta, K., Thiem, J. Eds. Springer (Berlin). pp. 2441–2470.
22. Lucas, A. H., Reason, D. C. (1999) *Immunol. Rev.* **171**, 89–104.
23. Kuberan, B., Linhardt, R. J. (2000) *Curr. Org. Chem.* **4**, 653–677.
24. Danishefsky, S. J., Allen, J. R. (2000) *Angew. Chem. Int. Ed.* **39**, 836–863.
25. Pozsgay, V. (2001) *Adv. Carbohydr. Chem. Biochem.* **56**, 153–199.
26. Galonic, D. P., Gin, D. Y. (2007) *Nature* **446**, 1000–1007.
27. Seeberger, P. H., Werz, D. B. (2007) *Nature* **446**, 1046–1051.
28. Demchenko, A. V. Ed. (2008) *Handbook of Chemical Glycosylation: Advances in Stereoselectivity and Therapeutic Relevance.* Wiley-VCH (Weinheim).
29. Lemieux, R. U. (1954) *Adv. Carbohydr. Chem. Biochem.* **9**, 1–57 and references therein.
30. Capon, B. (1969) *Chem. Rev.* **69**, 407–496.
31. Gervay, J., Nguyen, T. N., Hadd, M. J. (1997) *Carbohydr. Res.* **300**, 119–125.
32. Nukada, T., Berces, A., Zgierski, M. Z., Whitfield, D. M. (1998) *J. Am. Chem. Soc.* **120**, 13291–13295.
33. Nguyen, H. M., Chen, Y. N., Duron, S. G., Gin, D. Y. (2001) *J. Am. Chem. Soc.* **123**, 8766–8772.
34. Ayala, L., Lucero, C. G., Romero, J. A. C., Tabacco, S. A., Woerpel, K. A. (2003) *J. Am. Chem. Soc.* **125**, 15521–15528.
35. Crich, D., Chandrasekera, N. S. (2004) *Angew. Chem. Int. Ed.* **43**, 5386–5389 and references therein.
36. Boebel, T. A., Gin, D. Y. (2005) *J. Org. Chem.* **70**, 5818–5826.
37. Li, Z., Gildersleeve, J. (2006) *J. Am. Chem. Soc.* **128**, 11612–11619.
38. Jensen, H. H., Bols, M. (2006) *Acc. Chem. Res.* **39**, 259–265.
39. Whitfield, D. M. (2009) *Adv. Carbohydr. Chem. Biochem.* **62**, 83–159.
40. Mydock, L. K., Demchenko, A. V. (2010) *Org. Biomol. Chem.* **8**, 497–510.
41. Beaver, M. G., Woerpel, K. A. (2010) *J. Org. Chem.* **75**, 1107–1118.
42. Crich, D. (2010) *Acc. Chem. Res.* **43**, 1144–1153.
43. Pedersen, C. M., Marinescu, L. G., Bols, M. (2010) *C. R. Chimie* **14**, 17–43.

44. Nokami, T., Shibuya, A., Manabe, S., Ito, Y., Yoshida, J. (2009) *Chem. Eur. J.* **15**, 2252–2255.
45. Huang, M., Retailleau, P., Bohe, L., Crich, D. (2012) *J. Am. Chem. Soc.* **134**, 14746–14749.
46. Huang, M., Garrett, G. E., Birlirakis, N., Bohe, L., Pratt, D. A., Crich, D. (2012) *Nature: Chemistry* **4**, 663–667.
47. Crich, D. (2011) *J. Org. Chem.* **76**, 9193–9209.
48. Kaeothip, S., Yasomanee, J. P., Demchenko, A. V. (2012) *J. Org. Chem.* **77**, 291–299.
49. Kononov, L. O., Malysheva, N. N., Orlova, A. V., Zinin, A. I., Laptinskaya, T. V., Kononova, E. G., Kolotyrkina, N. G. (2012) *Eur. J. Org. Chem.* 1926–1934.
50. Whitfield, D. M. (2012) *Carbohydr. Res.* **356**, 180–190.
51. Whitfield, D. M. (2012) *Carbohydr. Res.* **356**, 191–195.
52. Ranade, S. C., Demchenko, A. V. (2013) *J. Carbohydr. Chem.* **32**, 1–43.
53. Smoot, J. T., Demchenko, A. V. (2009) *Adv. Carbohydr. Chem. Biochem.* **62**, 161–250.
54. Mootoo, D. R., Konradsson, P., Udodong, U., Fraser-Reid, B. (1988) *J. Am. Chem. Soc.* **110**, 5583–5584.
55. Fraser-Reid, B., Udodong, U. E., Wu, Z. F., Ottosson, H., Merritt, J. R., Rao, C. S., Roberts, C., Madsen, R. (1992) *Synlett*, 927–942 and references therein.
56. Nicolaou, K. C., Ueno, H. (1997) In *Preparative Carbohydrate Chemistry*. Hanessian, S. Ed. Marcel Dekker, Inc. (New York). pp. 313–338.
57. Williams, L. J., Garbaccio, R. M., Danishefsky, S. J. (2000) In *Carbohydrates in Chemistry and Biology, Vol. 1*. Ernst, B., Hart, G. W., Sinay, P. Eds. Wiley-VCH (Weinheim, New York). pp. 61–92.
58. Kahne, D., Walker, S., Cheng, Y., van Engen, D. (1989) *J. Am. Chem. Soc.* **111**, 6881–6882.
59. Roy, R., Andersson, F. O., Letellier, M. (1992) *Tetrahedron Lett.* **33**, 6053–6056.
60. Boons, G. J., Isles, S. (1994) *Tetrahedron Lett.* **35**, 3593–3596.
61. Kanie, O., Ito, Y., Ogawa, T. (1994) *J. Am. Chem. Soc.* **116**, 12073–12074.
62. Kanie, O. (2000) In *Carbohydrates in Chemistry and Biology*, Vol. 1. Ernst, B., Hart, G. W., Sinay, P. Eds. Wiley-VCH (Weinheim, New York). pp. 407–426.
63. Huang, L., Wang, Z., Huang, X. (2004) *Chem. Commun.* 1960–1961.
64. Huang, X., Huang, L., Wang, H., Ye, X. S. (2004) *Angew. Chem. Int. Ed.* **43**, 5221–5224.

65. Wang, Z., Zhou, L., Ei-Boubbou, K., Ye, X. S., Huang, X. (2007) *J. Org. Chem.* **72**, 6409–6420.
66. Parameswar, A. R., Demchenko, A. V. (2009) In *Progress in the Synthesis of Complex Carbohydrate Chains of Plant and Microbial Polysaccharides.* Nifantiev, N. E. Ed. Transworld Res. Network (Kerala). pp. 463–488.
67. Kaeothip, S., Demchenko, A. V. (2011) *Carbohydr. Res.* **346**, 1371–1388.
68. Premathilake, H. D., Demchenko, A. V. (2011) In *Topics in Current Chemistry: Reactivity Tuning in Oligosaccharide Assembly, Vol. 301.* Fraser-Reid, B., Lopez, J. C. Eds. Springer-Verlag (Berlin-Heidelberg). pp. 189–221.
69. Plante, O. J., Palmacci, E. R., Seeberger, P. H. (2001) *Science* **291**, 1523–1527.
70. Seeberger, P. H. (2008) *Chem. Soc. Rev.* **37**, 19–28.
71. Plante, O. J., Palmacci, E. R., Seeberger, P. H. (2003) *Adv. Carbohydr. Chem. Biochem.* **58**, 35–54.
72. Fruchtel, J. S., Jung, G. (1996) *Angew. Chem. Int. Ed. Engl.* **35**, 17–42.
73. Winter, M. (1996) In *Combinatorial Peptide and Nonpeptide Libraries: A Handbook.* Jung, G. Ed. VCH (Wienheim, New York, Basel, Cambridge, Tokyo). pp. 465–510.
74. Hermkens, P. H. H., Ottenheijm, H. C. J., Rees, D. (1996) *Tetrahedron* **52**, 4527–4554.
75. Brown, R. C. D. (1998) *J. Chem. Soc., Perkin Trans.* **1**, 3293–3320.
76. Merrifield, B. (1984) *Br. Polym. J.* **16**, 173–178.
77. Toy, P. H., Lam, Y. Eds. (2012) *Solid-Phase Organic Synthesis* John Wiley & Sons, Inc. (Hoboken)
78. Merrifield, R. B. (1963) *J. Am. Chem. Soc.* **85**, 2149–2154.
79. Frechet, J. M., Schuerch, C. (1971) *J. Am. Chem. Soc.* **93**, 492–496.
80. Schmidt, R. R., Jonke, S., Liu, K. (2007) In *ACS Symp. Ser. (Frontiers in Modern Carbohydrate Chemistry)* Vol. 960. Demchenko, A. V. Ed. Oxford University Press (Oxford). pp. 209–237.
81. Seeberger, P. H. (2002) *J. Carbohydr. Chem.* **21**, 613–643.
82. Seeberger, P. H., Haase, W. C. (2000) *Chem. Rev.* **100**, 4349–4393.
83. Tanaka, K., Fukase, K. (2012) In *Solid-Phase Organic Synthesis.* Toy, P. H., Lam, Y. Eds. John Wiley & Sons, Inc. (Hoboken). pp. 489–530.
84. Zhu, T., Boons, G. J. (1998) *Angew. Chem. Int. Ed.* **37**, 1898–1900.
85. Zhu, T., Boons, G. J. (1999) *Angew. Chem., Int. Ed. Engl.* **38**, 3495–3497.
86. Plante, O. J., Palmacci, E. R., Seeberger, P. H. (2003) *Adv. Carbohydr. Chem. Biochem.* **58**, 35–54.
87. Seeberger, P. H. (2003) *Chem. Commun.* 115–1121.

88. Seeberger, P. H., Werz, D. B. (2005) *Nature Rev.* **4**, 751–763.
89. Lepenies, B., Yin, J., Seeberger, P. H. (2010) *Current Opinion in Chemical Biology* **14**, 404–411.
90. Andrade, R. B., Plante, O. J., Melean, L. G., Seeberger, P. H. (1999) *Org. Lett.* **1**, 1811–1814.
91. Palmacci, E. R., Plante, O. J., Hewitt, M. C., Seeberger, P. H. (2003) *Helvetica Chimica Acta* **86**, 3975–3990.
92. Melean, L. G., Love, K. R., Seeberger, P. H. (2002) *Carbohydrate Research* **337**, 1893–1916.
93. Kanemitsu, T., Seeberger, P. H. (2003) *Organic Letters* **5**, 4541–4544.
94. Eller, S., Collot, M., Yin, J., Hahm, H. S., Seeberger, P. H. (2013) *Angew. Chem. Int. Ed.* **52**, 5858–5861.
95. Esposito, D., Hurevich, M., Castagner, B., Wang, C. C., Seeberger, P. H. (2012) *Beilstein Journal of Organic Chemistry* **8**, 1601–1609.
96. Kandasamy, J., Hurevich, M., Seeberger, P. H. (2013) *Chem. Commun.* **49**, 4453–4455.
97. Werz, D. B., Castagner, B., Seeberger, P. H. (2007) *J. Am. Chem. Soc.* **129**, 2770–2771.
98. Hurevich, M., Seeberger, P. H. (2014) *Chem. Commun.* **50**, 1851–1853.
99. Codée, J. D. C., Kröck, L., Castagner, B., Seeberger, P. H. (2008) *Chemistry — A European Journal* **14**, 3987–3994.
100. Walvoort, M. T. C., van den Elst, H., Plante, O. J., Krock, L., Seeberger, P. H., Overkleeft, H. S., van der Marel, G. A., Codee, J. D. C. (2012) *Angew. Chem. Int. Ed.* **51**, 4393–4396.
101. Hakomori, S. (1981) *Ann. Rev. Biochem.* **50**, 733–764.
102. Hakomori, S. (1984) *Annu. Rev. Immunol.* **2**, 103–126.
103. Hakomori, S. (1998) *Acta Anat.* **161**, 79–90.
104. Hakomori, S. (2008) *Biochim. Biophys. Acta* **1780**, 325–346.
105. Guo, Z., Wang, Q. (2009) *Curr. Op. Chem. Biology* **13**, 608–617.
106. Allen, J. R., Allen, J. G., Zhang, X. F., Williams, L. J., Zatorski, A., Ragupathi, G., Livingston, P. O., Danishefsky, S. J. (2000) *Chem. Eur. J.* **6**, 1366–1375.
107. Bosse, F., Marcaurelle, L. A., Seeberger, P. H. (2002) *J. Org. Chem.* **67**, 6659–6670.
108. Jeon, I., Iyer, K., Danishefsky, S. J. (2009) *J. Org. Chem.* **74**, 8452–8455.
109. Love, K. R., Seeberger, P. H. (2004) *Angew. Chem. Int. Edit.* **43**, 602–605.
110. Grubbs, R. H., Miller, S. J., Fu, G. C. (1995) *Acc. Chem. Res.* **28**, 446–452.

111. Krock, L., Esposito, D., Castagner, B., Wang, C. C., Bindschadler, P., Seeberger, P. H. (2012) *Chem. Sci.* **3**, 1617–1622.
112. Kim, K. S., Kim, J. H., Lee, Y. J., Park, J. (2001) *J. Am. Chem. Soc.* **123**, 8477–8481.
113. Calin, O., Eller, S., Seeberger, P. H. (2013) *Angew. Chem. Int. Ed.* **52**, 5862–5865.
114. Egusa, K., Kusumoto, S., Fukase, K. (2003) *Eur. J. Org. Chem.* **2003**, 3435–3445.
115. Hanashima, S., Manabe, S., Ito, Y. (2005) *Angew. Chem. Int. Ed.* **44**, 4218–4228.
116. Ando, H., Manabe, S., Nakahara, Y., Ito, Y. (2001) *J. Am. Chem. Soc.* **123**, 3848–3849.
117. Wu, B., Hua, Z., Warren, J. D., Ranganathan, K., Wan, Q., Chen, G., Tan, Z., Chen, J., Endo, A., Danishefsky, S. J. (2006) *Tetrahedron Lett.* **47**, 5577–5579.
118. Vijaya Ganesh, N., Fujikawa, K., Tan, Y. H., Stine, K. J., Demchenko, A. V. (2012) *Org. Lett.* **14**, 3036–3039.
119. Nigudkar, S. S., Parameswar, A. R., Pornsuriyasak, P., Stine, K. J., Demchenko, A. V. (2013) *Org. Biomol. Chem.* **11**, 4068–4076.
120. Fraser-Reid, B., Lopez, J. C. (2011) In *Topics in Current Chemistry,* Vol. 301. Springer-Verlag (Berlin-Heidelberg).
121. Jensen, H. H., Pedersen, C. M., Bols, M. (2007) *Chem. Eur. J.* **13**, 7576–7582.
122. Pedersen, C. M., Nordstrom, L. U., Bols, M. (2007) *J. Am. Chem. Soc.* **129**, 9222–9235.
123. Mydock, L. K., Demchenko, A. V. (2008) *Org. Lett.* **10**, 2103–2106.
124. Premathilake, H. D., Mydock, L. K., Demchenko, A. V. (2010) *J. Org. Chem.* **75**, 1095–1100.
125. Zhang, Z., Ollmann, I. R., Ye, X. S., Wischnat, R., Baasov, T., Wong, C. H. (1999) *J. Am. Chem. Soc.* **121**, 734–753.
126. Hsu, C. H., Hung, S. C., Wu, C. Y., Wong, C. H. (2011) *Angew. Chem. Int. Ed.* **50**, 11872–11923.
127. Huang, T. Y., Zulueta, M. M. L., Hung, S. C. (2011) *Org. Lett.* **13**, 1506–1509.
128. Huang, Y. L., Hung, J. T., Cheung, S. K., Lee, H. Y., Chu, K. C., Li, S. T., Lin, Y. C., Ren, C. T., Cheng, T. J., Hsu, T. L., Yu, A. L., Wu, C. Y., Wong, C. H. (2013) *Proc. Nat. Acad. Sci.* **110**, 2517–2522.
129. Wilson, B. G., Fraser-Reid, B. (1995) *J. Org. Chem.* **60**, 317–320.
130. Grice, P., Ley, S. V., Pietruszka, J., Priepke, H. W. M., Walther, E. P. E. (1995) *Synlett* 781–784.

131. Baeschlin, D. K., Chaperon, A. R., Charbonneau, V., Green, L. G., Ley, S. V., Lucking, U., Walther, E. (1998) *Angew. Chem. Int. Edit.* **37**, 3423–3428.
132. Douglas, N. L., Ley, S. V., Lucking, U., Warriner, S. L. (1998) *J. Chem. Soc., Perkin Trans.* **1**, 51–65.
133. Tanaka, H., Yamada, H., Takahashi, T. (2007) *Trends Glycosci. Glycotechnol.* **19**, 183–193.
134. Takahashi, T., Adachi, M., Matsuda, A., Doi, T. (2000) *Tetrahedron Lett.* **41**, 2599–2603.
135. Tanaka, H., Matoba, N., Tsukamoto, H., Takimoto, H., Yamada, H., Takahashi, T. (2005) *Synlett* 0824–08282.
136. Miura, T., Goto, K., Waragai, H., Matsumoto, H., Hirose, Y., Ohmae, M., Ishida, H., Satoh, A., Inazu, T. (2004) *J. Org. Chem.* **69**, 5348–5353.
137. Carrel, F. R., Geyer, K., Codée, J. D. C., Seeberger, P. H. (2007) *Org. Lett.* **9**, 2285–2288.
138. Jaipuri, F. A., Pohl, N. L. (2008) *Org. Biomol. Chem.* **6**, 2686–2691.
139. Pornsuriyasak, P., Ranade, S. C., Li, A., Parlato, M. C., Sims, C. R., Shulga, O. V., Stine, K. J., Demchenko, A. V. (2009) *Chem. Commun.* 1834–1836.
140. Ganesh, N. V., Fujikawa, K., Tan, Y. H., Nigudkar, S. S., Stine, K. J., Demchenko, A. V. (2013) *J. Org. Chem.* **78**, 6849–6857.
141. Amatore, C., Jutand, A., Mallet, J. M., Meyer, G., Sinay, P. (1990) *J. Chem. Soc., Chem. Commun.* 718–719.
142. Balavoine, G., Berteina, S., Gref, A., Fischer, J. C., Lubineau, A. (1995) *J. Carbohydr. Chem.* **14**, 1217–1236.
143. Yamago, S., Kokubo, K., Yoshida, J. I. (1997) *Chem. Lett.* **26**, 111–112.
144. Yamago, S., Kokubo, K., Hara, O., Masuda, S., Yoshida, J. (2002) *J. Org. Chem.* **67**, 8584–8592.
145. France, R. R., Compton, R. G., Davis, B. G., Fairbanks, A. J., Rees, N. V., Wadhawan, J. D. (2004) *Org. Biomol. Chem.* **2**, 2195–2202.
146. Nokami, T., Shibuya, A., Tsuyama, H., Suga, S., Bowers, A. A., Crich, D., Yoshida, J. (2007) *J. Am. Chem. Soc.* **129**, 10922–10928.
147. Nokami, T., Hayashi, R., Saigusa, Y., Shimizu, A., Liu, C. Y., Mong, K. K. T., Yoshida, J. I. (2013) *Org. Lett.* **15**, 4520–4523.

Part 4:

Carbohydrate-Based Compounds for Medicinal Chemistry Applications

Chapter 11
Iminosugars: The Potential of Carbohydrate Analogs

Naoki Asano

BioApply Co., Ltd., 1–95 Tsuchishimizu, Kanazawa 920-0955, Japan

11.1 Introduction

Approximately five decades have passed since the first iminosugar, nojirimycin (1), was discovered from the cultured broth of the *Streptomyces* species [1]. Iminosugars are sugar analogs with a nitrogen in place of the ring oxygen of the corresponding sugar. Such sugar mimetics inhibit glycosidases because of a structural resemblance to the sugar moiety of the natural substrate. Glycosidases are involved in a wide range of important biological processes, such as intestinal digestion, post-translational processing of the sugar chain of glycoproteins, quality-control systems in the endoplasmic reticulum (ER) and ER-associated degradation mechanisms, the lysosomal catabolism of glycoconjugates, and further cell–cell and cell–virus recognition [2]. Therefore, glycosidase inhibitors have great potential for the treatment of diabetes, viral infection, lysosomal storage disorders, and metastatic cancer [3]. The basis of their pharmaceutical applications have been founded by three natural products: 1-deoxynojirimycin (DNJ) (2), swainsonine (3), and castanospermine (4). However, the number of glycosidase inhibitors that have progressed to practical use beyond the discovery phase are restricted due to the lack of adequate potency and selectivity. Despite these drawbacks, the first-generation iminosugars have already demonstrated their potential with two marketed drugs, Glyset® (5) and Zavesca® (6), and some others in clinical development.

In the last decade, it has become clear that sugar mimetics do not need to be glycosidase inhibitors for biological effects and can also act as immune modulators and pharmacological chaperones of misfolded proteins without inhibiting glycosidases [4–6]. The ability of iminosugars to bind to lysosomal glycosidases and enhance the stability of the mutant enzymes, enabling the enzyme to attain its maturation and trafficking to the lysosome, is well documented [7,8]. The challenge of pharmacological chaperones for new therapeutic possibility was opened up by 1-deoxygalactonojirimycin (DGJ) (7) [9]. Very recently, some of the first-generation iminosugars are finding many new pharmaceutical applications by novel bioassays.

Most reports on natural and synthetic iminosugars have focused on D-iminosugars with D-*glucose*, D-*mannose*, and D-*galactose* configurations because their target enzymes are D-glycosidases, except for α-L-fucosidase and α-L-rhamnosidase, and are essential for the survival and existence of all living organisms [2,3,10]. The ability of enzymes to discriminate between enantiomers is vital for organisms. The enzyme can distinguish between the desired substrate and its mirror image (the enantiomer); this is because the enzyme has an active site that will only accept one of the enantiomers, but reject the other. Therefore, D-iminosugars fit the active site of D-glycosidases and show excellent inhibition for them. This may be the reason why synthetic L-iminosugars have attracted little attention. However, in recent years, Fleet *et al.* have published amazing reports stating that some synthetic five-membered L-iminosugars are much more potent inhibitors of D-glycosidases than their D-enantiomers, and further show improved selectivity in D-glycosidase inhibition [11–14]. Thus, the importance for L-iminosugars is being rapidly reconsidered [15]. There is no doubt that many natural and synthetic iminosugars will find new therapeutic applications once they are evaluated in novel bioassays without getting caught up on "glycosidase inhibitors" and "structural similarity" with the corresponding sugars.

The scope of this article is to summarize and re-evaluate the first generation D-iminosugars and their derivatives for pharmaceutical purposes, and outline new findings on L-iminosugars for new biological activities.

11.2 1-Deoxynojirimycin and Its Derivatives

11.2.1 *Historical background*

Nojirimycin (NJ) (Fig. 11.1) (1) was the first iminosugar discovered from the cultured broth of the *Streptomyces* species [1]. The novel structural feature of NJ is the simple replacement of the ring oxygen in D-glucose by nitrogen. This simplicity opened the way to a remarkable glucosidase inhibitor. DNJ (2) was originally prepared by catalytic hydrogenation of NJ with a platinum catalyst or by chemical reduction with sodium borohydride ($NaBH_4$) [16]. It was later iso-

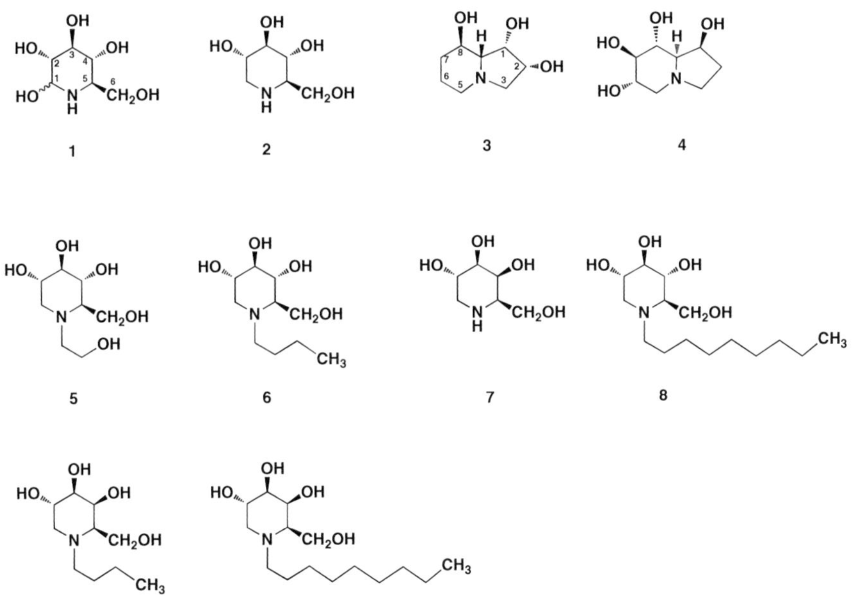

1 2 3 4

5 6 7 8

9 10

Figure 11.1 Structures of Compounds **1–10**.

lated from the root bark of mulberry trees and called moranoline [17] and has now been shown to be present in many disparate species of plants and microorganisms [18]. Despite the excellent α-glucosidase inhibitory activity *in vitro*, the efficacy of DNJ *in vivo* was only moderate [19]. Miglitol (Glyset®) (5) was selected as the most favorable inhibitor out of a large number of synthetic DNJ derivatives and launched in Germany as the first market in 1998, for the treatment of patients with type 2 diabetes. This is the first example of iminosugars as therapeutic agents.

The viral envelope glycoproteins are often essential for virion assembly and secretion and/or infectivity. Compounds that interfere with the glycosylation processes of viral glycoproteins can be expected to be antiviral agents. A number of enzymes, such as processing α-glucosidases, processing α-mannosidases, and a variety of glycosyltransferases, participate in the formation of the sugar chains in the viral envelope glycoproteins. In particular, inhibitors of processing α-glucosidase I, such as *N*-butyl-DNJ (6), exhibit anti-human immunodeficiency virus (HIV) activities *in vitro* at subcytotoxic concentrations [20]. The *in vivo* data obtained so far also promise practical use of processing α-glucosidase I inhibitors as anti-HIV agents. Although *N*-butyl-DNJ (SC 49483) was evaluated in a Phase II clinical trial as an anti-HIV agent, further clinical trials were abandoned because of side effect issues and less efficacy in maintaining high serum levels of the drug. *N*-Nonyl-DNJ (8) is 100–200 times more potent than *N*-butyl-DNJ in inhibiting the hepatitis B virus (HBV) in cell-based assays [21]. In the absence of a reliable *in vitro* infectivity assay for HCV, potential antiviral drugs initially have been tested by using a related pestivirus, bovine viral diarrhea virus (BVDV) [22]. It has been shown that the antiviral effect of the long-alkyl-chain derivative *N*-nonyl-DNJ is more pronounced than that of the short-alkyl-chain derivative *N*-butyl-DNJ [22]. Interestingly, *N*-butyl-DGJ (9) shows no antiviral activity towards BVDV, whereas *N*-nonyl-DGJ (10) is a potent inhibitor [23]. Processing α-glucosidase I inhibition is ruled out as this antiviral mechanism, since both compounds do not inhibit the enzyme. Pavlović *et al.* focused on HCV p7 protein that presents a potential target for antiviral therapy [24]. HCV p7 has been suggested to be a member of a group of small proteins known as viroporin, which mediate cation permeability across membranes and are important for virion release or maturation [25–27]. HCV and BVDV both contain the p7 protein, but flaviviruses such as dengue virus (DENV) do not contain. The p7 protein is capable of forming ion channels in black lipid membranes using an artificial lipid bilayer system. The long-alkyl-chain iminosugar derivatives, such as *N*-nonyl-DNJ and *N*-nonyl-DGJ, inhibit the formation of the p7 ion channel [24].

Lysosomes are membrane-bound acidic organelles that contain over 50 different acid hydrolases and break down macromolecules for recycling in eukaryotic cells. Among macromolecules, glycosphingolipids (GSLs) are an essential cell membrane component containing polysaccharide chains. The deficiency of any individual lysosomal enzyme leads to a lysosomal storage disorder (LSD) and organ damage arises from progressive accumulation of the substrate for the deficient enzyme. The goal of current therapy for LSDs is the reduction of accumulated substrate in lysosomes. The successful treatment for such diseases to date is enzyme replacement therapy (ERT) and another option is substrate reduction therapy (SRT). ERT is based on the intravenous administration of a recombinant form of the deficient enzyme, and is now available for Gaucher disease, Fabry disease, Pompe disease, and mucopolysaccharidoses (MPSs) I, II, and VI. The aim of SRT is to reduce GSL substrate influx into the lysosomes by small-molecule inhibitors of GSL synthesis. *N*-Butyl-DNJ (Zavesca®) is an inhibitor of glucosylceramide synthase and is used for the treatment of type I Gaucher disease and Niemann–Pick type C disease [28]. This is the second example of iminosugars as therapeutic agents. *N*-Butyl-DNJ is also being evaluated in patients with neuronopathic Gaucher disease, but a recent report has shown no significant benefit for neurological manifestations of type III patients [29]. In addition, many patients treated with *N*-butyl-DNJ have experienced side effects, including diarrhea, weight loss, tremors, and peripheral neuropathy [30].

11.2.2 *Future prospects*

Dengue viral infection has become a major public health threat to 2.5 billion people worldwide at risk of infection. Dengue fever is a mosquito-borne disease caused by any one of four closely related dengue viruses (DENVs). A vaccine for dengue must be effective against all four DENV serotypes. Although development for suitable vaccines has been ongoing, no effective vaccines are currently available. Considering the difficulty in developing DENV vaccines, anti-DENV drug discovery is becoming increasingly important. Several DENV proteins are glycoproteins with *N*-linked glycans. Removing glucose residues from the glycan chains attached to nascent glycoproteins are required for proper folding and function [31]. These steps are catalyzed by ER-resident α-glucosidases I and II, and inhibitors of these enzymes as described above have been known to show a potent antiviral effect. In addition, it was described above that *N*-butyl-DNJ had failed in an HIV clinical trial because of low potency and difficulty in reaching therapeutic concentration. Recently, Qu *et al.* found that an oxygenated *N*-alkyl DNJ derivative (11) (Fig. 11.2) (CM–10–18) significantly improves antiviral potency compared to *N*-butyl-DNJ [32]. Compound CM–10–18 was further found to protect mice from lethal DENV infection [33]. Very recently, Yu *et al.* focused on a structural exploration of the tail group in CM–10–18 to further improve its antiviral profile and synthesized the CM–10–18 derivatives with a tertiary alcohol functional group or an ether bond in its alkylated side chain [34]. Among

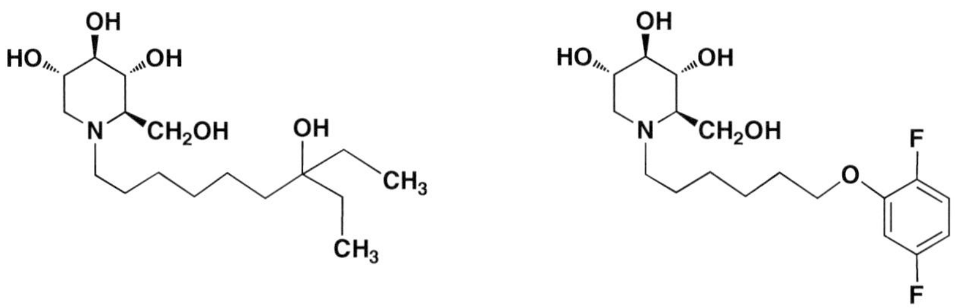

11 **12**

Figure 11.2 Structures of Compounds **11** and **12**.

them, compound (12), which has a 6-(2,5-difluorophenoxy)hexyl group as the side chain, was found to be one of the most potent antiviral compounds against DENV with a median effective concentration (EC_{50}) value of 0.4 μM, while the EC_{50} value of CM–10–18 was 6.5 μM. Furthermore, compound (12) was well tolerated in male Sprague–Dawley rats at a dose of up to 200 mg/kg and displayed desirable pharmacokinetics profiles, with much improved bioavailability compared to CM–10–18. Thus, iminosugars and their derivatives still have great potential as antiviral agents.

Over the past decade, pharmacological chaperone (PC) therapy, which was first suggested by Fan *et al.* [35], has been proposed as a potential treatment for LSDs. Small-molecule PCs are designed to selectively bind and stabilize mutant enzymes in the ER, thereby facilitating correct folding and trafficking to lysosomes, and increasing total cellular enzyme activity [36,37]. Recently, a number of small-molecule compounds have been identified as PCs for the mutant enzymes associated with LSDs, including Fabry, Gaucher, Pompe, Tay–Sachs, Sandhoff, and GM1–gangliosidosis [38]. The most advanced PC is migalastat hydrochloride (migalastat HCl, DGJ HCl, Amigal®) developed for the treatment of Fabry disease. Fabry disease is an X-linked LSD caused by mutations in the gene encoding lysosomal α-galactosidase A (α-Gal A). Migalastat HCl selectively binds some mutant forms of α-Gal A and increases enzyme stability and trafficking to the lysosomes. It is in Phase III of clinical studies to evaluate its safety and efficacy as a potent PC. Furthermore, migarastat HCl is also in Phase II development for the treatment of Fabry disease as a combination therapy with already existing ERT. In February 2013, Amicus Therapeutics presented results from an open-label Phase II trial, which showed that in patients who took an oral dose of migalastat HCl two hours before receiving an infusion of α-Gal A, active enzyme levels in plasma increased in 22 out of 23 instances compared to ERT alone.

Pompe disease (glycogen storage disease type II) is an LSD caused by mutations of acid α-glucosidase (GAA), and its deficiency results in generalized tissue glycogen accumulation and secondary cardiac and skeletal muscle destruction [39,40]. ERT is currently the primary treatment for Pompe disease, and Myozyme® and Lumizyme®, which are recombinant human GAA (rhGAA), are the only two approval products. ERT is effective in the non-neurological symptoms of lysosomal storage diseases, but is not beneficial in the diseases that primarily affect the central nervous system since enzymes administrated do not efficiently cross the blood–brain barrier. The DNJ HCl (AT2220, duvoglustat HCl) has been shown to increase endogenous levels of many different mutant forms of GAA [41]. Hence, the PCs have been proposed as a potential therapy for Pompe disease as well [41,42]. Data from Phase I studies demonstrated that AT2220 was generally safe and well tolerated at all doses evaluated with no drug-related serious adverse events; however, at the high dose administrated at the start of Phase II clinical trials as a monotherapy treatment in adults with Pompe disease, serious adverse effects were seen in several participants in February 2009 and, as a result, AT2220 was put on clinical hold. Khanna *et al.* have recently demonstrated that AT2220 stabilizes rhGAA *in vitro*, minimizing protein denaturation and inactivation at neutral pH and physiological temperature [43]. Their studies using rats and GAA knock-out mice indicated that oral pre-administration of AT2220 increases the circulating half-life of rhGAA and leads to significant increases in rhGAA activity in disease-relevant tissues, and that AT2220-mediated increases in rhGAA tissue levels translate to greater glycogen reduction, compared to administration of rhGAA alone. After careful review of additional and Phase I studies, Amicus decided to shift AT2220 research from monotherapy to co-administration and AT2220 was released from clinical hold in March 2011 after data submission and discussions with the Food and Drug Administration (FDA). Thus, the use of PCs as a combination therapy to boost the efficacy of existing ERTs appears to be much more promising.

11.3 (+)-Lentiginosine and (−)-Lentiginosine

11.3.1 *Historical backgrounds*

Lentiginosine (Fig. 11.3), a dihydroxyindolizidine alkaloid, was first isolated from the leaves of *Astragalus lentiginosus* [44]. This

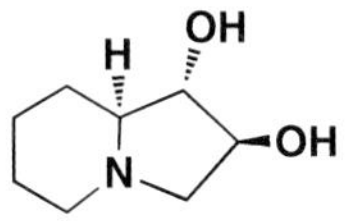

13

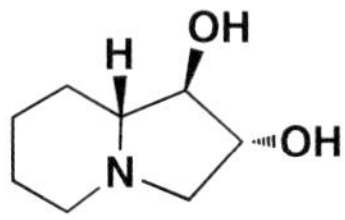

14

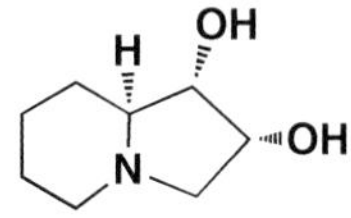

15

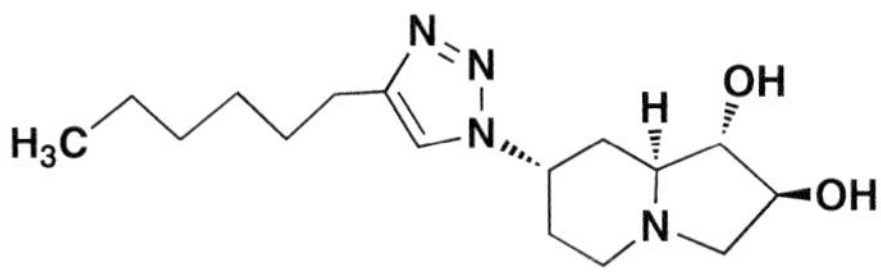

16

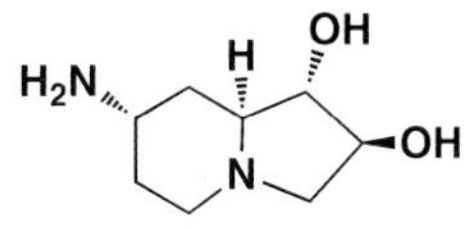

17

Figure 11.3 Structures of compounds **13–17**.

plant is widely distributed in the western United States and co-produces swainsonine (3). The ingestion by range animals of certain species of locoweeds (*Astragalus* and *Oxytropis* spp.) results in a chronic neurological disease known as locoism [45]. Swainsonine is a powerful inhibitor of lysosomal [46] and Golgi α-mannosidase II [47], while lentiginosine is a reasonably good inhibitor of fungal amyloglucosidase, with an inhibitory constant (K_i) value of 10 μM, but it does not inhibit other α-glucosidases, or any other glycosidases [44]. In 1995, on the basis of the synthesis of both (+)-lentiginosine (13) and (–)-lentiginosine (14) and their inhibition of amyloglucosidase, the natural alkaloid was assigned as (+)-lentiginosine [48]. Synthetic (+)-lentiginosine displayed amyloglucosidase inhibition (K_i = 2 μM) five times stronger than that reported for natural lentiginosine, and 35 times stronger than that measured for (–)-lentiginosine [48]. Pastuzak *et al.* also have reported the natural occurrence of 2-epilentiginosine (15) in the leaves of *A. lentiginosus* and its lack of glycosidase inhibitory activity [44]. However, 2-epilentiginosine, which was isolated from *Ipomoea carnea* together with swainsonine, was proven to be a potent inhibitor of rat lysosomal α-mannosidase, with a median inhibitory concentration (IC_{50}) value of 4.6 μM [49]. (+)-Lentiginosine and (–)-swainsonine have been postulated to be biosynthesized from (1*R*, 8*S*)-1-hydroxyindolizidine by hydroxylation at the C-2 position [44].

11.3.2 *Future prospects*

Heat shock protein 90 (Hsp90) is a highly conserved and essential stress protein (or molecular chaperone) that is expressed in all eukaryotic cells. Its functions result from the ability of Hsp90 to chaperone a wide range of client proteins that play a central pathogenic role in human diseases including cancer, neurodegenerative diseases, and viral infections [50,51]. Hsp90 plays an important role in the maintenance of multiple oncogenic pathways and is essential for maintaining the folding, stability, and conformation of several oncoproteins to help retain aberrant activity and sustain the malignant state [52]. Furthermore, it has also been observed that Hsp90 is over-expressed in a variety of human malignancies [53]. Hence, Hsp90 has emerged as an exciting target for the development of cancer chemotherapeutics, and its inhibitors have become an attractive chemotherapeutic approach. The druggability of Hsp90 was established using the natural products geldanamycin and radicicol mimicking the unusual adenosine triphosphate (ATP) structure adopted in an ATP binding pocket within the N-terminal domain of Hsp90 [54–57]. Thus, they lead to potent and selective inhibition of ATP binding and hydrolysis and, in turn, to the depletion of oncogenic clients through ubiquitin-mediated proteasomal degradation [58,59]. Although preclinical data with these natural products have heightened interest in Hsp90 as a drug target, results from initial studies indicated a lack of efficacy in most tumor types and the utility of these inhibitors was also limited by their lack of specificity, problems with solubility, and toxicity, particularly hepatotoxicity [52,60]. However, these natural products provided the chemical basis for successors that subsequently entered clinical trials. Many clinical and translational researches on Hsp90 have been based on the geldanamycin derivatives, such as 17-allylamino-17-demethoxygeldanamycin (17-AAG) and 17-dimethylaminoethylamino-17-demethoxygeldanamycin (17-DMAG). However, despite the promising results seen in certain cancers, development of 17-AAG (tanespimycin) as an anticancer drug was subsequently discontinued.

Very recently, Piaz *et al.* focused on natural (+)-lentiginosine with a simple structure, high water solubility, and significant biological activity, from screening to identify novel Hsp90 inhibitors among several natural compounds [61]. They characterized Hsp90 inhibition using a panel of chemical approaches by a surface plasmon resonance (SPR) based on a binding assay, and biological approaches including limited proteolysis, and biochemical and cellular assays. While synthetic (−)-lentiginosine did not bind to Hsp90, (+)-lentiginosine and the derivative (16), bearing a hydrophobic side chain on C-7, showed higher affinity towards Hsp90, with K_d values of 24.7 and 45.9 nM, respectively. On the other hand, introduction of a primary amino group to C-7 to give (+)-7-aminolentiginosine (17) worsened both the kinetics and the thermodynamics of its interaction with Hsp90. (+)-Lentiginosine inhibited the ATPase activity of Hsp90 in a concentration-dependent manner and showed a more potent inhibitory activity than 17-AAG. The effect of (+)-lentiginosine on the chaperone activity of Hsp90 was also investigated by monitoring the citrate synthase-thermal-induced aggregation in the presence of Hsp90. Upon incubation at elevated temperatures, citrate synthase undergoes quantitative aggregation, but the presence of stoichiometric amounts of Hsp90 significantly changes the protein aggregation kinetics [62]. Addition of a four-fold molar excess of (+)-lentiginosine to the reaction mixture gave a curve slope almost comparable to that observed without the chaperone, whereas the addition of the same amount of (−)-lentiginosine did not perturb the citrate synthase+Hsp90 curve. To identify the Hsp90 region involved in (+)-lentiginosine binding, Piaz *et al.* used limited proteolysis–mass spectrometry, the efficiency of which was recently demonstrated in the investigation of Hsp90–inhibitor complexes [63]. The proteolytic patterns observed on the Hsp90–(+)-lentiginosine complex confirmed a direct interaction occurring between (+)-lentiginosine and Hsp90 in a free native state, and suggested an interaction mainly involving the Hsp90 middle domain. Thus, the middle domain of Hsp90, but not an ATP binding pocket within the N-terminal domain, was suggested to be a promising binding site for new classes of Hsp90 inhibitors with multi-target anticancer potential.

In 2010, Macchi *et al.* reported an interesting biological activity of (−)-lentiginosine [64]. Both enantiomers of lentiginosine slightly increased cell proliferation in human mixed lymphocytes at low concentration (5 μM), but at higher concentration (0.5–1 mM), their effect was remarkably different. Both enantiomers inhibited cell proliferation, but the effect of (−)-lentiginosine was much more potent than (+)-enantiomer. (−)-Lentiginosine exhibited a proapoptotic activity towards tumor cells, but it was less proapoptotic towards normal cells and less cytotoxic. Furthermore, this apoptosis was caspase-dependent, as shown by the increased expression and activity of caspase-3 and -8 in treated cells. Recently, (−)-lentiginosine has been reported to increase caspase-9 expression by 1.5–3.1 folds in all tumor cell lines and to increase the cytochrome c level in the cytoplasm by 2.3–2.6 folds in treated cells, indicating a remarkable collapse of the mitochondrial membrane potential [65]. In addition, whether the p53 tumor suppressor is involved in (−)-lentiginosine-mediated cell death was investigated. In about half of all human cancers, p53 is either lost or mutated in a way that compromises its function, and normal p53 acts as a safeguard against cancer by regulating the expression of a series of genes that mediate cell cycle arrest or apoptosis [66]. Apoptosis by (−)-lentiginosine was induced in p53-deficient cell lines as well, indicating that it is p53-independent. These data obtained by (−)-lentiginosine encourage further investigation on similar compounds to make them an interesting platform for a new generation of anticancer drugs.

11.4 1,4-Dideoxy-1,4-imino-L-arabinitol (LAB) and Its Derivatives

11.4.1 *Historical background*

L-Iminosugars (Fig. 11.4), except L-1-deoxyfuconojirimycin (18) [67] and L-1-deoxyrhamnojirimycin (19) [68] mimicking natural L-sugars, have attracted little attention because of their presumed lack of biological activity profile. However, there are interesting early reports on the biological activities of one L-iminosugar. 1,4-Dideoxy-1,4-imino-D-arabinitol (DAB) (20), which was isolated from the seeds of *Angylocalyx pynaertii* (Leguminosae), is a powerful inhibitor of yeast α-glucosidase with an IC_{50} value of 0.18 μM [69, 70], while its synthetic L-enantiomer, 1,4-dideoxy-1,4-imino-L-arabinitol (LAB) (21), is a much more potent inhibitor of mammalian α-glucosidases than DAB [71]. Interestingly, DAB is a competitive inhibitor of α-glucosidases, but LAB was found to inhibit them in a noncompetitive manner [72]. The K_i values of DAB and LAB for rat intestinal isomaltase were 2.4 and 0.07 μM, respectively. Introduction of the hydroxymethyl group to the α-1-*C* position of DAB leads to 2,5-dideoxy-2,5-imino-D-mannitol (D-DMDP) (22), which was first found in the leaves of the legume *Derris elliptica* [73]. D-DMDP is now known as one of the most widespread of secondary metabolite sugar

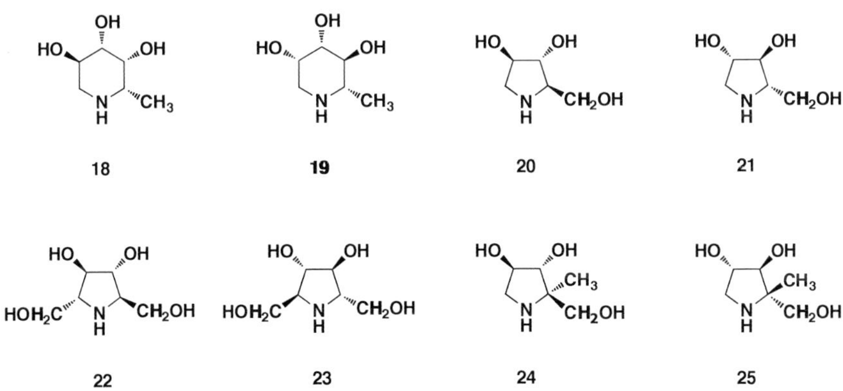

Figure 11.4 Structures of Compounds **18–25**.

mimics [74]. D-DMDP has long been reported as a good inhibitor of glycosidases, but the inhibitory properties of its enantiomer L-DMDP (23) have not been studied. D-DMDP is a potent inhibitor of α-glucosidase, β-glucosidase, and β-galactosidase [75]. Synthetic enantiomers of iminosugars frequently inhibit the same enzymes as the natural products [72,76–78]. L-DMDP was a highly specific inhibitor of plant and mammalian α-glucosidases, abolishing its inhibitory activity towards β-glucosidase and β-galactosidase [79, 80]. While the IC_{50} values of L-DMDP towards rat sucrase and isomaltase were 0.1 and 0.05 μM, respectively, those of D-DMDP were 55 and 75 μM, respectively. In 2011, da Cruz *et al.* have synthesized 4-*C*-Me-DAB (24) and 4-*C*-Me-LAB (25) and investigated their glycosidase inhibitory properties [81]. 4-*C*-Me-DAB was a competitive inhibitor of rat intestinal sucrase with a K_i value of 0.89 μM, whereas 4-*C*-Me-LAB was a noncompetitive inhibitor of the enzyme with a K_i value of 0.95 μM.

11.4.2 *Future prospects*

Tay–Sachs and Sandhoff diseases are LSDs resulting from decreased lysosomal β-hexosaminidase A (Hex A) activity. Hex A deficiency leads to the neuronal storage of GM2 ganglioside (GM2) and causes progressive neurodegenerative disease [82]. In humans, β-hexosaminidase exists as two major isoforms: Hex A and Hex B. The lysosomal β-hexosaminidases are dimers composed of α and/or β subunits. Hex A is a heterodimer of α (encoded by the *HEXA* gene) and β (encoded by the *HEXB* gene), whereas Hex B is a homodimer of β subunits. Hex S is a minor isoform and a homodimer of α subunits. Only Hex A has a unique physiological function, since Hex B and Hex S are unable to hydrolyze GM2 *in vivo* [83]. Tay–Sachs disease results from mutations of the *HEXA* gene with inherited deficiency of Hex A activity and normal expression of Hex B, while Sandhoff disease results from mutations of the *HEXB* gene, with resulting deficiency of both Hex A and Hex B [84]. The critical threshold level for disease has been estimated to be 5–10% of normal Hex A levels [85]. Currently, there is no therapy for both diseases. One promising therapy for these diseases involves the use of Hex A inhibitors as pharmacological chaperones to enhance the enzyme activity at subcritical levels. In 2004, Tropak *et al.* reported that, by growing adult Tay–Sachs and Sandhoff fibroblasts in culture medium containing hexosaminidase inhibitors, the residual protein and activity levels of intracellular Hex A elevated well above the critical 10% of normal levels [86]. Four Hex inhibitors used in this study were 2-acetamido-2-deoxynojirimycin (NJNAc) (26), 2-acetamido-1,2-dideoxynojirimycin (DNJNAc) (27), 6-acetamido-6-deoxycastanospermine (CASNAc) (28), and *N*-acetyl-glucosamine-thiazoline (NGT) (29), and their K_i values were 0.005 (for Jack bean Hex), 0.7 (for Jack bean Hex), 0.5 (for human placenta Hex), and 0.28 μM (for Jack bean Hex), respectively. In conclusion, the authors propose that these Hex inhibitors function as pharmacological chaperones, enhancing the stability of the native conformation of the enzyme, and increasing the amount of Hex A capable of exiting the ER for transport to the lysosome.

The finding that synthetic LAB and L-DMDP are more powerful and are specific inhibitors of α-glucosidase than natural DAB and D-DMDP, led to the design and synthesis of Hex inhibitors. Rountree *et al.* synthesized 2-acetamido-1,4-imino-1,2,4-trideoxy-D-arabinitol (DABNAc) (30), 2-acetamido-1,4-imino-1,2,4-trideoxy-L-arabinitol (LABNAc) (31), and their *N*-benzyl derivatives, *N*-Bn-DABNAc (32) and *N*-Bn-LABNAc (33), and evaluated their inhibitory activities towards a variety of glycosidases, including Hex [87]. As a result, DABNAc and *N*-Bn-DABNAc exhibited only weak inhibition towards some Hexs, but LABNAc and *N*-Bn-LABNAc were found to be potent inhibitors of these enzymes, with IC_{50} values in the micromolar to submicromolar range. None of these compounds showed inhibition towards glycosidases other than Hexs. *N*-Bn-LABNAc was a 2.5-fold more potent inhibitor of human placenta Hex than DNJNAc, with a K_i value of 3.7 μM, but LABNAc was 1.5-fold weaker than DNJNAc (K_i = 9.5 μM). LABNAc and *N*-Bn-LABNAc were also found to inhibit Hex in a noncompetitive manner, suggesting a similarity with LAB and L-DMDP. The cell line GM3461, which was obtained from a Tay–Sachs patient with the adult-onset form of the disease, was used to investigate the chaperoning effect of

N-Bn-LABNAc on the α subunit. The Hex activities were evaluated with the synthetic substrates, 4-methylumbelliferyl-β-D-*N*-acetylglucosamine (MUG), which is hydrolyzed by all Hex isoenzymes, and MUG-6-sulfate (MUGS), which is hydrolyzed only by Hex isoenzymes containing an α subunit, such as Hex A and Hex S [88]. *N*-Bn-LABNAc at a concentration of 50 μM increased intracellular Hex activity two-fold using MUG and, at 5 μM, it resulted in a similar increase using MUGS. These data support the enhancement of Hex B activity at higher concentrations of inhibitor, whereas Hex A and/or Hex S activity are increased at much lower concentrations. Thus, noncompetitive inhibitors of Hex, such as *N*-Bn-LABNAc, have been shown to be effective in raising Hex activity to levels that might be of therapeutic value to adult Tay–Sachs and Sandhoff patients. This is the second report on a chaperoning activity by a noncompetitive inhibitor. It is the first report that L-isofagomine acted as a pharmacological chaperone in Gaucher cells [89]. D-Isofagomine (D-IFG) (34) is a very strong inhibitor of human β-glucocerebrosidase

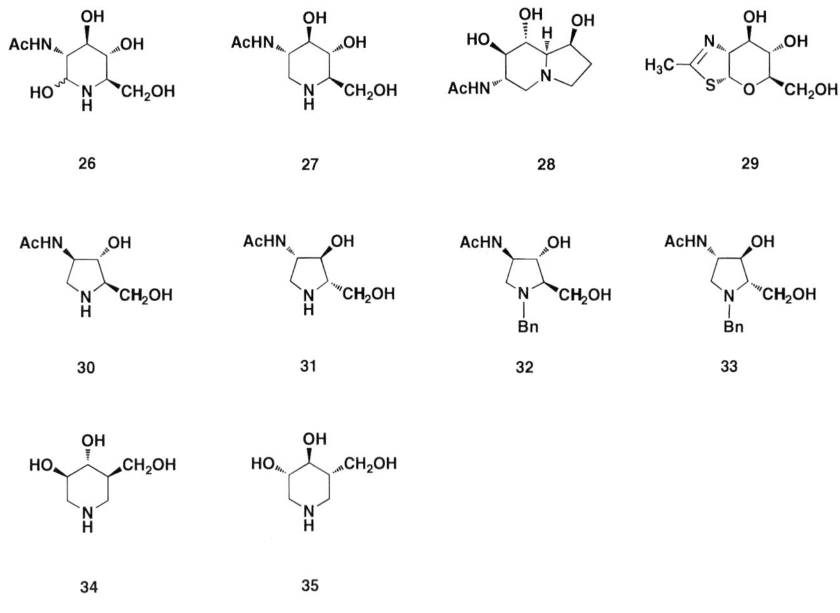

Figure 11.5 Structures of Compounds **26–35**.

(GCase) and inhibits the enzyme in a competitive manner, with a K_i value of 0.016–0.025 μM [72,90]. It is known to be an effective pharmacological chaperone for some Gaucher variants [91–93]. Surprisingly, L-IFG (36), a noncompetitive inhibitor of GCase with a K_i value of 5.7 μM, showed a chaperoning activity in a dose-dependent manner in N370S Gaucher fibroblasts, with a 1.6-fold increase at 500 μM [89]. On the other hand, treatment with 10 μM D-IFG caused a two-fold increase in GCase activity. These findings that noncompetitive inhibitors showed a chaperoning activity are very important, since all successful pharmacological chaperones for LSDs to date have been active-site-directed competitive inhibitors. Noncompetitive inhibitors of lysosomal glycosidases may be effective towards the variant with a mutation that destabilizes a domain distinct from the catalytic domain. A number of ligands with pharmacological specificity may have potential for the chemotherapeutic treatment of human genetic disorders.

11.5 Conclusions

Iminosugars are sugar analogs with a nitrogen in place of the ring oxygen of the corresponding sugar, and naturally occurring iminosugars are one of the most interesting discoveries and the most fascinating nature arts in the fields of natural product chemistry. The first natural iminosugar, NJ, showed novel antibiotic and glucosidase inhibitory activities [1,16]. This gave chemists and biologists of those days an impact. DNJ, which was originally prepared by the removal of the anomeric hydroxyl group of NJ [16], was later isolated from the cultured broth of bacteria and plants [18]. A large-scale production of DNJ with bacteria made it possible to introduce Glyset® and Zavesca® to market. The initial pharmaceutical interest in iminosugars was related to their properties as glycosidase inhibitors, but it has become clear that they do not need to be glycosidase inhibitors for expression of bioactivities, as seen in this review. In addition, high potency and selectivity in D-glycosidase inhibition seen in synthetic five-membered L-iminosugars may represent one of the key elements for an effective medicinal use of L-iminosugars.

There is no doubt that a number of natural and synthetic iminosugars will find new therapeutic applications once they are evaluated in novel and sometimes unexpected bioassays without getting caught up on "glycosidase inhibitors" and "structural similarity" with the corresponding sugars. Current state of our knowledge on iminosugars seems to be just the tip of the iceberg. Iminosugars and related compounds are waiting that their potential ability is developed by scientists' intelligence.

References

1. Inoue, S., Tsuruoka, T., Niida, T. (1966) *J. Antibiot.* **19**, 288–292.
2. Asano, N. (2003) *Glycobiology* **13**, 93R–104R.
3. Compain, P., Martin, O. R. Eds. (2007) *Iminosugars: From Synthesis to the Therapeutic Applications.* John Wiley & Sons Ltd. (West Sussex).
4. Horne, G., Wilson, F. X., Tinsley, J., Williams, D. H., Storer, R. (2011) *Drug Discov. Today* **16**, 107–118.
5. Horne, G., Wilson, F. X. (2011) *Prog. Med. Chem.* **50**, 135–176.
6. Nash, R. J., Kato, A., Yu, C. Y., Fleet, G. W. J. (2011) *Future Med. Chem.* **3**, 1513–1521.
7. Fan, J. Q. (2003) *Trends Pharmacol. Sci.* **24**, 355–360.
8. Fan, J. Q., Ishii, S. (2007) *FEBS J.* **274**, 4962–4971.
9. Fan, J. Q., Ishii, S., Asano, N., Suzuki, Y. (1999) *Nat. Med.* **5**, 112–115.
10. Stuetz, A. E. Ed. (1999) *Iminosugars as Glycosidase Inhibitors: Nojirimycin and Beyond.* Wiley-VCH (New York).
11. Yu, C. Y., Asano, N., Ikeda, K., Wang, M. X., Butters, T. D., Wormald, M. R., Dwek, R. A., Winters, A. L., Nash, R. J., Fleet, G. W. J. (2004) *Chem. Commun.* **2004**, 1936–1937.
12. Asano, N., Ikeda, K., Yu, L., Kato, A., Takebayashi, K., Adachi, I., Kato, I., Ouchi, H., Takahata, H., Fleet, G. W. J. (2005) *Tetrahedron: Asymmetry* **49**, 223–229.
13. Rountree, J. S. S., Butters, T. D., Wormald, M. R., Dwek, R. A., Asano, N., Ikeda, K., Evinson, E. L., Nash, R. J., Fleet, G. W. J. (2007) *Tetrahedron Lett.* **48**, 4287–4291.
14. Rountree, J. S. S., Butters, T. D., Wormald, M. R., Boomkamp, S. D., Dwek, R. A., Asano, N., Ikeda, K., Evinson, E. L., Nash, R. J., Fleet, G. W. J. (2009) *ChemMedChem* **4**, 378–392.
15. D'Alonzo, D., Guaragna, A., Palimbo, G. (2009) *Curr. Med. Chem.* **16**, 473–505.

16. Inoue, S., Tsuruoka, T., Ito, T., Niida, T. (1968) *Tetrahedron* **24**, 2125–2144.
17. Yagi, M., Kouno, T., Aoyagi, Y. (1976) *Nippon Nogeikagaku Kaishi* **50**, 571–572.
18. Watson, A. A., Fleet, G. W. J., Asano, N., Molyneux, R. J. (2001) *Phytochemistry* **56**, 265–295.
19. Junge, B., Matzke, M., Stoltefuss, J. (1996) In *Handbook of Experimental Pharmacology*. Kuhlmann, J., Pols, W. Eds. Springer (Berlin, Heidelberg, New York). pp. 411–482.
20. Van den Broek, L. A. G. M. (1997) In *Carbohydrates in Drug Design*. Witczak, Z. J., Nieforth, K. A. Eds. Marcel Dekker (New York). pp. 471–493.
21. Mehta, A., Zitzmann, N., Rudd, P. M., Block, T. M., Dwek, R. A. (1998) *FEBS Lett.* **430**, 17–22.
22. Zitzmann, N., Mehta, A. S., Carrouée, S., Butters, T. D., Platt, F. M., McCauley, J., Blumberg, B. S., Dwek, R. A., Block, T. M. (1999) *Proc. Natl. Acad. Sci. USA* **96**, 11878–11882.
23. Durantel, D., Branza-Nichita, N., Carrouée, S., Butters, T. D., Dwek, R. A., Zitzmann, N. (2001) *J. Virol.* **75**, 8987–8998.
24. Pavlović, D., Neville, D. C., Argaud, O., Blumberg, B., Dwek, R. A., Fischer, W. B., Zitzmann, N. (2003) *Proc. Natl. Acad. Sci. USA* **100**, 6104–6108.
25. Harada, T., Tautz, N., Thiel, H. J. (2000) *J. Virol.* **74**, 9498–9506.
26. Carrère-Kremer, S., Montpellier-Pala, C., Cocquerel, L., Whchowski, C., Penin, F., Dubuisson, J. (2002) *J. Virol.* **76**, 3720–3730.
27. Carrasco, L. (1995) *Adv. Virus Res.* **45**, 61–112.
28. Wraith, J. E., Vecchio, D., Jacklin, E., Abel, L., Chadha-Boreham, H., Luzy, C., Giorgino, R., Patterson, M. C. (2010). *Mol. Genet. Metab.* **99**, 351–357.
29. Schiffmann, R., FitzGibbon, E. J., Harris, C., DeVile, C., Davies, E. H., Abel, L., Schaik, I. N. V., Benko, W. S., Timmons, M., Ries, M., Vellodi, A. (2008) *Ann. Neurol.* **64**, 514–522.
30. Zimran, A., Elstein, D. (2003) *Philos. Trans. R. Soc. London, B: Biol. Sci.* **358**, 961–966.
31. Helenius, A., Aebi, M. (2004) *Annu. Rev. Biochem.* **73**, 1019–1049.
32. Qu, X., Pan, X., Weidner, J., Yu, W., Alonzi, D., Xu, X., Butters, T., Block, T., Guo, J. T., Chang, J. (2011) *Antimicrob. Agents Chemother.* **55**, 1036–1044.
33. Chang, J., Schul, W., Yip, A., Xu, X., Guo, J. T., Block, T. M. (2011) *Antiviral Res.* **92**, 369–371.

34. Yu, W., Gill, T., Wang, L., Du, Y., Ye, H., Qu, X., Guo, J. T., Cuconati, A., Zhao, K., Block, T. M., Xu, X., Chang, J. (2012) *J. Med. Chem.* **55**, 6061–6075.
35. Fan, J. Q., Ishii, S., Asano, N., Suzuki, Y. (1999) *Nat. Med.* **5**, 112–115.
36. Fan, J. Q. (2003) *Trend Pharmacol. Sci.* **24**, 355–360.
37. Fan, J. Q., Ishii, S. (2007) *FEBS J.* **274**, 4962–4971.
38. Valenzano, K. J., Khanna, R., Powe, A. C., Boyd, R., Lee, G., Flanagan, J. J., Benjamin, E. R. (2011) *Assay Drug Dev. Technol.* **9**, 213–235.
39. Kishnani, P. S., Howell, R. R. (2004) *J. Pediatr.* **144**, 535–543.
40. Van der Ploeg, A. T., Reuser, A. J. J. (2008) *Lancet* **372**, 1342–1353.
41. Parenti, G., Zuppaldi, A., Pittis, M. G., Tuzzi, M. R., Annunziata, I., Meroni, G., Porto, C., Donaudy, F., Rossi, B., Rossi, M., Filocamo, M., Donati, A., Bembi, B., Ballabio, A., Andria, G. (2007) *Mol. Ther.* **15**, 508–514.
42. Flanagan, J. J., Rossi, B., Tang, K., Wu, X., Mascioli, K., Donaudy, F., Tuzzi, M. R., Fontana, F., Cubellis, V., Porto, C., Benjamin, E., Lockhart, D. J., Valenzano, K. J., Andria, G., Parenti, G., Do, H. V. (2009) *Hum. Mutat.* **30**, 1683–1692.
43. Khanna, R., Flanagan, J. J., Feng, J., Soska, R., Frascella, M., Pellegrino, L. J., Lun, Y., Guillen, D., Lockhart, D. J., Valenzano, K. J. (2012) *PLoS One* **7**, e40776.
44. Pastuszak, I., Molyneux, R. J., James, L. F., Elbein, A. D. (1990) *Biochemistry* **29**, 1886–1891.
45. Molyneux, R. J., James, L. F. (1982) *Science* **216**, 190–191.
46. Dorling, P. R., Huxtable, C. R., Colegate, S. M. (1980) *Biochem. J.* **191**, 649–651.
47. Fuhrmann, U., Bause, E., Legler, G., Ploegh, H. (1984) *Nature* **307**, 755–758.
48. Brandi, A., Cicchi, S., Cordero, F. M., Fringnoli, R., Goti, A., Picasso, S., Vogel, P. (1995) *J. Org. Chem.* **60**, 6806–6812.
49. Haraguchi, M., Gorniak, S. L., Ikeda, K., Minami, Y., Kato, A., Watson, A. A., Nash, R. J., Molyneux, R. J., Asano, N. (2003) *J. Agric. Food Chem.* **51**, 4995–5000.
50. Xu, W., Neckers, L. (2007) *Clin. Cancer Res.* **13**, 1625–1629.
51. Mahalingal, D., Swords, R., Carew, J. S., Nawrocki, S. T., Bhalla, K., Giles, F. J. (2009) *Brit. J. Cancer* **100**, 1523–1529.
52. Wang, Y., Trepel, J. B., Neckers, L. M., Giaccone, G. (2010) *Curr. Opin. Investig. Drugs* **11**, 1466–1476.
53. Calderwood, S. K., Ciocca, D. R. (2008) *Int. J. Hyperther.* **24**, 31–39.
54. Uehara, Y. (2003) *Curr. Cancer Drug Targets* **3**, 325–330.

55. Shulte, T. W., Akinaga, S., Soga, S., Sullivan, W., Stensgard, B., Toft, D., Neckers, L. M. (1998) *Cell Stress Chaperones* **3**, 100–108.
56. Grenert, J. P., Sullivan, W. P., Fadden, P., Haystead, T. A., Clark, J., Mimnaugh, E., Krutzsch, H., Ochel, H. J., Schulte, T. W., Sausville, E., Neckers, L. M. *J. Biol. Chem.* **272**, 23843–23850.
57. Roe, S. M., Prodromou, C., O'Brien, R., Ladbury, J. E., Piper, P. W., Pearl, L. H. (1999) *J. Med. Chem.* **42**, 260–266.
58. Mimnaugh, E. G., Chavany, C., Neckers, L. (1996) *J. Biol. Chem.* **271**, 22796–22801.
59. Schneider, C., Sepp-Lorenzino, L., Nimmesgern, E., Ouerfelli, O., Danishefsky, S., Rosen, N., Hartl, F. U. (1996) *Proc. Natl. Acd. Sci. USA* **93**, 14536–14541.
60. Kim, Y. S., Alarcon, S. V., Lee, S., Lee, M. J., Giaccone, G., Neckers, L., Trepel, J. B. (2009) *Curr. Top. Med. Chem.* **9**, 1479–1492.
61. Dal Piaz, F., Vassallo, A., Chini, M. G., Cordero, F., Pisano, C., Bifulco, G., Tommasi, N. D., Brandi, A. (2012) *PLoS One* **7**, e43316.
62. Jacob, U., Lilie, H., Meyer, I., Buchner, J. (1995) *J. Biol. Chem.* **270**, 7288–7294.
63. Giommarelli, C., Zuco, V., Favini, E., Pisano, C., Dal Piaz, F., De Tommasi, N., Zunino, F. (2010) *Cell. Mol. Life Sci.* **67**, 995–1004.
64. Macchi, B., Minutolo, A., Grelli, S., Cardona, F., Cordero, F. M., Mastino, A., Brandi, A. (2010) *Glycobiology* **20**, 500–506.
65. Minutolo, A., Grelli, S., Marino-Merlo, F., Cordero, F. M., Brandi, A., Macchi, B., Mastino, A. (2012) *Cell Death Dis.* **3**, e358.
66. Muller, P. A., Vousden, K. H., Norman, J. C. (2011) *J. Cell Biol.* **192**, 209–218.
67. Fleet, G. W. J., Shaw, A. N., Evans, S. V., Fellows, L. E. (1985) *J. Chem. Soc. Chem. Commun.* **1985**, 841–842.
68. Zhou, P., Salleh, M. H., Chan, P. C. M., Lajoie, G., Honek, J. F., Chandra Nambiar, P. T., Ward, O. P. (1993) *Carbohydr. Res.* **239**, 155–166.
69. Nash, R. J., Bell, E. A., Williams, J. M. (1985) *Phytochemistry* **24**, 1620–1622.
70. Fleet, G. W. J., Nicholas, S. J., Smith, P. W. (1985) *Tetrahedron Lett.* **26**, 3127–3130.
71. Scofield, A. M., Fellows, L. E., Nash, R. J., Fleet, G. W. J. (1986) *Life Sci.* **39**, 645–650.
72. Asano, N., Ikeda, K., Yu, L., Kato, A., Takebayashi, K., Adachi, I., Kato, I., Ouchi, H., Takahata, H., Fleet, G. W. J. (2005) *Tetrahedron Asymmetry* **16**, 223–229.

73. Welter, A., Jadot, J., Dardenne, G., Marlier, M., Casimir, J. (1976) *Phytochemisry* **15**, 747–749.
74. Alison, A. A., Fleet, G. W. J., Asano, N., Molyneux, R. J., Nash, R. J. (2001) *Phytochemistry* **56**, 265–295.
75. Asano, N., Oseki, K., Kizu, H., Matsui, K. (1994) *J. Med. Chem.* **37**, 3701–3706.
76. Kato, A., Kato, N., Kano, E., Adachi, I., Ikeda, K., Yu, L., Okamoto, T., Banba, Y., Ouchi, H., Takahata, H., Fleet, G. W. J. (2005) *J. Med. Chem.* **48**, 2036–2044.
77. Clinch, K., Evans, G. B., Fleet, G. W. J., Furneaux, R. H., Johnson, S. W., Lenz, D. H., Mee, S. P. H., Rands, P. R., Schramm, V. L., Taylor, R. E. A., Tyler, P. C. (2006) *Biomol. Chem.* **16**, 1131–1139.
78. D'Alonzo, D., Guaragna, A., Palumbo, G. (2009) *Curr. Med. Chem.* **16**, 473–505.
79. Yu, C. Y., Asano, N., Ikeda, K., Wang, M. X., Butters, T. D., Wormald, M. R., Dwek, R. A., Winters, A. L., Nash, R. J., Fleet, G. W. J. (2004) *Chem. Comm.* **2004**, 1936–1937.
80. Ayers, B., Ngo, N., Jenkinson, S. F., Martínez, R. F., Shimada, Y., Adachi, I., Weymouth-Wilson, A. C., Kato, A., Fleet, G. W. J. (2012) *J. Org. Chem.* **77**, 7777–7792.
81. da Cruz, F. P., Newberry, S., Jenkinson, S. F., Wormald, M. R., Butters, T. D., Alonzi, D. S., Nakagawa, S., Becq, F., Norez, C., Nash, R. J., Kato, A., Fleet, G. W. J. (2011) *Tetrahedron Lett.* **52**, 219–223.
82. Kolodny, E. H. (2001) *Adv. Genet.* **44**, 1010–1126.
83. Kolodny, E. H. (2007) In *The Molecular and Genetic Basis of Neurologic and Psychiatric Disease. 4.* Rosenberg, R. N., DiMauro, S., Paulson, H., Ptacek, L., Nestler, E. Eds. Lippincott, Williams & Wilkins (Philadelphia). pp. 221–229.
84. Gravel, R. A., Clarke, J. T. R., Kaback, M. M., Mahuran, D., Sandhoff, K., Suzuki, K. (1995) In *The Metabolic and Molecular Bases of Inherited Diseases. 7.* Scriver, C. R., Beaudet, A. L., Sly, W. S., Valle, D. Eds. McGraw-Hill (New York) 2, pp. 2839–2879.
85. Conzelmann, E., Sandhoff, K. (1984) *Dev. Neurosci.* **6**, 58–71.
86. Tropak, M. B., Reid, S. P., Guiral, M., Withers, S. G., Mahuran, D. (2004) *J. Biol. Chem.* **279**, 13478–13487.
87. Rountree, J. S. S., Butters, T. D., Wormald, M. R., Boomkamp, S. D., Dwek, R. A., Asano, N., Ikeda, K., Evinson, E. L., Nash, R. J., Fleet, G. W. J. (2009) *ChemMedChem* **4**, 378–392.
88. Kytzia, H. J., Sandhoff, K. (1985) *J. Biol. Chem.* **260**, 7568–7572.

89. Kuriyama, C., Kamiyama, O., Ikeda, K., Sanae, F., Kato, A., Adachi, I., Imahori, T., Takahata, H., Okamoto, T., Asano, N. (2008) *Bioorg. Med. Chem.* **16**, 7330–7336.
90. Zhu, X., Sheth, K. A., Li, S., Chang, H. H., Fan, J. Q. (2005) *Angew. Chem. Int. Ed.* **44**, 7450–7453.
91. Chang, H. H., Asano, N., Ishii, S., Ichikawa, Y., Fan, J. Q. (2006) *FEBS J.* **273**, 4082–4092.
92. Yu, Z., Sawker, A. R., Whalen, L. J., Wong, C. H., Kelly, J. W. (2007) *J. Med. Chem.* **50**, 94–100.
93. Steet, R. A., Chung, S., Wustman, B., Powe, A., Do, J., Kornfeld, S. A. (2006) *Proc. Natl. Acad. Sci. USA* **103**, 13813–13818.

Chapter 12

Vaccines against Bacterial Pathogens: A General Overview

Darielys Santana Medero, Dagmar Garcia Rivera, Yury Valdes Balbin and Vicente Verez Bencomo

Center for Biomolecular Chemistry, Calle 21 y 200, Atabey, Playa, La Habana, Cuba

With the exception of hygienic water, nothing has probably been more important in the history of infectious diseases and medicine in general than vaccines. According to the World Health Organization, some 20 different vaccines are recommended for routine use in humans [1], characterized by a high diversity. For example, the antituberculous *Mycobacterium bovis* Bacillus Calmette–Guerin (BCG) and smallpox vaccines are made from antigenically related strains belonging to the same bacterial species or viral family [2]. Other vaccines are living attenuated viruses, such as for measles, mumps, or rotavirus. In some cases, a single protein may be used as a vaccine: for example, purified inactivated toxins from diphtheria and tetanus or recombinant hepatitis B antigen.

From the list of 15 vaccines ranked by sales and representing the 15 top-selling vaccines of 2012, six are carbohydrate-based vaccines [3]. As a common factor, they are based on bacterial capsular polysaccharide (CPS) and all of them target meningitis and pneumonia caused by three pathogens: *Streptococcus pneumoniae*, *Haemophilus influenzae* type b and *Neisseria meningitidis*. Despite the sales, a big effort is undertaken worldwide for the introduction of conjugate vaccines against these three pathogens as there is sufficient evidence of their potential in preventing child morbidity and mortality.

The first clinical evaluation of a pneumococcal vaccine was conducted more than a century ago in 1911 by Sir Almroth E. Wright, who tested whole-cell heat-killed pneumococcal vaccines among several thousand gold miners in Johannesburg, South Africa [4]. The development of pneumococcal vaccines illustrates very well the big progress achieved in a century, from the vague understanding of the germ theory of disease to the current success of conjugate vaccines. It was originally based on the awareness that the cell surface of a pathogenic microorganism is rich in CPS molecules that confer virulence by virtue of their ability to shield the bacterium from the actions of complements. Strong evidence of the role of anti-CPS antibodies in the protection from infectious diseases paved the way to vaccine design based on purified capsular polysaccharide or their polysaccharide–protein conjugate.

Today, it is accepted as a constituted knowledge that the best way to induce antibodies against a bacterial polysaccharide is through carbohydrate–protein conjugate vaccines. The first example of glycoconjugate vaccines introduced in clinical practice was against *H. influenzae* type b (Hib). The studies conducted for the licensure of Hib glycoconjugate vaccines were extended for several years with different types of glycoconjugates [5–8]. Later in 2004, a vaccine containing a synthetic oligosaccharide fragment was also licensed and introduced [9].

The wide range of different structural models applied to the development of Hib conjugate vaccines (Table 12.1) has lead to effective vaccines, showing that polysaccharide–protein conjugation is a generic methodology for the production of novel bacterial vaccines against pathogens for which no other successful prophylaxis exist. Hib conjugate vaccines have had an enormous impact on reducing the burden of infectious disease worldwide, leading to 50–95% decreases in the incidence of Hib disease and to the prevention of 78% of all cases of meningitis in children aged 0–4 years in the developed world [10].

Table 12.1 Hib conjugate vaccines licensed for use among children.

Hib

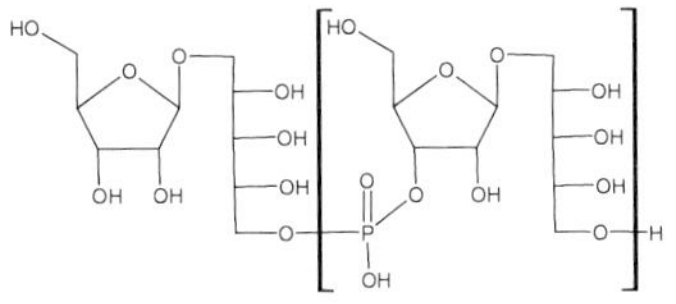

Polysaccharide	Protein carrier	Hib Conjugation chemistry	Manufacturer	Trade-name	Vaccine
Small	CRM_{197}	Reductive amination	Wyeth-Lederle	HibTITER	HbOC
Small	CRM_{197}	CDI	Chiron	VaxemHib	HbOC
Medium	Meningococcus group B OMP	Thiol-bromoacetate	Marck	PedvaxHib	PRP-OMP
Large	Tetanus toxoid	CDI	Sanofi-Pasteur/GSK	ActHIB/OmniHIB	PRP-T
Synthetic oligosaccharide	Tetanus toxoid	Thiol-maleimido	Heber biotec	QuimiHib	PRP-T

Notes: CRM_{197} diphtheria toxin mutant protein, Meningococcus group B OMP *Neisseria meningitidis* group B outer membrane protein, CDI carbodiimide, GSK GlaxoSmithKline.

A monovalent conjugate vaccine for *N. meningitidis* serogroup C was introduced in the United Kingdom (UK) with a dramatic effect on serogroup C disease [11]. Another monovalent glycoconjugate vaccine against serogroup A, named MenAfriVac®, was developed and introduced in Africa to combat epidemics in the meningitis belt [12].

Hib is rather exceptional, but for the other pathogens, several serogroups or serotypes differing in their capsular polysaccharides are involved in the disease. Therefore, the success of monovalent glycoconjugate vaccines was followed by the introduction of more complex multivalent vaccines, containing several components individually conjugated to a carrier protein. The introduction of the first pneumococcal conjugate vaccine PCV7 (Prevnar®, Wyeth, PA, USA) marked a breakthrough in the development of glycoconjugate vaccines. Seven individual *S. pneumoniae* CPS from serotypes 4, 6B, 9V, 14, 18C, 19F, and 23F were individually submitted to periodic oxidation, conjugated to CRM197 carrier protein, combined, and adsorbed on aluminum phosphate. Their doses are as low as 2 or 4 μg of conjugate polysaccharide [13], compared to 25 μg of the polysaccharide in unconjugated polysaccharide vaccines. The vaccine led to a decrease in invasive pneumococcal disease by 75% in children younger than 5 years and by 94% for the pneumococcal serotypes contained in the vaccine [14, 15].

The need to include serotypes 1 and 5 is more prevalent in Latin America, Asia, and Africa, and the reemergence in North America and Europe of some serotypes absent from the initial formulation prompted the development of even more complex glycoconjugate vaccine PCV10, adding serotypes 1, 5, and 7F [16] (Synflorix™, GlaxoSmithKline Biologicals, Belgium), and PCV13, including those three and additionally serotypes 3, 6A, and 19A in a new version of Prevnar [17] (Pfizer, USA). The development of pneumococcal conjugate vaccines illustrates very well the state of the art and the level that could be attained with present technology, despite the diversity of structure of capsular polysaccharides (Fig. 12.1).

For *N. meningitidis*, a more complex tetravalent conjugate vaccine containing the CPS from serogroups A, C, W135, and Y conjugate to carrier protein was also licensed and is in use in several countries (Fig. 12.2), where the diversity of capsular polysaccharides is also important [18].

Conjugate vaccines were also successful in preventing typhoid fever, fever an important cause of morbidity and mortality in developing countries [19]. Infection occurs when *S. enterica* sv. Typhi colonizes the intestine and is rapidly disseminated intracellularly. Surprisingly, even though it is an enteric pathogen, protection correlates with the level of serum immunoglobulin G (IgG) antibodies and is consistent with the hypothesis that mucosally-secreted IgG antibodies should prevent intestinal colonization [20], similar to that observed with conjugate vaccines for meningitis, pneumonia, and nasopharyngeal colonization.

There are not commercial vaccines available against other important encapsulated bacteria affecting human beings; examples are *S. aureus* 5 and 8, *N. meningitidis* serogroup X, *E. coli* K types 1, 2, 5, 12 and 13, and Group B *Streptococcus* types I, II, and III.

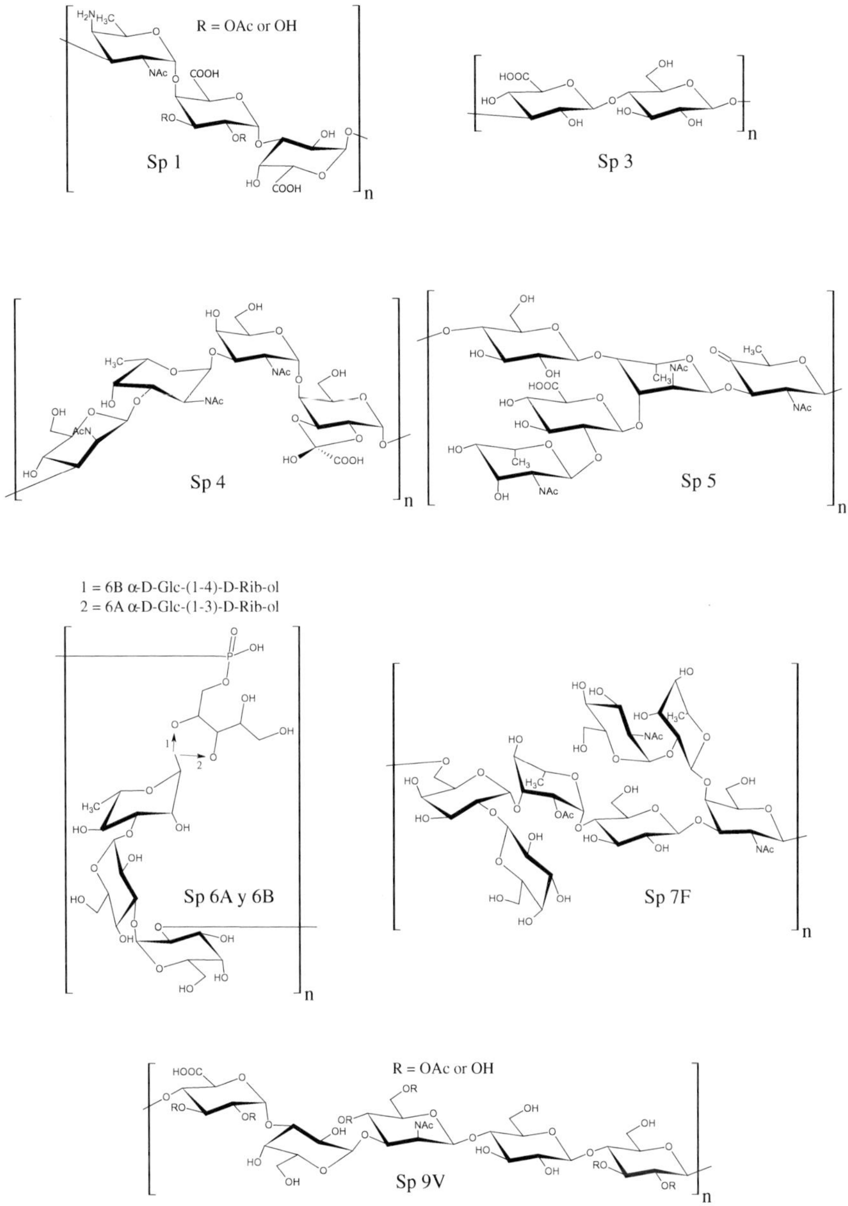

Figure 12.1 Structure of capsular polysaccharides included in licensed pneumococcal conjugate vaccines.

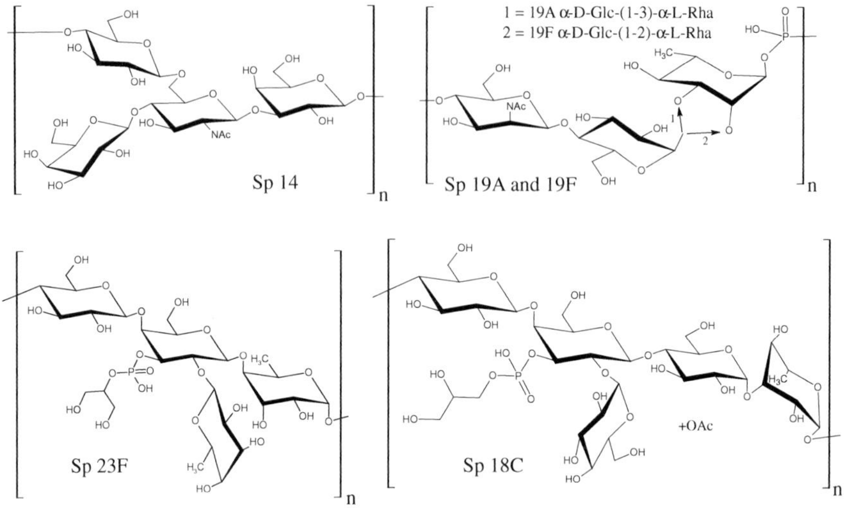

Figure 12.1 (*Continued*)

For CPS, the identification and isolation of the antigenic molecule [21, 22, 23], the preparation of carbohydrate–protein conjugate and the proof of protection could be performed in a relatively straightforward way. Technologies are now available for isolation of a pure CPS from the surface of bacteria and also for the conjugation to a protein [24] to produce the desired immunogen [25]. The feasibility of use of a synthetic oligosaccharide–protein conjugate in specific cases was also demonstrated for Hib [9].

The development of a conjugate vaccine against the aforementioned pathogens is not a matter of technology. Group B *Streptococcus*, for example, is actually the leading cause of neonatal sepsis and meningitis, associated with significant morbidity and mortality [26]. Protection of newborns requires a protective level of type-specific antibodies in pregnant women [27], and the problems presented by vaccination of pregnant women limit clinical trials of conjugate vaccines in this population [28].

The role of CPS as a virulence factor and the protective role of anti-CPS antibodies have been shown to be more complex in some cases, such as the 'microencapsulated' *S. aureus* types 5 and 8.

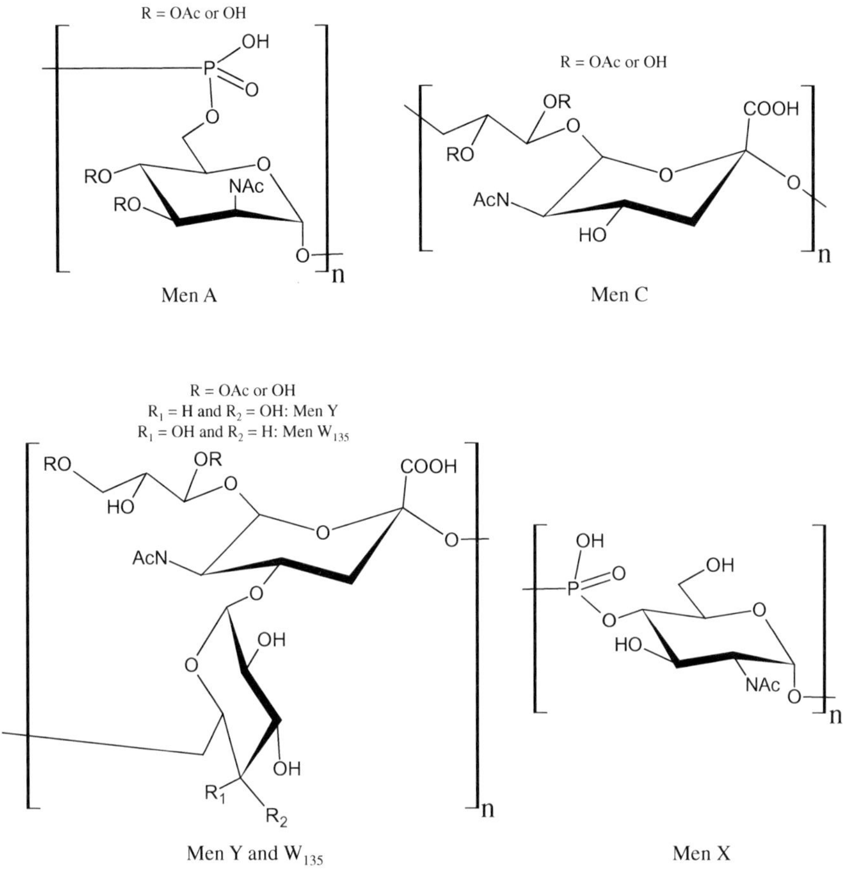

Figure 12.2 Structure of capsular polysaccharides from serogroups A, C, W135, and Y included in licensed meningococcal conjugate vaccines. Currently, meningococcal X (Men X) vaccine is in development stage.

About 70–80% of *S. aureus* strains produce one of two CPS antigens, CPS5 or CPS8 [29], and more than 90% also produce another surface polysaccharide, poly-*N*-acetyl glucosamine (PNAG) [30]. The use of capsular polysaccharide–conjugate vaccine targeting specific serotypes 5 and 8 failed to show efficacy in preventing staphylococcal bacteremia in hemodialysis patients [31], and also in preventing nasopharyngeal colonization [32]. Bacterial production of the capsule is under genetic and environmental control [33, 34]. While encapsulation promotes infection in most models and the antibodies against the CPS provide protection, in other models the encapsulation reduces virulence [35]. A promising alternative could be the induction of antibodies against the PNAG of bacterial biofilms [36]. In this particular case, the use of PNAG could be a promising alternative [37].

Other bacteria, such as *N. meningitidis* serogroup B and *E. coli* K1, have polysaccharides that are not immunogenic in humans, probably due to a great structural similarity of the CPS to some structures in the host cells [38]. Even conjugation to tetanus toxoid (TT) did not improve the immunogenicity of this polysaccharide [39]. In this case, the human immune system fails to elicit an adaptive immune response to the α-2,8-linked sialic acid homopolymeric capsule of serogroup B meningococci, which structurally resembles saccharides on human neural cell adhesion molecules [40] and oligosaccharides on fetal tissues [41]. In an attempt to overcome the problem, the *N*-acetyl group was replaced by *N*-propionyl, obtaining conjugates to TT with improved immunogenicity. Unfortunately, the project was stepped down by the perceived risk of autoimmunity, the poor performance in the clinical trial [42], and the recent success of a vaccine base on selected protein of *N. meningitidis* group B licenced as Bexsero [43].

N. meningitidis group X became only recently an important cause of meningitis in Africa [44]. Conjugates of medium size polysaccharide fragments to CRM197 [45] and of small fragments to TT and diphtheria toxin (DT) were recently reported [46]. In both cases, the conjugates induced the production of antibodies with bactericidal activity. A synthetic approach was also undertaken with mono-, di-, and trisaccharides [47]. Certainly, a vaccine against Men X will be developed and ready to enter clinical evaluation in the next future.

Several important human pathogens like *Vibrio cholera*, *Shigella*, or *Salmonella* are unencapsulated. For *Shigella*, the ethiological agent of shigellosis, a prospective study demonstrated an association between serum antibodies to lipopolysaccharide (LPS)-carbohydrates and the attack rate of the disease [48]. Several other relevant finding supported the initial assumption that serum IgG anti-LPS of enteric nonencapsulated bacteria could confer immunity by killing the inoculums of homologous bacteria on the epithelial surface of the small intestine [49].

However, at present, the development of conjugate vaccines with the LPS *O*-polysaccharide is not as straightforward as that for CPS. The needed *O*-polysaccharides could be obtained from the purified LPS after acid or base hydrolysis [50,51]. They are usually shorter than the CPS, but after conjugation to a protein, their immunological properties resemble those of the CPS.

Several conjugates of *Shigella sonnei* and *Shigella flexneri* 2a were obtained with bacterial toxoids [52] and tested in clinical trials with adults [53]. After improvement of the conjugates by selection of *Pseudomonas aeruginosa* recombinant mutant exoprotein A and its prior succinylation [54], the conjugates were evaluated in a Phase III study conducted with 2,799 healthy children 1–4 years old. Two 25-μg doses of the conjugate were injected and the children were followed for two years for episodes of acute diarrhea. As a result, Shigella conjugates were demonstrated to be safe and immunogenic in children 1–4 years old, and the *S. sonnei* conjugate elicited 71.1% efficacy in children 3–4 years old and can be predicted to be efficacious in individuals older than 3 years of age. The purification of *O*-polysaccharides from LPS is more complex than from the CPS; a lot of effort was dedicated to the chemical synthesis of large oligosaccharide fragments and to the study of their protein conjugates from *Shigella dysenteriae* type 1 [55] and *S. flexneri* 2a [56]. Some of these conjugates could enter in clinical trials in the future.

Another excellent example to illustrate the state of the art in this particular field is *Vibrio cholerae*. There are two serotypes of *V. cholera* causing the disease: O1 and O139. *V. cholerae* O1 is an unencapsulated bacterium, which depends on its LPS for colonization. *V. cholerae* O139 has, in contrast, a capsular polysaccharide. The role of anti-carbohydrate antibodies in both cases were considered relevant since the seventies [57].

The O1 serogroup can be further classified into the Ogawa and Inaba serotypes, by a difference in the terminal nonreducing D-perosamine residue that is *O*-methylated in the Ogawa

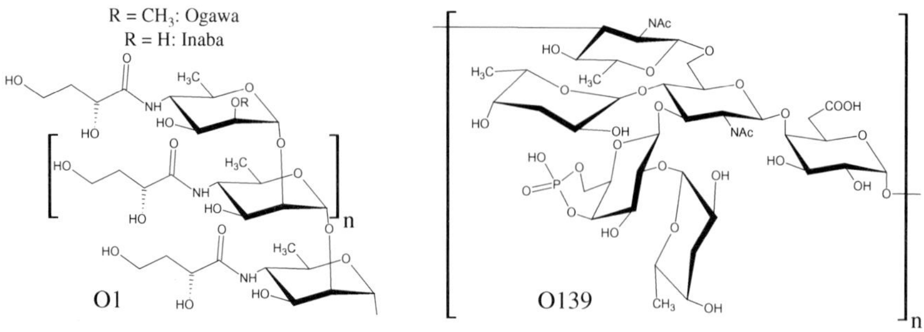

Figure 12.3 Structure of the O1 and O139 polysaccharides from serotypes of *Vibrio cholera.*

serotype [58] (Fig. 12.3). Both polysaccharides were employed in an attempt to prepare protein conjugates as potential candidates [59]. The *O*-polysaccharides were obtained by their treatment with sodium hydroxide (NaOH) or hydrazine, to result in de-*O*-acylation. It should be noted that any de-*N*-acylation, which could take place along with de-*O*-acylation, would be undesirable, as it could also result in partial cleavage of the tetronic acid side chains present in the *O*-specific polysaccharide (*O*-SP).

Inaba *O*-polysaccharide conjugated to cholera toxin was compared in a clinical trial with a whole cell vaccine. The conjugates induced a vibriocidal response, nevertheless, four-fold lower than that for the cellular vaccine [60]. As interest in preventing cholera continues today, the preparation of better immunogens is pursued by exploring conjugation methods and carrier proteins and by changing the isolation methods of the *O*-polysaccharide to acid hydrolysis [61].

Ogawa polysaccharides were also conjugated to human serum albumin (HSA) using β-glucan as a matrix [62]. The conjugates induce vibriocidal activity and improve IgG and IgA antibody responses [63]. The comparison of methods for purification of the polysaccharide from the LPS also favor acid hydrolysis vs. hydrazinolysis [64].

As the isolation of the polysaccharide from the LPS is complex, the synthesis of several fragments from Ogawa and Inaba polysaccharides, as well as their protein conjugates, was accomplished [65]. The glycoconjugates of synthetic hexasaccharides induce a strong antibody response that, surprisingly, was vibriocidal only in the case of Ogawa [66,67].

One dose of currently available oral cholera vaccines is more immunogenic than conjugate vaccines, but their shortcoming is the relatively short-term protection against cholera. An important role in the future for LPS conjugate vaccines could be in a so-called "prime-boost strategy", consisting of priming with a cellular vaccine, followed by a booster with a conjugate. Preliminary results in mice showed an improvement in protection [68].

V. cholerae O139 was approached in a similar way. Conjugates were obtained either from the CPS [69] or from *O*-polysaccharides isolated by acid hydrolysis of the LPS [70]. Both the LPS and the CPS share the same epitopes. Conjugates of the CPS, with recombinant DT or with the *O*-polysaccharide with TT, elicited high levels of serum anti-CPS IgG in mice with vibriocidal activity. Further clinical development of some of these glycoconjugate vaccines is expected in the future.

The *O*-polysaccharide from the LPS of *Salmonella* Paratyphi A was also recently explored for conjugate vaccines, as the bacterium represents a significant proportion of enteric fever in Asia. The *O*-polysaccharide was released directly from the culture supernatant by acetic acid hydrolysis through the 3-deoxy-D-manno-octulosonic acid (KDO) moiety of the LPS core. Their conjugates to CRM197 protein induce a strong bactericidal activity [71]. Further development of a bivalent conjugate vaccine, including *S.* Typhi and Paratyphi A, could be expected in the near future.

Moraxella catarrhalis is an increasing cause of otitis media and sinusitis in children, and lower respiratory tract infections in adults [72]. The reduction of colonization of other common respiratory pathogens as a consequence of the introduction of glycoconjugate vaccines is probably the reason for this increase and justifies the need of a vaccine for *M. catarrhalis.* The bacterium is unencapsulated with a lipooligosaccharide as the major surface antigen targeted by bactericidal antibodies. Serotypes A and B account for 90% of clinical isolates [73].

The oligosaccharide fraction from types A [74] and B [75] were obtained by hydrazinolysis and conjugated to carrier proteins. The conjugates were immunogenic for the parent serotypes in mice and rabbits with elevated complement-mediated bactericidal activity against the homologous strain. Synthetic oligosaccharides up to undecamer and their conjugates were recently prepared and these oligosaccharide–protein conjugates elicit a cross-reactive protection, as they could also be considered potential vaccine candidates [76]. Further development into clinical evaluation is expected for the near future.

The success of glycoconjugate vaccines against the bacteria representing the three most important respiratory pathogens together with recent advances in glycosciences concerning screening, detection, isolation, and synthesis of surface carbohydrates, has allowed an upsurge of new potential glycoconjugate vaccines in the last 4 years. Conjugate vaccines that were impossible to conceive 15 years ago are in progress against *Clostridium difficile* nosocomial infection [77,78], anthrax [79,80], candidiasis [81,82], and group A *streptococcus* [83], only to cite some of them.

References

1. WHO. (2014) WHO recommendations for routine immunization — summary tables. [Online] Available at: http://www.who.int/immunization/policy/immunization_tables/en/index.html [Accessed 20 September 2014].
2. WHO. (2004) *Weekly Epidemiological Record* **79**, 25–40.
3. GEN Genetic Engineering & Biotechnology News (2013) Top 15 Vaccines of 2012. [Online] Available at: http://www.genengnews.com/insight-and-intelligenceand153/top-15-vaccines-of-2012/77899844/?page=1 [Accessed 1 November 2013].
4. Wright, A. E., Morgan, W. P., Colebrook, L., Dodgson, R. W. (1914) *Lancet* **10**, 87–95.
5. Anderson, P., Insel, R. A., Smith, D. H., Cate, T. R., Couch, R. B., Glezen, W. P. (1981) *J. Infect. Dis.* **144**, 530.
6. Jennings, H. J., Lugowski, C. (1981) *J. Immunol.* **127**, 1011.

7. Tai, J. Y., Vella, P. P., McLean, A. A., Woodhour, A. F., McAleer, W. J., Sha, A., Dennis-Sykes, C., Hilleman (1987) *M. R. Proc. Soc. Exp. Biol. Med.* **184**, 154–161.
8. Schneerson, R., Barrera, O., Sutton, A., Robbins, J. B. (1980) *J. Exp. Med.* **152**, 361.
9. Verez-Bencomo, V., Fernandez-Santana, V., Hardy, E., Toledo, M. E., Rodriguez, M. C., Heynngezz, L., Rodriguez, A., Baly, A., Herrera, L., Izquierdo, M., Villar, A., Valdes, Y., Cosme, K., Deler, M. L., Montane, M., Garcia, E., Ramos, A., Aguilar, A., Medina, E., Torano, G., Sosa, I., Hernandez, I., Martinez, R., Muzachio, A., Carmenates, A., Costa, L., Cardoso, F., Campa, C., Diaz, M., Roy, R. (2004) *Science* **305**, 522.
10. Peltola, H. (2000) *Clin. Microbiol. Rev.* **13**, 302.
11. Trotter, C. L., Andrews, N. J., Kaczmarski, E. B., Miller, E., Ramsay, M. E. (2004) *Lancet* **364**, 365.
12 Frasch, C. E., Preziosi, M.-P., LaForce, F. M., (2012) *Hum. Vaccin. Immunother.* **8**, 715–724.
13. Oosterhuis-Kafeja, F., Beutels, P., Van Damme, P. (2007) *Vaccine* **25**, 2194–2212.
14. Centers for Disease Control and Prevention (2005) *Morb. Mortal. Wkly Rep.* **54**, 893.
15. Black, S., Shinefield, H., Fireman, B., Lewis, E., Ray, P., Hansen, J. R., Elvin, L., Ensor, K. M., Hackell, J., Siber, G., Malinoski, F., Madore, D., Chang, I., Kohberger, R., Watson, W., Austrian, R., Edwards, K.(2000) *Pediatr. Infect. Dis. J.* **19**, 187–195.
16. Prymula, R., Schuerman, L. (2009) *Expert Review of Vaccines* **8**, 1479.
17. [Online] Available at: http://www.fda.gov/BiologicsBloodVaccines/Vaccines/ApprovedProducts/ucm201667.htm [Accessed 1 November 2013].
18. Halperin, S. A., Bettinger, J. A., Greenwood, B., Harrison, L. H., Jelfs, J., Ladhani, S. N., McIntyre, P., Ramsay, M. E., Safadi, M. A. (2012) *Vaccine* **30**, B26.
19. Crump, J. A., Luby, S. P., Mintz, E. D. (2004). *Bull. World Health Organ.* **82**, 346–353.
20. Robbins, J. B., Schneerson, R., Szu, S. C. (1996) In *Novel Strategies in Design and Production of Vaccines.* Cohet, S., Shafferman, A. Eds. Plenum Press (New York). pp. 169–182.
21. Calix, J., Nahm, M., Zartler, E. (2011) *J. Bacteriol.* **193**, 5271–5278.
22. Pinheiro, T., Barbosa, A., Godinho da Silva, J. (2006) *J. Chromatogr. B* **832**, 262–267.

23. Lepenies, B., Seeberger, P. (2010) *Immunopharmacol. Immunotoxicol.* **32**, 196–207.
24. Guo, Z., Jennings, H. (2001) *Methods Mol. Med.* **66**, 49–54.
25. Frasch, C. E., Preziosi, M. P., LaForce, F. M. (2012) *Hum. Vaccin. Immunother.* **8**, 715–724.
26. Koenig, J. M., Keenan, W. J. (2009) *Pediatr. Clin. North. Am.* **56**, 689–708.
27. Backer, C. J., Kasper, D. L. (1976) *N. Eng. J. Medicine* **294**, 753–756.
28. Munoz, F. M., Ferrieri, P. (2013) *Vaccine* **31**, D46–D51.
29. Verdier, I., Durand, G., Bes, M., Taylor, K. L., Lina, G., Vandenesch, F., Fattom, A. I., Etienne, J. (2007) *J. Clin Microbiol.* **45**, 725–729.
30. Maira-Litran, T., Kropec, A., Goldmann, D., Pier, G. B. (2004) *Vaccine* **22**, 872–879.
31. Shinefield, H., Black, S. (2006) *Pediatr. Infect. Dis. J.* **25**, 167–168.
32. Creech, C. B., Johnson, B. G., Alsentzer, A. R., Hohenboken, M., Edwards, K. M., Talbot, T. R. (2010) *Vaccine* **28**, 256–260.
33. Weiser, J. N., Bae, D., Epino, H., Gordon, S. B., Kapoor, M., Zenewicz, L. A., Shchepetov, M. (2001) *Infect. Immun.* **69**, 5430–5439.
34. Luong, T., Lee, C. (2006) *Microbiol.* **152**, 3123–3131.
35. O'Riordan, K., Lee, J. C. (2004) *Clin. Microbiol. Rev.* **17**, 218–234.
36. Maira-Litran, T., Kropec, A., Goldmann, D. A., Pier, G. B. (2005) *Infect. Immun.* **73**, 6752.
37. Gening, M. L., Maira-Litran, T., Kropec, A., Skurnik, D., Grout, M., Tsvetkov, Y. E., Nifantiev, N. E., Pier, G. B. (2010) *Infect. Immun.* **78**, 764.
38. Jennings, H. J. (1990) *Curr. Top. Microbiol. Immunol.* **150**, 97–127.
39. Jennings, H. J., Lugowski, C. (1981) *J. Immunol.* **127**, 1011–1018.
40. Suzuki, M., Angata, K., Nakayama, J., Fukuda, M. (2003) *J. Biol. Chem.* **278**, 49459–49468.
41. Finne, J., Leinonen, M., Makela, P. H. (1983) *Lancet* **2**, 355–357.
42. Bruge, J., Bouveret-Le-Cam, N., Danve, B., Rougon, G., Schulz, D. (2004) *Vaccine* **22**, 1087–1096.
43. Vesikari, T., Esposito, S., Prymula, R., Ypma, E., Kohl, I., Toneatto, D., Dull, P., Kimura, A. (2013) *Lancet* **38**, 1825–1835.
44. Boisier, P., Nicolas, P., Djibo, S., Taha, M.-K., Jeanne, I., Boubacar H., Maínassara, B., Tenebray, K., Kairo, K., Giorgini, D., Chanteau, S. (2007) *Clin. Infect. Dis.* **44**, 657.
45. Micoli, F., Romano, M. R., Tontini, M., Cappelletti, E., Gavini, M., Proietti, D., Rondini, S., Swennen, E., Santini, L., Filippini, S., Balocchi, C., Adamo, R., Pluschke, G., Norheim, G., Pollard, A., Saul, A., Rappuoli,

R., MacLennan, C. A., Berti, Fr., Costantino, P. (2013) *PNAS* **110**, 19077–19082.

46. Glez-Miró, M., Garrido, R., Rodríguez, L., Espinosa, C., Cardoso, F., Sanchez, Y., Santana, D., Chang, J., Valdés, Y., García, D., Acevedo, R., González, D., Cardoso, D., Vérez, V. Poster in: International Congress on Meningeal Disease Vaccines. Cuba, 19–24 May 2013.
47. Morelli, L., Lay, L. (2013) *ARKIVOC* **166**,166–184.
48. Cohen, D., Green, M. S., Block, C., Slepon, R., Ofek, I. (1991) *J. Clin. Microbiol.* **29**, 386–389.
49. Robbins, J. B., Chu, C. Y., Schneerson, R. (1982) *Clin. Infect. Dis.* **15**, 346.
50. Volk, W. A., Salomonsky, N. L., Hunt, D. (1972) *J. Biol. Chem.* **247**, 3881–3887.
51. Holst, O. Ed. (2000). *Methods in Molecular Biology, Bacterial Toxins: Methods and Protocols.* Humana Press, Inc., (Totowa, NJ) **145**, 345–353.
52. Taylor, D. N., Trofa, A. C., Sadoff, J., Chu, C., Bryla, D., Shiloach, J., Cohen, D., Ashkenazi, S., Lerman, Y., Egan, W., Schneerson, R., Robbins, J. B. (1993) *Infect. Immun.* **61**, 3678–3687.
53. Cohen, D., Ashkenazi, S., Green, M. S., Gdalevich, M., Robin, G., Slepon, R., Yavzori, M., Orr, N., Block, C., Ashkenazi, I., Shemer, J., Taylor, D. N., Hale, T. L., Sadoff, J. C., Pavliakova, D., Schneerson, R., Robbins, J. B. (1997) *Lancet* **349**, 155–159.
54. Passwell, J. H., Ashkenzi, S., Banet-Levi, Y., Ramon-Saraf, R., Farzam, N., Lerner-Geva, L., Even-Nir, H., Yerushalmi, B., Chu, C., Shiloach, J., Robbins, J. B., Schneerson, R. (2010) *Vaccine* **28**, 2231–2235.
55. Pozsgay, V., Chu, C., Pannell, L., Wolfe, J., Robbins, J. B., Schneerson, R. (1999) *Proc. Natl. Acad. Sci. USA* **96**, 5194.
56. Phalipon, A., Costachel, C., Grandjean, C., Thuizat, A., Guerreiro, C., Tanguy, M., Nato, F., Vulliez-Le Normand, B., Bélot, F., Wright, K., Marcel-Peyre, V., Sansonetti, P. J., Mulard, L. A. (2006) *J. Immunol.* **176**, 1686.
57. Levine, M. M., Nalin, D. R., Craig, J. P., Hoover, D., Bergquist, E. J., Waterman, D., Holley, H. P., Hornick, R. B., Pierce, N. P., Libonati, J. P. (1979) *Trans. R. Soc. Trop. Med. Hyg.* **73**, 3–9.
58. Ito, T., Higuchi, T., Hirobe, M., Hiramatsu, K., Yokota, T. (1994) *Carbohydr. Res.* **256**, 113.
59. Kabir, S. (1987) *J. Med. Microbiol.* **23**, 9–18.
60. Gupta, R. K., Taylor, D. N., Bryla, D. A., Robbins, J. B., Szu, S. S. C. (1998) *Infect. Immun.* **66**, 3095–3309.

61. Grandjean, C., Wade, T K., Ropartz, D., Ernst, L., Wade, W. F. (2013) *Pathogens and Disease* **67**, 136–158.
62. Machová, E., Bystrick, S., Gáliková A., Kogan, G. (2000) *Eur. J. Med. Chem.* **237**, 681–687.
63. Paulovicová, E., Machová, E., Hostacká, A., Bystrick, S. (2006) *Clin. Experim. Imm.* **144**, 521–527.
64. Paulovicova, E., Korcova, J., Farkas, P., Bystricky, S. (2010) *J. Med. Microbiology* **59**, 1440–1448.
65. Zhang, J., Kovac, P. (1999) *Carbohydr. Res.* **321**, 157–167.
66. Chernyak, A., Kondo, S., Wade, T. K., Meeks, M. D., Alzari, P. M., Fournier, J. M., Taylor, R. K., Kovác, P., Wade, W. F. (2002) *J. Inf. Dis.* **185**, 950–962.
67. Saksena, R., Ma, X., Wade, T. K., Kovác, P., Wade, W. F. (2006) *FEMS Immunol. Med. Mic.* **47**, 116–128.
68. Tarique, A., Kalsy, A., Arifuzzaman, M., Rollins, S. M., Charles, R. C., Leung, D. T., Harris, J. B., LaRocque, R. C., Sheikh, A., Bhuiyan, M. S., Saksena, R., Clements, J. D., Calderwood, S. B., Qadri, F., Kovác, P., Ryan, E. T. (2012) *Clin. Vaccine Immunol.* **19**, 594.
69. Kossaczka, Z., Shiloach, J., Johnson, V. D., Taylor, N. R., Finkelstein, A. J., Robbins, B., Szu, S. C. (2000) *Infect. Imm.* **68**, 5037–5043.
70. Boutonnier, A., Villeneuve, S., Nato F., Dassy, B., Fournier, J. M. (2001) *Infect. Immun.* **69**, 3488–3493.
71. Micoli, F., Rondini, S., Gavini, M., Lanzilao, L., Medaglini, D., Saul, A., Martin, L. B. (2012) *PLoS One* **7**, e47039.
72. Murphy, T. F., Brauer, A. L., Grant, B. J., Sethi, S. (2005) *Am. J. Respir. Crit. Care Med.* **172**, 195.
73. Vaneechoutte, M., Verschraegen, G., Claeys, G., Van Den Abeele, A. M. (1990) *J. Clin. Microbiol.* **28**, 182–187.
74. Gu, X. X., Chen, J., Barenkamp, S. J., Robbins, J. B., Tsai, C. M., Lim, D. J., Battery, J. (1998) *Infect. Immun.* **66**, 1891.
75. Yu, S., Gu, X. X. (2005) *Infect. Immun.* **73**, 2790.
76. Campbell, A., Lohman, G., Plante, O. J. (2011) Patent WO 137181.
77. Adamo, R., Roman, M. R., Berti, F., Leuzzi, R., Tontini, M., Danieli, E., Cappelletti, E., Cakici, O. S., Swennen, E., Pinto, V., Brogioni, B., Proietti, D., Galeotti, C. L., Lay, L., Monteiro, M. A., Scarselli, M., Costantino, P. (2012) *ACS Chem. Biol.* 7, 1420.
78. Martin, C. E., Broecker, F., Oberli, M. A., Komor, J., Mattner, J., Anish, C., Seeberger, P. H. (2013) *J. Am. Chem. Soc.* **135**, 9713.

79. Tamborrini, M., Werz, D. B., Frey, J., Pluschke, G., Seeberger, P. H. (2006) *Angew. Chem. Int. Ed.* **45**, 6581.
80. Metha, A. S., Saile, E., Zhong, W., Buskas, T., Carlson, R., Kannenberg, E., Reed, Y., Quinn, C., Boons, G. J. (2006) *Chem. Eur. J.* **12**, 9136.
81. Torosantucci, A., Chiani, P., Bromuro, C., De Bernardis, F., Palma, A. S., Liu, Y., Mignogna, G., Maras, B., Colone, M., Stringaro, A., Zamboni, S., Feizi, T., Cassone, A. (2009) *PLoS One* **4**, e5392.
82. Xin, H., Cartmell, J., Bailey, J. J., Dziadek, S., Bundle, D. R., Cutler, J. E. (2012) *PLoS One* **7**, e35106.
83. Kabanova, A., Margarit, I., Berti, F., Romano, M. R., Grandi, G., Bensi, G., Chiarot, E., Proietti, D., Swennen, E., Cappelletti, E., Fontani, P., Casini, D., Adamo, R., Pinto, V., Skibinski, D., Capo, S., Buffi, G., Gallotta, M., Christ, W. J., Campbell, A. S., Pena, J., Seeberger, P. H., Rappuoli, R., Costantino, P. (2011) *Vaccine* **29**, 104–114.

Chapter 13

Carbohydrate-Based Antibacterial Vaccines: Current Progress and Future Outlook

Sachin S. Shivatare and Chung-Yi Wu

The Genomic Research Center, Academia Sinica, Taipei, Taiwan

13.1 Introduction

Despite the advent of antibiotics, implementation of national vaccination campaigns, and intensive care support, bacterial meningitis continues to be an important cause of morbidity and mortality among high-risk groups (infants, children under two years of age, and immunocompromised patients), especially in developing countries. Some of the leading causes of bacterial meningitis include *Haemophilus influenzae* (Hib; most often caused by type b), *Streptococcus pneumoniae*, group B *Streptococcus*, *Listeria monocytogenes*, and *Neisseria meningitidis*. Early administration of an optimal antibiotic therapy for bacterial meningitis has been shown to be essential to minimize lethality and morbidity; however, the steady increase in antibiotic resistance has become an issue of great concern.

Carbohydrates, in the form of oligosaccharides and glycoconjugates, are associated with protein stabilization, cancer metastasis, viral host interactions, bacterial pathogenicity, and signal transduction [1–4], besides their role in metabolism and as structural building blocks. The structurally diverse carbohydrate epitopes are being adopted for use in vaccine design against human pathogens and cancers. Specially, bacterial polysaccharides are usually associated with the cell surface of pathogenic bacteria in the form of lipopolysaccharides, peptidoglycans, and capsular polysaccharides [5,6]. These immense polysaccharides help bacteria escape from the host immune system and prevent antibody-mediated complement activation and phagocytosis. The common feature of the organisms that causes most of the bacterial meningitis is a carbohydrate capsule with different oligosaccharide patterns that act as both a virulence determinant and target of protective antibody [7]. Therefore, antibodies specifically against bacterial surface polysaccharides may enhance the elimination of pathogenic bacteria.

13.2 Overview of Bacterial Capsular Polysaccharides

Bacterial glycoconjugates consist of a diverse range of macromolecules that include peptidoglycans, lipopolysaccharides, periplasmic glucans, and capsular polysaccharides, secreted by pathogenic bacteria as part of their cell wall components. A thick, mucous-like layer of extracellular polysaccharides are high molecular-weight polymers of repetitive sugar units in such a large amount, that they encapsulate microorganisms. A dense array of polysaccharides that surrounds the bacterial capsule represents a unique feature of the particular species and also the strain of bacteria considered. The highly conserved capsular polysaccharides of pathogenic bacteria are the most exposed surface to host defense against host immune system and are an important virulence determinant, and thus form an attractive target for vaccine development [8,9]. However, the development of effective carbohydrate-specific vaccines against bacterial pathogens has been hindered by the considerable complexity and heterogeneity of capsular polysaccharides. Furthermore, some of the bacterial polysaccharides (e.g. polysialic acid glycans of *meningococcus* group B) mimic the native host glycans, and therefore, may be tolerated by the host immune system, and lead to poor immunogenicity [10].

Glycoconjugate vaccines containing fragments of capsular polysaccharide attached to immunogenic carriers have been made against many bacterial infections, such as *H. influenza, S. pneumonia* and *N. meningitidis* [11–15]. Despite their structural diversity, certain common sugars form the basis for different groups or serotypes. For example, *Escherichia coli* K1 and *Meningitidis* group B both produce sialic-acid-containing capsules.

13.2.1 *Haemophilus influenza*

Haemophilus influenza (Hi), formerly known as *bacillus influenza,* are Gram-negative bacteria belonging to the *Pasteurellaceae* family. On the basis of their distinct capsular polysaccharide antigens, *H. influenzae* bacteria are classified into six serotypes: a, b, c, d, e, and f (Table 13.1) [16]. Specifically, the capsule of *H. influenzae* type a (Hia), which is responsible for up to 10% of *Haemophilus* infections, consists of repeating units of β-D-*N*-acetylglucosamine-D-ribitol residues joined through phosphoric diester linkages. Type b *H. influenzae* (Hib), which causes severe and permanent neurological defects among high-risk groups, consists of β-D-ribose-D-ribitol-5-phosphate that has been included in the first commercial vaccine (trade name QuimiHib) developed in Cuba against Hib [17]. Type c and f consist of the repeating units of D-GlcNAc-(1→3)-β-D-Gal and D-GalNAc-(1→4)-β-3-acetyl-GalNAc, respectively. Their repeating units are joined by α-phosphoric diester linkages. Type d capsular polysaccharides consist of D-GlcNAc-β-(1→3)-linked to the unusual 2-acetamido-2-deoxy-D-mannose uronic acid, where C6 of uronic acid forms an amide linkage with amino acid residues, including L-serine, L-threonine, or L-alanine, that has been proposed to be of phylogenetic significance. The type e capsular polysaccharides consist of disaccharide repeating units of D-GlcNAc-(1→4)-β-2-acetamido-2-deoxy-D-mannose uronic acid, whereas the additional fructose substitution at the 3-*O* position further classified it into type e. All the *H. influenza* derived capsular polysaccharides are acidic due to the presence of a carboxylic acid functionality on sugar residues and the phosphoric diester linkages.

Table 13.1 Structures of serotype specific polysaccharide antigens of *H. influenzae.*

Serotypes	Structure
a	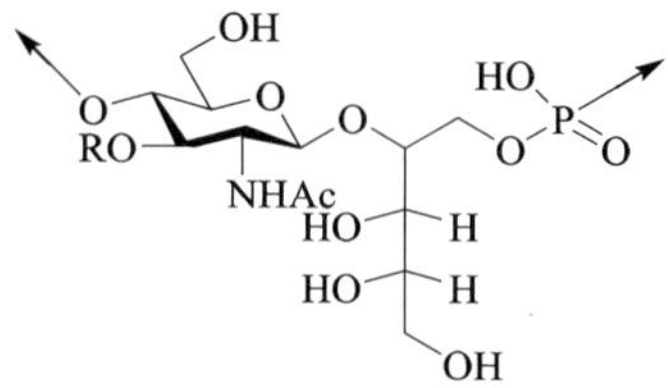 R = –OAc, –H
b	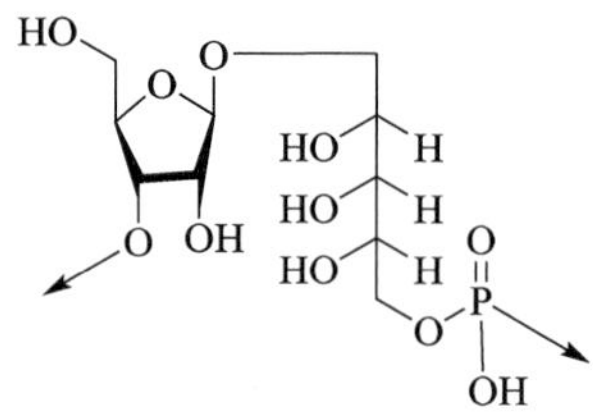
c	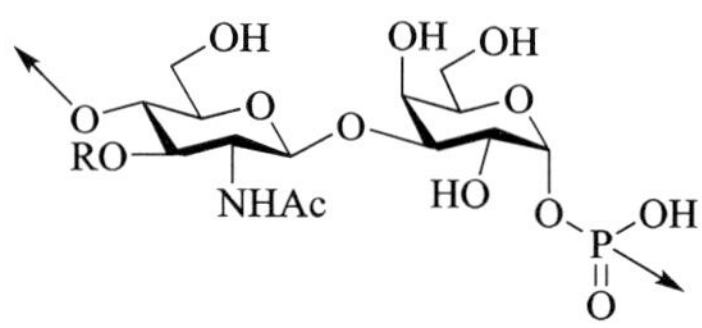 R = –OAc, –H
d	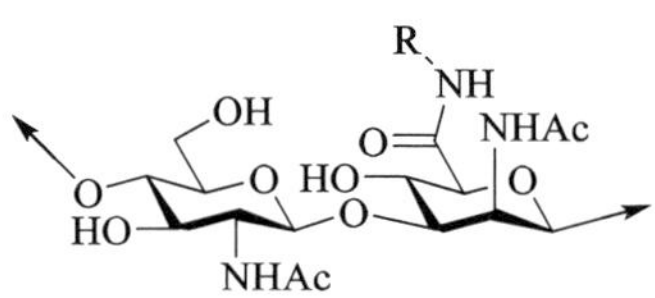 R = L-serine L-threonine L-alanine
e, e′	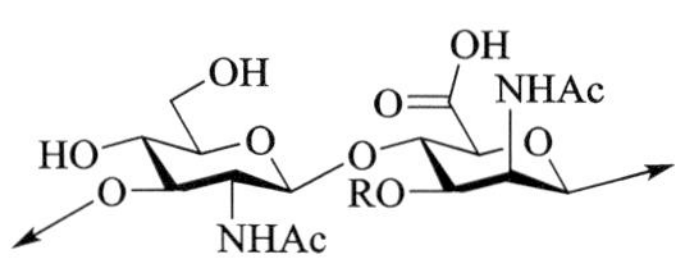 Type R = e –H e' –β–D–Frucotse
f	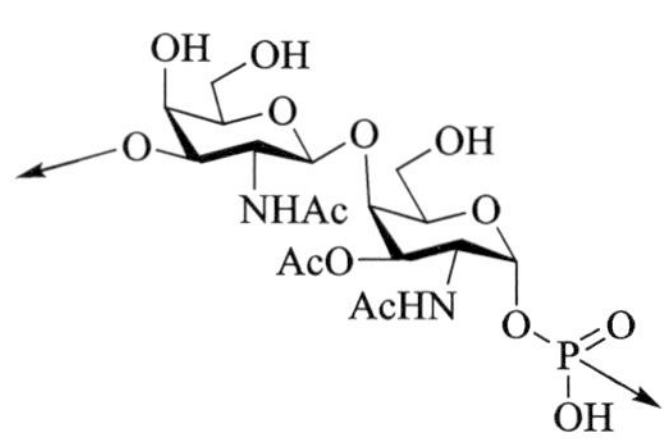

The *O*-acetylated polysaccharides of type a, c, and f are of particular importance because of their immunodominant properties.

13.2.2 *Streptococcus pneumoniae*

An estimate indicated that more than four million deaths occur annually, mostly among children under the age of five, due to acute lower respiratory infections caused by pneumococci [18–20]. The *S. pneumoniae*, a Gram-positive organism with a well-formed polysaccharide capsule, is the most common cause of community-acquired pneumonia (CAP), bacterial meningitis, bacteremia, and otitis media in adults and young children [21,22]. Up to the present date, 90 different serotypes of *S. pneumoniae* have been identified; however, only a few of these are common causes of pneumococcal disease. Each type has a structurally distinct capsular polysaccharide (CPS), composed of repeating oligosaccharide units joined by glycosidic linkages. The development of a protective vaccine against *S. pneumoniae* was particularly challenging due to a large number of pneumococcal serotypes. The currently available pneumococcal polysaccharide vaccines are composed of the 23 most common pneumococcal serotypes, which account for 85–90% of the serotypes that are responsible for almost 88% of invasive infections [23,24].

The large number of pneumococcal serotypes leads to a huge structural diversity and complexity; therefore, only the polysaccharide structures of pneumococci that are the major causes of morbidity and mortality are discussed here (Table 13.2). The capsular polysaccharide from *S. pneumoniae* type 1 consists of a non-branched trisaccharide repeating unit [→3)-α-(2→4)-dideoxy-4-amino-D-FucNAc-(1→4)-α-D-GalA*p*-(1→3)-α-D-GalA*p*-(1→] that carries an alternating single positive charge of free amine and two negative charges of two carboxylates, responsible for the Zwitterionic nature of polysaccharides [25]. The type 2 capsular polysaccharide consists of singly branched hexasaccharide repeating units, each containing [→4)-β-D-Glc*p*-(1→3)-α-L-Rha*p*-{α-D-Glc*p*A-(1→6)-α-D-Glc*p*-(1→2)-}, {(1→3)-α-L-Rha*p*-(1→3)-β-L-Rha*p*-(1→)}] [26]. The structural homologies and extensive serological cross-reactivities favor the development of multivalent vaccines against invasive pneumococcal infections.

Table 13.2 Capsular polysaccharides of *Streptococcus pneumonia.*

Serotypes	Structure
1	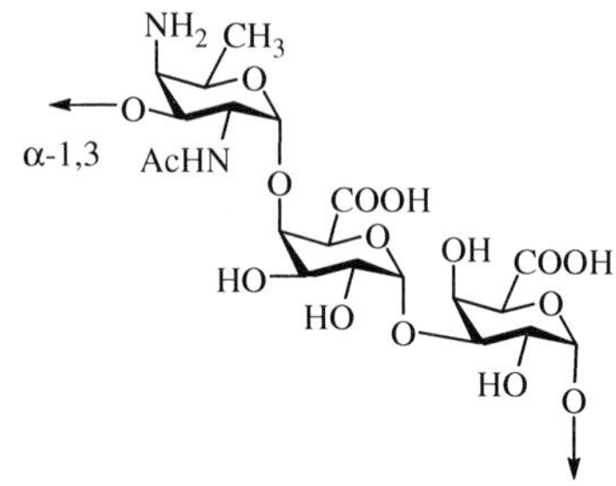
2	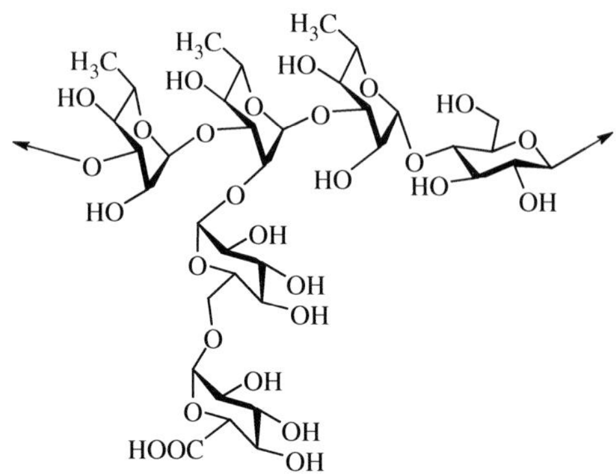
3	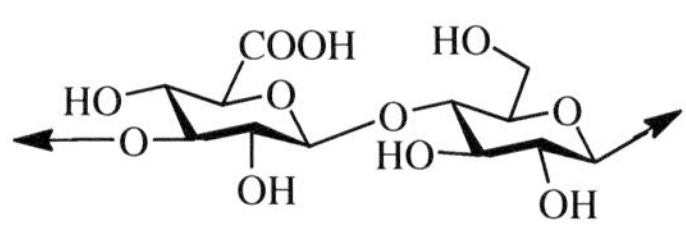
4	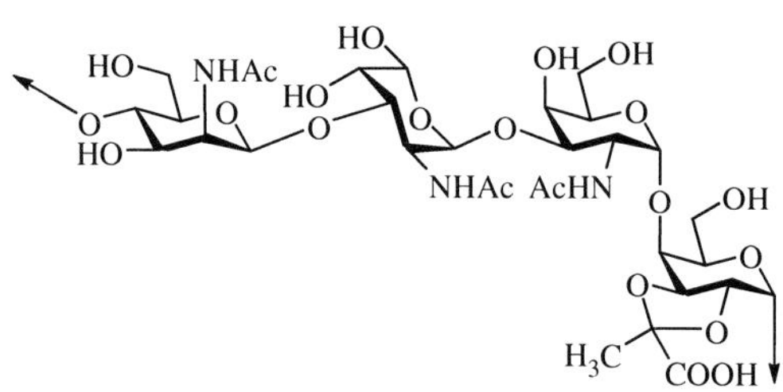

(*Continued*)

Table 13.2 (*Continued*)

Serotypes	Structure
6A	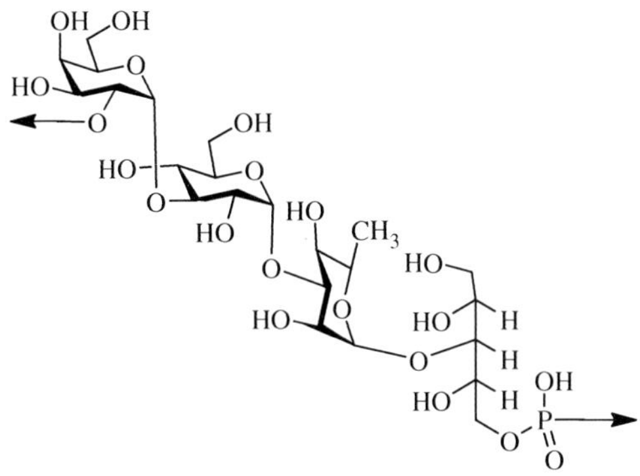
6B	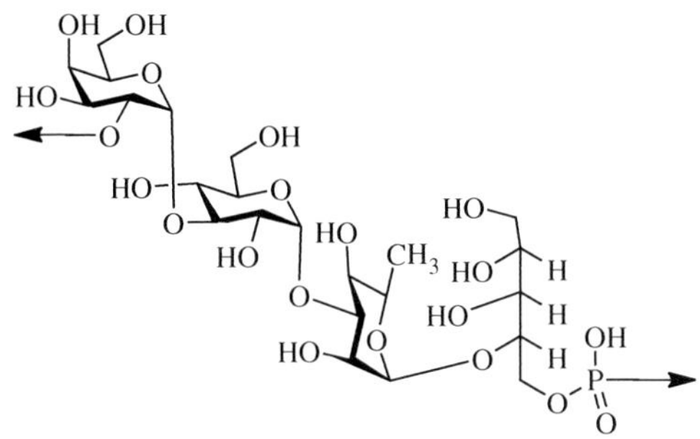
9A	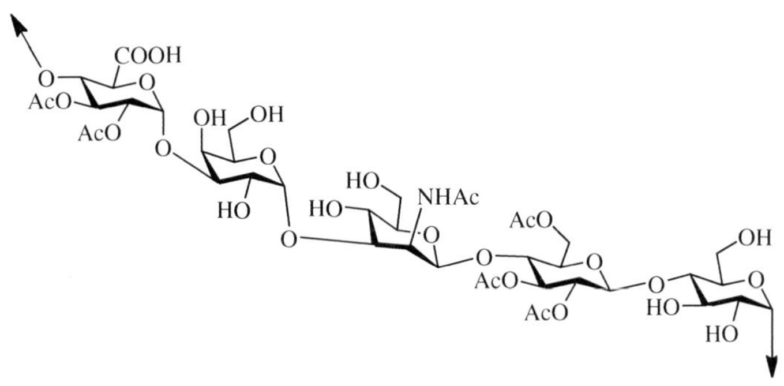

(*Continued*)

Table 13.2 (*Continued*)

Serotypes	Structure
9n	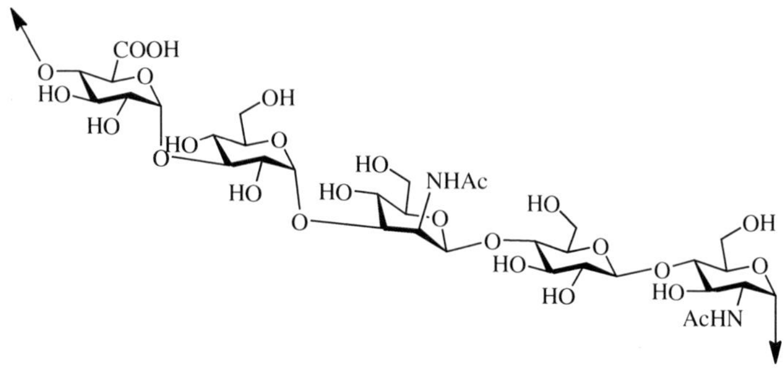
14	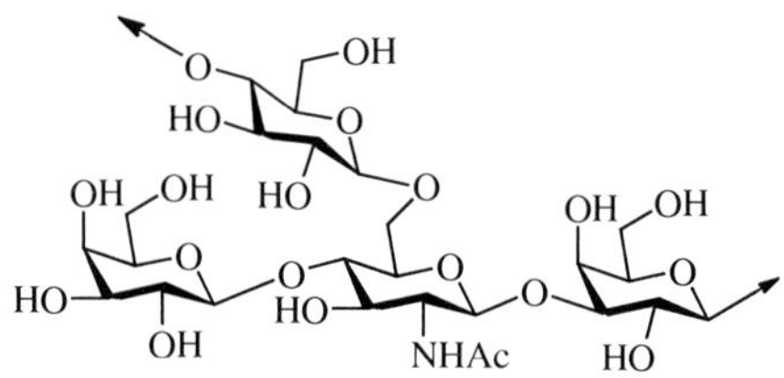
18C	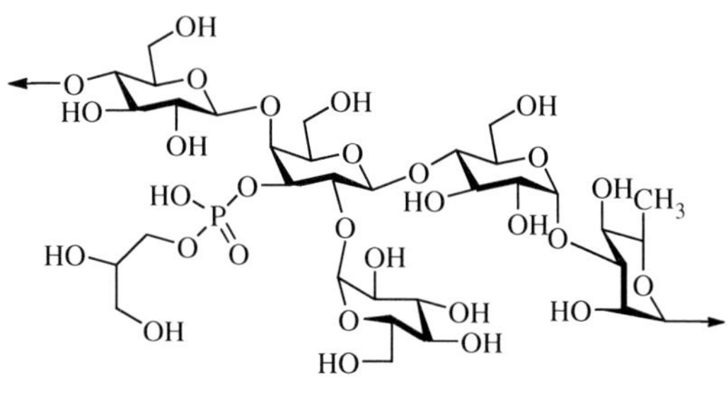
19A	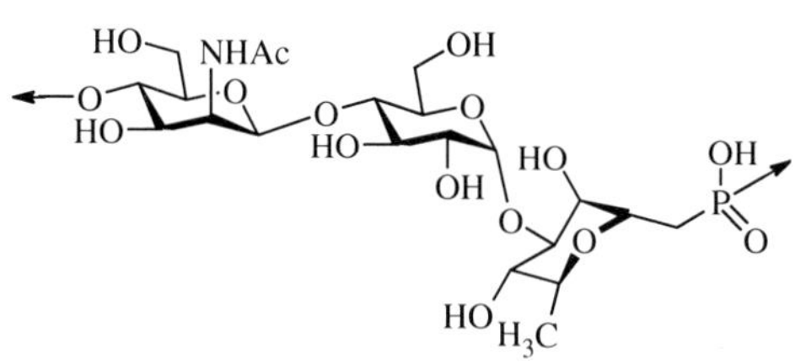

(*Continued*)

Table 13.2 (*Continued*)

Serotypes	Structure
19F	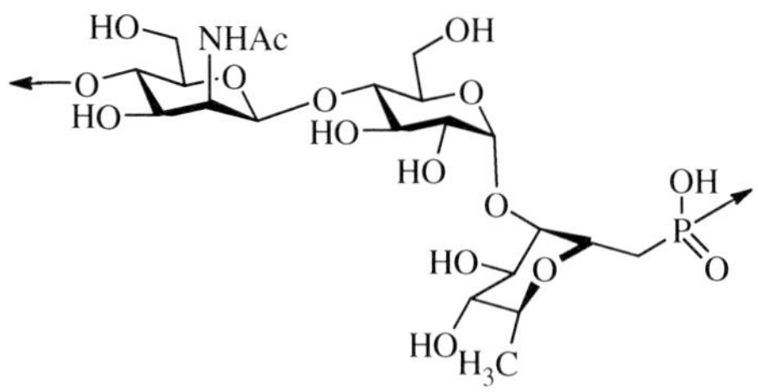
23F	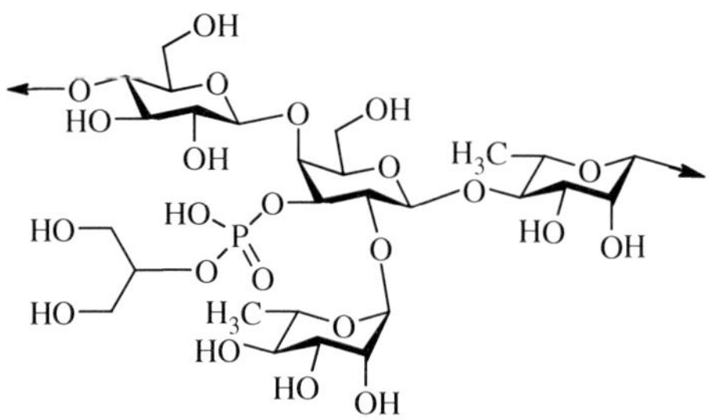

13.2.3 *Neisseria meningitidis*

The Gram-negative encapsulated organism *N. meningitidis* is the main cause of septicemia and bacterial meningitidis worldwide in children and adults. Bacteria *N. meningitidis* colonize in the mucous membranes of humans as pathogens, causing many epidemics across the world [27,28]. Based on the structure of the capsular polysaccharides, 13 serogroups of *N. meningitidis* have so far been identified including A, B, C, 29E, H, I, K, L, W135, X, Y, and Z [29, 30]. Approximately 90% of invasive meningococcal infections are caused by strains belonging to serotypes A, B, C, and W135 [31]. Although all *N. meningitidis* serotypes can cause epidemics, the group A strains are the causative agents for repeating epidemics in Sub-Saharan Africa [32], whereas serogroups B and C are the most prevalent in Europe and Latin/North America, respectively. The group A polysaccharides are homopolymers of 2-acetamido-2-deoxy-D-mannopyranosyl phosphate, linked through 1, 6 linkages with partial 3-*O*-acetylation (Table 13.3) [33]. The polysaccharides of

Table 13.3 Structures of polysaccharide antigens of *Neisseria meningitidis*.

Groups	Structure
A	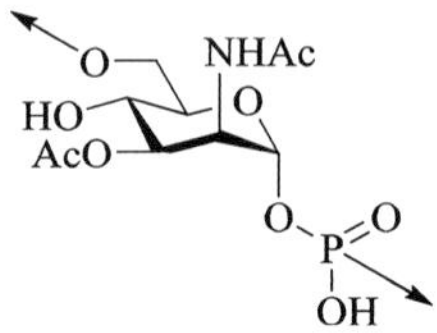
B	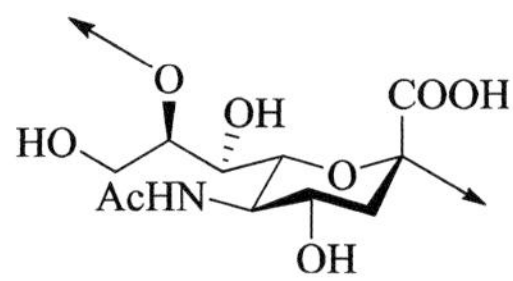
C	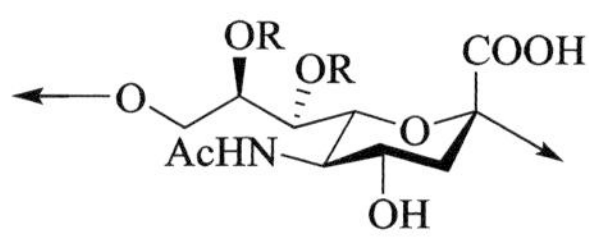 R = –OAc, –H
W-135	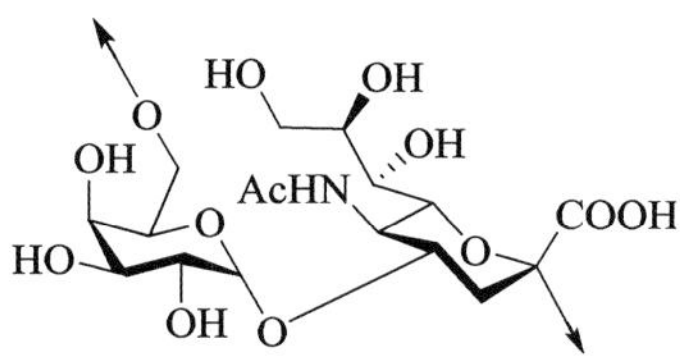
X	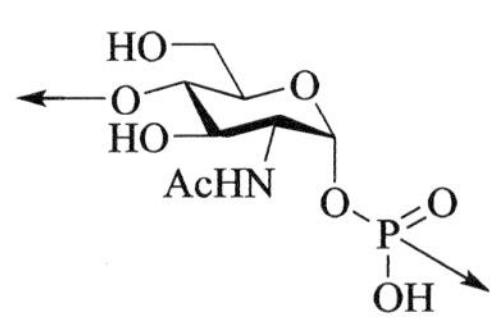
Y	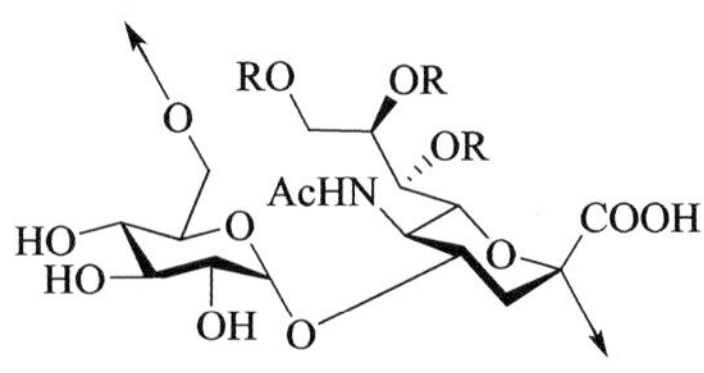 R = –OAc, –H

(*Continued*)

Table 13.3 (*Continued*)

Groups	Structure
Z	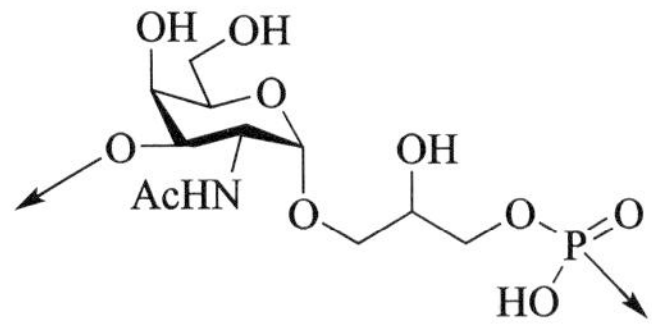
29e	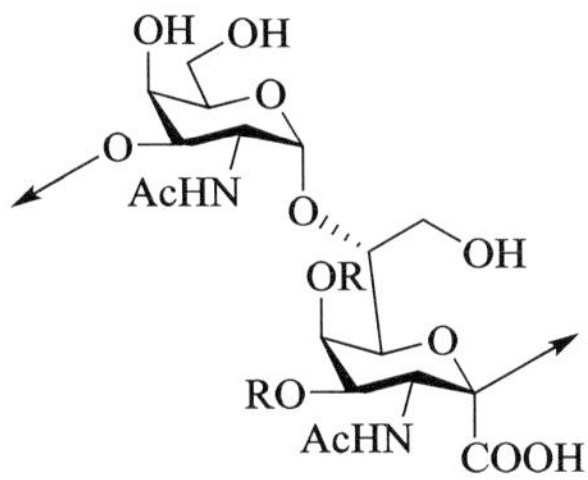 R = –OAc, –H

group B and C are repeating units of *N*-acetyl neuraminic acid (polysialic acid) linked via α (2→8) and α (2→9) linkages, respectively. However, the meningococcal group B polysaccharide is a "self"-antigen because of its expression by a number of host tissues during early fetal development, which further led to a decrease in immunogenicity, even after conjugation to a carrier protein [34]. Group W-135 capsular polysaccharide is composed of repeating →6)-α-D-Gal-(1→4)-α-D-Neu5Ac-(2→disaccharides [35]. The serogroup X polysaccharide is a homopolymer of (α1→4)-linked *N*-acetylglucosamine (GlcNAc)-1-phosphate [36]. The proportion of meningococcal cases caused by serogroup Y has increased dramatically from 2% to 37% during the nineties in the United States (US). The polysaccharide capsule of serogroup Y consists of repeating units of →6)-α-D-Glc-(1→4)-α-D-Neu5Ac-(2→ disaccharide [35]. The careful examination of nearly identical structures of groups W-135 and Y polysaccharide indicated that the only difference is the orientation of the hydroxyl group at the C4 position, equatorial in Y and axial in W-135. Apart from the pathogenic serotypes A, B, C, and W-135, the continuing occurrence of meningococcal cases caused by serogroups X, Z, and 29E emphasize the need to develop the potential vaccine against these meningococcal serogroups. The capsular polysaccharides of serogroup 29E and Z are structurally similar, consisting of copolymers of galactosamine conjugated to either 2-keto-3-deoxyoctulosonic acid (KDO; group 29E) or glycerol phosphate (group Z) [37].

13.2.4 *Group B streptococcus*

Group B streptococci are Gram-positive bacteria commonly found in the intestinal, vaginal, and rectal areas in 15–45% of all healthy women, and are a leading cause of infections that can affect non-pregnant adults, pregnant women, and their newborn infants [38–41]. Based on their capsular polysaccharide structures, group B streptococci are classified into at least nine different serotypes, such as Ia, Ib, II, III, IV, V, VI, VII, and VIII, depending on their capsular polysaccharide structures (Table 13.4) [42–45]. Before 1990, most of the group B *streptococcus* infections were caused by serotypes Ia, Ib, II, and III. The first reports of serotype V causing infections in newborns occurred in the early nineties. Based on the serotype distributions, it has been postulated that an effective vaccine composed of types Ia, Ib, II, III, and V may provide protection from 85–100% of invasive diseases. Several group B *streptococcus* capsular serotypes consist of sialic acid primarily attached through an α (2→3) linkage to the inner galactose residue. The terminal sialic acid is amenable to conjugation to protein carriers via simple periodic oxidation or reductive amination, which constitutes the general strategies of glycoconjugate synthesis.

13.2.5 *Salmonella typhi*

According to the World Health Organization (WHO) report, a worldwide estimation of 16 million people contracted typhoid and about 600,000 die from the disease every year [46–50]. Typhoid fever, also known as "typhoid", is a potentially fatal multisystemic illness caused primarily by the bacterium *Salmonella typhi,* which is

Table 13.4. Structures of polysaccharide antigens of Group B *streptococcus*.

Type	Structure
Ia	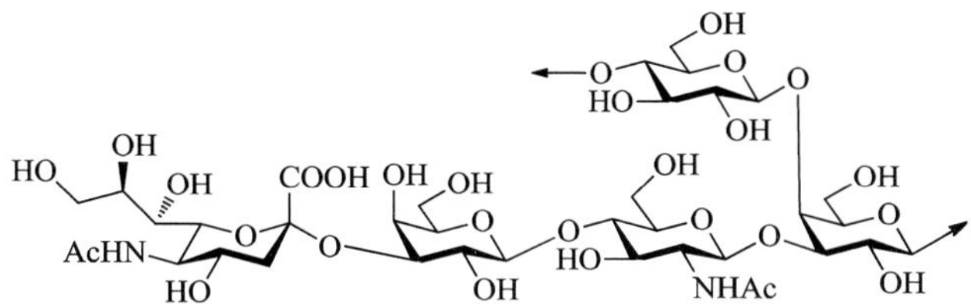
Ib	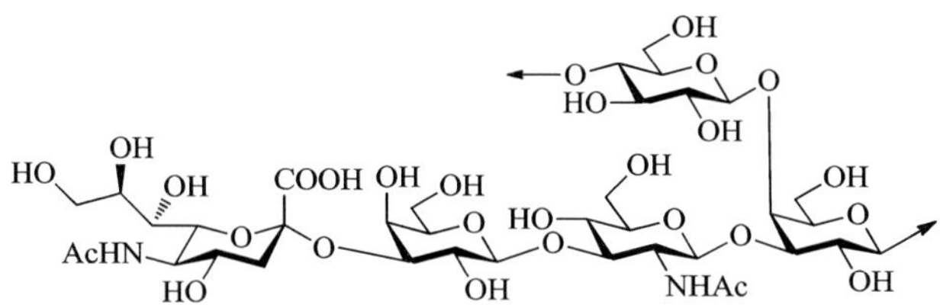
II	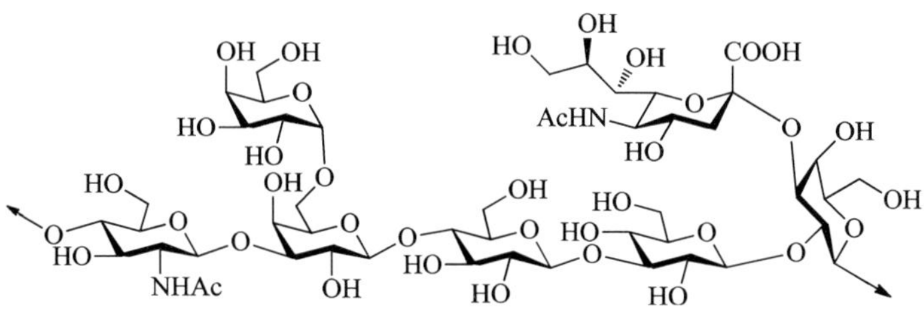
III	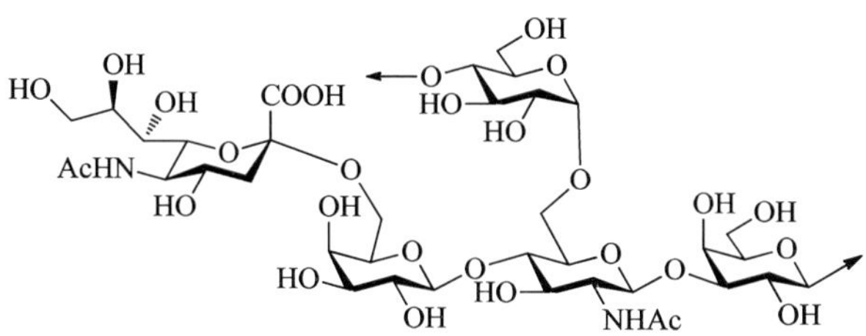
V	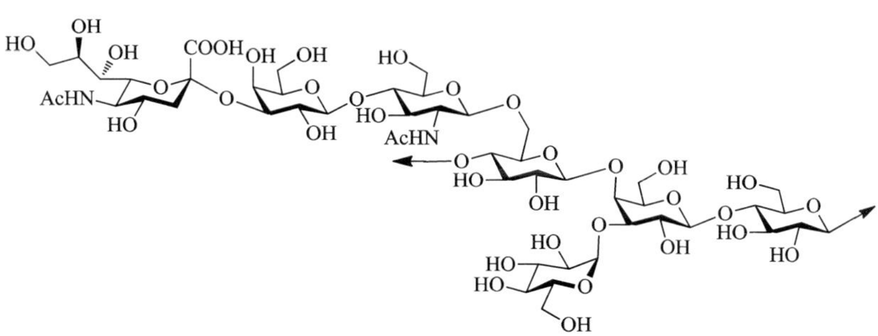

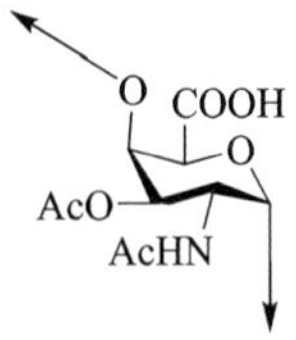

Figure 13.1 Structure of Vi antigen from *Salmonella typhi.*

transmitted through contaminated food and water [51]. *S. typhi* has been a major human intracellular pathogen for thousands of years, thriving in poor public sanitation conditions. The use of antibiotics has significantly reduced the frequency of typhoid fever in developed countries; however, typhoid fever remains endemic in the poor world. Moreover, in recent years, *S. typhi* has acquired resistance to many of the antibiotics used for its treatment.

The capsular polysaccharide of *S. typhi*, also known as Vi antigen, is a linear homopolymer of (1→4)-linked-α-D-*N*-acetyl galactosamine uronic acid, variably *O*-acetylated at the C-3 position (Fig. 13.1) [52]. The Vi capsular antigen is the main determinant of virulence factors for typhoid fever and, thus, continues to be the most important target for the development of effective treatment and prophylaxis for this disease. One of the two licensed vaccines that were developed against *S. typhi* was primarily based on the Vi antigen (Typhim Vi), while the other one was a live attenuated oral vaccine from the Ty21 strain.

13.3 Licensed Carbohydrate-Based Antibacterial Vaccines

In 1923, Heidelberger and Avery established a correlation between capsular polysaccharides and pneumococcal serotypes [53]. Later in 1930, Francis and Tillet demonstrated that vaccination with serotype-specific polysaccharides generated an immune response against heterologous types of pneumococci species [54]. Thereafter, in 1950, Heidelberger *et al.* showed that pneumococcal capsular polysaccharide vaccine could induce high titers of serotype-specific antibodies lasting for 5–8 months [55]. The first polysaccharide vaccine, PneumoVax®, composed of unconjugated capsular polysaccharides, was isolated from 14 *pneumonia* serotypes in 1977, and by 1983, this was modified to 23 valent vaccines by Merck and Co. in 1983 [56]. Although this polysaccharide vaccine is effective in healthy adults, it fails to provide adequate antibody responses in high-risk groups, such as infants, young children, and the elderly, the populations that are the most vulnerable to bacterial infections due to immature or compromised immune systems [57–60]. For example, pneumococcal vaccine PneumoVax, Vi antigen-based vaccine against typhoid and the four component *N. meningitidis* groups, A, C, Y, and W135 vaccine, failed to provide adequate protection for children younger than two years old. In general, naturally occurring carbohydrate molecules are usually large repetitive units, these are not processed by antigen presenting cells (APCs), but interact directly with B cells to trigger the immune response that is independent of $CD4^+$ T helper cells [61]. Such T-cell independent responses are instant, less robust, and primarily consist of low-affinity immunoglobulin M (IgM) antibodies, in contrast to T-cell dependent responses that are typically long lasting against peptide backbones [62, 63]. The strong, durable, and protective antibody response to the vast majority of oligosaccharides could be generated by providing exogenous $CD4^+$ T-cell epitopes, usually through attachment of isolated polysaccharides to protein scaffolds [64]. T-cell epitope peptides incorporated into carrier proteins facilitate the uptake and processing of glycoconjugates by APCs to trigger the activation of helper T-cells. Thus, high-affinity, long-lived, and class-switched antibodies can be generated even among persons in high-risk groups.

As early as in 1929, Goebel and Avery first demonstrated that the poor immunogenicity of purified *S. pneumoniae* type 3 polysaccharide could be enhanced by covalent conjugation of the polysaccharide to an immunogenic protein carrier [65]. The distinct advantage of conjugate vaccines is the ability to induce affinity-matured, isotype-switched, and T-cell dependent protective immunoglobulin G (IgG) responses, providing long-lasting immunity, particularly in newborns.

The success of glycoconjugate vaccines opened new avenues in the field of vaccine immunology and accelerated vaccine development. A great wealth of information about the structure–immunological activity relationship of polysaccharides and their protein conjugates has been accumulated in the course of commercial development of glycoconjuagte vaccines to prevent bacterial infections caused by *H. influenzae* type b, *N. meningitidis*, *S. enterica serovar typhi*, and *S. pneumoniae* (Table 13.5) [12,66,67]. The first licensed glycoconjugate vaccine against *H. influenzae* type b has been introduced in routine vaccination programs in developed countries. Consequently, meningitis in adults, as well as among high-risk groups, has completely disappeared and, eventually, there are no more cases in the whole population as a result of "herd immunity". Subsequently, the successful glycoconjugate vaccine against *N. meningitidis* type C containing meningococcal group C polysaccharide conjugated to CRM_{197} was developed. The tetravalent meningococcal conjugate vaccine containing the A, C, Y, and W135 serotypes of *N. meningitidis* is currently marketed under the trade name Menactra® by Sanofi Pasteur and Menveo® by Novartis. Moreover, some of the carbohydrate-based vaccines are under investigation against *Staphylococcus aureus*, *Shigella sonnei*, *Shigella flexneri*, and many others.

13.3.1 *Haemophilus influenza type b*

Hib is the leading cause of severe diseases, including meningitidis, pneumonia, septicemia, epiglottitis, and septic arthritis in younger children, particularly in developing countries. The first generation Hib vaccine (HbPV) was attempted with purified polyribosyl–ribitol phosphate, a component of the polysaccharide capsule of type b organisms. In preliminary studies, this vaccine was found to be effective in adults and older children, but it failed to provide adequate immunity in infants, the population more vulnerable to invasive Hib diseases [68]. However, due to the severity of Hib related cases, the vaccine was commercialized for children older than two years of age in the United States in 1985, despite its poor immunogenicity in high risk groups.

Table 13.5 Licensed polysaccharide and conjugate vaccines.

Organism	Trade name	Manufacturer	Components	Current status
Haemophilus influenza (Type b)	Pedvax HIB	Merck & Co.	CPS to OMPs	Licensed (1990)
	ActHIB	Sanofi Pasteur	CPS with TT	Licensed (1993)
	Hiberix	GlaxoSmithKline Biologicals	CPS with TT	Licensed EU (1998) and USA (2009)
	QuimHib	Heber Biotech	Synthetic oligosaccharides to TT	Licensed in Cuba (2004)
	HibTiter, Vaxem-Hib	Wyeth	CPS to CRM_{197} (12 mers)	Licensed (1990)
Neisseria meningitidis A, C, Y and W-135	Menactra	Sanofi Pasteur	Group A, C, Y and W-135 CPS to DT	Licensed (2005)
	Menveo	Novartis	Group A, C, Y and W-135 CPS to DT	Licensed (2010)
	Menomune	Sanofi Pasteur	Group A, C, Y and W-135 CPS	Licensed (1981)
N. meningitidis (group C)	Meningitec, Menjugate	Pfizer Chiron	Meningococcal group C CPS to CRM_{197}	Licensed in UK (1981)
	Neis-Vac-C	Baxter	De-O-Ac CPS to TT	Licensed in UK (1999)
Salmonella typhi	TYPHIM Vi	Sanofi Pasteur	Vi CPS	Licensed in France (1988)
Streptococcus pneumonia	Pneumovax 23	Merck & Co.	Pneumococcal polysaccharides 23-valent	Licensed (1983)
	Prevnar 13	Wyeth	Pneumococcal polysaccharides 7-valent to CRM_{197}	Licensed (2000)

In December of 1987, Aventis Pasteur obtained the license for the first Hib conjugate vaccine (PRP-D or ProHIBIT®), which was prepared from the purified capsular polysaccharide, consisting of a polymer of ribose, ribitol, and phosphate (PRP) of the *H. influenzae* type b strain, covalently attached to diphtheria toxoid (DT) [69]. Studies conducted with ProHIBIT® suggest that induced antibody levels in children of 18- to 20-months-old are 30-fold higher than those induced by polysaccharide vaccines in the same age group [70]. Following the success of this vaccine, three more glycoconjugate vaccines formulated with various protein carriers with distinct immunogenicities were licensed for routine Hib vaccination programs in developing countries (Fig. 13.2). Among these second generation vaccines, the 1990 PedvaxHIB® is marketed by Merck and Co. for infants and children 2 to 71 months of age. This vaccine is composed of highly purified polyribosylribitol phosphate polysaccharide of Hib covalently conjugated to an outer membrane protein

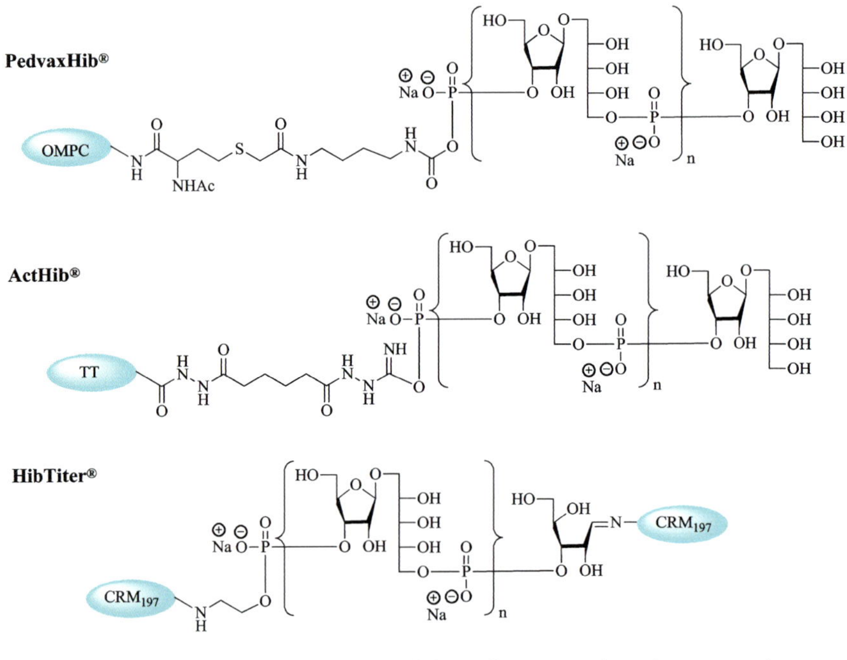

Figure 13.2 Currently licensed *Haemophilus influenza type b* conjugate vaccines.

complex (OMPC). Although the vaccine was recommended in a two-dose regime in the United States, apparently, protective immune response was induced after only one dose in infants, probably due to the direct adjuvant properties of the carrier protein, OMPC [71]. The vaccines containing PRP conjugated to tetanus toxoid (ActHib®–Sanofi Pasteur) and a non-toxic mutant of diphtheria toxin, CRM_{197} (HibTiter®–Pfizer), have shown significant immunogenicity after three doses in infants [72]. Among these Hib vaccines, PedvaxHIB® is the most potent, where >10 mg/L of antibody titer could be induced after the first administration. Interestingly, HibTiter®, the least immunogenic of the conjugates, has been shown to be effective in reducing the incidence of the invasive Hib infections in Finland [73].

13.3.2 *Neisseria meningitidis*

Among the various serogroups of *N. meningitidis*, serogroups A, B, C, Y, and W135 account for almost 90% of invasive infections worldwide. The meningococcal polysaccharide vaccine, Menomune® (MPSV4), consisting of serogroups A, C, Y, and W135-specific polysaccharide antigens, was made available by Sanofi Pasteur in the seventies for people 2 to 55 years of age [74]. Other examples of meningococcal serogroups A, C, Y, and W135-specific polysaccharide vaccines, such as Mencevax® by GlaxoSmithKline and NmVac4-A/C/Y/W134 by JN-International Medical Corporation, have not been licensed in the United States, but are routinely used worldwide. The use of these meningococcal plain polysaccharide vaccines in worldwide national public health vaccination programs was hindered by several key issues: (1) These vaccines elicit only primary immune responses of low-affinity IgM antibodies [75]; (2) booster doses do not increase antibody levels, but also, they can result in hyporesponsiveness [76]; (3) no memory response was generated in infants; and (4) meningococcal serogroup B, a major cause of infections (up to 90% in the United Kingdom), has not been included in polysaccharide tetravalent formulation, thus rendering them ineffective against group B infections.

In 2005, Sanofi Pasteur obtained the first US license for meningococcal conjugate vaccine, Menactra® (MCV4), which was composed of 4 μg of each of meningococcal A, C, Y, and W135 polysaccharides conjugated to ~48 μg DT [77]. The other formulation, Menveo® (MCV4), in which the carrier protein was replaced by diphtheria CRM197, was licensed to Novartis in 2010 for people 2 to 55 years of age and later received US Food and Drug Administration (FDA) approval for use in children as young as 2 months in August of 2013 [77]. In April of 2011, Menactra was approved by FDA for children as young as nine months old. The vaccination age got pushed down to six weeks old when meningococcal bivalent vaccine, MenHibrix® (GlaxoSmithKline Biologicals), was approved by FDA in 2012 to prevent the disease caused by *N. meningitidis* serogroups C and Y, and *H. influenzae* type b. Other efforts include the Meningitidis Vaccine Project (MVP) funded by the Bill and Melinda Gates Foundation for the development of an affordable serogroup A conjugate vaccine for Sub-Saharan Africa, where the group A is responsible for periodic, large-scale epidemics. The resultant conjugate vaccine (MenAfriVac®), consisting of serogroup A polysaccharide conjugated to tetanus-toxin through reductive amination, was tested and introduced into mass vaccination program for widespread use in African meningitis belt countries since 2010 [78].

Despite these impressive successes of conjugate vaccines against invasive diseases caused by meningococcal serogroups A, C, Y, and W135, serogroup B remains a major cause of endemic meningitidis in both developed and developing countries [34]. The main limitation of the development of vaccines against group B is structural similarity between the group B polysaccharide capsule and normal human brain tissue composed of α-(2,8)-*N*-acetyl neuraminic acid repeating units, resulting in immune tolerance. Attempts to break the immune tolerance through slight modification on the polysaccharide moiety, particularly by replacing the *N*-acetyl group with *N*-propionyl and conjugation to protein (NprPSA-TT), showed elevated bactericidal IgG antibodies without detection of autoantibodies [79,80]. Overall, the group B capsular polysaccharide still remains a promising target for further investigations. The Bexsero®, a new type of group B vaccine developed by Novartis is currently licensed in Europe, Canada, and Australia, which is a unique multicomponent vaccine containing four main immunogenic components: there, recombinant proteins combine with outer membrane vesicles derived from the meningococcal NZ98/254 strain.

13.3.3 *Streptococcus pneumonia*

The *S. pneumonia* polysaccharide capsule represents an important virulence factor, and also interferes with phagocytosis by preventing C3b opsonization of the bacterial cells. The most commonly used polysaccharide vaccine, PPV23 (Pneumovax®, Merck), formulated with 23 different pneumococcal serotypes (1, 2, 3, 4, 5, 6b, 7F, 8, 9N, 9V, 10A, 11A, 12F, 14, 15B, 17F, 18C, 19F, 19A, 20, 22F, 23F, and 33F) has been available since 1983 in certain age categories, but is poorly immunogenic in young children and has no effect on nasopharyngeal carriage [57,58,81]. To improve the immunogenicity for children under two years of age, a heptavalent conjugate vaccine, PCV7 (Prevnar®, Wyeth), formulated with serotypes 4-, 6B-, 9V-, 14-, 18C-, 19F-, and 23F-specific polysaccharides individually conjugated to DT mutant CRM197, was licensed in 2000 for a routine national vaccination campaign in US Studies with Prevnar®, in which >37,000 children showed 94% prevention of invasive pneumococcal diseases [82,83]. Prevnar® was also recommended for active immunization in infants and toddlers against otitis media caused by serotypes that are part of the vaccine. However, for such serotypes, the protection against otitis media is lesser than the protection against invasive diseases. Synflorix®, a decavalent conjugate vaccine consisting of antigen from ten pneumococcal serotypes (seven in Prevnar®, plus serotypes 1, 5, and 7F), was approved by the European Medicines Agency for use in the European Union in January of 2009. In February of 2010, a tri-decavalent pneumococcal vaccine (Prevnar13®, Pfizer), which protects against an additional six serotypes, was approved for children 6 weeks through 17 years of age for the prevention of invasive disease caused by 13 *S. pneumonia* serotypes (1, 3, 4, 5, 6A, 6B,

7F, 9V, 14, 18C, 19A, 19F, and 23F), and for children 6 weeks through 5 years of age for the prevention of otitis media caused by seven of the 13 serotypes (4, 6B, 9V, 14, 18C, 19F, and 23F) [84]. Recently, the development of serotype specific monoclonal antibodies have been shown to reduce bacterial load, suppress host inflammatory response, and further prolong the survival time in mouse models of pneumococcal infection [85]. The research for developing an effective vaccine against broad pneumococcal serotypes both for adults and children under the age of two is currently underway.

13.4 Development of Synthetic Antibacterial Conjugate Vaccine

In spite of the impressive success of glycoconjugate vaccines, FDA was not quick to accept vaccines consisting of the polysaccharide component isolated from bacterial sources, especially via acid hydrolysis and subsequent size exclusion fractionation, usually as complex mixtures, because the isolation and purification of the polysaccharides from natural sources is often troublesome, leading to problematic quality control, such as batch-to-batch variability, residual contamination, and toxicity with lipopolysaccharide-derived polysaccharides. Moreover, the attachment of carbohydrate components to carrier proteins through various chemical processes (e.g. reductive amination) generally destroys the reducing end, leading to a loss of some immunogenicity and resulting in conjugates with ambiguities in composition and structural orientation. All these problems can be avoided using totally synthetic methods to obtain polysaccharide components. The recent advancement in oligosaccharide synthesis, including "regioselective one-pot protection methodologies" [86], "programmable one-pot glycosylation" [87], and "automated solid-phase oligosaccharide synthesis" [88,89], facilitate the production of homogeneous and well-characterized glycoconjugate vaccines in a safe and reproducible manner.

13.4.1 *Haemophilus influenza type b*

In 2004, the breakthrough in synthetic vaccines was made by a research team in Cuba via large-scale good manufacturing practice (GMP) production of fully synthetic glycoconjugate vaccines (trade name QuimiHib®) against *H. influenza* type b. The successful large-scale synthesis of ribosylribitol oligomers was achieved using low-cost starting materials, such as ribitol derivative (1), and ribose acetate (2) (Fig. 13.3) via one-step polycondensation reactions using H-phosphonate chemistry [17]. A vaccine prototype was made by conjugation of synthetic antigen to thiolated human serum albumin (HSA) with an average epitope ratio of eight repeating units. The resulting conjugate vaccine (sPRP-HSA) was evaluated in enzyme-linked immunosorbent assays (ELISAs) and found to be highly antigenic, being recognized by anti-Hib sera of mice immunized with commercially available vaccines (Vaxem-Hib and Hiberix). In another vaccine prototype, synthetic antigen was coupled to the more immunogenic carrier protein, tetanus toxoid (TT). The clinical evaluation of this vaccine showed induction of long-term protective antibody titers with a 99.7% success rate in children, subsequently QuimiHib becoming Cuba's national vaccination program in 2004 [90].

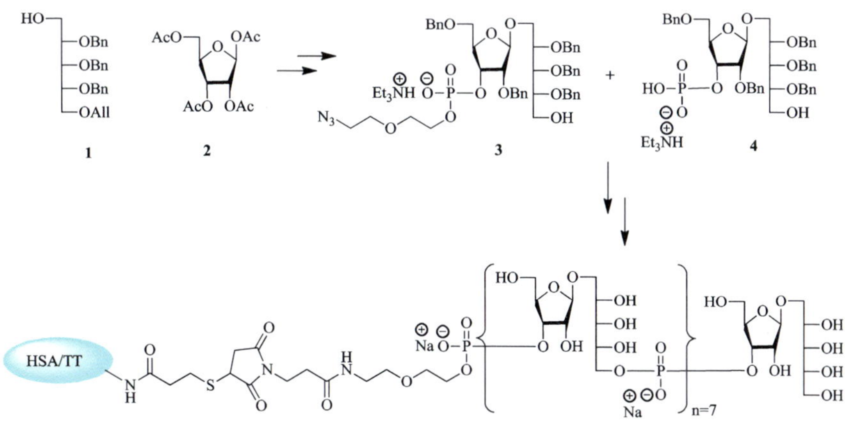

Figure 13.3 Synthesis of QuimiHib.

13.4.2 *The Shigella group bacteria S. dysenteriae*

Shigella is the genus of Gram-negative, rod-shaped bacteria that causes shigellosis (bacillary dysentry) throughout the world. *Shigella* are classified into four serogroups, namely *S. dysenteriae, S. flexneri, S. boydii,* and *S. sonnei,* where each group is further classified into several serotypes, except for *S. sonnei,* which contains only one serotype. The *S. dysenteriae* type 1 is the leading cause of diarrhoea and dysentery, in the absence of any licensed vaccine against this bacterium [91,92]. The *S. dysenteriae* expressed lipopolysaccharide is composed of tetrasaccharide repeating units of α-linked D-galactose, *N*-acetyl-D-glucosamine, and *L*-ramnose residues. Pozsgay's group has reported the first synthesis of the tetrasaccharide unit and its longer oligomers are up to four repeating units long [93,94]. The synthetic antigens were conjugated to HSA through heterobifunctional linkers to form a vaccine construct (Fig. 13.4). Immunological evaluation of glycoconjugate constructs in mice showed that hexadecasaccharide constructs with a molar ratio of 9:1 (carbohydrate: protein) gave the highest IgG titers [95].

13.4.3 *Neisseria meningitidis*

Group B meningococci remain a major global health concern, and the poor immunogenicity of the group B meningococcal

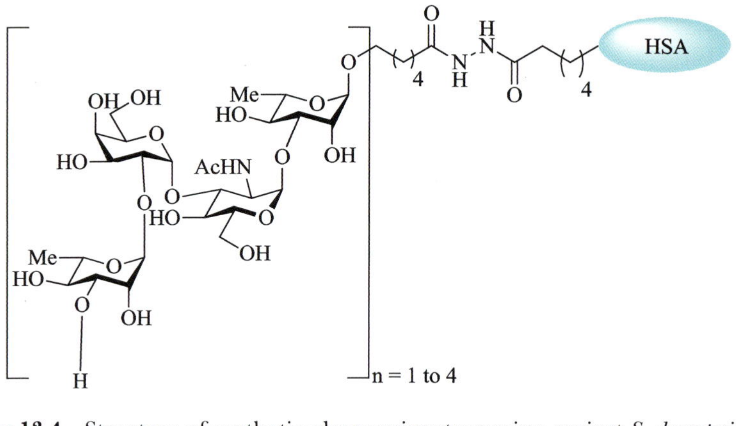

Figure 13.4 Structure of synthetic glycoconjugate vaccine against *S. dysenteriae.*

polysaccharide prevents the development of polysaccharide-based vaccines against *N. meningitis*. Although the polysaccharideb–protein conjugate vaccines were able to enhance polysaccharide-specific antibody levels, including antibodies of the IgG isotype, these levels were still generally low, and no bactericidal activity was reported. The poor immunogenicity of polysaccharides, quite possibly due to its similarity to certain human tissues, can be improved by creating a chemically modified serogroup B antigen. Jennings and co-workers first achieved this goal by replacing the *N*-acetyl groups of the group B meningococcal polysaccharide with *N*-propionyl groups and further conjugated to TT and HSA at the reducing or non-reducing end to form synthetic glycoconjugate vaccines (Fig. 13.5) [79, 96, 97]. Immunological studies showed that the synthetic constructs induced high titers of polysaccharide-specific antibodies in mice, which were highly bactericidal for group B meningococcal and passively protective against *E. coli* K1 [98].

The fully synthetic glycopeptidolipid construct (Fig. 13.6) was developed by Geert-Jan Boons and co-workers for investigation as a vaccine against *N. meningitidis* [99]. A glycoconjugate consists of three major components: L-*glycero*-D-*manno*-heptose, a saccharide

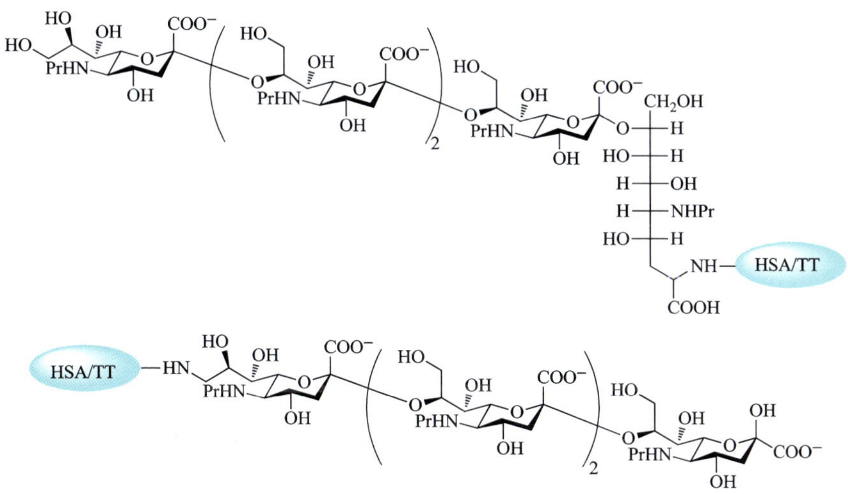

Figure 13.5 Polysialic acid-based glycoconjugate constructs against *N. meningitidis group B.*

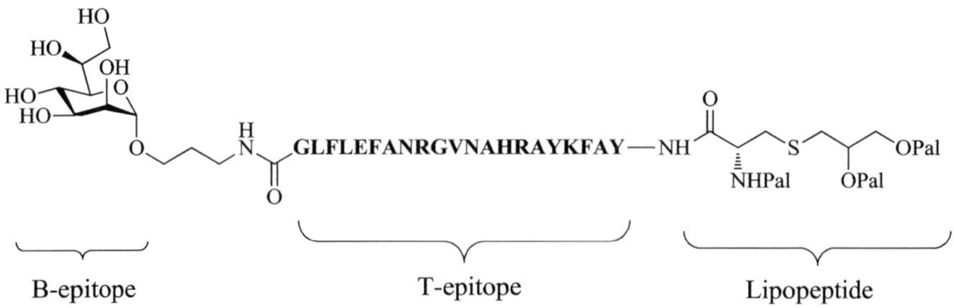

Figure 13.6 Structure of L-*glycero*-D-*manno*-heptose antigen–T-epitope–lipopeptide conjugate vaccine.

representing a partial structure of the inner-core oligosaccharide, acting as a B epitope; T-cell epitope, a peptide sequence identified as a major histocompatibility complex (MHC) class II restricted site for human T-cells and derived from an outer-membrane protein of *N. meningitidis*; and N-terminal of T-cell epitope attached to the lipopeptide, Pam3Cys, which acts as an immunostimulant and serves as a membrane anchor in the lipid layer of cell membranes to ensure efficient and long-term delivery of the antigen.

Group A strains of *N. meningitidis* are the main cause of explosive epidemics in Sub-Saharan Africa. The development of an efficient conjugate vaccine against *N. meningitidis* A is greatly hampered by the poor hydrolytic stability of its capsular polysaccharide, which is made up of (1→ 6)-linked 2-acetamido-2-deoxy-alpha-D-mannopyranosyl phosphate repeating units. Pozsgay *et al.* reported the synthesis of fragments, including mono- with or without 6-*O*-phosphate, bi-, and trisaccharide units [100]. Then, the saccharides with amino alkyl spacers at the reducing end were *N*-acylated with hepta-4,6-dienoic acid and the resulting conjugated diene-equipped saccharides were subjected to a Diels–Alder type addition with maleimidobutyryl-group functionalized HSA to form covalent conjugates containing up to 26 saccharide haptens per albumin molecule (Fig. 13.7). Vaccine evaluation of the neoglycoconjugates indicated that the polyclonal meningococcus group A antiserum could recognize a fragment as small as a monosaccharide, however, the *O*-acetyl group(s) present in the natural capsular material is not essential for antigenicity.

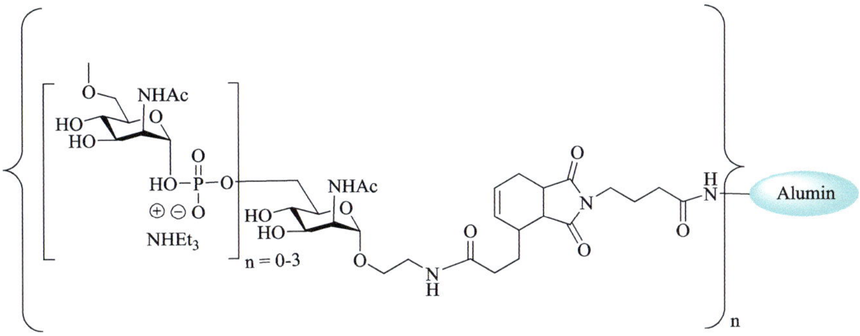

Figure 13.7 Structure of *N. meningitidis* group A glycoconjugate vaccines.

During 2001 and 2002, the first large-scale epidemics of serogroup W135 meningitis in Sub-Saharan Africa were reported from Burkina Faso. By the beginning of 2012, more than 4,000 cases of meningococcal disease, caused mainly by serogroup W135, were reported in countries of the African meningitis belt. Therefore, an affordable supply of serogroup W135-containing polysaccharide vaccine for epidemic control is of great priority. The Wu group has recently reported the first total synthesis of *N. meningitidis* serogroup W135 capsular disaccharide to decasaccharide, which consists of →6)-α-D-Gal-(1→4)-α-D-Neu5Ac(9-OAc)-(2→ repeating units. The resulting saccharides were chemically attached to CRM_{197} using thiol-maleimide-based Michael-type addition to form conjugates with various carbohydrate:epitope ratios (Fig. 13.8). Immunological results showed that induced IgG antibody titers were much higher than IgM antibody titers. Moreover, the results showed that a tetrasaccharide is the minimum length of oligosaccharide required to make effective antibodies that possess potent bactericidal activity [101].

13.4.4 *Vibrio cholerae*

Vibrio cholerae, the Gram-negative bacterium, causes cholera, leading to severe diarrhoea [102]. Currently, the *V. cholerae* is classified into 206 "O" serogroups based on the differences in the sugar composition of the heat-stable surface somatic "O" antigen. The epidemic

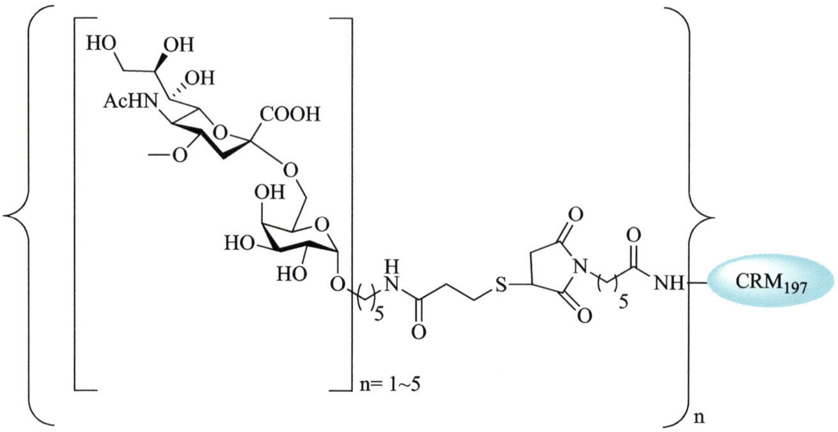

Figure 13.8 Structure of *N. meningitidis W135* glycoconjugate vaccines.

cholera was majorly caused by *V. cholerae* strains of the O1 serogroup; however, the latest pathogenic serotype O139 was discovered in 1992. *V. cholerae* strain O1 occurs in two serotypes, namely Ogawa and Inaba, the only structural difference being the 2-*O*-methylation of the terminal 4-amino-4,6-dideoxy-D-mannose unit in the Ogawa serotype (Fig. 13.9) [103,104].

Recently, Szu *et al.* reported the chemical synthesis of hexa-, octa-, and decasaccharide fragments of the *O*-specific polysaccharide of *V. cholerae* O1, serotype Ogawa. These oligosaccharides with a carboxylic acid at the reducing end were conjugated to TT by a novel linking scheme consisting of 17-atom linkers of hydrazide and alkyl bonds (Fig. 13.9). The immune evaluation of glycocojugates showed that decasaccharide conjugates elicited the highest anti-LPS IgG. Moreover, apart from chain length, a conjugate with a molar ratio of 230 ± 10 monosaccharide units per TT induced high antibody response [105]. Furthermore, the conjugates also elicited cross-reactive antibodies to the serotype Inaba. Meeks *et al.* reported the preparation of neoglycoconjugates from *V. cholerae* O1, serotype Inaba (disaccharide, tetrasaccharide, and hexasaccharide conjugated to bovine serum albumin (BSA)) using squaric acid chemistry, at different levels of substitution. BALB/c mice responded to the

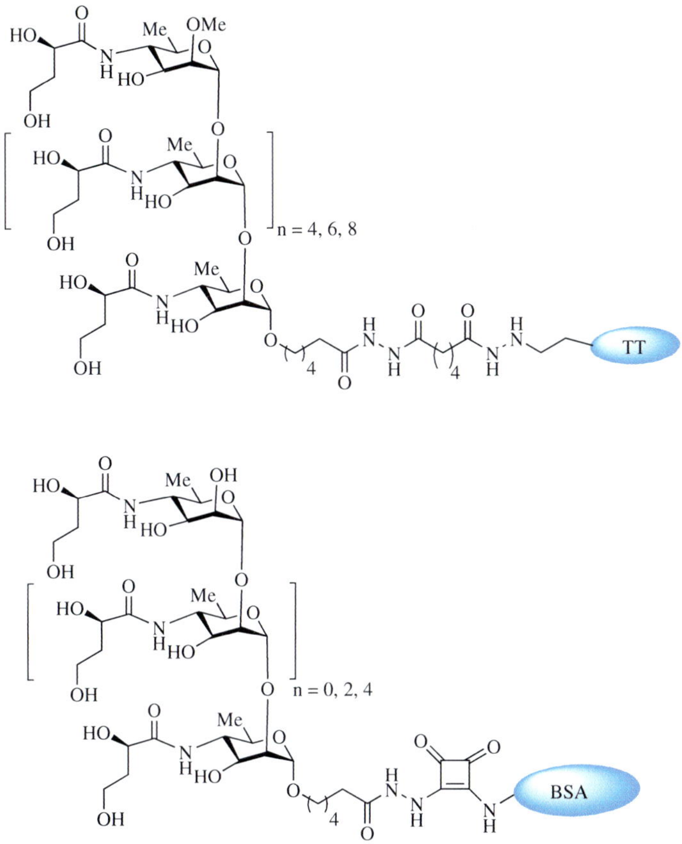

Figure 13.9 Chemical composition of conjugate vaccine against *Vibrio cholerae* serotype Ogawa (Upper panel) and Inaba (Lower panel).

Inaba-specific conjugates with similar titers of antibodies of those obtained from mice immunized with Ogawa-specific conjugates with BSA. However, the Inaba-specific antibodies (IgM and IgG1) were neither vibriocidal nor protective in the infant mouse cholera model [106,107].

13.5 Perspectives and Future Challenges

The significant progress achieved in immunology in recent years has resulted in the development of several sophisticated vaccines against invasive bacteria and other human pathogens. Most of the licensed and under development polysaccharide-based bacterial vaccines meet the necessary prophylactic requirement including: (i) safety and efficacy; (ii) specificity for the polysaccharide capsule that is the major bacterial virulence determinant; (iii) restriction of transmission and thus, induction of herd immunity; and (iv) providing T-cell dependent long-lasting immunity with immunological memory in adults, children, and high-risk groups. However, despite the extensive research in this field, the precise nature of the molecular events that permit polysaccharides conjugated to protein carriers to be processed as T-dependent antigens remains unclear, and more efforts are certainly needed.

The commercial success of pure synthetic conjugate vaccine (Quimi Hib) clearly highlighted the utility and safety of chemically well-defined oligosaccharide-based vaccine over the natural polysaccharide counterpart, which is usually characterized by significant heterogeneity that may lead to jeopardizing reproducibility of production batches and contamination by highly toxic components. The development in key technologies, such as carbohydrate microarrays, automated oligosaccharide syntheses, and biophysical and computational methods for studying antigenb–antibody interactions, rapidly accelerated the access to a wide range of oligosaccharides and allowed researchers to gain a deeper understanding of their functions. However, to produce some uncommon sugars that are part of bacterial capsules, glycochemists have to focus on the development of rapid and efficient methodologies including suitable, selective protection and deprotection steps, together with proper glycosylating agents. Progress has been made in the synthesis of metabolically stable analogs of bacterial polysaccharide antigens (e.g. C- or S-glycosides) to improve immunogenicity of conjugate vaccines. The studies of the vaccine obtained from such modified analogs may improve our understanding of antigen presentation and processing to B-cells and stimulation of the immune system to induce protective antibodies.

It is becoming increasingly apparent that the modes of glycotope presentation in the context of vaccines affect how they are undertaken and processed by the body to raise immune responses against them. New synthetic and microarray technologies are being developed to enable us to explore the interaction of the immune system with self and non-self glycans. Overall, such key discoveries can be expected to translate into new therapeutics and vaccines to combat invasive bacterial diseases and improve human health.

References

1. Varki, A. (1993) *Glycobiology* **3**, 97–130.
2. Dwek, R. A. (1996) *Chem. Rev.* **96**, 683–720.
3. Bertozzi, C. R., Kiessling, L. L. (2001) *Science* **291**, 2357–2364.
4. Ohtsubo, K., Marth, J. D. (2006) *Cell* **126**, 855–867.
5. Hancock, I. C. (1997) *Biochem. Soc. Trans.* **25**, 183–187.
6. Bugg, T. D. Brandish, P. E. (1994) *FEMS Microbiol. Lett.* **119**, 255–262.
7. Daffe, M., Etienne, G. (1999) *Tuber. Lung Dis.* **79**, 153–169.
8. Lesinski, G. B., Westerink, M. A. (2001) *Curr. Drug Targets* **1**, 325–334.
9. Lesinski, G. B. Westerink, M. A. (2001) *J. Microbiol. Method* **47**, 135–149.
10. Coquillat, D., Bruge, J., Danve, B., Latour, M., Hurpin, C., Schulz, D., Durbec, P., Rougon, G. (2001) *Infect. Immun.* **69**, 7130–7139.
11. Finn, A. (2004) *Brit. Med. Bull.* **70**, 1–14.
12. Goldblatt, D. (1998) *J. Med. Microbiol.* **47**, 563–567.
13. Huang, Y. L., Wu, C. Y. (2010) *Expert Rev. Vaccines* **9**, 1257–1274.
14. Ada, G., Isaacs, D. (2003) *Clin. Microbiol. Infect.* **9**, 79–85.
15. Jones, C. (2005) *An. Acad. Bras. Cienc.* **77**, 293–324.
16. Ulanova, M., Tsang, R. S. (2009) *Infect. Genet. Evol.* **9**, 594–605.
17. Verez-Bencomo, V., Fernandez-Santana, V., Hardy, E., Toledo, M. E., Rodriguez, M. C., Heynngnezz, L., Rodriguez, A., Baly, A., Herrera, L., Izquierdo, M., Villar, A., Valdes, Y., Cosme, K., Deler, M. L., Montane, M., Garcia, E., Ramos, A., Aguilar, A., Medina, E., Torano, G., Sosa, I., Hernandez, I., Martinez, R., Muzachio, A., Carmenates, A., Costa, L., Cardoso, F., Campa, C., Diaz, M., Roy, R. (2004) *Science* **305**, 522–525.
18. Meszner, Z. (2014) *Orv. Hetil.* **155**, 243–247.
19. Schwartz, B., Van Beneden, C., Whitney, C. G., Levine, O. S. (2000) *MMWR. Recommendations and reports: Morbidity and mortality weekly report. Recommendations and reports/CDC* **49**, 1–35.

20. Nuorti, J. P., Whitney, C. G. (2010) *MMWR. Recommendations and reports: Morbidity and mortality weekly report. Recommendations and reports/ CDC* **59**, 1–18.
21. Dagan, R. (2010) *Vaccine* **19**, S9–S16.
22. Dowell, S. F., Butler, J. C., Giebink, G. S., Jacobs, M. R., Jernigan, D., Musher, D. M., Rakowsky, A., Schwartz, B. (1999) *Pediatr. Infect. Dis. J.* **18**, 1–9.
23. WHO (2008) *Releve epidemiologique hebdomadaire/Section d'hygiene du Secretariat de la Societe des Nations = Weekly epidemiological record/Health Section of the Secretariat of the League of Nations* **83**, 373–384.
24. Whitney, C. G., Moore, M., Pilishvili, T. (2012) *MMWR. Morbidity and mortality weekly report* **61**, 816–819.
25. Lindberg, B., Lindqvist, B., Lonngren, J., Powell, D. A. (1980) *Carbohydrate Res.* **78**, 111–117.
26. Jansson, P. E., Lindberg, B., Anderson, M., Lindquist, U., Henrichsen, J. (1998) *Carbohydrate Res.* **182**, 111–117.
27. Jodar, L., Feavers, I. M., Salisbury, D., Granoff, D. M. (2002) *Lancet* **359**, 1499–1508.
28. Lo, H., Tang, C. M., Exley, R. M. (2009) *Lancet Infect. Dis.* **9**, 418–427.
29. Broud, D. D., Griffiss, J. M., Baker, C. J. (1979) *J. Infect. Dis.* **140**, 465–470.
30. Craven, D. E., Shen, K. T., Frasch, C. E. (1982) *Infect. Immun.* **37**, 132–137.
31. Cohen R., Levy, C. (2012) *Arch. Pediatr.* **19**, S61–S64.
32. Elias, J. (2014) *Lancet* **383**, 6–8.
33. Swartley, J. S., Liu, L. J., Miller, Y. K., Martin, L. E., Edupuganti, S., Stephens, D. S. (1998) *J. Bacter.* **180**, 1533–1539.
34. Zimmer, S. M., Stephens, D. S. (2006) *Curr. Opin. Invest. Drugs* **7**, 733–739.
35. Ram, S., Lewis, L. A., Agarwal, S. (2011) *J. Biol. Chem.* **286**, 8297–8307.
36. Muindi, K. M., McCarthy, P. C., Wang, T., Vionnet, J., Battistel, M., Jankowska, E., Vann, W. F. (2014) *Glycobiology* **24**, 139–149.
37. Griffiss, J. M., Brandt, B. L. (1983) *J. Gen. Microbiol.* **129**, 447–452.
38. Baker, C. J., Edwards, M. S. (2003) *Arch. Dis. Child.* **88**, 375–378.
39. Decheva, A. (2013) *Akush. Ginekol.* **3**, 52, 55–60.
40. Money, D., Allen, V. M. (2013) *J. Obstet. Gynaecol. Can.* **35**, 939–951.
41. Schneeberger, C., Kazemier, B. M., Geerlings, S. E. (2014) *Curr. Opin. Infect. Dis.* **27**, 108–114.
42. Jennings, H. J., Katzenellenbogen, E., Lugowski, C., Kasper, D. L. (1983) *Biochem.* **22**, 1258–1264.

43. Schifferle, R. E., Jennings, H. J., Wessels, M. R., Katzenellenbogen, E., Roy, R., Kasper, D. L. (1985) *J. Immunol.* **135**, 4164–4170.
44. Wessels, M. R., Benedi, W. J., Jennings, H. J., Michon, F., DiFabio, J. L., Kasper, D. L. (1989) *Infect. Immun.* **57**, 1089–1094.
45. Wessels, M. R., DiFabio, J. L., Benedi, V. J., Kasper, D. L., Michon, F., Brisson, J. R., Jelinkova, J., Jennings, H. J. (1991) *J. Biol. Chem.* **266**, 6714–6719.
46. Bhutta, Z. A., Threlfall, J. (2009) *J. Am. Med. Association* **302**, 898–899.
47. Buckle, G. C., Walker, C. L., Black, R. E. (2012) *J. Glob. Health* **2**, 010401.
48. Crump, J. A., Luby, S. P., Mintz, E. D. (2004) *Bull World Health Organ.* **82**, 346–353.
49. Guzman, C. A., Borsutzky, S., Griot-Wenk, M., Metcalfe, I. C., Pearman, J., Collioud, A., Favre, D., Dietrich, G. (2006) *Vaccine* **24**, 3804–3811.
50. Martin, L. B. (2012) *Curr. Opin. Infect. Dis.* **25**, 489–499.
51. Sinha, A., Sazawal, S., Kumar, R., Sood, S., Reddaiah, V. P., Singh, B., Rao, M., Naficy, A., Clemens, J. D., Bhan, M. K. (1999) *Lancet* **354**, 734–737.
52. Szu, S. C., Li, X. R., Stone, A. L., Robbins, J. B. (1991) *Infect. Immun.* **59**, 4555–4561.
53. Heidelberger, M., Avery, O. T. (1923) *J. Exp. Med.* **38**, 73–79.
54. Tillett, W. S., Francis, T. (1929) *J. Exp. Med.* **50**, 687–701.
55. Heidelberger, M., Dilapi, M. M., Siegel, M., Walter, A. W. (1950) *J. Immunol.* **65**, 535–541.
56. Whitney, C. G. (2010) *MMWR. Morbidity and mortality weekly report* **59**, 1102–1106.
57. Shapiro, E. D., Berg, A. T., Austrian, R., Schroeder, D., Parcells, V., Margolis, A., Adair, R. K., Clemens, J. D. (1991) *N. Engl. J. Med.* **325**, 1453–1460.
58. Melegaro, A., Edmunds, W. J. (2004) *Eur. J. Epidemiol.* **19**, 353–363.
59. Mitchell, R., Truck, J., Pollard, A. J. (2013) *Expert Opin. Biol. Ther.* **13**, 1451–1465.
60. Vila-Corcoles, A., Ochoa-Gondar, O., Guzman, J. A., Rodriguez-Blanco, T., Salsench, E., Fuentes, C. M. (2010) *BMC Infect. Dis.* **10**, 73.
61. Cobb, B. A., Wang, Q., Tzianabos, A. O., Kasper, D. L. (2004) *Cell* **117**, 677–687.
62. Snapper, C. M., Mond, J. J. (1996) *J. Immunol.* **157**, 2229–2233.
63. Vos, Q., Lees, A., Wu, Z. Q., Snapper, C. M., Mond, J. J. (2000) *Immunol. Rev.* **176**, 154–170.

64. Avery, O. T., Goebel, W. F. (1931) *J. Exp. Med.* **54**, 437–447.
65. Avery, O. T., Goebel, W. F. (1929) *J. Exp. Med.* **50**, 533–550.
66. Maiden, M. C. (2013) *Philos. Trans. R. Soc. Lond. Series B, Biological Sciences* **368**, 20120147.
67. Trotter, C. L., McVernon, J., Ramsay, M. E., Whitney, C. G., Mulholland, E. K., Goldblatt, D., Hombach, J., Kieny, M. P. (2008) *Vaccine* **26**, 4434–4445.
68. Kayhty, H., Karanko, V., Peltola, H., Makela, P. H. *Pediatrics* **74**, 857–865.
69. Schneerson, R., Barrera, O., Sutton, A., Robbins, J. B. (1980) *J. Exp. Med.* **152**, 361–376.
70. Peter, C. G. (1998) *Pediatrics* **102**, 252–254.
71. Latz, E., Franko, J., Golenbock, D. T., Schreiber, J. R. (2004) *J. Immunol.* **172**, 2431–2438.
72. Anderson, P. W., Pichichero, M. E., Insel, R. A., Betts, R., Eby, R., Smith, D. H. (1986) *J. Immunol.* **137**, 1181–1186.
73. Eskola, J., Peltola, H., Kayhty, H., Takala, A. K., Makela, P. H. (1992) *J. Infect. Dis.* **165**, S137–S138.
74. Armand, J., Arminjon, F., Mynard, M. C., Lafaix, C. (1982) *J. Biol. Stand.* **10**, 335–339.
75. Reingold, A. L., Broome, C. V., Hightower, A. W., Ajello, G. W., Bolan, G. A., Adamsbaum, C., Jones, E. E., Phillips, C., Tiendrebeogo, H., Yada, A. (1985) *Lancet* **2**, 114–118.
76. Bilukha, O. O., Rosenstein, N. (2005) *MMWR. Recommendations and reports: Morbidity and mortality weekly report. Centers for Disease Control* **54**, 1–21.
77. Pilishvili, T. (2011) *MMWR. Morbidity and mortality weekly report* **60**, 72–76.
78. LaForce, F. M., Okwo-Bele, J. M. (2011) *Health affairs (Project Hope)* **30**, 1049–1057.
79. Jennings, H. J., Roy, R., Gamian, A. (1986) *J. Immunol.* **137**, 1708–1713.
80. Fusco, P. C., Michon, F., Tai, J. Y., Blake, M. S. (1997) *J. Infect. Dis.* **175**, 364–372.
81. Bricks, L. F. (1994) *J. Pediatr.* **70**, 75–81.
82. Fritzell, B., Fletcher, M. A. (2011) *Expert Rev. Vaccines* **10**, 263–290.
83. Plosker, G. L. (2013) *Paediatric Drugs* **15**, 403–423.
84. Whitney, C. G. (2010) *MMWR. Morbidity and mortality weekly report* **59**, 258–261.

85. Burns, T., Abadi, M., Pirofski, L. A. (2005) *Infect. Immun.* **73**, 4530–4538.
86. Wang, C. C., Zulueta, M. M., Hung, S. C. (2011) *Chimia* **65**, 54–58.
87. Wu, C. Y., Wong, C. H. (2011) *Top. Curr. Chem.* **301**, 223–252.
88. Hsu, C. H., Hung, S. C., Wu, C. Y., Wong, C. H. (2011) *Angew. Chem. Intl. Ed.* **50**, 11872–11923.
89. Seeberger, P. H. (2008) *Chem. Soc. Rev.* **37**, 19–28.
90. Torano, G., Toledo, M. E., Baly, A., Fernandez-Santana, V., Rodriguez, F., Alvarez, Y., Serrano, T., Musachio, A., Hernandez, I., Hardy, E., Rodriguez, A., Hernandez, H., Aguilar, A., Sanchez, R., Diaz, M., Muzio, V., Dfana, J., Rodriguez, M. C., Heynngnezz, L., Verez-Bencomo, V. (2006) *Clin. Vaccine Immunol.* **13**, 1052–1056.
91. Kotloff, K. L., Winickoff, J. P., Ivanoff, B., Clemens, J. D., Swerdlow, D. L., Sansonetti, P. J., Adak, G. K., Levine, M. M. (1999) *Bull. World Health Organ.* **77**, 651–666.
92. Liu, B., Knirel, Y. A., Feng, L., Perepelov, A. V., Senchenkova, S. N., Wang, Q., Reeves, P. R., Wang, L. (2008) *FEMS Microbiol. Rev.* **32**, 627–653.
93. Pozsgay, V. (1998) *J. Org. Chem.* **63**, 5983–5999.
94. Pozsgay, V., Ekborg, G., Sampathkumar, S. G. (2006) *Carbohydrate Res.* **341**, 1408–1427.
95. Kubler-Kielb, J., Vinogradov, E., Mocca, C., Pozsgay, V., Coxon, B., Robbins, J. B., Schneerson, R. (2010) *Carbohydrate Res.* **345**, 1600–1608.
96. Jennings, H. J. (1989) *Contrib. Microbiol. Immunol.* **10**, 151–165.
97. Jennings, H. J., Lugowski, C., Ashton, F. E. (1984) *Infect. Immun.* **43**, 407–412.
98. Krug, L. M., Ragupathi, G., Hood, C., George, C., Hong, F., Shen, R., Abrey, L., Jennings, H. J., Kris, M. G., Livingston, P. O. (2012) *Cancer Immunol. Immunother.* **61**, 9–18.
99. Zhang, Y., Gaekwad, J., Wolfert, M. A., Boons, G. J. (2008) *Chem. Eur. J.* **14**, 558–569.
100. Berkin, A., Coxon, B., Pozsgay, V. (2002) *Chem. Eur. J.* **8**, 4424–4433.
101. Wang, C. H., Li, S. T., Lin, T. L., Cheng, Y. Y., Sun, T. H., Wang, J. T., Cheng, T. J., Mong, K. K., Wong, C. H., Wu, C. Y. (2013) *Angew. Chem. Intl. Ed.* **52**, 9157–9161.
102. Moore, S., Thomson, N., Mutreja, A., Piarroux, R. (2014) *Clin. Microbiol. Infect.* **20**(5), 373–379.
103. Levine, M. M., Noriega, F. (1995) *Papua New Guinea Med.* **38**, 325–331.

104. Mandal, S., Mandal, M. D., Pal, N. K. (2011) *Asian Pac. J. Trop. Med.* **4**, 573–580.
105. Ftacek, P., Nelson, V., Szu, S. C. (2013) *Glycoconjugate J.* **30**, 871–880.
106. Chernyak, A., Kondo, S., Wade, T. K., Meeks, M. D., Alzari, P. M., Fournier, J. M., Taylor, R. K., Kovac, P., Wade, W. F. (2002) *J. Infect. Dis.* **185**, 950–962.
107. Meeks, M. D., Saksena, R., Ma, X., Wade, T. K., Taylor, R. K., Kovac, P., Wade, W. F. (2004) *Infect. Immun.* **72**, 4090–4101.

Chapter 14

Synthetic Vaccines against Cancers

Michele Fiore, Baptiste Thomas, Gour Chand Daskhan
and Olivier Renaudet

University Grenoble Alpes, DCM, 38000 Grenoble, France

14.1 Introduction

Cancer immunotherapy has recently emerged as a promising alternative to antitumoral treatments, such as surgery, chemo- and/or radiotherapies that often provoke side effects. In this context, major efforts are being made to design synthetic vaccines capable of triggering immunity against cancers [1–7]. These antitumoral synthetic vaccines are first composed of tumor-associated carbohydrate antigens (TACAs), such as Tn, Thomsen–Friedenreich (TF) (or T), sialyl Tn (sTn), sialyl TF, Lewis y (Le^y), GM2, or Globo H (Fig. 14.1), that are overexpressed at the surface of a large variety of tumors, where they play crucial roles in invasion and metastasis processes.

The presence of TACAs in a vaccine formulation is indeed necessary to activate B lymphocytes; however, it is insufficient to produce high titers of antibodies and to induce reproducible and long-lasting immunity against tumors, due to their inability to activate T-helper lymphocytes. To achieve this purpose and induce T-cell dependent response, two approaches have been proposed. The semi-synthetic strategy (Fig. 14.2) requires the covalent conjugation of a TACA to a protein carrier, such as keyhole limpet

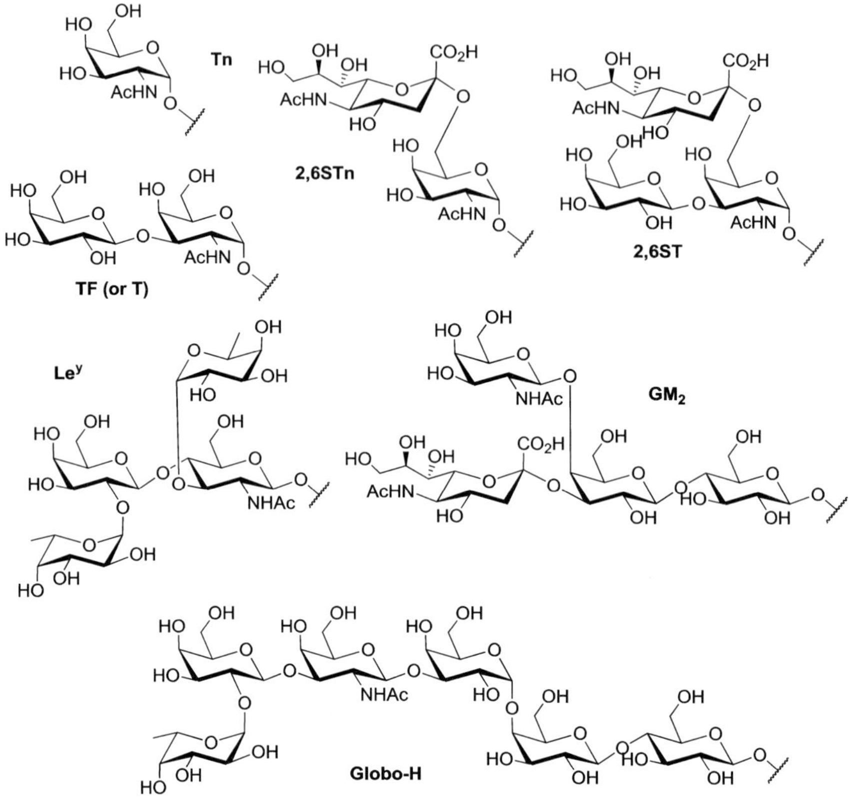

Figure 14.1 Chemical structures of several TACAs.

hemocyanin (KLH), bovine serum albumin (BSA), or tetanus toxin (TTox), which provides peptide fragments that can activate T-helper lymphocytes. However, their utilization remains debatable because of side effects and problems of reproducibility due to their uncertain molecular composition. As an alternative, well-defined fully synthetic vaccines have been designed. They combine, within a single molecule, T-helper ($CD4^+$) peptide antigen to stimulate T-helper lymphocytes and clusters of TACAs to promote a B-cell response (Fig. 14.2). In this chapter, we describe these two approaches through the selection of various promising synthetic vaccines described so far.

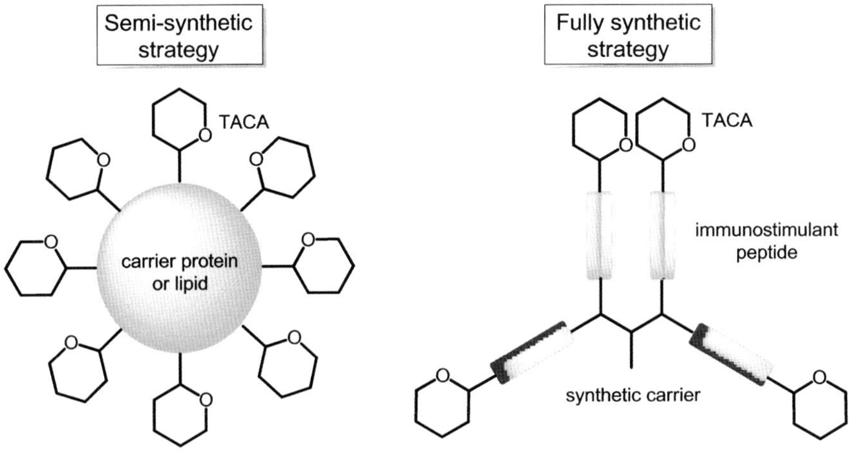

Figure 14.2 Two main classes of antitumoral vaccines.

The preparation of TACAs remains the prerequisite to the development of both semi-synthetic and fully synthetic vaccines. These topics being out of the scope of this chapter, we suggest readers to refer to excellent reviews that cover recent advances in carbohydrate chemistry [8–9], and chemoenzymatic [10] and biotechnological synthesis [11].

14.2 The Semi-Synthetic Approach

14.2.1 *Protein carriers*

Mucins are heavily glycosylated glycoproteins bearing *O*-linked oligosaccharides associated to serine (Ser) or threonine (Thr) residues. Among these glycoproteins, mucin 1 (MUC1) is composed of numerous tandem repeats of the sequence histamine (His)–glycine (Gly)–valine (Val)–Thr–Ser–alanine (Ala)–proline (Pro)–asparagine (Asp)–Thr–arginine (Arg)–Pro–Ala–Pro–Gly–Ser–Thr–Ala–Pro–Pro–Ala. They are overexpressed with different glycosylation profiles to the membrane of almost all epithelial cells, as well as on tumors, which make these structures attractive targets for the development of anticancer vaccines. A paper published in 1997 was at the origin of a long-lasting and exciting story by the group of H. Kunz. In this study, the authors described, for the first time, a fully-supported synthesis of tumor-associated antigen glycopeptides from the tandem repeat sequence MUC1 [12], and later from of the epithelial mucin, MUC4 [13]. This group has further demonstrated that the conjugation of monovalent MUC1, bearing either STn (1) [14] or TF (2) [15] antigens to BSA or TTox through a PEGylated squarate linker (Fig. 14.3), stimulate strong and highly selective immune responses. In particular, the production of high titers of immunoglobulin G (IgG), which strongly binds to epithelial tumor cells of the MCF-7 cell line, has been observed [15].

In another report, new synthetic MUC1-based vaccines containing two immunodominant domains have been elaborated [16]. As nuclear magnetic resonance (NMR) spectroscopic analyses revealed, the glycan side chains in the Ser–Thr–Ala–Pro–Pro–Ala sequence strongly influence the peptide conformation and tumor selectivity, and the authors decided to incorporate the STn antigen at the serine or the threonine at positions [17] and [18] (3) (Fig. 14.3). In addition, the peptide sequence was elongated to contain the second immunodominant domain, Ala–Pro–Asp–Thr–Arg–Pro. The resulting MUC1 glycopeptide was next coupled to carrier proteins, such as BSA or TTox, and the conjugate was used for immunization of

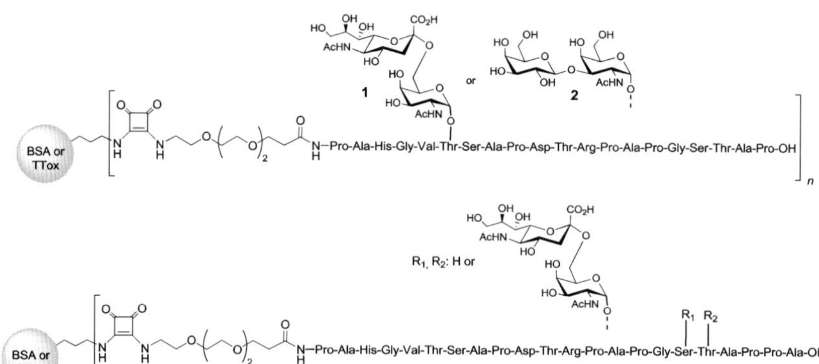

3

Figure 14.3 Example of monovalent MUC1-conjugate.

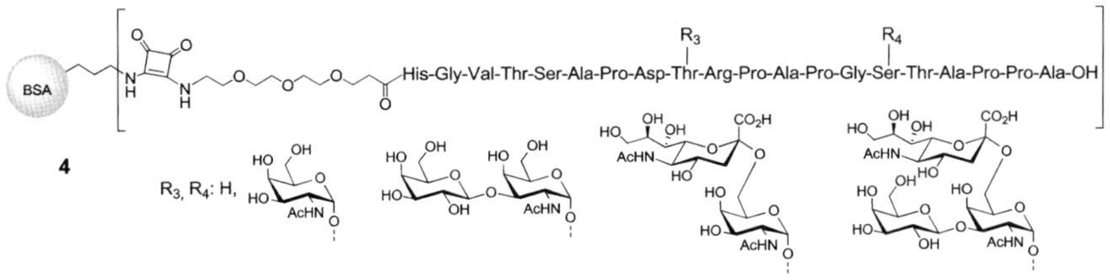

Figure 14.4 Examples of divalent MUC1-based vaccines.

BALB/cJ mice. Strong antibody (Ab) responses were observed against breast tumor tissues, thus suggesting that those Abs can be used as valuable diagnostic tools.

In a subsequent paper, the same group investigated the influence of the glycosylation pattern in MUC1 on the immune response [17]. Following the same approach, BSA-conjugated vaccines (4) have been prepared with various divalent combinations of sialylated and unsialylated antigens (Fig. 14.4). Immunological data have shown the production of high titers of IgG antibodies capable of binding strongly to the Michigan Cancer Foundation-7 (MCF-7) breast tumor cells. Furthermore, this study revealed that variations on the glycosylation pattern in MUC1-conjugates significantly influence the immune response.

Elegant contributions have been made by Danishefsky and coworkers, who have described several methods to install amino acids and linkers to TACAs. Briefly, the "cassette approach" [18], based on the iterative utilization of glycals [19], was developed to synthesize complex carbohydrate antigens, such as Tn, TF, STn, Le^y, or Globo-H with high efficiency. Coupling of these structures was performed onto fluorenylmethoxycarbonyl (Fmoc)-protected serine or threonine to provide clusters of peptide containing mucin-related antigens [20–21]. The incorporation of allyl or pentenyl alcohols provided other key building blocks for the conjugation to proteins, such as KLH [22–25]. The process required both ozonolysis of the alkene and reductive amination with amino groups of the protein to provide conjugates with 350 (5) to 716 (6) carbohydrate residues per protein when bifunctional linker 4-(4-*N*-maleimidomethyl)

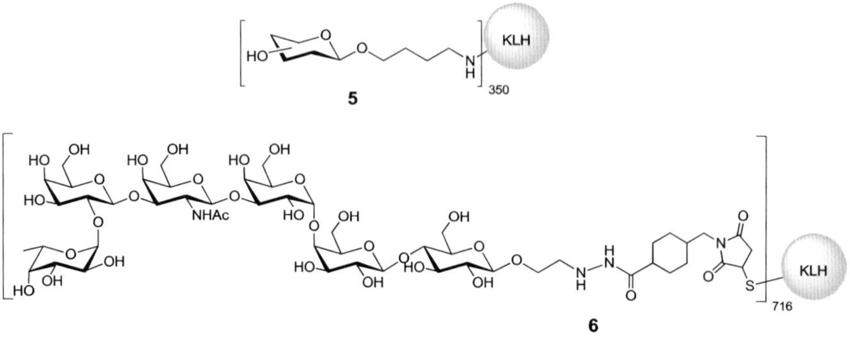

Figure 14.5 Examples of KLH-based vaccines from the Danishefsky group.

cyclohexane-1-carbonylhydrazide was used [26] (Fig. 14.5). In the latter case, the KLH carrier should be treated with 2-iminothiolane to convert amino to thiol groups. When such vaccine-containing Globo-H (6) was co-administrated by subcutaneous vaccinations with the QS-21 immunoadjuvant to patients having prostate cancer, the induction of high-titer immunoglobulin M (IgM) Abs was observed. More interestingly, complete immunological study in humans has indicated that the treatment effect was still observed three months after the vaccination protocol [26]. It should be mentioned that Horner–Emmons olefination from *n*-pentenyl glycosyls and glycine-derived phosphonate has also been described by the same group to provide unnatural glycosylamino acids containing the Tn, Globo-H, and Le^y antigens with excellent diastereos electivity [27].

Attachment of a trivalent glycopeptide to KLH or BSA was also reported by the same research group. To this end, a thiodiaminopropyl linker was conjugated to at the *C*-terminal end of the glycopeptides, and a Michael-type coupling was performed with KLH or BSA (7) derivatized with *m*-maleimimidobenzoyl group (Fig. 14.6). The resulting vaccine (7) was thus obtained with a grafting ratio of about 317 clusters per protein for KLH and only seven for BSA [21]. When immunization was performed with this vaccine in the presence of the adjuvant QS-21, high IgM and IgG titers were observed against the cluster of Tn antigen by enzyme-linked

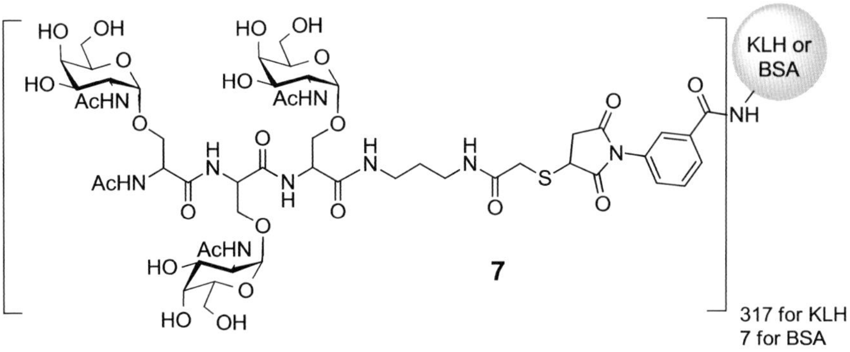

Figure 14.6 Structure of a trivalent glycopeptide-protein conjugate.

immunosorbent assay (ELISA). In addition, these antibodies were found to be strongly reactive with Tn-positive human colon cancer cell lines by flow cytometry.

With the aim to mimic the heterogeneous tumor surface glycocalix, the same coupling strategy was employed to prepare unimolecular pentavalent conjugates that included different carbohydrate antigens (i.e. Tn, TF, STn, Le^y, and Globo-H) within a single carrier protein [20, 28, 29]. The authors thus expected to improve the immunological properties against prostate cancer by triggering a "multifaceted" immunological response. For this purpose, several glycosylamino acids have been prepared, incorporated into a peptide backbone, and linked covalently to a carrier protein to provide vaccines, such as (8) (Fig. 14.7), with excellent grafting ratio (228:1). Preclinical ELISA-based evaluation has shown that the presence of multiple TACAs do not suppress the IgM and IgG Abs responses against any of the carbohydrate antigens [28].

A more recent study by the same group described an improved conjugation protocol leading to (8), which provides a higher epitope ratio (i.e. 505:1) [30]. Immunization of mice in the presence of QS-21 first revealed a high IgG and IgM antibody response against each of the five individual carbohydrate antigens, suggesting that these antibodies can recognize the MCF-7 breast cancer cell lines expressing these five carbohydrate antigens. This study demonstrated

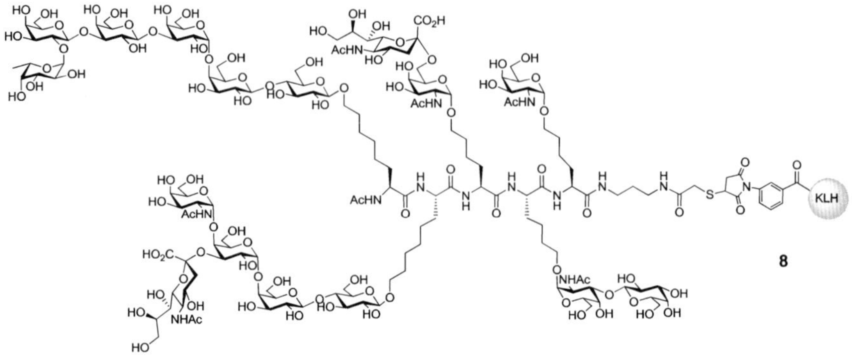

8

Figure 14.7 Unimolecular pentavalent KLH-based vaccine.

the effectiveness of clusters of carbohydrate antigens in generating an improved immune response against tumors for the first time.

Despite the large variety of vaccine candidates developed so far, clinical evaluations still remain unsatisfactory due to side effects and the sensitivity of TACAs towards endogenous enzymes, such as glycosidases. Therefore, structural modifications of native TACAs, including *C*- or *S*-glycosides have been recently proposed as antigen analogs [31, 32]. Most of these antigen analogs were found to provide similar, and sometimes stronger, immunological responses with improved resistance against enzymatic degradation and bioavailability. For example, a structurally simple thioether-bridged mimetic [33] of the GM3 lactone is a hydrolytically stable mimetic of a metabolite of the GM3 antigen and was described by Nativi and coworkers (Fig. 14.8). When conjugated to KLH (9), this antigen mimetic was proved to be a potent immunogen capable of eliciting specific anti-melanoma Abs.

Interestingly, the substitution of a sugar hydroxy group by a more electronegative fluorine atom was found to enhance the bioavailabilities of antigenic structures. Recently, Hoffmann-Röder *et al.* have developed hydrolysis-resistant mucin analogs [15] with a fluorine substituent at C-6, C2′, or C6′ that was incorporated into the tandem repeat sequence of MUC1 (Fig. 14.8). Immunological studies of vaccine candidates composed of the TTox carrier and the MUC1 glycopeptide displaying the native T-antigen or a fluorinated analog

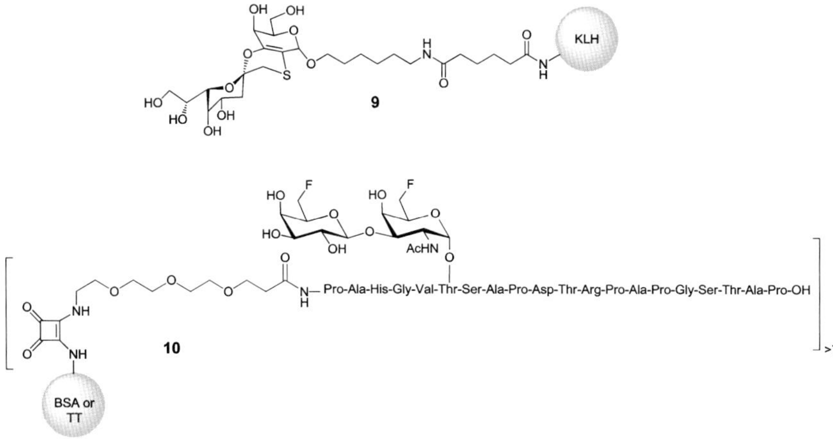

Figure 14.8 Examples of vaccines containing antigen analogs.

(10) have revealed that the fluoride substitution at C-6 improved the vaccine stability without reducing the immunogenicity.

14.2.2 *Lipid carriers*

Conjugation of carbohydrate and peptide antigens to lipid tails can be classified into semi-synthetic vaccines due to the self-assembling potency of these structures to form liposomes, which are the bioactive forms of the vaccine. The first significant contribution in this field was reported by Singhal *et al.*, who have shown that the association of dimeric Tn antigens to the tripalmitoyl-S-glycerylcysteinylserine (Pam_3Cys) from *Escherichia coli*, a ligand for toll-like receptors 2 (TLR-2), promotes a significant IgM and IgG response without requiring carrier proteins [34]. Later, more sophisticated constructions (11), containing Pam_3Cys and multivalent heptapeptide presenting the Le^y antigen (Fig. 14.9), have been designed and synthesized by the group of Danishefsky [35,36]. Here again, the presence of multiple copies of the Le^y-antigen was essential to induce a stronger Abs response.

G. J. Boons and co-workers have certainly reported the most impressive examples in the class of semi-synthetic structures. Self-

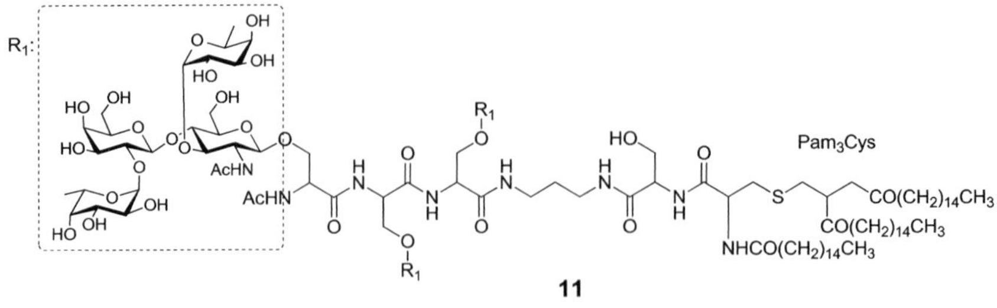

Figure 14.9 Pam_3Cys–Le^y vaccine designed by Danishefsky.

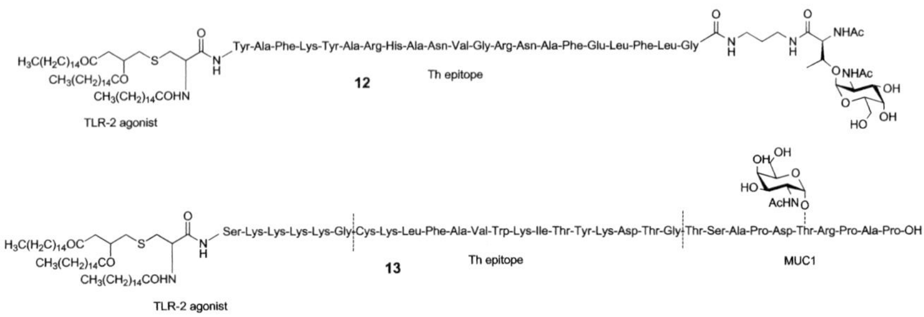

Figure 14.10 Self-adjuvating three-component vaccine candidates.

adjuvating three-component vaccine candidates were synthesized using a convergent strategy (Fig. 14.10). A first construction (12) was composed of Pam_3Cys, the Tn antigen, and the T-cell epitope peptide "YAF" derived from an outer-membrane protein of *Neisseria meningitidis* [37].

After having prepared the glycolipopeptide (12) by successive coupling of the lipid and carbohydrate moieties to the *N*- and *C*-termini of the protected peptide, respectively, phospholipid-based liposomes have been prepared by hydration of a lipid film containing (12), cholesterol, phosphatidylcholine, and phosphatidyl-ethanolamine. After extrusion through 100-nm membranes, transmission electron microscopy confirmed the liposomes' diameter of approximately 100 nm. Further immunological studies have shown that the presence of Pam_3Cys to TACA alone only induces low titers of Abs, whereas the peptide T epitope is required to promote a class switch to IgG. The same group has designed a more sophisticated glycolipopeptide containing a MUC1 fragment (13) that was prepared using a liposome-mediated native chemical ligation (NCL) [38]. The synthetic process involved a cysteine-terminated MUC1 epitope containing Tn, the helper T-epitope YAF, and the Pam_3CysSK_4 moiety. The latter lipid moiety was directly coupled into liposomes to avoid solubility problems. Several other compounds were synthesized using the same protocol [39,40]. Interestingly, immunological studies have led to the conclusion that the three components are essential to induce strong therapeutic responses in mouse models of mammary cancer [39,41]. In particular, it was demonstrated that this tripartite combination allows the activation of cytotoxic T-lymphocytes (CTLs) and antibody-dependent cellular cytotoxicity-mediating antibodies specific for tumors expressing MUC1, whereas unglycosylated structures were significantly less effective.

More recently, the group of Payne has proposed an elegant strategy to prepare similar self-adjuvating multicomponent vaccine candidates (14) combining a MUC1 glycopeptide with a T-helper peptide and the lipopeptide immunoadjuvant Pam_3Cys–Ser [42] via pentafluorophenyl ester condensation (Fig. 14.11).

Immunological evaluation in murine models has shown the potency of these structures to produce high levels of IgG antibodies in the absence of an external adjuvant, liposomal preparation, or carrier protein. Moreover, these Abs were capable of recognizing MCF-7 breast cancer cell line epitopes.

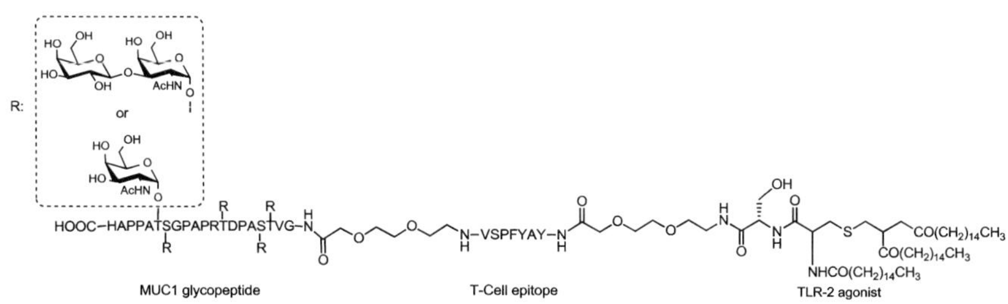

14

Figure 14.11 Three-component vaccine candidate designed by Payne.

14.3 The Fully Synthetic Approach

In 1997, the group of C. Leclerc (Institut Pasteur, Paris) designed fully synthetic multiple antigen glycopeptides (15) (MAGs) that combine clustered Tn antigens to improve the delivery to antigen presenting cells by means of cell–receptor cross-linking and endocytosis [43], and $CD4^+$ peptide epitope from the type I-poliovirus protein (PV) within a well-defined dendritic core (15), [44] (Fig. 14.12). Immunization of mice with MAG (15) stimulates the production of anti-Tn IgG Abs and increases the survival rate of mice without undesired immune responses against the carrier itself [45]. This result prompted the authors to design new structures containing combinations of human leukocyte antigen (HLA)-restricted T-helper epitopes, i.e. PV, TTox, or universal Pan DR (PADRE) with trimeric Tn antigens (16) linked to serine, threonine [46], or homoserine [47] (Fig. 14.12). Immunological evaluation in non-human

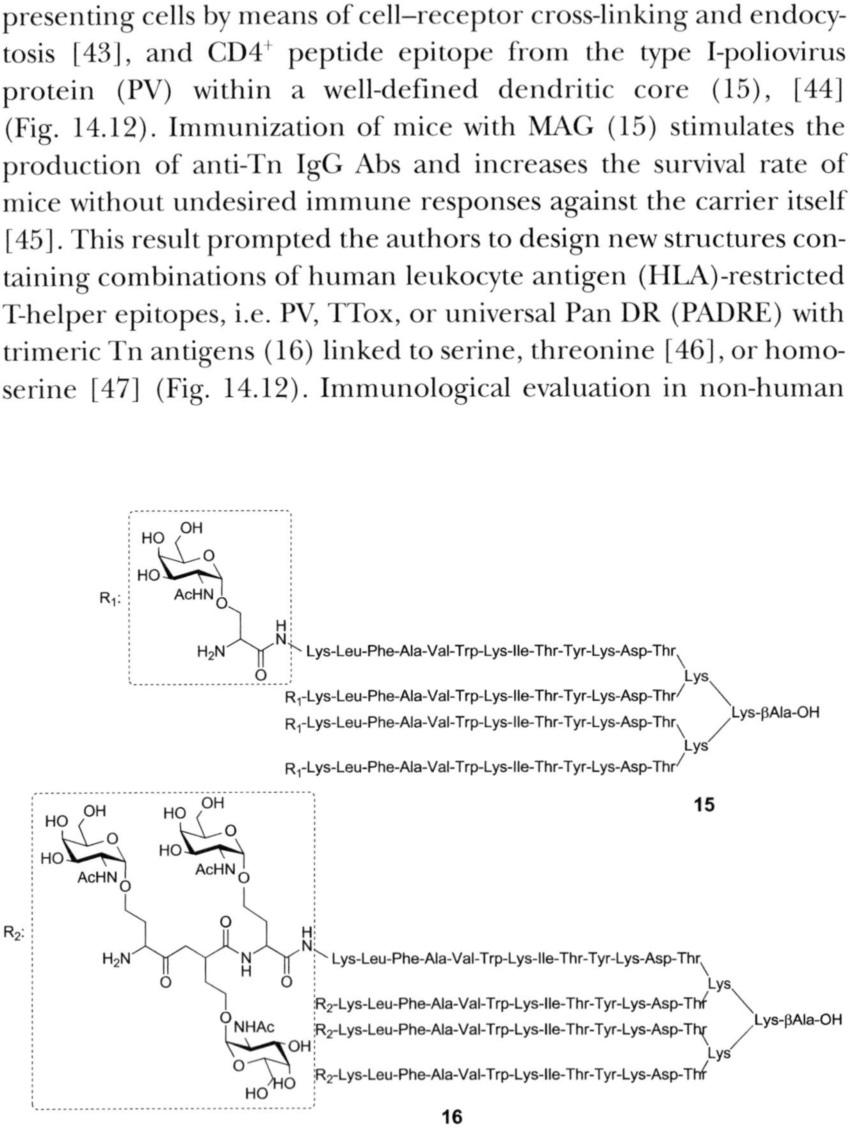

Figure 14.12 Multiple antigenic glycopeptide (MAG) designed by Leclerc.

primates has revealed that MAGs elicit stronger anti-carbohydrate IgG responses against the tumor expressing the Tn antigen than traditional KLH-based vaccines, and more importantly, that these antibodies can mediate antibody-dependent cell cytotoxicity against human tumors. More recently, the same group has demonstrated that both the Tn-cluster density and the peptide backbone significantly affect the binding of anti-Tn Abs, thus highlighting the influence of the TACA presentation on the humoral response [48].

In 2001, Kunz and co-workers synthesized glycosylated amino acids with pre-defined anomer configurations, which were used as building blocks for the preparation of the MUC-1 glycopeptide antigen containing the immunodominant Pro–Asp–Thr–Arg–Pro motif (Fig. 14.13) [49]. As tumor-associated glycopeptides derived from mucins should be associated to T-cell epitopes in order to induce a robust immune response, the authors next coupled them to T-cell peptide epitope TToxs (17) to stimulate a tumor-selective response [50]. Whereas incomplete structures only induce the proliferation of CD3-positive, instead of cytotoxic CD8-positive T-cells, the complete vaccine (17) was shown to induce a stronger cytotoxic T-cell response.

Later on, the same group synthesized MUC1 sequences containing one STn alone or associated with one or two Tn antigens. These structures were either linked to the $CD4^+$ ovalbumin fragment $OVA_{323-339}$ (18) using a fragment condensation strategy [51], or conjugated to BSA (19) (Fig. 14.14). The latter were found to display an average loading between seven and ten, and were used for coating to microtiter plates to detect the Abs induced after immunizations.

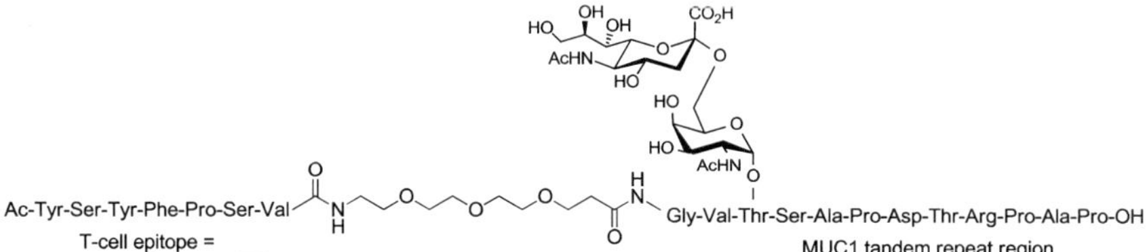

17

Figure 14.13 Synthetic vaccine containing TT fragment and MUC1 tandem repeat sequence.

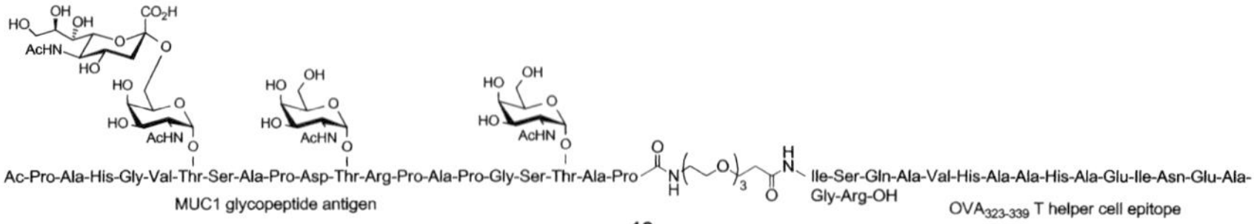

18

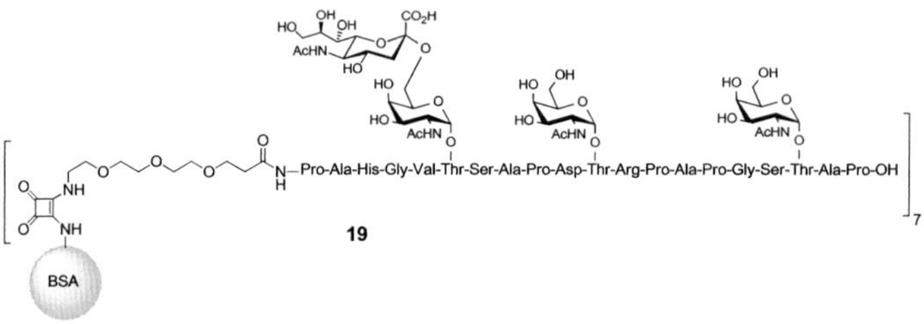

19

Figure 14.14 Strucure of MUC1–OVA conjugates.

Interestingly, when compared to partially glycosylated structures, Abs induced by a monoglycosylated sialyl Tn-MUC1 vaccine was found to provide the highest binding affinity to the BSA conjugate.

On the basis of the model of Leclerc, the group of Dumy and Renaudet have selected cyclopeptide carriers to construct synthetic vaccines containing B- and T-cell antigens displayed within regioselectively addressable domains of this scaffold. Due to its stable conformation, these antigens are oriented in well-defined spatial orientations, which offer the advantage of preventing steric hindrance between them [52]. To avoid problems that may occur during final deprotection, the authors selected oxime ligation [53] to assemble each element to the scaffold. A first generation of vaccine [20] combining clustered Tn antigens and the helper T-cell epitope derived from PV has been prepared (Fig. 14.15) [53]. For this purpose, aminooxy αGalNac as a Tn antigen analog [54], and the PV peptide functionalized with N-terminal aldehyde were synthesized [55]. These structures were successfully linked to a cyclopeptide scaffold containing aldehyde and aminooxy functionalities by using two successive oximation reactions. Immunological evaluation in mice showed the production of Abs that can recognize native Tn on Jurkat cells, and demonstrated the utility of cyclopeptides as non-immunological carriers for antigenic structures for the development

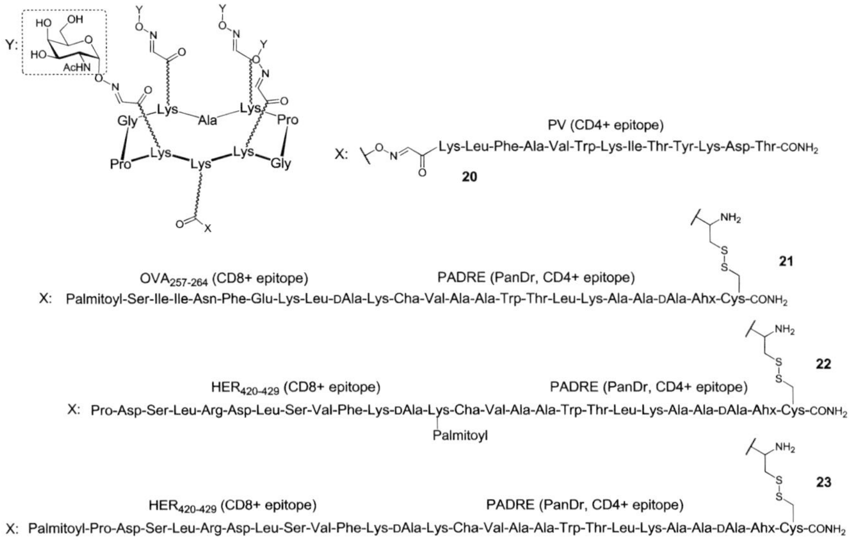

Figure 14.15 Structure of cyclopeptide-based synthetic vaccines designed by Renaudet.

of antitumor vaccines [53]. Subsequently, the same group developed four-component vaccines (21), bearing for the first time within the same scaffold a cluster of Tn antigens (as B-cell epitope), a PADRE peptide (as $CD4^+$ T-helper peptide epitope), a fragment of ovalbumin $OVA_{257-264}$ (as $CD8^+$ CTL peptide epitope), and a palmitic acid (as immunoadjuvant) [56]. They reasoned that combining these four elements might ensure both humoral and cellular responses, and self-adjuvanting properties to the vaccines. The Tn antigens were attached to the scaffold by using oxime ligation while the chimeric palmitoyl–OVA–PADRE lipopeptide was linked by disulfide bridge formation. Both mass spectroscopy and reverse phase high-performance liquid chromatography (RP-HPLC) analysis confirmed the accurate molecular definition of this structure. Further immunological responses were investigated *in vitro* and *in vivo* with both BALB/c and B6 mice inoculated with MO5 carcinoma tumors. Expectedly, it has been observed that the glycolipopeptide vaccine stimulates both B- and T-cell responses without co-administration with an external adjuvant [57]. More remarkably, safety, improved survival, and protection were also observed, thus suggesting that such constructions combining these four essential components represent attractive tools for antitumoral immunotherapy.

Other constructions have been prepared to evaluate the influence of the position of the lipid adjuvant in the magnitude of B- and T-cell immunogenicity and the protective efficacy [58]. Glycolipopeptides bearing the chimeric peptide antigen (22–23) composed of the PADRE peptide and the $CD8^+$ T-cell epitope from human epidermal growth factor receptor 2 (HER-2) ($HER_{420-429}$) have been prepared by following the same synthetic route. In one structure (22), a palmitic acid was attached at the N-terminus (linear glycolipopeptide), while the other one (23) displayed the lipid in the middle of the peptide (branched glycolipopeptide). The molecular uptake, the processing and the cross-presentation pathways were next evaluated. Interestingly, the studies showed that the linear glycolipopeptide (22) was captured by dendritic cells (DCs) more efficiently than the branched structure (23); therefore DC maturation, TLR-2-dependent T-cell response, and tumoral regression were improved. By contrast, (23) stimulated a stronger tumor-specific IgG response than (22).

Danishefsky and co-workers have reported other promising synthetic vaccines based on similar cyclopeptide carriers [59]. Cyclopeptides bearing both the Tn and sTn antigens, or cross-linked Tn antigen, has been designed and prepared using a convergent synthetic strategy (Fig. 14.16). After conjugation to KLH, the authors expected to mimic the heterogenicity of tumor glycocalix using compound (24) and to enhance a cluster-recognizing Ab response by retaining an appropriate distance between TACAs through the rigidified presentation, as shown in (25). However, immunological evaluation has not been reported so far.

Chemoselective ligation was also utilized by the group of A. Delmas which has prepared glycopeptides combining MUC1 and PADRE peptides [60,61]. Two MUC1 peptides carrying the Tn (2) and TF (27) antigens on the threonine residue of the immunodominant Pro–Asp–Thr–Arg–Pro sequence was attached to the PADRE peptide by successive oxime ligations (Fig. 14.17). Immunological assays have revealed that this vaccine prototype

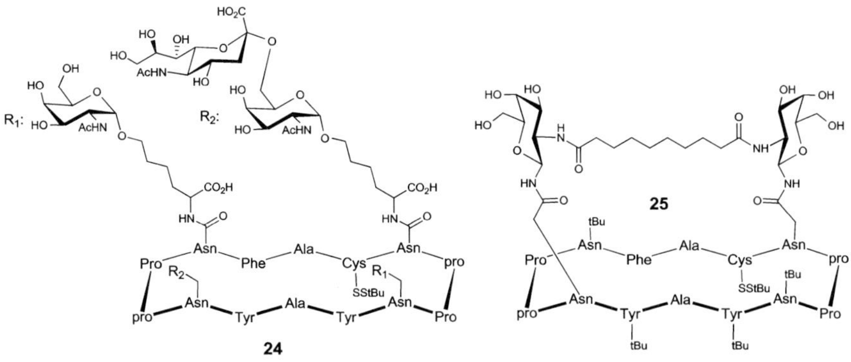

Figure 14.16 Structure of cyclopeptide-based synthetic vaccines developed by Danishefsky.

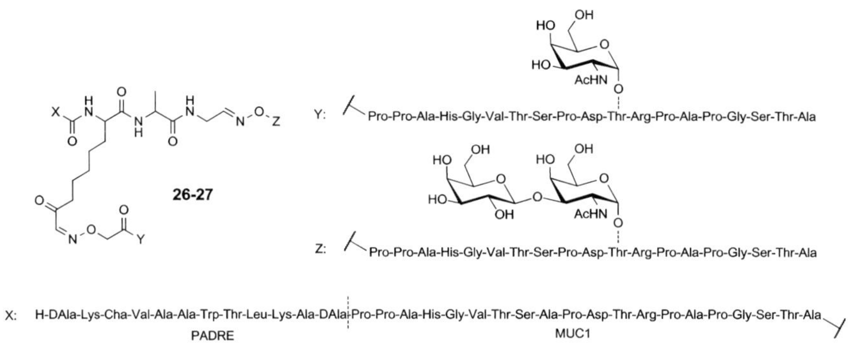

Figure 14.17 Structure of a fully synthetic vaccine prepared by oxime ligation.

induces the production of Abs that recognized native MUC-1 expressed on human breast tumors.

Other significant studies in this field have been demonstrated by the group of F. Peri with stable mimics of the Tn antigen. The *C*-glycosydic analog of Tn was synthesized from allyl-*C*-glucopyranoside [62] and was conjugated into an aminooxy, containing T-cell peptide antigens from ovalbumin (i.e. $OVA_{328–340}$ and $OVA_{327–339}$) to provide monovalent (28) and divalent (29) neoglycopeptides (Fig. 14.18). Biological studies have revealed that such structures are stable towards enzymatic degradation and can be presented by DCs without loss of antigen specificity [63]. Moreover, the vaccine (29)

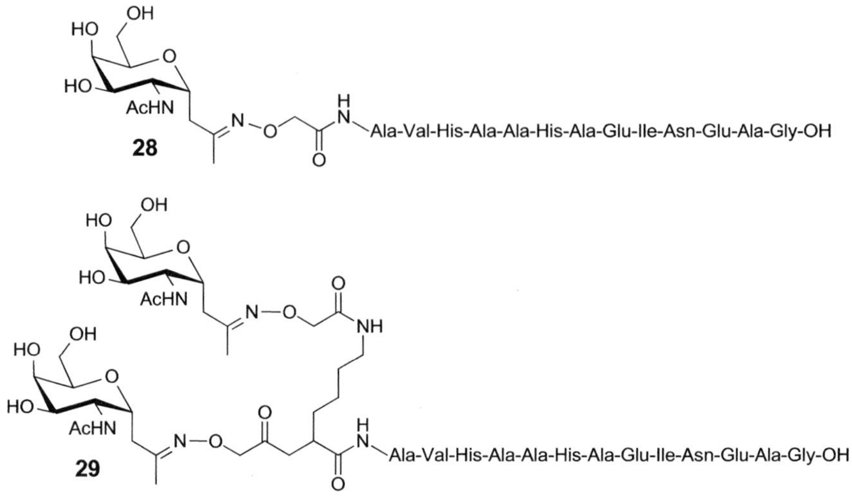

Figure 14.18 Synthetic vaccines designed by Peri and co-workers.

bearing two *C*-glycosides was presented to T-cells with a higher efficiency than the molecule (28) having only one sugar residue [64]. Once again, this study highlights the importance of the multivalent carbohydrate antigen presentation.

14.4 Conclusion

This chapter highlights the recent progresses made in the design of sophisticated antitumoral vaccines with the aim to stimulate more specific and efficient immune responses against tumors. Particularly, new generations of structures associating three and four components that are carbohydrate antigens, $CD4^+$ and/or $CD8^+$ peptide antigens, and TLR agonists have been found to promote remarkable humoral and cellular responses, protective efficacy, as well as self-adjuvanting properties. Furthermore, unimolecular multiantigenic constructions displaying multiple carbohydrate antigens undoubtedly represent another essential basis for further optimization of vaccine candidates. Together with the recent utilization of glyconanoparticles [65], these progresses open up promising perspectives in the development of selective immunotherapy against human cancers.

References

1. Busksas, T., Thompson, P., Boons, G. J. (2009) *Chem. Commun.* **36**, 5335–5349.
2. Guo, Z., Wang, Q. (2009) *Curr. Opin. Chem. Biol.* **13**, 608–617.
3. Morelli, L., Poletti, L., Lay, L. (2011) *Eur. J. Org. Chem.* **29**, 5723–5777.
4. Shiao, T. C., Roy, R. (2012) *New. J. Chem.* **36**, 324–339.
5. Peri, F. (2013) *Chem. Soc. Rev.* **42**, 4543–4556.
6. Gaidzik, N., Westerlind, U., Kunz, H. (2013) *Chem. Soc. Rev.* **42**, 4421–4442.
7. Wilson, R. M., Danishefsky, S. J. (2013) *J. Am. Chem. Soc.* **135**, 14462–14472.
8. Wang, C. C., Lee, J. C., Luo, S. Y., Kulkarni, S. S., Huang, Y. W., Lee, C. C., Chang, K. L., Hung, S. C. (2007) *Nature* **446**, 896–899.
9. Seeberger, P. H., Werz, D. B. (2007) *Nature* **446**, 1046–1051.
10. Koeller, K. M., Wong, C. H. (2000) *Chem. Rev.* **100**, 4465–4493.
11. Endo, T., Koizumi, S. (2000) *Curr. Opin. Struct. Biol.* **10**, 536–541.
12. Liebe B., Kunz H. (1997) *Angew. Chem. Int. Ed.* **36**, 618–620.
13. Brocke, C., Kunz, H. (2004) *Synthesis* **4**, 525–542.
14. Kaiser, A., Gaidzik, N., Westerlind, U., Kowalczyk, D., Hobel, A., Schmitt, E., Kunz, H. (2009) *Angew. Chem. Int. Ed.* **48**, 7551–7555.
15. Hoffmann-Röder, A., Kaiser, A., Wagner, S., Gaidzik, N., Kowalczyk, D., Westerlind, U., Gerlitzki, B., Schmitt, E., Kunz, H. (2010) *Angew. Chem. Int. Ed.* **49**, 8498–8503.
16. Gaidzik, N., Kaiser, A., Kowalczyk, D., Westerlind, U., Gerlitzki, B., Sinn, H. P., Schmitt, E., Kunz, H. (2011) *Angew. Chem. Int. Ed.* **50**, 9977–9981.
17. Cai, H., Huang, Z. H., Shi, L., Sun, Z. Y., Zhao, Y. F., Kunz, H., Li, Y. M. (2012) *Angew. Chem. Int. Ed.* **51**, 1719–1723.
18. Chen, X. T., Sames D., Danishefsky, S. J. (1998) *J. Am. Chem. Soc.* **120**, 7760–7769.
19. Seeberger, P. H., Danishefsky, S. J. (1998) *Acc. Chem. Res.* **31**, 685–695.
20. Schwarz, J. B., Kuduk, S. D., Chen, X. T., Sames, D., Glunz, P. W., Danishefsky, S. J. (1999) *J. Am. Chem. Soc.* **121**, 2662–2673.
21. Kuduk, S. D., Schwarz, J. B., Chen, X. T., Glunz, P. W., Sames, D., Ragupathi, G., Livingston, P. O., Danishefsky, S. J. (1998) *J. Am. Chem. Soc.* **120**, 12474–12485.
22. Bilodeau, M. T., Park, T. K., Hu, S., Randolph, J. T., Danishefsky, S. J., Livingston, P. O., Zhang, S. (1995) *J. Am. Chem. Soc.* **117**, 7840–7841.

23. Danishefsky, S. J., Behar, V., Randolph, J. T., Lloyd, K. O. (1995) *J. Am. Chem. Soc.* **117**, 5701–5711.
24. Allen, J. R., Allen, J. G., Zhang, X. F., Williams, L. J., Zatorski, A., Ragupathi, G., Livingston, P. O., Danishefsky, S. J. (2000) *Chem. Eur. J.* **6**, 1366–1375.
25. Ragupathi, G., Park, T. K., Zhang, S. L., Kim, I. J., Graber, L., Adluri, S., Lloyd, K. O., Danishefsky, S. J., Livingston, P. O. (1997) *Angew. Chem. Int. Ed.* **36**, 125–128.
26. Slovin, S. F., Ragupathi, G., Adluri, S., Ungers, G., Terry, K., Kim, S., Spassova, M., Bornmann, W. G., Fazzari, M., Dantis, L., Olkiewicz, K., Lloyd, K. O., Livingston, P. O., Danishefsky, S. J., Scher H. I. (1999) *Proc. Natl. Acad. Sci. USA* **96**, 5710–5715.
27. Allen, J. R., Harris, C. R., Danishefsky, S. J. (2001) *J. Am. Chem. Soc.* **123**, 1890–1897.
28. Ragupathi, G., Coltart, D. M., Williams, L. J., Koide, F., Kagan, E., Allen, J., Harris, C., Glunz, P. W., Livingston, P. O., Danishefsky, S. J. (2002) *Proc. Natl. Acad. Sci. USA* **99**, 13699–13704.
29. Keding, S. J., Danishefsky, S. J. (2004) *Proc. Natl. Acad. Sci. USA* **101**, 11937–11942.
30. Zhu, J., Wan, Q., Lee, D., Yang, G., Spassova, M. K., Ouerfelli, O., Ragupathi, G., Damani, P., Livingston, P. O., Danishefsky, S. J. (2009) *J. Am. Chem. Soc.* **131**, 9298–9303.
31. Bundle, D. R., Rich, J. R., Jacques, S., Yu, H. N., Nitz, M., Ling, C. C. (2005) *Angew. Chem. Int. Ed.* **44**, 7725–7729.
32. Awad, L., Madani, R., Gillig, A., Kolympadi, M., Philgren, M., Muhs, A., Gérard, C., Vogel, P. (2012) *Chem. Eur. J.* **18**, 8578–8582.
33. Arcangeli, A., Toma, L., Contiero, L., Crociani, O., Legnani, L., Lunghi, C., Nesti, E., Moneti, G., Richichi, B., Nativi, C. (2010) *Bioconjugate Chem.* **21**, 1432–1438.
34. Toyokuni, T., Dean, B., Cai, S., Boivin, D., Hakomori, S. I., Singhal, A. K. (1994) *J. Am. Chem. Soc.* **116**, 395–396.
35. Glunz, P. W., Hintermann, S., Williams, L. J., Schwarz, J. B., Kuduk, S. D., Kudryashov, V., Lloyd, K. O., Danishefsky, S. J. (2000) *J. Am. Chem. Soc.* **122**, 7273–7279.
36. Kudryashov, V., Glunz, P. W., Williams, L. J., Hintermann, S., Danishefsky, S. J., Lloyd, K. O. (2001) *Proc. Natl. Acad. Sci. USA* **98**, 3264–3269.
37. Buskas, T., Ingale, S., Boons, G. J. (2005) *Angew. Chem. Int. Ed.* **44**, 5985–5988.
38. Ingale, S., Buskas, T., Boons, G. J. (2006) *Org. Lett.* **8**, 5785–5788.

39. Ingale, S., Wolfert, M. A., Gaekwad, J., Buskas, T., Boons, G.-J. (2007) *Nature Chem. Biol.* **3**, 663–667.
40. Ingale S., Wolfert, M. A., Buskas, T., Boons, G. J. (2009) *ChemBioChem* **10**, 455–463.
41. Lakshminarayanan, V., Thompson, P., Wolfert, M. A., Buskas, T., Bradley, J. M., Pathangey, L. B., Madsen, C. S., Cohen, P. A., Gendler, S. J., Boons, G. J. (2012) *Proc. Natl. Acad. Sci. USA* **109**, 261–266.
42. Wilkinson, B. L., Day, S., Malins, L. R., Apostolopoulos, V., Payne, R. J. (2011) *Angew. Chem. Int. Ed.* **50**, 1635–1639.
43. Ota, S., Ono, T., Morita, A., Uenaka, A., Harada, M., Nakayama, E. (2002) *Cancer Res.* **62**, 1471–1476.
44. Bay, S., Lo-Man, R., Osinaga, E., Nakada, H., Leclerc., C, Cantacuzene, D. (1997) *J. Pept. Res.* **49**, 620–625.
45. Lo-Man, R., Bay, S., Vichier-Guerre, S., Dériaud, E., Cantacuzene, D., Leclerc, C. (1999) *Cancer Res.* **59**, 1520–1524.
46. Lo-Man, R., Vichier-Guerre, S., Perraut, R., Dériaud, E., Huteau, V., Ben Mohamed, L., Diop, O. M., Livingston, P. O., Bay, S., Leclerc, C. (2004) *Cancer Res.* **64**, 4987–4994.
47. Vichier-Guerre, S., Lo-Man, R., Huteau, V., Dériaud, E., Leclerc, C., Bay, S. (2004) *Bioorg. Med. Chem. Lett.* **14**, 3567–3570.
48. Mazal, D., Lo-Man, R., Bay, S., Pritsch, O., Dériaud, E., Ganneau, C., Medeiros, A., Ubillos, L., Obal, G., Berois, N., Bollati-Fogolin, M., Leclerc, C., Osinaga, E. (2013) *Cancer Immunol. Immunother.* **62**, 1107–1122.
49. Keil, S., Claus, C., Dippold, W., Kunz, H. (2001) *Angew. Chem. Int. Ed.* **40**, 366–369.
50. Dziadek, S, Hobel, A, Schmitt, E, Kunz, H. (2005) *Angew. Chem. Int. Ed.* **44**, 7630–7635.
51. Westerlind, U., Hobel, A., Gaidzik, N., Schmitt E., Kunz, H. (2008) *Angew. Chem. Int. Ed.* **47**, 7551–7556.
52. Galan, M. C., Dumy, P., Renaudet, O. (2013) *Chem. Soc. Rev.* **42**, 4599–4612.
53. Ulrich, S., Boturyn, D., Marra, A., Renaudet, O., Dumy, P. *Chem. Eur. J.* Grigalevicius, S., Chierici, S., Renaudet O., Lo-Man R., Dériaud, E., Leclerc, C., Dumy, P. (2005) *Bioconjugate Chem.* **16**, 1149–1159.
54. Renaudet, O., Dumy, P. (2001) *Tetrahedron Lett.* **42**, 7575–7578.
55. Geoghegan, K. F., Stroh, J. G. (1992) *Bioconjugate Chem.* **3**, 138–146.
56. Renaudet, O., Ben Mohamed, L., Dasgupta, G., Bettahi, I., Dumy, P. (2008) *ChemMedChem* **3**, 737–741.

57. Bettahi, I., Dasgupta, G., Renaudet, O., Chentoufi, A. A., Zhang, X., Carpenter, D., Yoon, S., Dumy, P., Ben Mohamed, L. (2009) *Cancer Immunol. Immunother.* **58**, 187–200.
58. Renaudet, O., Dasgupta, G., Bettahi, I., Shi, A., Nesburn, A. B., Dumy, P., Ben Mohamed, L. (2010) *PLoS One* **5**, e11216.
59. Jeon, I., Lee, D., Krauss, I. J., Danishefsky, S. J. (2009) *J. Am. Chem. Soc.* **131**, 14337–14344.
60. Cremer, G. A., Bureaud, N., Piller, V., Kunz, H., Piller, F., Delmas, A. F. (2006) *ChemMedChem* **1**, 965–968.
61. Cremer, G. A., Bureaud, N., Lelièvre, D., Piller, V., Piller, F., Delmas A. (2004) *Chem. Eur. J.* **10**, 6353–6360.
62. Cipolla, L., La Ferla, B., Lay, L., Peri, F., Nicotra, F. (2000) *Tetrahedron: Asym.* **11**, 295–303.
63. Cipolla, L., Rescigno, M., Leone, A., Peri, F., La Ferla, B., Nicotra, F. (2002) *Bioorg. Med. Chem.* **10**, 1639–1646.
64. Peri, F., Cipolla, L., Rescigno, M., La Ferla, B., Nicotra, F. (2001) *Bioconjugate Chem.* **12**, 325–328.
65. Marradi, M., Chiodo, F., Garcia, I., Penadés, S. (2013) *Chem. Soc. Rev.* **42**, 4728–4745.

Chapter 15

DC-SIGN as a Target for Drug Development Based on Carbohydrates

Sara Sattin,* Franck Fieschi[†,‡,§] and Anna Bernardi*

*Università degli Studi di Milano, Dipartimento di Chimica, via Golgi, 19, 20133, Milano

[†]Université Grenoble Alpes, Institut de Biologie Structurale, 6 rue Jules Horowitz, Grenoble, F-38000, France

[‡]CNRS, UMR 5075, Grenoble, F-38000, France

[§]CEA, DSV UMR 5075, Grenoble, F-38000, France

15.1 Introduction

Dendritic cell-specific intercellular adhesion molecule-3 grabbing non-integrin (DC-SIGN) is a tetrameric calcium dependent (C-type) lectin, expressed by dermal dendritic cells (DCs), which specifically recognize highly-glycosylated structures displayed at the surface of several pathogens [1–4]. Recognition by DC-SIGN is known to play a key role in human immunodeficiency virus (HIV) transmission and is involved in a number of other viral infections; as such, it is currently regarded as an interesting target for the design of anti-viral agents [5]. Immature DCs that express DC-SIGN are located in vaginal, cervical, and rectal mucosae, and are among the first cells encountered by HIV during sexual transmission. DC-SIGN captures the virus at their surface by binding high-mannose glycans present on its envelope glycoprotein, gp120. The detailed molecular mechanisms by which this receptor operates have not been fully elucidated; however it seems clear that, following this initial interaction, whole virus particles are transmitted to $CD4^+$ T lymphocytes, a process called infection in *trans*. It has also been shown that blocking HIV recognition by DC-SIGN effectively abrogates in *trans* infection. Thus, a number of groups are working to design DC-SIGN antagonists [6–11], with the goal of unveiling a new strategy to prevent sexual transmission of HIV. Effective modulators of DC-SIGN are also needed to help clarify the different biological processes in which this receptor is involved.

The main natural ligand recognized by the carbohydrate recognition domain of DC-SIGN is the high mannose glycan $(Man)_9$-$(GlcNAc)_2$, also known as Man_9, a branched oligosaccharide presented in multiple copies by several pathogen glycoproteins and specifically by the gp120 envelope protein of HIV. DC-SIGN can also recognize branched fucosylated structures bearing terminal galactose residues, such as the Lewis antigens. The primary interaction of oligosaccharides and DC-SIGN occurs by coordination of a residue of the oligosaccharide (often the one at the non-reducing end) to a calcium (Ca^{2+}) binding site exposed to the surface of the protein. X-ray data are available for complexes of DC-SIGN carbohydrate recognition domains (CRDs) with both mannose oligosaccharides and Lewis-X [12–14].

A handful of unnatural ligands have been described so far, both mono- and multivalent, and the subject has been recently reviewed [6,7,15].

Our groups have been involved for the past six to seven years in a collaborative project at the European level directed towards the discovery of novel DC-SIGN antagonists. In this context, we have developed the design and synthesis of both mannose-based and fucose-based ligands. Our results are summarized below.

15.2 Fucose-Based Ligands

The interaction of DC-SIGN with fucosylated oligosaccharides occurs mainly with branched fucosylated structures bearing terminal galactose residues, such as the Lewis antigens.

Fucosylated dendrimers [16] and fucose-based ligands [17–19] have attracted attention because they promise to deliver selective DC-SIGN antagonists that do not interfere with the protective action of Langerin, a mannose-selective C-lectin, towards the HIV infection process [20]. Langerin is expressed by Langerhans cells, a subset of dendritic cells of the epidermis and mucosa. Langerin binds HIV by interaction with gp120 glycans and sequesters it into Birbeck's granules where it can be degraded [20]. Thus, in the quest towards novel antiviral compounds and materials, it is considered to be essential to develop agents that are selective for DC-SIGN versus Langerin.

The X-ray structure of the DC-SIGN–Lewis-X complex [14] shows the fucose residue of Lewis-X (Fig. 15.1) interacting with the primary Ca^{2+} site, while the galactose residue is stabilized by hydrogen (H)-bonding in a nearby region of the DC-SIGN CRD (Fig. 15.6A). It has long been established that the Lewis-X trisaccharide (Gal[Fucα(1–3)β(1–4)GlcNAc) possesses limited conformational flexibility and it is mostly described by one major tweezers-like conformation that brings the α-face of fucose in hydrophobic contact with the β-face of galactose (see [21] and references therein). This conformation, which is adopted by the sugar in solution, is generally conserved in lectin complexes, including in the DC-SIGN–Lewis-X X-ray structure. Using the three-dimensional (3D) structure of Lewis-X as a template, we designed a mimic (1) (Fig. 15.1) as an unnatural fucose-based ligand for DC-SIGN [17]. The ligand is based on an α-fucosylamide anchor, which can drive the molecule to the DC-SIGN primary binding site. This anchor is connected via

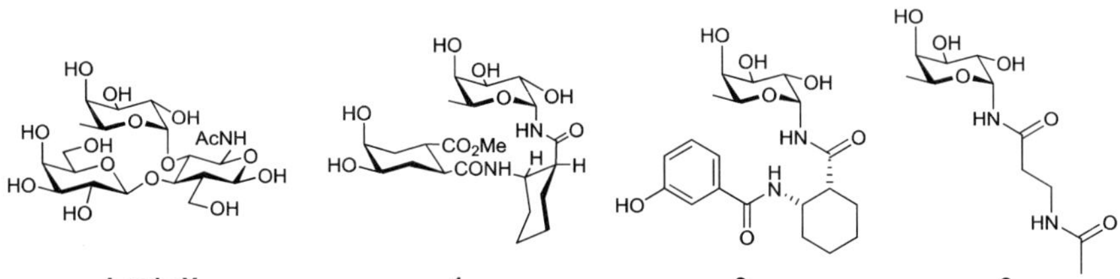

Lewis-X 1 2 3

Figure 15.1 Fucose-based DC-SIGN ligands.

amide bonds to a galactose mimic through a cyclic *cis*-β-aminoacid linker, which defines the molecular shape, forcing the fucose residue and the galactose stand-in to overlap, like they do in the Lewis trisaccharide.

The synthesis of mimic (1) relies on a one-pot Staudinger reduction/aza-Wittig sequence described by De Shong [22] and leading to the key intermediate (4) (Fig. 15.2). The oxazoline (4) can be efficiently acylated by pyridylthiolesters, in a process that we have exploited for the synthesis of (1) (Fig. 15.2) and of a number of derivatives with different amino acids [18]. Acylation of the resulting *N*-glycosyl amino acid (5) with the galactose mimic (6), followed by deacetylation, afforded (1).

DC-SIGN binding studies performed by a surface plasmon resonance (SPR) biosensor showed that (1) rivals the affinity of the natural ligand, but additional structure–activity relationship (SAR) studies revealed that the molecule is probably over-designed. Indeed, both replacement of the galactose mimic with a variety of fragments and change of the central building block with cyclic and acyclic 2-amino acids resulted in minor variations of the affinity towards DC-SIGN. As an example, compound (2) (Fig. 15.1), identified as the most potent inhibitor, is roughly twice as active as

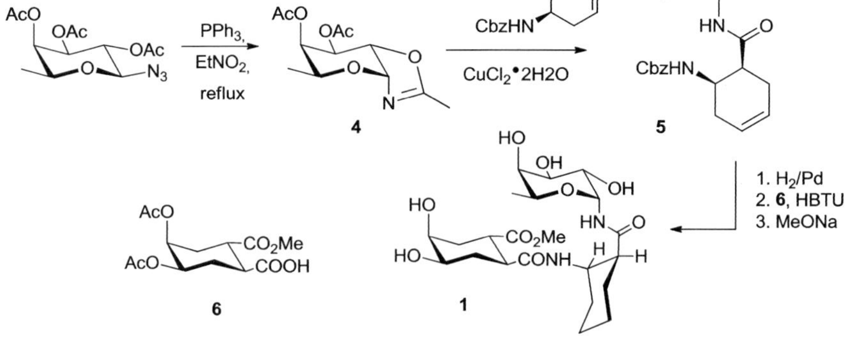

Figure 15.2 Synthesis of **1**.

Lewis-X, whereas the synthetically more accessible compound (3) (Fig. 15.1) was found to be almost as active as the natural ligand [18]. These small differences suggested that the galactose mimic fragment in (1) gave a limited contribution to the binding interaction, a hypothesis that was supported by nuclear magnetic resonance (NMR) studies [23].

From a practical point of view, a well-validated approach to compensate for weak protein–carbohydrate interactions is the synthesis of multivalent ligands that present several copies of the monomeric epitope [see Chapter 17 in this book]. So, we developed appropriate functionalizations for (2) and (3) to obtain linker-armed molecules (8) and (9) (Fig. 15.3) that could be used for the synthesis of glycodendrimers [19].

A tetravalent presentation of the fucose-based ligands was obtained by click reaction on pentaerythritol-based tetra-alkyne (10) (Fig. 15.4) and the relative affinity of (11) and (12) for DC-SIGN was examined in the SPR competition assay. Modest amplification factors were obtained ($\beta = 1.3$ for (12) and $\beta = 2$ for (11)) [24], most likely through statistical rebinding [25], since the dendrimer is too small to bridge two binding sites with a distance of almost 4 nm in the DC-SIGN tetramer [26]. Nonetheless, an order of magnitude improvement was achieved and compound (11) displayed a median inhibitory concentration (IC_{50}) value of 39 μM.

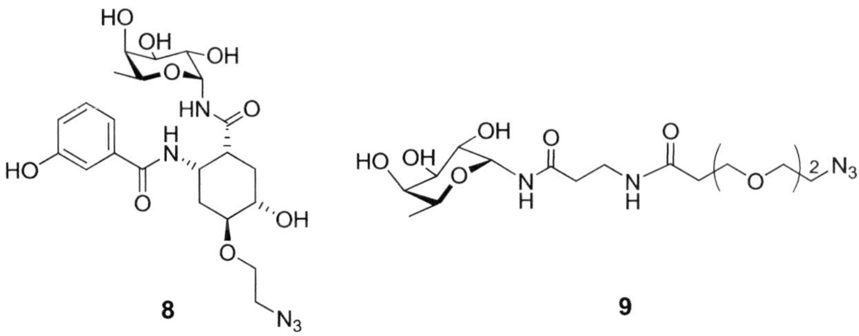

8 9

Figure 15.3 Linker-armed fucose-based ligands **8** and **9**.

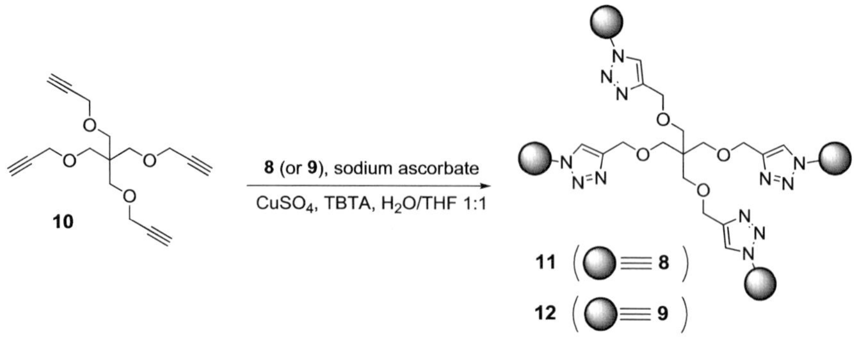

Figure 15.4 Synthesis of the tetravalent dendrimers **11** and **12**.

15.3 Mannose-Based Ligands

Mannose-based DC-SIGN ligands have been studied more extensively than those based on fucose, probably due to the fact that most of the natural DC-SIGN ligands contain mannose or higher mannose structures.

In 2007, a group of 2-C-substituted branched D-mannose analogs was reported as better DC-SIGN binders than D-mannose [27]. Garber *et al.* used shikimic acid as a mannose mimic, since the hydroxyl groups at position 2, 3, and 4 have the same configurations [28]. The same authors also obtained multivalent presentation of their mimic by conjugation with polymeric scaffolds [28]. The functionalized polymer carrying 29 copies of the monovalent ligand has an IC_{50} value of 2.9 μM and its length should be enough to reach two CRDs of DC-SIGN.

For our design of mannose-based antagonists, we have used the high-mannose oligosaccharide Man9 ((13), Fig. 15.5) as a lead structure, and the observation, reported by the Seeberger's lab [29], was that high density arrays of 1,2-manobioside (Man-α-1,2-Man) bind to DC-SIGN almost as effectively as the entire Man_9. To mimic 1,2-mannobioside, the pseudo-1,2-mannobioside (14) (Fig. 15.5) was designed, which contains a mannose unit connected to a conformationally locked cyclohexanediol [30].

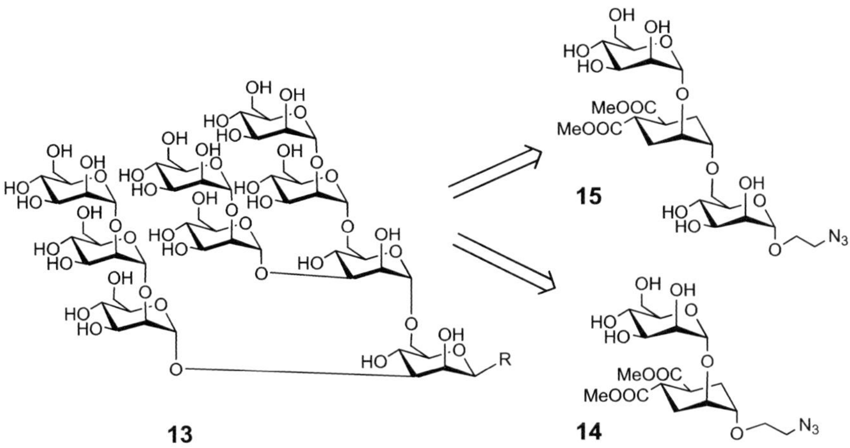

Figure 15.5 Glycomimetics structures **14** and **15** based on the high-mannose oligosaccharide **13** (Man9).

The latter acts as a mimic of a reducing end mannose residue and features a spacer-arm terminated with azido or amino functionality, useful to generate multivalent DC-SIGN ligands. The design of the mimic (14) was supported by modelling, showing that the molecule replicates the conformational behavior of the natural disaccharide, but mostly populates its extended conformation (E) [31]. Furthermore, this molecule is more stable than the natural counterpart to jack-bean mannosidase-catalyzed hydrolysis [31]. Saturation transfer difference (STD)–NMR experiments proved that (14) interacts with DC-SIGN and infection studies of DC-SIGN expressing Jurkat cells by a pseudo-typed Ebola virus showed that (14) acts as an inhibitor of viral entry with an IC_{50} value of 0.6 mM. This value was approximately three times lower than that of the natural disaccharide Man-α-1,2-Man, which also showed a marked cytotoxicity not exhibited by (14). SPR studies showed that (14) inhibits DC-SIGN binding to mannosylated bovine serum albumin (BSA, Man-BSA) with an IC_{50} value of 1 mM, about half of the value measured for mannose itself [32] and very similar to Man-α-1,2-Man [33]. However, the latter binds to Langerin with the same affinity, while (14) displays a modest, but measurable selectivity [33].

As mentioned above, selectivity against Langerin is a key concern in the development of DC-SIGN antagonists.

The pseudo-trisaccharide (15), mimicking the linear Man-α-1,2-Man-α-1,6-Man trisaccharide of the D3 arm of Man9, was designed following the same concept [34]. This compound inhibits DC-SIGN binding to Man-BSA with an IC_{50} value of 130 μM (SPR) [32]. This corresponds to *ca.* a 15-fold improvement over mannose and a 10-fold improvement over (14) in the same inhibition experiment. Perhaps even more interestingly, (15) exhibits a 20-fold selectivity for DC-SIGN compared to Langerin in an SPR-based assay [35] and it was, therefore, the first antagonist that we formulated in polyvalent presentations able to inhibit DC-SIGN mediated HIV infection both in cellular and cervical explant models [32,35].

Recently, the high-resolution X-ray structure of the (14)/DC-SIGN complex [33] has become available. The structure revealed that the glycomimetic interacts with the protein in a different orientation, compared to the natural mannobioside. In this pose (Fig. 15.6B), the cyclohexane ring of the mimic interacts with the V351 residue of the protein, in a very exposed area of the binding site. NMR studies confirmed that this binding mode is represented in solution and that it is probably unique. Interestingly, the binding epitope of (14) on the protein is the same used by Lewis-X [14] (Fig. 15.6A) and

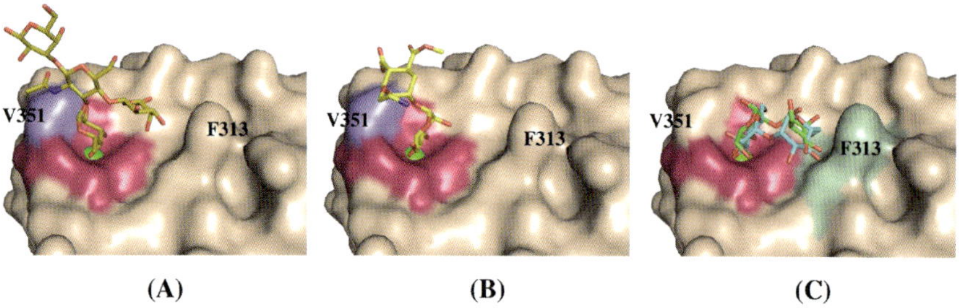

Figure 15.6 Binding mode analysis of DC-SIGN ligands by X-ray crystallography. (A) X-ray structure of the DC-SIGN/Lewis-X complex [14]; (B) X-ray structure of the DC-SIGN/**14** complex [33]; (C) Overlayed structure of major (green) and minor (blue) binding mode of Man-α-1,2-Man with DC-SIGN [13, 36]. Common residues involved in all binding epitopes are in pink; V351 common to Lewis-X and **14** is in blue; residues common to the multiple binding modes of Man-α-1,2-Man are in green.

markedly different from the ones observed for the natural Man-α-1,2-Man (Fig. 15.6C) [13]. Considering the selectivity of Lewis-X for DC-SIGN versus Langerin, this finding may provide a structural basis for the lectin selectivity of (14) and (15).

Analysis of the structure shows that the methyl ester groups of (14) in the complex are exposed to the solvent and extend towards the surface of DC-SIGN. Larger groups in this position can be expected to reach the protein surface and make additional contacts. Based on these and other (synthetic) considerations, we attempted to optimize this lead and prepared a library of amide derivatives of (14), following the approach shown in Fig. 15.7 [37,38]. The optimized synthesis gave fast and efficient access to a group of bis-amides (16), decorated with an azido-terminated linker, which allowed for easy conjugation to aglycons via copper catalyzed azide–alkyne cycloadditions.

SPR inhibition tests (Fig. 15.8) showed that many bis-benzylamido derivatives are more active by a factor of 3–4, compared to the parent pseudo-mannobioside (14). They approach the activity of (15) in the same test and display similar or better selectivity against Langerin [38], at a fraction of the synthetic cost. NMR studies showed relatively high saturation transfer for the aromatic amide

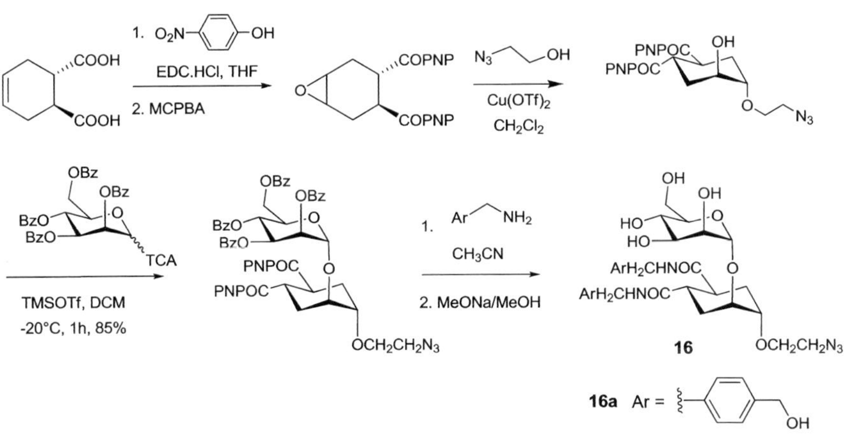

Figure 15.7 Synthesis of a library of secondary amide derivatives of general structure **16** of pseudo-dimannoside **14**.

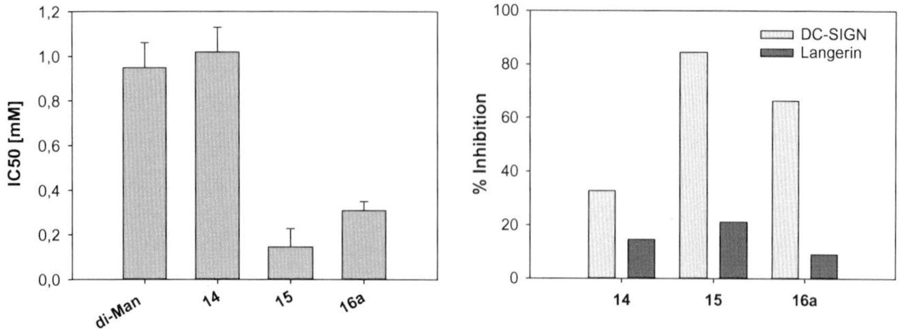

Figure 15.8 Inhibition of lectin binding to Man-BSA (SPR). Left: Inhibition of DC-SIGN binding: IC_{50} of Man-α-1,2-Man (**di-Man**), **14**, **15** and **16a**. Right: DC-SIGN and Langerin inhibition levels measured for a 0.5 mM concentration of the compounds.

moieties in the STD spectrum, suggesting a close proximity between the aromatic groups of the ligand and the binding site of DC-SIGN. Among the molecules prepared here, ligand (16a) (Fig. 15.7) was selected for further studies, due to a combination of good solubility, synthetic accessibility, activity, and selectivity. Polyvalent presentations of this ligand on dendrimeric scaffolds are described below.

The affinity for DC-SIGN of glycomimetic monovalent ligands obtained so far is such that their therapeutic potential is limited. However, appropriate levels of affinity have been obtained when the ligands have been presented in a multimeric form using dendritic scaffolds [32, 35, 39, 40]. Glycodendritic structures and glycopolymers have been used for the development of multivalent DC-SIGN antagonists and are partly described in Chapter 17 of this book (for a review, see [8]; not covered by the review: [16, 41]). The first multimeric presentation of glycomimetics (14) and (15) were synthesized by conjugation of the monovalent ligands to tetra- and multivalent scaffolds based on bis-hydroxymethylpropionic acid as building block. Tetravalent presentation of (15) in dendron (18) (Fig. 15.9) was shown to inhibit *trans* infection of T lymphocytes by DC-SIGN expressing B-cells, which had been pre-incubated with HIV in the presence of (18). Infection was abrogated almost totally by (18) in 100 μM concentration, and an IC_{50} *ca.* 10 μM could be estimated.

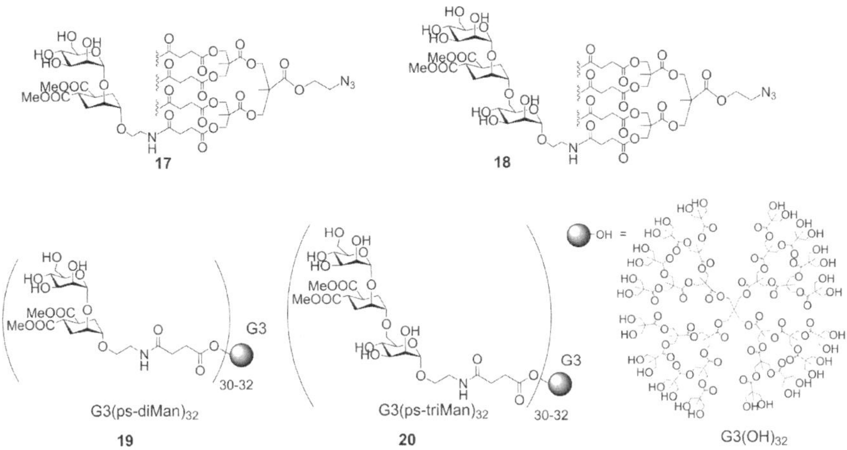

Figure 15.9 Multivalent presentation of glycomimetics **14** and **15** using dendron (**17**,**18**) and dendrimer (**19**,**20**) scaffolds.

The same dendron was tested in a human cervical explant model to evaluate the potential of developing a topical microbiocide against sexually-transmitted viral infections [35]. The results showed that (18) strongly reduces the infection of cervical explants by different HIV-1 R5-topic strains and is also effective, at higher doses, against X-4 topic virus. The dendron also induces immune activation and proinflammatory responses, which contribute to its antiviral activity.

Other bis-hydroxymethylpropionic acid based dendrimers (Boltorn-type) were synthesized to obtain multivalent glycomimetics conjugates with four to 32 copies of ligands (14) and (15) ((17)–(20), Fig. 15.9). These tetra- and multivalent systems were tested *in vitro* using an infection model based on pseudotyped viral particles with the Ebola virus envelope glycoprotein GP1 [39], which is exclusively dependent on DC-SIGN [42]. In these experiments, the tetravalent systems (17) and (18) were very active in the low micromolar range, and the multivalent systems (19) and (20) showed a very strong inhibition effect with IC_{50} values in the nanomolar range. Despite the different affinities of the monovalent spearheads (14) and (15), relatively small differences were observed between equivalent constructs obtained from the two ligands. In other words, the pseudo-trimannoside (15), which at the monovalent level is more active than (14) by an order of magnitude, does not fully profit from avidity effects allowed by polyvalency. This puzzling behavior appears to depend on the loss of alternative binding modes available to (15) only in the monovalent state, a feature that we have investigated in great detail and recently characterized [43]. As a consequence, despite the improved affinity of the pseudomannotrioside at the monovalent level, our data suggest that the pseudo-mannobioside (14) is a better lead for the structural optimization of ligands.

Indeed, we have recently described dendrimer (21) [40] carrying six copies of ligand (16a), and have shown that this novel material provides 100% inhibition of the HIV infection at 10 μM concentration, a clear improvement from the previous candidate (18) (Fig. 15.10). Additionally, (21) was found to block DC-SIGN mediated uptake of Dengue virus (DENV) by Raji cells with a low micromolar IC_{50}. Like HIV, DENV is critically dependent on DC-SIGN for host infection [44, 45]. No clinical treatments are available for DENV infection and there is a clear requirement for novel antiviral agents in this field [46]. The promising results obtained in this study establish the bases for the preparation of improved materials with even higher antiviral activities.

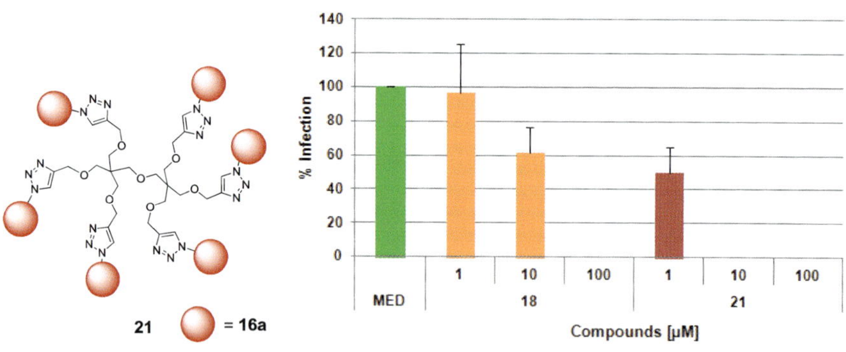

Figure 15.10 (Left) Dendrimer **21**; (Right) HIV *trans* infection levels after treatment with dendrimers **18** and **21**. Experiments were performed on CD4+ T cells purified from 3 healthy donors. Data are represented as percentage of untreated control infection level. Values represent the mean ±SD.

15.4 Conclusions

A majority of pathogens that infect humans use the mucosal entry pathway and recognize DC-SIGN as their primary target. These pathogens include respiratory viruses and those responsible for sexually transmitted diseases, such as HIV and human papilloma virus (HPV). Therefore, the design of molecules and materials capable of interacting efficiently with DC-SIGN and to block DC-SIGN dependent infections is a topic of great interest.

We have shown that glycomimetic molecules capable of antagonizing this lectin can be designed by taking advantage of the 3D structure of known oligosaccharide ligands and of available structural information on the lectin–ligand complexes. The small-molecule monovalent ligands obtained are often endowed with limited protein affinity, but display improved drug-like properties compared to natural sugars. This approach has allowed the identification of promising leads, which are giving encouraging results in terms of selectivity as well. Multivalent presentation on polymeric scaffolds of these ligands has afforded high-affinity antagonists, but the selectivity of these materials against different C-lectins with mannose or fucose selectivity still needs to be studied [47]. Much work remains to be done to allow selection of polyvalent scaffolds and ligands optimal in size, shape, and valency, and finely tuned to the supramolecular architecture of individual C-type lectins, but it is clear already that this work has the potential to represent a major step forward in the quest for effective antiviral therapeutics.

References

1. Geijtenbeek, T. B. H., Kwon, D. S., Torensma, R., van Vliet, S. J., van Duijnhoven, G. C. F., Middel, J., Cornelissen, I. L. M. H. A., Nottet, H. S. L. M., KewalRamani, V. N., Littman, D. R., Figdor, C. G., van Kooyk, Y. (2000) *Cell* **100**, 587–597.
2. Geijtenbeek, T. B. H., Torensma, R., van Vliet, S. J., van Duijnhoven, G. C. F., Adema, G. J., van Kooyk, Y., Figdor, C. G. (2000) *Cell* **100**, 575–585.

3. Mitchell, D. A., Fadden, A. J., Drickamer, K. (2001) *J. Biol. Chem.* **276**, 28939–28945.
4. van Kooyk, Y., Geijtenbeek, T. B. (2003) *Nat. Rev. Immunol.* **3**, 697–709.
5. Ernst, B., Magnani, J. L. (2009) *Nat. Rev. Drug Discov.* **8**, 661–677.
6. Anderluh, M., Jug, G., Svajger, U., Obermajer, N. (2012) *Curr. Med. Chem.* **19**, 992–1007.
7. Reina, J. J., Bernardi, A., Clerici, M., Rojo, J. (2010) *Future Med. Chem.* **2**, 1141–1159.
8. Lepenies, B., Lee, J., Sonkaria, S. (2013) *Adv. Drug Delver. Rev.* **65**, 1271–1281.
9. Prost, L. R., Grim, J. C., Tonelli, M., Kiessling, L. L. (2012) *ACS Chem. Biol.* **7**, 1603–1608.
10. Xiao, L., Hung, K.-C., Takahashi, T. T., Joo, K.-I., Lim, M., Roberts, R. W., Wang, P. (2013) *ACS Chem. Biol.* **8**, 967–977.
11. Zhang, Q., Collins, J., Anastasaki, A., Wallis, R., Mitchell, D. A., Becer, C. R., Haddleton, D. M. (2013) *Angew. Chem. Int. Ed. Engl.* **52**, 4435–4439.
12. Feinberg, H., Mitchell, D. A., Drickamer, K., Weis, W. I. (2001) *Science* **294**, 2163–2166.
13. Feinberg, H., Castelli, R., Drickamer, K., Seeberger, P. H., Weis, W. I. (2007) *J. Biol. Chem.* **282**, 4202–4209.
14. Guo, Y., Feinberg, H., Conroy, E., Mitchell, D. A., Alvarez, R., Blixt, O., Taylor, M. E., Weis, W. I., Drickamer, K. (2004) *Nat. Struct. Mol. Biol.* **11**, 591–598.
15. Reina, J. J., Bernardi, A. (2012) *Mini-Rev. Med. Chem.* **12**, 1434–1442.
16. Garcia-Vallejo, J. J., Koning, N., Ambrosini, M., Kalay, H., Vuist, I., Sarrami-Forooshani, R., Geijtenbeek, T. B. H., van Kooyk, Y. (2013) *Int. Immunol.* **25**, 221–233.
17. Timpano, G., Tabarani, G., Anderluh, M., Invernizzi, D., Vasile, F., Potenza, D., Nieto, P. M., Rojo, J., Fieschi, F., Bernardi, A. (2008) *ChemBioChem* **9**, 1921–1930.
18. Andreini, M., Doknic, D., Sutkeviciute, I., Reina, J. J., Duan, J., Chabrol, E., Thepaut, M., Moroni, E., Doro, F., Belvisi, L., Weiser, J., Rojo, J., Fieschi, F., Bernardi, A. (2011) *Org. Biomol. Chem.* **9**, 5778–5786.
19. Doknic, D., Abramo, M., Sutkeviciute, I., Reinhardt, A., Guzzi, C., Schlegel, M. K., Potenza, D., Nieto, P. M., Fieschi, F., Seeberger, P. H., Bernardi, A. (2013) *Eur. J. Org. Chem.* **2013**, 5303–5314.

20. de Witte, L., Nabatov, A., Pion, M., Fluitsma, D., de Jong, M. A., de Gruijl, T., Piguet, V., van Kooyk, Y., Geijtenbeek, T. B. (2007) *Nat. Med.* **13**, 367–371.
21. Zierke, M., Smieško, M., Rabbani, S., Aeschbacher, T., Cutting, B., Allain, F. H. T., Schubert, M., Ernst, B. (2013) *J. Am. Chem. Soc.* **135**, 13464–13472.
22. Damkaci, F., DeShong, P. (2003) *J. Am. Chem. Soc.* **125**, 4408–4409.
23. Guzzi, C., Angulo, J., Doro, F., Reina, J. J., Thepaut, M., Fieschi, F., Bernardi, A., Rojo, J., Nieto, P. M. (2011) *Org. Biomol. Chem.* **9**, 7705–7712.
24. Mammen, M., Choi, S. K., Whitesides, G. M. (1998) *Angew. Chem. Int. Ed. Engl.* **37**, 2754–2794.
25. Krishnamurthy, V. M., Estroff, L. A., Whitesides, G. M. (2006) In *Fragment-Based Approaches in Drug Discovery*, Jahnke, W., Erlanson, D. A., Eds. Wiley (New York). pp. 11–53.
26. Tabarani, G., Thepaut, M., Stroebel, D., Ebel, C., Vives, C., Vachette, P., Durand, D., Fieschi, F. (2009) *J. Biol. Chem.* **284**, 21229–21240.
27. Mitchell, D. A., Jones, N. A., Hunter, S. J., Cook, J. M. D., Jenkinson, S. F., Wormald, M. R., Dwek, R. A., Fleet, G. W. J. (2007) *Tetrahedron-Asymmetr.* **18**, 1502–1510.
28. Garber, K. C. A., Wangkanont, K., Carlson, E. E., Kiessling, L. L. (2010) *Chem. Commun.* **46**, 6747–6749.
29. Adams, E. W., Ratner, D. M., Bokesch, H. R., McMahon, J. B., O'Keefe, B. R., Seeberger, P. H. (2004) *Chem. Biol.* **11**, 875–881.
30. Reina, J. J., Sattin, S., Invernizzi, D., Mari, S., Martinez-Prats, L., Tabarani, G., Fieschi, F., Delgado, R., Nieto, P. M., Rojo, J., Bernardi, A. (2007) *ChemMedChem* **2**, 1030–1036.
31. Mari, S., Posteri, H., Marcou, G., Potenza, D., Micheli, F., Cañada, F. J., Jimenez-Barbero, J., Bernardi, A. (2004) *Eur. J. Org. Chem.* 5119 – 5125.
32. Sattin, S., Daghetti, A., Thépaut, M., Berzi, A., Sánchez-Navarro, M., Tabarani, G., Rojo, J., Fieschi, F., Clerici, M., Bernardi, A. (2010) *ACS Chem. Biol.* **5**, 301–312.
33. Thepaut, M., Guzzi, C., Sutkeviciute, I., Sattin, S., Ribeiro-Viana, R., Varga, N., Chabrol, E., Rojo, J., Bernardi, A., Angulo, J., Nieto, P. M., Fieschi, F. (2013) *J. Am. Chem. Soc.* **135**, 2518–2529.
34. Mari, S., Sánchez-Medina, I., Mereghetti, P., Belvisi, L., Jiménez-Barbero, J., Bernardi, A. (2007) *Carb. Res.* **342**, 1859–1868.

35. Berzi, A., Reina, J. J., Ottria, R., Sutkeviciute, I., Antonazzo, P., Sanchez-Navarro, M., Chabrol, E., Biasin, M., Trabattoni, D., Cetin, I., Rojo, J., Fieschi, F., Bernardi, A., Clerici, M. (2012) *AIDS* **26**, 127–137.
36. Angulo, J., Díaz, I., Reina, J. J., Tabarani, G., Fieschi, F., Rojo, J., Nieto, P. M. (2008) *ChemBioChem* **9**, 2225–2227.
37. Obermajer, N., Sattin, S., Colombo, C., Bruno, M., Švajger, U., Anderluh, M., Bernardi, A. (2011) *Mol. Div.* **15**, 347–360.
38. Varga, N., Sutkeviciute, I., Guzzi, C., McGeagh, J., Petit-Haertlein, I., Gugliotta, S., Weiser, J., Angulo, J., Fieschi, F., Bernardi, A. (2013) *Chem. Eur. J.* **19**, 4786–4797.
39. Luczkowiak, J., Sattin, S., Sutkeviciute, I., Reina, J. J., Sanchez-Navarro, M., Thepaut, M., Martinez-Prats, L., Daghetti, A., Fieschi, F., Delgado, R., Bernardi, A., Rojo, J. (2011) *Bioconjug. Chem.* **22**, 1354–1365.
40. Varga, N., Sutkeviciute, I., Ribeiro-Viana, R., Berzi, A., Ramdasi, R., Daghetti, A., Vettoretti, G., Amara, A., Clerici, M., Rojo, J., Fieschi, F., Bernardi, A. (2014) *Biomaterials* **35**, 4175–4184.
41. Dehuyser, L., Schaeffer, E., Chaloin, O., Mueller, C. G., Baati, R., Wagner, A. (2012) *Bioconjug. Chem.* **23**, 1731–1739.
42. Lasala, F., Arce, E., Otero, J. R., Rojo, J., Delgado, R. (2003) *Antimicrob. Agents Chemother.* **47**, 3970–3972.
43. Sutkeviciute, I., Thepaut, M., Sattin, S., Berzi, A., McGeagh, J., Grudinin, S., Weiser, J., Le Roy, A., Reina, J. J., Rojo, J., Clerici, M., Bernardi, A., Ebel, C., Fieschi, F. (2014) *ACS Chem. Biol*, **9**, 1377–1385.
44. Tassaneetrithep, B., Burgess, T. H., Granelli-Piperno, A., Trumpfheller, C., Finke, J., Sun, W., Eller, M. A., Pattanapanyasat, K., Sarasombath, S., Birx, D. L., Steinman, R. M., Schlesinger, S., Marovich, M. A. (2003) *J. Exp. Med.* **197**, 823–829.
45. Lozach, P.-Y., Burleigh, L., Staropoli, I., Navarro-Sanchez, E., Harriague, J., Virelizier, J.-L., Rey, F. A., Desprès, P., Arenzana-Seisdedos, F., Amara, A. (2005) *J. Biol. Chem.* **280**, 23698–23708.
46. Hidari, K., Abe, T., Suzuki, T. (2013) *Viruses* **5**, 605–618.
47. Scharenberg, M., Schwardt, O., Rabbani, S., Ernst, B. (2012) *J. Med. Chem.* **55**, 9810–9816.

Chapter 16

Carbohydrate, Biomaterials, and Tissue Engineering Applications

Laura Russo,* Antonella Sgambato,* Davide Bini,* Ilaria Calloni,* Davide Origgi,* Fatma Cetin Telli*,† and Laura Cipolla*

**Department of Biotechnology and Biosciences, University of Milano-, Bicocca, Italy*

†Ege University, Faculty of Science, Department of Chemistry, Bornova/Izmir 35100, Turkey

16.1 Introduction

The increasing importance of biomaterials in our society over the past decades can be tracked in a number of ways, including the growth of tissue engineering applications, both as an academic discipline and as an important industry [1]. Initially, the main goal of scientists was the design of materials largely performing mechanical functions, being preferably "inert" without interaction with the biology of the host organism in order to prevent rejection [2,3]. Early research and fortuitous accidents linking materials chemistry to biological responses provided a rational basis for developing new biomaterials [1]. The molecular biology revolution of the seventies, and advances in genomics, proteomics, and more recently in glycomics, significantly affected the ways in which biomaterials are designed, fabricated and used nowadays. It is well known that a series of interactions occur between the surface of biomaterials and the biological environment after they have been implanted into the human body. Therefore, the biomaterials surface plays an extremely relevant role in the response of artificial medical devices to the biological environment. Bioactive materials, in which the material scaffold has been functionalized with biomolecules in order to promote the desired biological response, increase the chances of local regeneration [4]. Different class of biomolecules have been exploited toward cell-responsive biomaterials, including: peptides, like the adhesive cue arginine (Arg)–glycine (Gly)–asparagine (Asp), targeting cell adhesion receptors (such as integrin or syndecan) [5]; other regulatory molecules (i.e. cytokines, growth factors) and small molecules, such as kartogenin (KGN) promoting chondrocyte differentiation for cartilage repair [6] or acetylcholine targeting guanine nucleotide-binding protein (G protein)-coupled receptor used for neural repair [7]; and several growth factors. In this scenario, one of the most relevant class of biomolecules involved in molecular recognition bioprocesses, the glycans [8], has been neglected and, therefore, should be included in the investigation on functional biomaterials [9].

It is now well established that glycan interactions with their receptors play a fundamental role in various critical intra- and intercellular events [10–12]. Moreover, carbohydrate structures encode information that modulates interactions between cells and the extracellular matrix (ECM) [13,14]. On the basis of these premises, saccharidic motifs are undoubtedly interesting cues to be used for the upgrading of synthetic or natural polymers to smart biomaterials with the ability to cross-talk with their biological environment.

Unlike most man-made materials, materials used in living systems are frequently multifunctional and dynamic [15]. In general, materials from natural sources (e.g. purified protein components, such as collagens from animal tissues) are advantageous because of their inherent properties of biological recognition, and susceptibility to cell-triggered proteolytic degradation and remodeling. While naturally occurring protein-based materials have numerous inherent advantageous properties, such as appropriate size, specificity, and affinity for their specific application, they are often difficult and expensive to manufacture and lack biostability. These and many other issues (e.g. purification, immunogenicity, and pathogen contamination) have egged on the design and development of synthetic biomaterials for tissue engineering applications that are potentially useful in a wide range of medical and non-medical applications [16]. One of the driving forces behind the use of synthetic materials is to simplify independent changes to the biochemical and biomechanical properties. Artificial components, such as polymers, offer an alternative as they are often easy to manufacture, feature low costs, and display high stability over a range of conditions. Independently from chemical and mechanical properties, these materials can be feathered with biologically active biomolecules (i.e. peptides or carbohydrates) that are able to drive and mimic the biochemical properties and activities of the natural ECM. Innovative biomaterials might be able to perform and stimulate a cross-talk mechanism with cells and biomolecules from the physiological milieu.

16.2 Glycoscience for Biomaterials Design

Glycans hold a broader framework of bioactivities: from those that are trivial, such as a source of energy or a purely structural role, to those that are crucial for the development, growth, function, or survival of an organism. This variety and complexity is not surprising if one keeps in mind their structural diversity. Moreover, carbohydrates have very strategic locations on the cell surface [17]. Despite their importance, glycans have not been given as much attention as signaling molecules in biomaterial design for tissue engineering and regenerative medicine applications [18–35]. One of the main successful exploitations of carbohydrates in biomaterial design is the creation of galactose-grafted scaffolds to support hepatocyte (liver cell) growth (Sections 16.2.4) [36,37].

In the following sections, examples of material functionalization ("biodecoration") with signaling and relevant glycidic scaffolds will be outlined. The sections will be organized based on the chemistry used for covalent attachment of the carbohydrates to the material surfaces.

16.2.1 *Material functionalization with glycidic structures via reductive amination*

Because of the higher accessibility surface, linear or fibrous proteins offer interesting properties as scaffolds for biomaterial design. One of the main representatives of the first family is collagen. Site-selective chemical modification of protein-based materials requires simple and efficient reactions, performable in high yield and under mild conditions. Different chemoselective reactions [38] are known for residue-specific modification of protein amino acids (cysteine [39], glutamine, serine, threonine, hydroxyproline, lysine [40], and arginine). Lysines bear a primary amine moiety and, although their side chains are protonated under physiological pH, they can react as nucleophiles. In this regard, reactions that have proven useful are the coupling of amino groups of lysines with aldehydes affording the corresponding Schiff base or a secondary amine after imine reduction. Collagen two-dimensional (2D) matrices have been decorated with galactose via reductive amination [41]. Collagen Type I from equine tendon was used for the preparation of patches by solvent casting method [42]. Neo-glycosylation of collagen (Fig. 16.1) was achieved, reacting lysine side-chain amino groups with lactose aqueous solution in sodium cyanoborohydride ($NaCNBH_3$) in citrate buffer.

The same procedure has been performed to prepare collagen matrices decorated with glucose moieties at their surface (Fig. 16.1) in order to investigate neuroblastoma F11 cell line behavior [43]. Images from a confocal microscope revealed a significantly higher frequency of cells with neuritic-like processes when plated on neoglucosylated collagen compared to pristine collagen and petri dishes, and functional analysis by the patch-clamp technique showed that neoglucosylated collagen matrices were able to drive cells to differentiate without the use of classical differentiating agents [44,45].

16.2.2 *Material functionalization with glycidic structures via thiol-ene reaction*

Another example of monosaccharides linkage to collagen based materials has been obtained via a thiol-ene approach [46], between

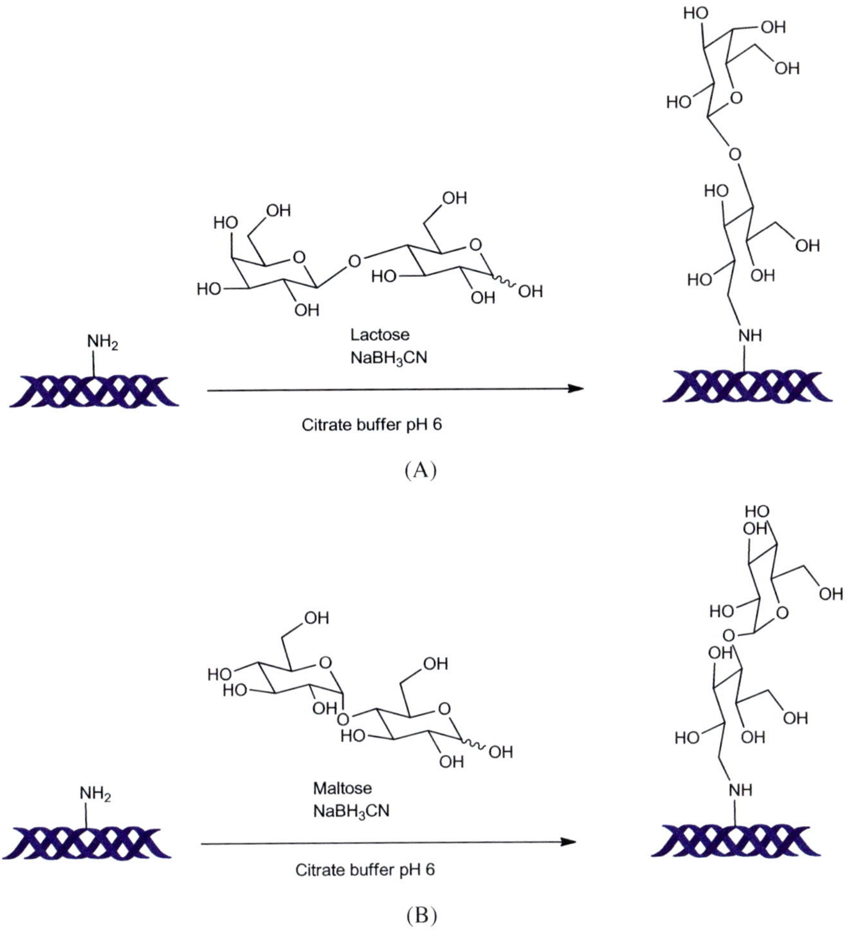

Figure 16.1 Neoglycosylation of collagen via reductive amination with lactose (A) and maltose (B), exposing galactose (A) and glucose (B) residues, respectively.

allyl glycosides and the thiol-functionalized material surface (Fig. 16.2). Since collagen contains few or no cysteine or cystine residues [47], thiol groups have been suitably introduced, reacting lysine side-chain amino groups with γ-thiobutyrolactone. Preliminary *in vivo* biological assays were performed, implanting unglycosylated and glycosylated collagen patches in osteoarthritic rats (OA) when the pathology was well established. The glycosylated collagen matrices resulted effective in improving mobility in OA models after implantation.

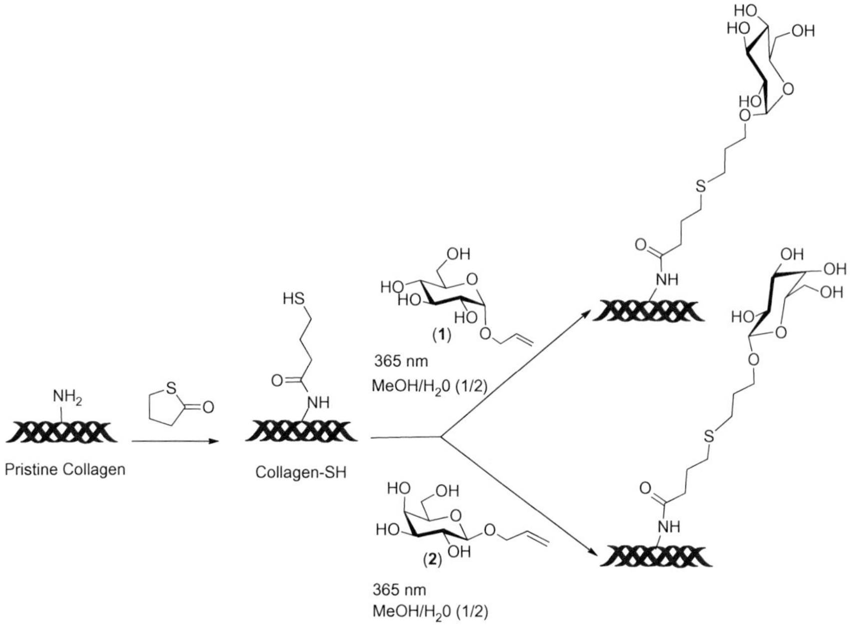

Figure 16.2 Thiol-ene reaction between thiolated collagen (collagen-SH) with allyl α-glucopyranoside (1) and allyl β-galactopyranoside (2).

Given the versatility of thiol-ene photoclick reactions on collagen matrices, the same procedure has also been used as a neoglycosylation method, allowing for the introduction of multi-antennary glycidic structures on the protein (Fig. 16.3). It must be pointed out, in fact, that most often binding events between glycans and their complementary receptors (i.e. lectins) [48] occur in a multivalent and cooperative manner [49]. The proposed strategy included thiolation of collagen, regioselective thiol-ene photoclick-reaction with a double-bond terminated heterobifunctional dendron, and final glycan conjugation by a second chemoselective reaction.

Thiol-ene functionalization has been used also for the functionalization of synthetic polymers. Glycopolymers have been shown to mimic the activity of natural glycoproteins through multivalent interactions with carbohydrate-binding proteins, e.g. lectins. Concerning synthetic glycopolymers, various strategies have been

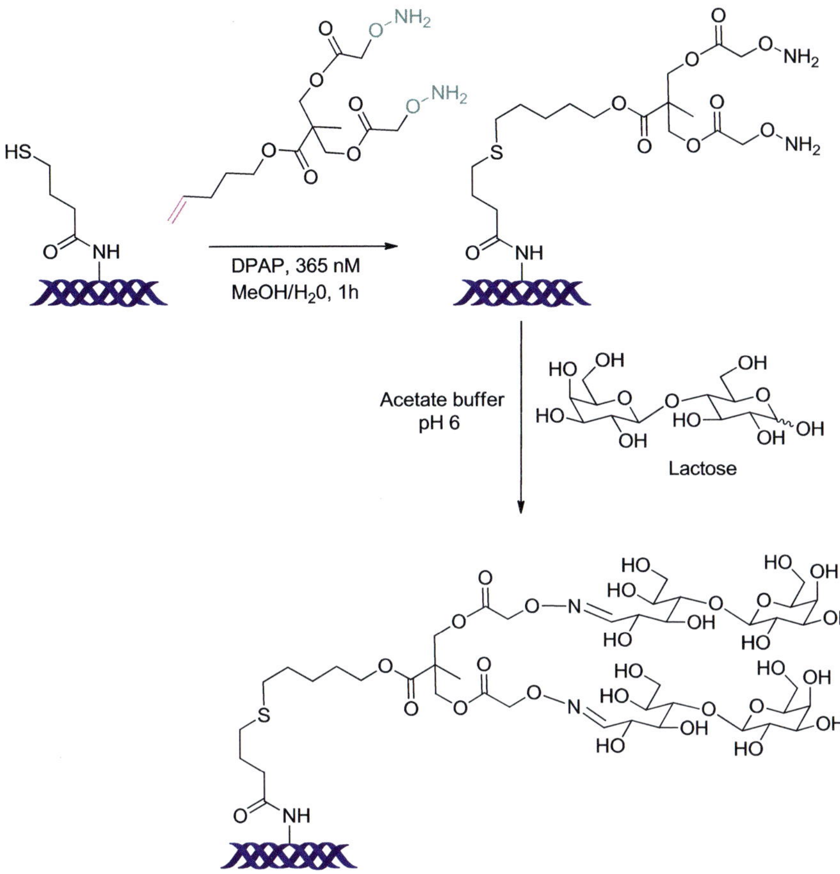

Figure 16.3 Photo-click reaction of collagen-SH with a bivalent dendron.

adopted until now, including free radical, controlled/living radical, anionic/cationic, and ring opening/ring opening metathesis polymerizations of carbohydrate-carrying monomers [50]. Roy *et al.* described an alternative approach to link more complex glycans using a combination of the reversible addition–fragmentation chain transfer (RAFT) polymerization and thiol-ene click chemistry. Trimannose has been used in the decoration step because their structural role in intracellular pathogens, such as *Mycobacterium tuberculosis* cell wall [51] and Leishmania parasite [52], and for its immunomodulator activity against these infectious diseases [53]. RAFT has been performed using allyl methacrylate (AMA), an asymmetrical divinyl monomer, copolymerized with *N*-(2-hydroxypropyl) methacrylamide (HPMA) to obtain multiple reactive pendant alkene groups along the polymer backbone. The conjugation step between the unprotected thiolated α(1,2)-trimannoside and the polymer (Fig. 16.4) was conducted with thiol-ene type "click" chemistries. These well-defined, trimannoside-containing copolymers could provide a useful alternative method for synthesizing polymeric glycan bioactive systems.

The glycan-modified polymer has been tested by ultraviolet-visible (UV-Vis) spectroscopy agglutination and surface plasmon resonance imaging (SPRi) analysis [54] for its ability to bind the mannose-specific lectin concanavalin A.

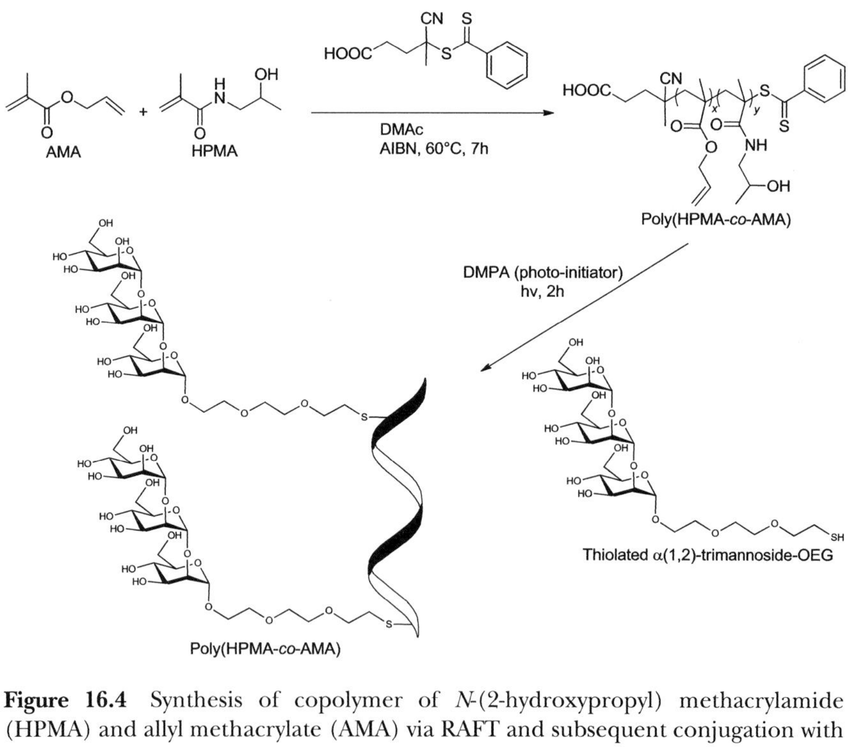

Figure 16.4 Synthesis of copolymer of *N*-(2-hydroxypropyl) methacrylamide (HPMA) and allyl methacrylate (AMA) via RAFT and subsequent conjugation with thiolated α(1,2)-trimannoside using thiol-ene chemistry.

16.2.3 *Material functionalization with glycidic structures via carbonyl–alkoxyamine chemoselective bioconjugation methods*

An alternative strategy to lysine functionalization, is the residue-specific modification of arginine [55]. Similar to lysine (5.91% of all residues in proteins), arginine is a relatively abundant residue in proteins (5.44%), though it has a lower tendency to be located on the surface of the protein itself [56]. Methylglyoxal modification of arginine has been proposed on collagen 2D scaffolds as a nature inspired novel bioorthogonal carbonyl–alkoxyamine chemoselective bioconjugation method, that can theoretically be extended to any protein. Methylglyoxal modification of arginine side-chains affords a methyl ketone that can be suitably conjugated to carbohydrates via a suitable linker (Fig. 16.5) [57].

Arginine modification was performed in a heterogeneous phase, immersing the collagen film in methylglyoxal solution [58]. In order to assess the presence of the methylketone functionality, chemical derivatization with suitable compounds for fluorescence analysis, X-ray photoelectron spectroscopy (XPS) and fourier transform infrared (FT-IR) analysis were performed.

The functionalization and, at the same time, the correct presentation of the galactose unit of lactose on the collagen film surface have been verified through a lectin-based assays [59].

16.2.4 *Material functionalization with glycidic structures via amide bond formation*

A different approach capitalizes on the amide bond formation between carboxyl and amino groups suitably located in a complementary position on the material and on the carbohydrate moieties, respectively. An example of a monosaccharide linkage to materials via amidation was proposed on hydroxyapatite (HA) [60]. Among various biocompatible materials, synthetic HA and carbonate hydroxyapatite (CHA) are widely used in many biomedical applications [61]. HA is the natural mineral ingredient of bones, teeth, and calcified tissues in vertebrates. Synthetic HA and CHA are used for

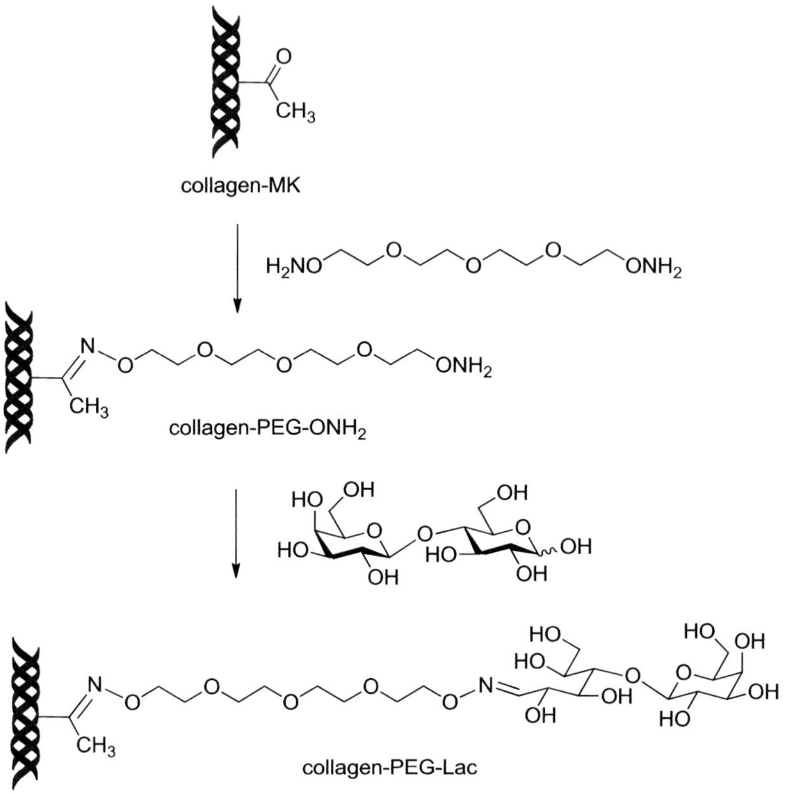

Figure 16.5 Methylketone (MK) introduced on arginine side-chain enables protein modification, where the carbonyl group reacts with an aminoxy molecule and collagen chemoselective bioconjugation with lactose.

human implant coatings, possessing a proven osteoconductivity, and its geometry may deeply influence specific tissue responses, as it happens for vascular in-growth [62]. Amide bond formation was also used for the functionalization of chitosan as the scaffold with lactobionic acid [63]. Among natural biomaterials, chitosan is a binary heteropolysaccharide consisting of (1→4)-linked 2-acetamido-2-deoxy-β-D-glucopyranose and 2-amino-2-deoxy-β-D-glucopyranose residues [64,65], which has been widely applied in clinic and exhibits essential advantages in biodegeneration, biocompatibility, and non-immunoreactions to the human body when compared to synthetic polymers. The aim of this study was to develop a novel natural nanofibrous scaffold with surface-bound galactose ligands to enhance the bioactivity and mechanical stability of primary hepatocytes in culture for liver regeneration.

Amide bond formation has also been used to functionalize synthetic nondegradable polymers like polystyrene-based porous polymers synthesized within high internal phase emulsions (polyHIPE) scaffolds. These scaffolds have shown very promising results with a range of different cell types due to their controllable morphology, high porosity, and suitable mechanical properties [66], but their hydrophobic nature represents one limitation for the potential biological applications. Hayward *et al.* described the inclusion of an ester comonomer into a styrene-based emulsion as a facile route to polyHIPE surface modification with aminoethyl glycosides (Fig. 16.6). They showed that a polystyrene-based polyHIPE containing pentafluorophenyl acrylate (PFPA) can be synthesized with a desirable morphology to permit three-dimensional (3D) cell growth [67]. The functionalization of this material was then employed downstrem the scaffold preparation, to immobilize pendent galactose residues via aminoethyl glycoside coupling with PFPA, taking inspiration from the strategy already described by Boyer and Davis for solution-based glycopolymer synthesis [68]. The selectivity and accessibility of the pendent galactose residues was verified using primary rat hepatocyte cell cultures, which contain a cell-surface asialoglycoprotein receptor (ASGPR31) that can specifically bind to galactose for enhanced cell adhesion and function [69].

Glucose-functionalized polyHIPEs were also prepared as a selectivity control for galactose during the primary rat cell adhesion experiments. Preliminary studies with primary rat hepatocytes on a galactose-functionalized polyHIPE showed that the carbohydrates are accessible and selective to cells entering the scaffold, seen by an enhanced functional activity of the cells as they initially adapt to the culture environment [70].

A different strategy has been used for carbohydrate grafting on polycaprolactone substrates [28]. Among synthetic polymers, aliphatic polyesters such as poly(ε-caprolactone) (PCL) have been

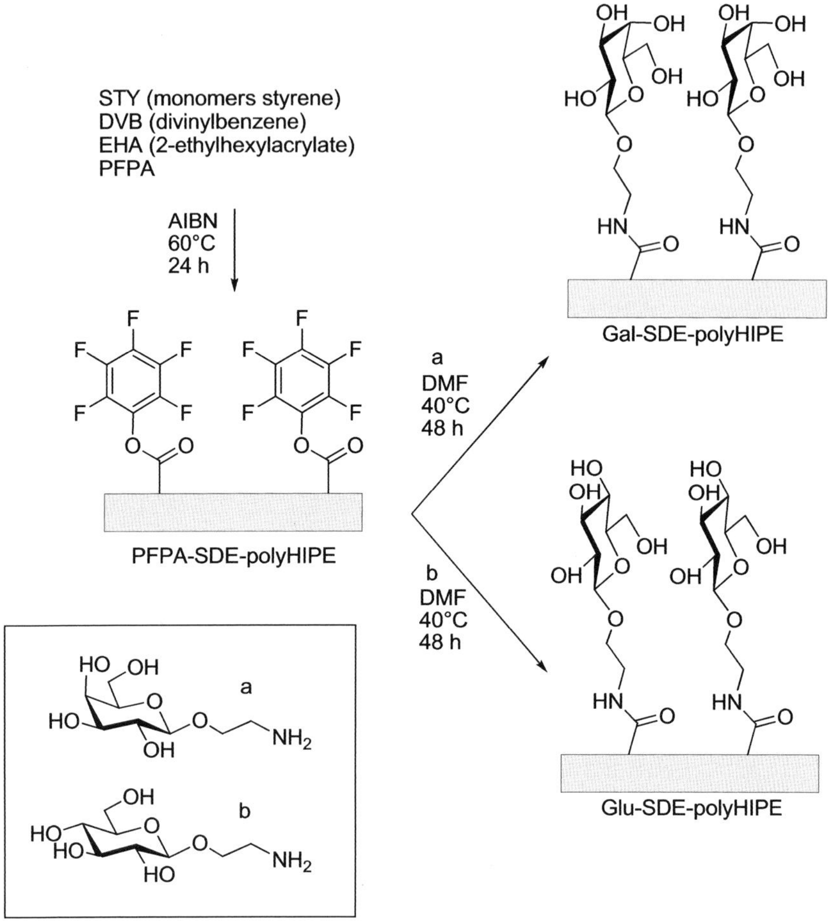

Figure 16.6 Synthetic strategy for the incorporation of carbohydrates onto the surface of a polystyrene-based polyHIPE (SDE-polyHIPE).

widely considered as biomaterials for designing scaffolds to support the regeneration of several tissue-engineered organs [71–73], due to a unique combination of biodegradability and biocompatibility properties. However, as any other synthetic polymer it does not present molecular motifs for cell biological recognition. Cell adhesion is strongly related to the surface properties of biomaterials, and it is influenced by substratum surface properties, such as surface charge, wettability, roughness, and topography. Modification of PCL-based

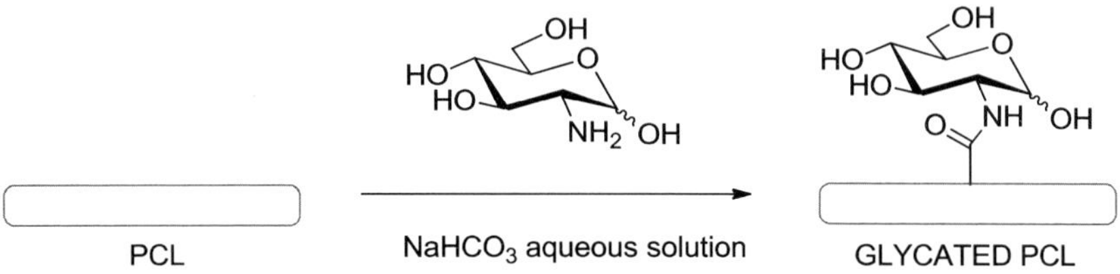

Figure 16.7 Aminolysis of PCL with glucosamine.

materials is desired in order to improve their hydrophilic properties and to achieve a friendly interface for living cells [74]. PCL functionalization was achieved by a one-step procedure of polymer aminolysis (Fig. 16.7), taking advantage of the amino group of a biologically relevant aminosugar, 2-amino-2-deoxyglucose (glucosamine). Glucosamine is an innovative way to render scaffolds more biocompatible by increasing hydrophilicity; in addition, it is worth noting that, since this monosaccharide is widespread in nature, including humans, and included in several biomedical formulations for osteoarthritis treatment, it should be a safe approach for biomaterial surface upgrades.

Preliminary biological assays evidence that the presence of glucosamine enhances both the hydrophilicity, as demonstrated by water contact angle measurements, and the biological performance of PCL, without negatively altering its mechanical behavior, thus upgrading it to an original biocompatible and bioactive polyester.

16.2.5 *Material grafting by azide-based click chemistry*

A high impact on the glyco-conjugation methodologies has been obtained by the introduction of the "azido-alkyne click-chemistry" approach [75, 76]. In actual state-of-the-art carbohydrate chemistry, the introduction of the alkyne or azide group in a saccharide is a routine process. This fact allows easy access to mono- as well as polyfunctionalized derivatives that can be rationally used for conjugation with other complementary functionalized synthetic and biological molecules or scaffolds by means of their intermolecular alkyne–azide 1,3-dipolar cycloaddition reaction [77]. In addition, click chemistry is also being explored by material scientists, both for the processing of bulk materials [78,79] and for the functionalization of material surfaces [80].

"Photo-click" reactions were also developed, affording several additional advantages, such as spatio-temporal reaction control via focused ultraviolet (UV)-light irradiation. An example of this class of reaction is the UV light-promoted grafting method of a molecule containing a photoreactive aryl azide group coated on a polymer surface [81,82]. In the case of organic azides, upon photolysis with ultraviolet light, a nitrene radical is generated. The nitrene radical formed is extremely reactive and can undergo a multitude of reactions, for example: insertion into carbon–hydrogen, nitrogen–hydrogen, and oxygen–hydrogen bonds; addition to olefins; proton abstraction reactions to give the corresponding amine; and in the case of aryl azides, a number of ring expansion reactions have been observed. Although the precise mechanism of surface modification is not fully understood, this method has been demonstrated to be very effective for a wide range of surfaces. This approach has been used recently to covalently link carbohydrates to polyethylene terephthalate (PET) fibers by Renaudie *et al.* [83], for example, by incorporating a sialyl–Lewis X saccharide, as it is known to interact specifically with L-selectin [84]. Thus, a new method for grafting a carbohydrate UV-reactive molecule, an azidophenyl lactamine (AzPhLac), Fig. 16.8, was proposed.

By reductive amination of lactose with an aminated phenyl azide molecule, the phenylazido group on lactose afforded the β-D-galactopyranosyl-(1→4)-1-*N*-[2-(4-azidopnenylamino)ethylamino]-1-deoxy-d-glucitol (AzPhLac); the PET fibers were then dipped (dip-coating method) in the AzPhLac aqueous solution, then dried and finally irradiated by UV light for the chemical fixation of the lactamine moiety via conversion of the phenylazido group to the highly reactive phenylnitrene. The complete surface modification is still not fully understood; the proposed mechanism by the authors also includes the formation of secondary amines. In addition, the highly

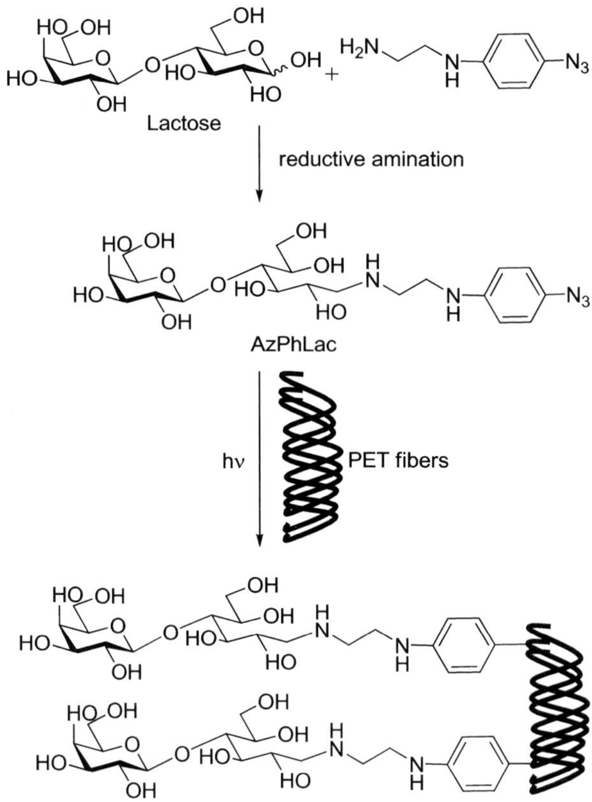

Figure 16.8 Grafting of azidated lactose to PET fibers by UV-irradiation.

reactive phenylnitrene can form covalent bonds not only with the hydrocarbons of the PET surface, but also with linkages of AzPhLac molecules in its immediate surroundings, leading to the formation of a polymeric multilayers, where the nitrene inserts onto the hydroxyl (–OH) groups.

A different strategy combines the azido–alkyne click-chemistry technique with plasma treatment (Fig. 16.9). Plasma treatment was used to introduce amino groups on polypropylene (PP) membrane surfaces (preactivation). A diazo-transfer reaction by triflyl azide transformed the amino groups into azido groups. Finally, the azido groups was reacted with propargyl by Huisgen to introduce cycloaddition 2-acetamido-2-deoxy-α-D-glucopyranoside.

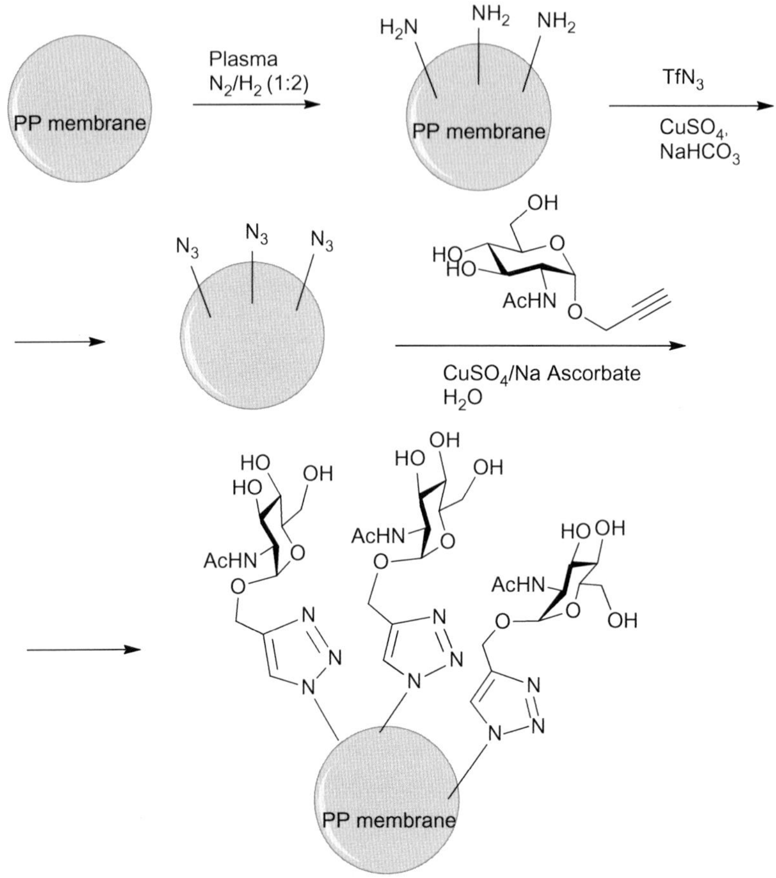

Figure 16.9 Diazo-transfer methodology on plasma-treated PP surfaces.

The group of Bertozzi [85] proposed a brilliant example of Copper (I)-catalyzed azide–alkyne cycloaddition (CuAAC) for the chemoselective ligation of azide-functionalized pyrene and glycan moieties to the alkyne-functionalized focal point and chain ends of a dendritic scaffold in order to "glycodecorate" single-walled carbon nanotubes (SWNTs) by ultrasonication in water. The structural, mechanical, electrical, and optical properties of SWNTs have stimulated considerable interest in their biological applications [86, 87], including biosensing, imaging, intracellular delivery, and cancer cell targeting. However, use of SWNTs in living systems requires strategies to diminish their cytotoxicity [88]. Thus, surface modifications that lower the toxicity of SWNTs, while simultaneously enabling specific biological recognition, are highly desirable [89,90]. A promising approach is to coat SWNTs with synthetic glycopolymers that are able to mimic the glycoproteins found on cell surfaces.

The proposed approach first involves the synthesis of dendritic structures, ensuring the mimicry of glycoproteins, and a suitable moiety for the anchoring to the nanotube surface. Thus, a pyrene tail was conjugated to the focal point of the dendrimer by CuAAC (Fig. 16.10). The resulting second generation dendrimer was further coupled with pent-4-ynoic anhydride to introduce additional alkyne groups. This third generation dendrimer was then reacted with a 2-azidoethyl mono- or disaccharide using CuAAC once again.

SWNTs coating with glycodendrimers resulted in a bioactive surface that was shown to mitigate their cytotoxicity. The synthetic method proposed by Bertozzi can be readily adapted to ligands for other receptor interactions, in view of novel applications, such as biosensors for carbohydrate-binding proteins and delivery agents that target specific cell surface receptors.

The Huisgen-type cycloaddition was also porposed for the glycodecoration of HA; in this example, a minimal spacer of difunctional polyethylene glycol (PEG) was used, possessing an azido group at one term for the Huisgen cycloaddition, and a carboxyl group on the other side for the covalent bonding to the hydroxyl groups present on the HA surface [91]. The Huisgen cycloaddition was performed onto HA–azide (N_3) with propargyl α-D-glucopyranoside. The "glycosylated" hydroxypatite was characterized by its ability to interact with glucose recognizing lectins.

16.3 Conclusions

Few examples of materials biofunctionalized with carbohydrates have appeared recently in the literature with the aim of ameliorating materials features, in terms of bioactivity, biocompatibility, and biostability. Despite the increasing development of glycomics and

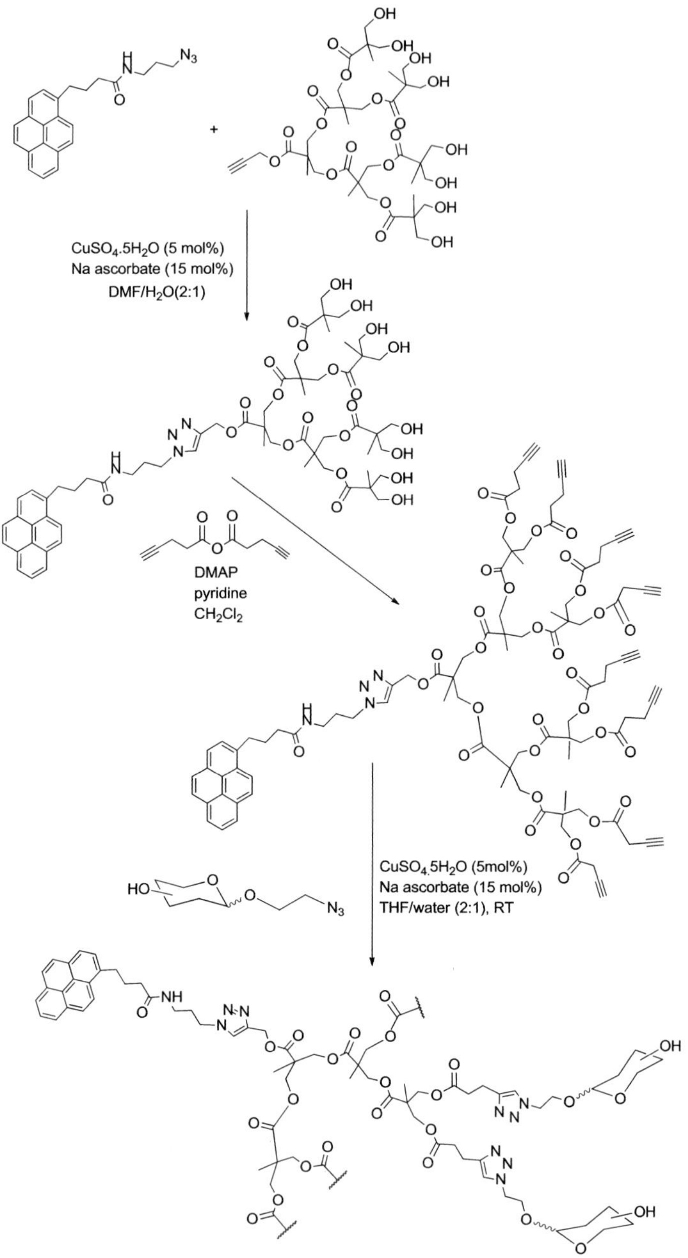

Figure 16.10 Generation of glycodendrimers for carbon nanotubes functionalization.

the methods at our disposal for oligossacharide synthesis, application of glycoscience for the design of biomaterials for tissue engineering is still scarce and a difficult challenge. We think that in the next few years, a great contribution by glycochemists will be given to this research field, which may experience a great explosion as observed for glycosylated nanoparticles and glycoarrays.

References

1. Huebsch, N., Mooney, D. J. (2009) *Nature* **462**, 426–432.
2. Ratner, B. D., Bryant, S. J. (2004) *Annu. Rev. Biomed. Eng.* **6**, 41–75.
3. Anderson, J. M., Rodriguez, A., Chang, D. T. (2008) *Semin. Immunol.* **20**, 86–100.
4. Mager, M. D., LaPointe, V., Stevens, M. M. (2011) *Nature Chem.* **3**, 582–589.
5. Hozumi, K., Otagiri, D., Yamada, Y., Sasaki, A., Fujimori, C., Wakai, Y., Uchida, T., Katagiri, F., Kikkawa, Y., Nomizu, M. (2010) *Biomaterials* **31**, 3237–3243.
6. Johnson, K., Zhu, S., Tremblay, M. S., Payette, J. N., Wang, J., Bouchez L. C., Meeusen, S., Althage, A., Cho, C. Y., Wu, X., Schultz, P. G. (2012) *Science* **336**, 717–721.
7. Zhou, Z., Yu, P., Geller, H. M., Ober, C. K. (2012) *Biomaterials* **33**, 2473–2481.
8. a) Service, R. F. (2012) *Science* **338**, 321–323; b) Stallforth, P., Lepenies, B., Adibekian, A., Seeberger, P. H. (2009) *J. Med. Chem.* **52**, 5561–5577; c) Hart, G. W., Copeland, R. J. (2010) *Cell* **143**, 672–676; d) Seeberger, P. H. (2009) *Nat. Chem. Biol.* **5**, 368–372.
9. a) Cipolla, L., Russo, L., Taraballi, F., Lupo, C., Bini, D., Gabrielli, L., Capitoli, A., Nicotra, F. (2012) In *Specialist Periodical Reports-Carbohydrate Chemistry* **38** Rauter, A. P., Lindhorst, T. Eds. Royal Society of Chemistry (London) Ch. 17; b) Chawla, K., Yu, T., Stutts, L., Yen, M., Guan, Z. (2012) *Biomaterials* **33**, 6052–6060; c) Sapsford, K. E., Algar, W. R., Berti, L., Gemmill, K. B., Casey, B. J., Oh, E. Stewart, M. H., Medintz, I. L. (2013) *Chem. Rev.* **113**, 1904–2074 d) Koepsel, J. T., Murphy, W. L. (2012) *Chem. Bio. Chem.* **13**, 1717–1724. e) Slaney, A. M., Wright, V. A., Meloncelli, P. J., Harris, K. D., West, L. J., Lowary, T. L., Buriak, J. M. (2011) *ACS Appl. Mater. Interfaces* **3**, 1601–1612;

f) Santoyo-Gonzalez, F., Hernandez-Mateo, F. (2009) *Chem. Soc. Rev.* **38**, 3449–3462.

10. Werz, D. B., Ranzinger, R., Herget, S., Adibekian, A., von der Lieth, C. W., Seeberger, P. H. (2007) *ACS Chem. Biol.* **2**, 685–691.
11. Kelley, W., Moremen, M. T., Nairn, A. V. (2012) *Nature Rev.* **13**, 448–462.
12. Cipolla, L., La Ferla, B., Airoldi, C., Zona, C., Orsato, A., Shaikh, N., Russo, L., Nicotra, F. (2010) *Future Med. Chem.* **2**, 587–599.
13. Ohtsubo, K., Marth, J. D. (2006) *Cell* **126**, 855–867.
14. Jurgensen, H. J., Madsen, D. H., Ingvarsen, S., Melander, M. C., Gårdsvoll, H., Patthy, L., Engelholm, L. H., Behrendt, N. A. J. (2011) *Biol. Chem.* **286**, 32736–32748.
15. Barish, R. D., Schulman, R., Rothemund, P. W., Winfree, E. (2009) *PNAS* **106**, 6054–6059.
16. Bryksin, A. V., Brown, A. C., Baksh, M. M., Finn, M. G., Barker, T. H. (2014) *Acta Biomateralia* **10,** 1761–1769.
17. Mager, M.D., LaPointe, V., Stevens, M.M. (2011) *Nature Chem.* **3**, 582.
18. Russo, L., Battocchio, C., Secchi, V., Magnano, E., Nappini, S., Taraballi, F., Gabrielli, L., Comelli, F., Papagni, A., Costa, B., Polzonetti, G., Nicotra, F., Natalello, A., Doglia, S. M., Cipolla, L. (2014) *Langmuir* **30**, 1336–1342.
19. Cipolla, L., Russo, L., Taraballi, F., Lupo, C., Bini, D., Gabrielli, L., Capitoli, A., Nicotra, F. (2012) In *SPR Carbohydrate Chemistry* 38, Rauter, A., Lindhorst, T. Eds. Royal Society of Chemistry (London). pp. 416–445.
20. Cipolla, L., Russo, L., Shaikh, N., Nicotra, F. (2013) In *Polymeric Biomaterials III Ed, Vol. I Polymers as Biomaterials CRC Press, Polymeric Biomaterials: 2 Volume Set,* 3rd ed. Dumitriu S., Popa, V. CRC Press (London). pp 715–736.
21. Chawla, K., Yu, T. B., Stutts, L., Yen, M., Guan, Z. (2012) *Biomaterials* **33**, 6052–6060.
22. Sun, X. L., Stabler, C. L., Cazalis, C. S., Chaikof, E. L. (2006) *Bioconjugate. Chem.* **17**, 52–57.
23. Freichels, H., Pourcelle, V., Auzély-Velty, R., Marchand-Brynaert, J., Jérôme (2012) *Biomacromolecules* **13**, 760–768.
24. Wendeln, C., Rinnen, S., Schulz, C., Arlinghaus, H. F., Ravoo, B. J. (2010) *Langmuir* **26**, 15966–15971.
25. Russo, L., Landi, E., Tampieri, A., Natalello, A., Doglia, S. M.; Gabrielli, L., Cipolla, L., Nicotra, F. (2011) *Carbohydr. Res.* **346**, 1564–1568.

26. Sandri, M., Natalello, A., Bini, D., Gabrielli, L., Cipolla, L., Nicotra, F. (2011) *Synlett* **13**, 1845–1848.
27. Russo, L., Zanini, S., Riccardi, C., Nicotra, F., Cipolla, L. (2011) *Mater. Today* **14**, 164–169.
28. Russo, L., Gloria, A., Russo, T., D'Amora, U., Taraballi, F., De Santis, R., Ambrosio, L., Nicotra, F., Cipolla, L. (2013) *RSC Adv.* **3**, 6286–6289.
29. Slaney, A. M., Wright, V. A., Meloncelli, P. J., Harris, K. D., West, L. J., Lowary, T. L., Buriak, J. M. (2011) *ACS Appl. Mater. Interfaces* **3**, 1601–1612.
30. Lancuški, A., Bossard, F., Fort, S. (2013) *Biomacromolecules* **14**, 1877–1884.
31. Bech, L., Lepoittevin, B., El Achhab, A., Lepleux, E., Teulé-Gay, L., Boisse-Laporte, C., Roger, P. (2007) *Langmuir* **23**, 10348–10352.
32. Acharya, C., Hinz, B., Kundu, S. C. (2008) *Biomaterials* **29**, 4665–4675.
33. Norberg, O., Deng, L., Yan, M., Ramström, O. (2009) *Bioconj. Chem.* **20**, 2364–2370.
34. Feng, Z. Q., Chu, X., Huang, N. P., Wang, T., Wang, Y., Shi, X., Ding, Y., Gu, Z. Z. (2009) *Biomaterials* **30**, 2753–2763.
35. Huang, L., Dolai, S., Raja, K., Kruk, M. (2010) *Langmuir* **26**, 2688–2693.
36. Oka, J. A., Weigel, P. H. (1986) *J. Cell. Biol.* **103**, 1055.
37. Griffith, L. G., Lopina, S. (1998) *Biomaterials* **19**, 979–986.
38. Lee, H., Pietrasik, J., Sheikoc, S. S., Matyjaszewski, K. (2010) *Progress in Polymer Science* **35**, 24–44.
39. a) Chalker, J. M., Bernardes, G. J. L., Lin, Y. A., Davis, B. G. (2009) *Chem. Asian J.* **4**, 630–640; b) Hudak, J. E., Barfield, R. M., de Hart, G. W., Grob, P., Nogales, E., Bertozzi, C. R., Rabuka, D. (2012) *Angew. Chem. Int. Ed.* **51**, 4161–4165; c) Esser-Kahn, A. P., Francis, M. B. (2008) *Angew. Chem. Int. Ed.* **47**, 3751–3754.
40. Tilley, S. D., Joshi, N. S., Francis, M. B. (2008) In *The Wiley Encyclopedia of Chemical Biology* Begley, T. Ed. Wiley-VCH (Weinheim). pp. 1–16.
41. Russo, L., Gautieri, A., Raspanti, M., Taraballi, F., Nicotra, F., Vesentini, S., Cipolla, L. (2014) *Carbohydr. Res.* **389**, 12–17.
42. Taraballi, F., Zanini, S., Lupo, C., Panseri, S., Cunha, C., Riccardi, C., Marcacci, M., Campione, M., Cipolla, L. (2013) *J. Colloid Interface Sci.* **394**, 590–597.
43. Russo, L., Sgambato, A., Lecchi, M., Pastori, V., Raspanti, M., Natalello, A., Doglia, S. M., Nicotra, F., Cipolla, L. (2014) *ACS Chem. Neurosci.* **5**, 261–265.
44. Edsjö, A., Holmquist, L., Påhlman, S. (2007) *Semin. Cancer. Biol.* **17**, 248–256.
45. Abemayor, E., Sidellt, N. (1989) *Environ. Health. Perspect.* **80**, 3–15.

46. a) Dondoni, A., Marra, A. (2012) *Chem. Soc. Rev.* **41**, 573–586; b) Hoyle, C. E., Bowman, C. N. (2010) *Angew. Chem. Int. Ed. Engl.* **49**, 1540–1573; c) Hoyle, C. E., Lowe, A. B., Bowman, C. N. (2010) *Chem. Soc. Rev.* **39**, 1355–1387; d) Azagarsamy, M. A., Anseth, K. S. (2013) *ACS MacroLett.* **2**, 5–9.
47. Miller, E. J. (1976) *Cell. Biochem.* **13**, 165–192.
48. a) Taylor, M. E., Drickamer, K. (2009) *Glycobiology* **19**, 1155–1162; b) Sharon, N., Lis, H. (2004) *Glycobiology* **14**, 53R–62R; c) Ambrosi, M., Cameron, N. R., Davis, B. G. (2005) *Org. Biomol. Chem.* **3**, 1593–1608.
49. Bertozzi, C. R., Kiessling, L. L. (2001) *Science* **291**, 2357.
50. a) Ladmiral, V., Melia, E., Haddleton, D. M. (2004) *Eur. Polym. J.* **40**, 431–449. b) Ting, S. R. S., Granville, A. M., Quemener, D., Davis, T. P., Stenzel M. H., Barner-Kowollik, C. (2007) *Aust. J. Chem.* **60**, 405–409. c) Hetzer, M., Chen, G., Barner-Kowollik, C., Stenzel, M. H. (2010) *Macromol. Biosci.* **10**, 119–126.
51. Turco, S. J., Descoteaux, A. (1992) *Annu. Rev. Microbiol.* **46**, 65–94.
52. Vercellone, A., Nigou, J., Puzo, G. (1998) *Front. Biosci.* **3**, 149–163.
53. Song, E. H., Osanya, A. O., Petersen, C. A., Pohl, N. L. (2010) *J. Am. Chem. Soc.* **132**, 11428–11430.
54. Roy, D., Ghosn, B., Song, E. H., Ratner, D. M., Stayton, P. S. (2013) *Polym. Chem.* **4**, 1153–1160.
55. Gauthier, M. A., Klok, H. A. (2011) *Biomacromolecules* **12**, 482–493.
56. Gauthier, M. A., Klok, H. A. (2008) *Chem. Commun.* **23**, 2591–2611.
57. Taraballi, F., Russo, L., Battocchio, C., Polzonetti, G., Nicotra, F., Cipolla, L. (2014) *Org. Biomol. Chem.* **12**, 4089–4092.
58. Gratzera, P. F., Santerrec, J. P., Lee, J. M. (2004) *Biomaterials* **25**, 2081–2094.
59. Slaney, A. M., Wright, V. A., Meloncelli, P. J., Harris, K. D., West, L. J., Lowary, T. L., Buriak, J. M. (2011) *ACS Appl. Mater. Interfaces* **3**, 1601–1612.
60. Sandri, M., Natalello, A., Bini, D., Gabrielli, L., Cipolla, L., Nicotra, F. (2011) *Synlett* **13**, 1845.
61. Dorozhkin, S. V. (2010) *Biomaterials* **31**, 1465.
62. Karageorgiou, V., Kaplan, D. (2005) *Biomaterials* **26**, 5474.
63. Feng, Z. Q., Chu, X., Huang, N. P., Wang, T., Wang, Y., Shi, X., Ding, Y., Gu, Z. Z. (2009) *Biomaterials* **30**, 2753.
64. Houseman, B.T., Mrksich, M. (2002) *Chem. Biol.* **9**, 443.
65. Houseman, B. T., Gawalt, E. S., Mrksich, M. (2003) *Langmuir* **19**, 1522.

66. a) Schutte, M., Fox, B., Baradez, M. O., Devonshire, A., Minguez, J., Bokhari, M., Przyborski, S., Marshall, D. (2011) *Assay Drug Dev. Technol.* **9**, 475–486. b) Bokhari, M., Carnachan, R. J., Cameron, N. R., Przyborski, S. A. (2007) *Biochem. Biophys. Res. Commun.* **354**, 1095–1100. c) Hayman, M. W., Smith, K. H., Cameron, N. R., Przyborski, S. A. J. (2005) *Biochem. Biophys. Methods* **62**, 231–224.
67. Kircher, L., Theato, P., Cameron, N. R. (2013) *Polymer* **54**, 1755–1761.
68. a) Boyer, C., Davis, T. P. (2009) *Chem. Commun.* 6029–6031, b) Beija, M., Li, Y., Lowe, A. B., Davis, T. P., Boyer, C. (2013) *Eur. Polym. J.* **49**, 3060–3071.
69. a) Ashwell, G., Morell, A. G. (1974) *Adv. Enzymol. Relat. Areas Mol. Biol.* **41**, 99–128. b) Cho, C. S., Hoshiba, T., Harada, I., Akaike, T. (2007) *React. Funct. Polym.* **67**, 1301–1310. c) Cho, C. S., Seo, S. J., Park, I. K., Kim, S. H., Kim, T. H., Hoshiba, T., Harada, I., Akaike, T. (2006) *Biomaterials* **27**, 576–585.
70. Hayward, A. S., Eissa, A. M., Maltman, D. J., Sano, N., Przyborski, S. A., Cameron, N. R. (2013) *Biomacromolecules* **14**, 4271–4277.
71. Gloria, A., De Santis, R., Ambrosio, L. (2010) *J. Appl. Biomater. Biomech.* **8**, 57–67.
72. Gloria, A., Russo, T., De Santis, R., Ambrosio, L. (2009) *J. Appl. Biomater. Biomech.* **7**, 141–152.
73. Causa, F., Battista, E., Della Moglie, R., Guarnieri, D., Iannone, M., Netti, P. (2010) *Langmuir* **26**, 9875–9878.
74. Gloria, A., Causa, F., Russo, T., Battista, E., Della Moglie, R., Zeppetelli, S., De Santis, R., Netti, P. A., Ambrosio, L. (2012) *Biomacromolecules* **13**, 3510–3521.
75. Remzi Becer, C., Hoogenboom, R., Schubert, U. S. (2009) *Angew. Chem. Int. Ed.* **48**, 4900.
76. Sletten, E. M., Bertozzi, C. R. (2009) *Angew. Chem. Int. Ed.* **48**, 6974–6998.
77. Santoyo-Gonzalez, F., Hernandez-Mateo, F. (2009) *Chem. Soc. Rev.* **38**, 3449–3462.
78. Heggli, M., Tirelli, N., Zisch A., Hubbell, A. (2003) *Bioconjugate Chem.* **14**, 967–973.
79. Fournier, D., Hoogenboom R., Schubert, U. S. (2007) *Chem. Soc. Rev.* **36**, 1369–1380.
80. van Delft, F. L., van Hest, J. C. M., Nolte, R. J. M., Rowan, A. E., Rutjes, F. P. J. T. (2007) *QSAR Comb. Sci.* **26**, 125–128.
81. Harmer, M. A. (1991) *Langmuir* **7**, 2010–2012.
82. Sugawara, T., Matsuda, T. (1994) *Macromolecules* **27**, 7809–7814.

83. Renaudie, L., Narvor, C. L., Lepleux E., Roger, P. (2007) *Biomacromolecules* **8**, 679–685.
84. Puksas, J. E., Chen, Y. (2004) *Biomacromolecules* **5**, 1141–1154.
85. Wu, P., Chen, X., Hu, N., Tam, U. C., Blixt, O., Zetti, A., Bertozzi, C. R. (2008) *Angew. Chem. Int. Ed.* **47**, 5022–5025.
86. Tasis, D., Tagmatarchis, N., Bianco, A., Prato, M. (2006) *Chem. Rev.* **106**, 1105–1136.
87. Kane, R. S., Stroock, A. D. (2007) *Biotechnol. Prog.* **23**, 316–319.
88. Colvin, V. L. (2003) *Nat. Biotechnol.* **21**, 1166–1170.
89. Chen, X., Lee, G. S., Zettl, A., Bertozzi, C. R. (2004) *Angew. Chem. Int. Ed.* **43**, 6111–6116.
90. Chen, X., Tam, U. C., Czlapinski, J. L., Lee, G. S., Rabuka, D., Zettl, A., Bertozzi, C. R. (2006) *J. Am. Chem. Soc.* **128**, 6292–6293.
91. Russo, L., Landi, E., Tampieri, A., Natalello, A., Doglia, S. M., Gabrielli, L., Cipolla, L., Nicotra, F. (2011) *Carb. Res.* **346**, 1564–1568.

Chapter 17

Carbohydrate Multivalent Systems: Synthesis and Therapeutic Opportunities

Jose J. Reina and Javier Rojo

Glycosystems Laboratory, Instituto de Investigaciones Químicas (IIQ) CSIC — Universidad de Sevilla, Americo Vespucio, 49, 41092 Sevilla, Spain

17.1 Introduction

Glycans are not only an important source of metabolic energy; they are also widely expressed as glycoconjugates (glycolipids and glycoproteins) on the surface of cells, forming a dense coat of carbohydrates, named the glycocalix, where they play a key role in many relevant biological processes [1–3]. During the last decade, the new field of chemical glycobiology has remarkably grown due to the increased appreciation for the ubiquity of glycans and their ability to encode chemical information [4,5]. One of the main goals in this field is to explain and understand how this information is encoded in sugar structures and how this information is read out by sugar binding proteins (lectins) [6,7]. Lectins are proteins that recognize glycans with high specificity; these carbohydrate–lectin interactions are implicated in several key biological events, such as cell–cell self-recognition processes, cell–extracellular matrix interactions, gamete fertilization, embryonic development, cell growth, cell differentiation, cell signaling, cell adhesion and migration, apoptosis, immunomodulation and inflammation, host–pathogen interactions, glycoprotein folding and routing, mitogenic induction, and homeostasis [1,8–10]. Thus, in principle, lectins can clearly be considered as potential targets for the development of new drugs. However, they have rarely been exploited for the discovery of novel therapeutic opportunities.

Carbohydrate–protein (specifically, carbohydrate–lectin) interactions are characterized by a high selectivity, metal ion dependence, and a low affinity compensated in nature by a multivalent presentation of glycans and receptors [11,12]. The carbohydrate recognition domains (CRDs) of lectins are often able to recognize complex oligosaccharides in a selective manner, but the affinity of lectins for monovalent carbohydrates is typically weak (dissociation constants are in the mM to the high μM range). However, most lectins are multimeric and, therefore, polyvalent presentations of saccharides acting as binding determinants for a given lectin are used in nature to establish multivalent interactions, increasing the affinity in comparison with the corresponding monovalent ligands [13–15]. In this context, a vast array of unnatural glycoconjugates (neoglycoconjugates) with different valencies and spatial arrangements of the ligand have been constructed (Fig. 17.1) as sugar binding protein antagonists, to prevent or treat diseases caused by pathogens [16] or to interfere in biological processes where carbohydrates are implicated. Scaffolds based on a large variety of structures, such as peptides and proteins [17–19], polymers [20,21], calixarenes [22,23], dendrimers [24,25], cyclodextrins [26–28], cyclopeptides [29], fullerenes [30], metallic nanoparticles, and quantum dots [31,32] provide nanoscale materials with anti-adhesive and cell targeting properties. In this chapter we will provide a brief overview with a selection of different examples based on the most common scaffolds used, including recent advances in the development of carbohydrate multivalent systems and the application of these systems as tools with potential therapeutic or diagnostic use.

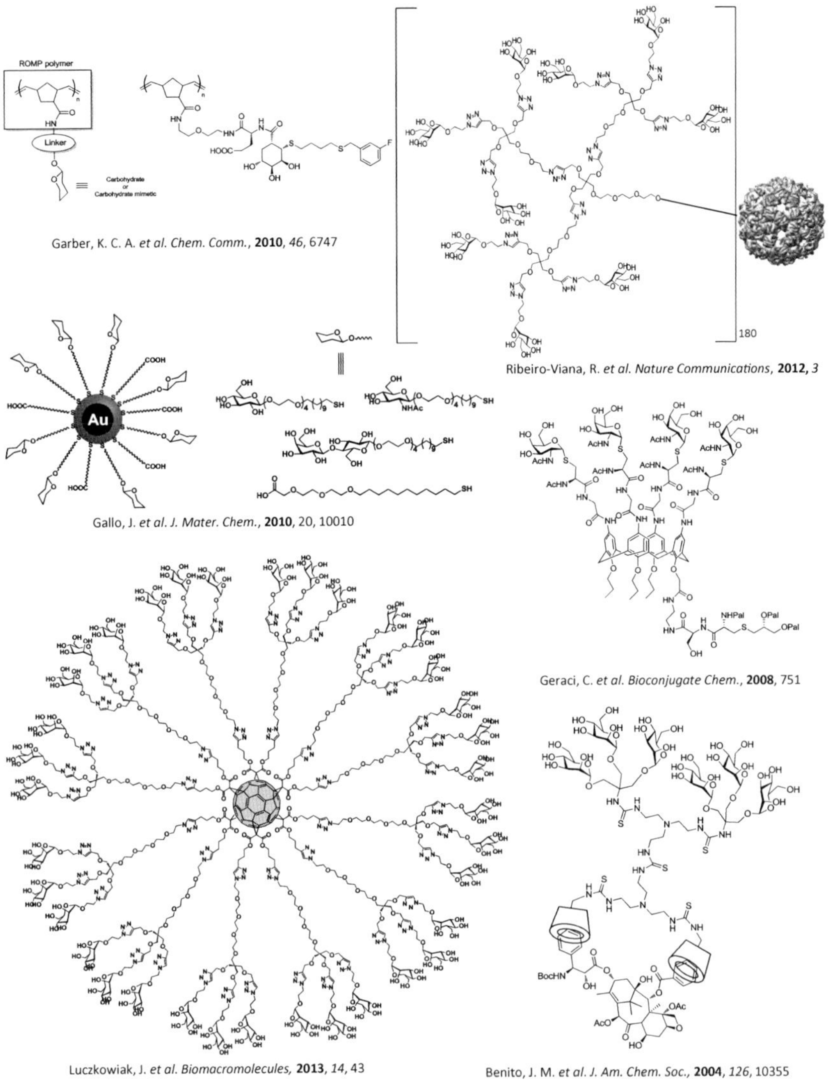

Figure 17.1 A selection of carbohydrate multivalent systems based on different scaffolds.

17.2 Glycopeptides and Glycoproteins

Natural glycoproteins and glycopeptides could be used to understand the vital role that carbohydrates play in nature [33–36] and, *a priori*, they could have potential applications in a clinical context for the development of therapeutics, diagnostics, and vaccines. However, the evaluation of these processes is complicated by the fact that natural expressed glycoproteins often emerge as heterogeneous mixtures [37]. Expression of glycoproteins in mammalian cell culture frequently leads to a mixture of glycoforms. The different properties demonstrated by each glycan within the resulting heterogeneous mixture not only present difficulties for therapeutic glycoproteins [38], but also problems in determining the exact structure–activity relationships. Therefore, synthetic glycoproteins and glycopeptides are good alternatives to obtain homogeneous systems. Several synthetic strategies have been developed for preparing glycoproteins using glycopeptides as model systems. These procedures encompass chemical, enzymatic, and molecular biological approaches, used independently or in combination. Artificial glycopeptides and glycoproteins require synthetic methods involving carbohydrate and peptide chemistry. Central to any of these synthetic strategies is the formation of the glycan–amino acid bond. Several reviews cover the strategies for the preparation of *N*-linked and *O*-linked glycoamino acids [17,39–41]. The resulting glycosylated amino acid is then typically incorporated into standard solid-phase peptide synthesis (SPPS) techniques following a linear assembly strategy. During the past decade, important advances have been achieved for the formation of glycopeptides, using linear and convergent strategies. The former ones mean the introduction of a non-natural amino acid in the peptide sequence using SPPS, followed by a chemoselective reaction to assemble the carbohydrate to the peptide chain. Additionally, these unnatural linkages can be used in mixed strategies where linear assembly and convergent elaborations are combined to provide large glycopeptides and glycoproteins [42]. A variety of methods have been developed for the preparation of unnatural glycan–peptide linkages [42]. Routine SPPS is typically limited to peptides containing less than 50 amino acids, but for glycoprotein synthesis, this technology is insufficient. However, linear coupling of such glycopeptide fragments allow for chain elongation to obtain glycoproteins. Native chemical ligation (NCL) provides a powerful tool for such glycopeptide coupling [43–48]. The process involves a chemoselective reaction of an N-terminal cysteine residue on one peptide with the C-terminal thioester of another. A different strategy to prepare synthetic glycoproteins is the use of chemoselective and site-specific glycosylation once the complete protein scaffold is prepared. One approach is the introduction of different tags, such as thiols of natural cysteine residues, and azides or alkynes in unnatural amino acids (azidohomoalanine (Aha) and homopropargylglycine (Hpg)). This last two tags have been used under the popular copper(I) catalyzed alkyne azide cycladdition (CuAAC) as a chemoselective reaction [49,50]. Synthetic glycopeptides and glycoproteins have potential applications as therapeutic agents, in particular, in the emergent field of synthetic vaccines [51]. Verez-Bencomo and co-workers have developed a glycoconjugate vaccine composed of a fully synthetic capsular polysaccharide antigen of *Haemophylus influenza* type b (HiB) [52], highlighting the fact that access to synthetic complex carbohydrate-based vaccines is feasible.

Recently, a novel strategy of "nested polyvalency" was reported to achieve a controlled design of highly polyvalent glycodendriproteins. The construction of these glycodendriprotein nanoparticles, mimicking the display in target pathogens, was based on the self-assembled (180-copy multimer) sphere-like protein icosahedron Qβ modified by site-directed mutagenesis to introduce a non-natural amino acid tag for the conjugation of glycodendrons via a CuAAC reaction [53,54]. These synthetic protein assemblies display the highest known number of carbohydrates (1,620 mannoses), yet present in a homogeneous manner. The glycoprotein particles showed high antiviral activity in the nanomolar to picomolar range, preventing dendritic cell infection by the Ebola pseudotyped virus through competitive blockage of the dendritic cell-specific intercellular adhesion molecule-3-grabbing non-integrin (DC-SIGN) receptor [53].

A particular example of glycopeptides are the glycocyclopeptides, which represent useful scaffolds for a multivalent presentation of glycans [29]. The first example was described by Renaudet and

Dumy in 2003 [55], using a cyclopeptide scaffold known as regioselectively addressable functionalized templates (RAFTs), composed of an antiparallel beta-sheet constrained by two L-proline–glycine beta-turns and stabilized into a locked and rigid conformation by intramolecular hydrogen bonds. Other cyclopeptides including two D-Proline–L-proline β turn inducers have been developed by Danishefsky and co-workers [56]. Multivalent anchoring sites for carbohydrate units are provided by lysine side chains that are oriented in one or both sites of the platform using an oxime-base conjugation strategy. The cyclopeptide template is prepared combining solid-phase and solution approaches. The key step is the cyclization of linear peptides, readily performed "head-to-tail" by treatment with (benzotriazol-1-yloxy)tripyrrolidinophosphonium hexafluorophosphate (PyBOP) and diisopropylethylamine (DIEA) at high dilution [55]. Additionally, the versatile synthetic methodology to prepare the cyclopeptide platform allows for the introduction of orthogonal functional groups in the amino acid side chains that permit the use of chemoselective ligations, providing a chemically controlled heterotopic presentation of glycocyclopeptides in combination with other molecules like peptides or DNA. An interesting example of this strategy was described using a glycocyclopeptide and a chemoselective ligation strategy (oxime and disulfide bond formation), where four essential components were combined within a single antitumoral synthetic vaccine candidate, producing immunity stimulation against cancer in a MO5/BALB/c tumor mouse model [29].

17.3 Glycopolymers

Glycopolymers are polymers bearing pendant carbohydrates. Nowadays, they have been considered as alternative structures to mimic the presentation of oligosaccharides in nature [5, 57]. They are becoming more and more important to understand biological interactions due to their unique recognition properties that are able to mimic oligosaccharides in binding processes to lectins [58–60]. Because they are synthetically tractable, these surrogates can be modified to optimize a desired activity or to probe a specific biological process [61]. Access to this structural diversity has been made possible by advances in polymer synthesis, which have provided the means to not only synthesize polymers with different structures, but also to control their properties. Accordingly, many polymerization strategies have now been applied to the preparation of glycopolymers. The classical radical polymerization reactions (free radical polymerization (FRP)) have now been complemented by controlled polymerization reactions, such as reversible addition–fragmentation chain-transfer (RAFT) polymerization, nitroxide-mediated polymerization (NMP), and Cu-catalyzed living radical polymerizations, like atom transfer radical polymerizations (ATRPs), transition metal-mediated living radical polymerization (TMM-LRP) and single-electron transfer living radical polymerization (SET-LRP) [62–64]. Non-radical polymerization strategies have also emerged, including the popular ring-opening metathesis polymerization (ROMP), which has been extensively used [65]. These approaches are relatively more tolerant with several functional monomers and, for this reason, they are almost as successful as ionic polymerizations in controlling the kinetic chain length [66]. Another major turning point in polymer science occurred in 2001, when Sharpless and co-workers [67] highlighted the importance of highly efficient organic reactions. These reactions have been called click reactions and, nowadays, around 20 organic reactions are claimed to be potential click reactions. Most of them have pros and cons, but more importantly, they have allowed polymer chemists to prepare tailor-made macromolecular structures in a well-defined manner [68]. Selected click reactions in combination with ionic or radical polymerization techniques have been used to prepare glycopolymers with controlled chain length, composition, and architecture [58].

Accordingly, glycopolymers have been used to mimic polysaccharides (i.e. glycosaminoglycans), glycoproteins, mucins, or even larger entities, such as viral particles or clustered glycans in cell surface microdomains [57, 69]. The research on glycopolymeric drugs mainly focuses on influenza hemagglutinin and neuraminidase inhibitors, human immunodeficiency virus (HIV), and Alzheimer's disease, affecting millions of people around the world [70]. However, the majority of these studies are still in preliminary stages, and extensive research is still required for clinical trials of glycopolymer-based drugs. Nevertheless, glycopolymers have a great potential to be used in the treatment of cancer and infectious diseases.

17.4 Glycodendritic Polymers (Glycodendrimers and Glycodendrons)

In deep contrast to glycopolymers, glycodendrimers and glycodendrons are structurally well-defined homogeneous monodisperse entities. They include a wide range of dendritic architectures that incorporate carbohydrates into their structures. Recently, a complete review of the state-of-the art of glycodendritic structures was published by Roy and Chambre [12]. Dendrimers and dendrons can be synthesized in a stepwise and controlled manner, following a convergent or a divergent strategy, providing increasing "generations" of structures that are homogeneous. The control of multimeric presentation of biologically relevant carbohydrates converts these systems in very interesting and useful multivalent carbohydrate scaffolds. A lot of examples of glycodendrimers with different cores, repetitive units, linkers, and oligosaccharide residues have been published in the literature during the past 25 years. The chemistry used to grow these dendrimer architectures has also been very different, from classical amide, ester, and ether bond formation, to Sonogashira coupling. More recently, the use of click chemistry reactions and, in particular, the CuAAC, have dominated the chemistry of glycodendrimers and glycodendrons [12].

Glycodendritic structures have been extensively used as tools to study and interfere in biological processes where carbohydrate–protein interactions are implicated. A relevant example was the multimeric presentation of high mannose oligosaccharide mimetics in multivalent scaffolds based on 2,2-bis(methylol)propionic acid (Bis-MPA) as building blocks to generate dendritic compounds with Boltorn-type structures, presenting high affinities for DC-SIGN, a lectin implied in the early stage of the infection processes of many pathogens [71]. These compounds were tested *in vitro* using an infection model based on pseudotyped viral Ebola particles expressing the GP1 viral envelope glycoprotein. The inhibitory concentration (IC_{50}) found in these experiments was in the nanomolar range, showing a very strong inhibition effect. Furthermore, the compounds were evaluated as suitable candidates for their development as topical HIV microbicides using a cervical explant HIV infection model, where inhibition activities around 80–90% were found [71].

Furthermore, glycodendrons are very versatile molecules due to the presence of an orthogonal functional group in the focal position that can be conjugated to other multivalent scaffolds like fullerenes, calixarenes, cyclodextrins, peptides, carbon nanotubes, etc. Some examples of these kinds of structures will be described in the next sections.

17.5 Glycocalixarenes and Glycocyclodextrins

Calixarenes, the cyclic oligomers obtained by the condensation of phenols and formaldehyde, are ideal scaffolds for the construction of multivalent glycosylated compounds with unique properties [72]. These carbohydrate multivalent systems have demonstrated an ability to interact with biological macromolecules and, consequently, to be potentially useful in interfering in many pathological phenomena. The valency of the calix[n]arene scaffold can be properly varied ($4 < n < 8$) by choosing the desired dimensions of the macrocycle. An important feature of calix[n]arenes is the dual possibility to keep them conformationally mobile to exploit an induce–fit binding, or, in contrast, to finely lock them in a specific structure more pre-organized for binding. Carbohydrates can be efficiently introduced, through a proper spacer, both at the phenolic oxygen atoms (lower rim) and/or the *para* position (with respect to the hydroxyl (–OH) group) on the anomeric nuclei (upper rim) of calixarenes, giving rise to glycoclusters with quite different characteristics. Currently, the chemistry of calixarenes is, in fact, well established, and different and effective methodologies are known for the introduction of chains, both at the lower and upper rims, of variable length and containing most of the functional groups (aldehyde (–CHO), carboxylic acid (–COOH), amine (–NH_2), *N*-chlorosuccinimide (–NCS), azide (–N_3), alkyl (C–C), etc.) that are frequently used for the conjugation of different ligands. Glycocalixarenes can also carry out other tasks potentially useful in nanomedicine [72]. Glycocalixarenes can form complexes with small hydrophobic molecules in water due to hydrophobic effects inside its lipophilic cavity. Moreover, the functionalization of the upper or lower rims with adequate polar groups also allows for the binding of polar guests, cations or anions [72]. It was demonstrated that glucosylthioureido calixarenes present high binding affinities towards a series of monoamine neurotransmitters (serotonin, norepinephrine, epinephrine, and dopamine) of relevant interest for neurodegenerative pathologies, such as schizophrenia and Parkinson's or Alzheimer's diseases.

A different alternative to load the cargo is to use a covalent binding, which takes advantage of the well-established synthetic methodologies devised mainly at the end of the last century, for the selective functionalization of calixarene scaffolds. The connections of the cargo should preferentially take place at the rim opposite to the one bearing the saccharide units in order to avoid disturbing protein–carbohydrate interactions. An interesting example of such an approach is represented by an anticancer vaccine candidate [73]. A cone calix(4)arene containing four tumor associated antigens (S-Tn) at the upper rim was covalently linked to the immunoadjuvant P_3CS (trihexadecanoyl-S-glycerylcysteinyl-serine) at the opposite lower rim. This compound presented a four-fold increase in antibody production in comparison with the monovalent analog exposing the same immunoadjuvant.

Cyclodextrins (CDs) are naturally occurring cyclic oligosaccharides derived from starch, composed of six (α), seven (β), or eight (γ) α(1-4)-linked glucoses. A variety of multivalent glycoconjugates with diverse architectures (CD–glycodendrimer conjugates, position selective CD–bis conjugates, face-selective glycoclusters, dual-face glycoclusters, and hyperbranched glycoclusters) have become accessible through appropriate ligation chemistries [28].

One of the most popular is the thiourea-forming coupling reaction between isothiocyanate-armed glycodendrons and mono-(C-6)-amino βCD. More recently, Seeberger and co-workers implemented the CuAAC reaction for the conjugation of an azide-armed glycodendrimer and a monosubstituted βCD derivative equipped with a terminal alkyne group [74].

Like calixarenes, cyclodextrins present an inner nanometric cavity in the cyclooligosaccharide core that remains essentially unaltered after sugar conjugation. It has been demonstrated that this cavity preserves the inclusion capabilities of CDs profusely exploited by the pharmaceutical industry for drug encapsulation purposes [28]. By taking advantage of the inclusion capabilities of CDs, receptor-targeted molecular delivery can be achieved through the guest–multivalent glyco CD–lectin ternary complex formation, provided that the corresponding supramolecular processes exhibit appropriate equilibrium constants. The magnitude of apparent stability constants for several drug–CD complexes ranges from 0 to 105 M^{-1}, the value of $K_{1:1}$ being most often between 50 and 2000 M^{-1} [75, 76]. Higher constants may lead to drug sequestering [77]. Regarding the affinity towards therapeutically useful lectins, association constants in the μM to nM range, which can be generally achieved by exploiting multivalency, may warrant efficient by eliciting receptor-mediated internalization mechanisms [78]. A particular example was described by Garcia-Fernandez and co-workers, wherein a molecular device was designed and purposely conceived to complex animitotic drug docetaxel (Taxotère®), consisting of a ditopic βCD host and an appended hexavalent mannose glycodendrimer. The systems were shown to bind specifically at the membrane of mouse alveolar nacrophages via the macrophage mannose receptor [79]. Although hydrophilic cyclodextrins are considered non-toxic at low to moderate oral dosages, parenteral administration may involve toxicity issues that need to be investigated in depth. In an additional level of complexity, chemical tailoring can promote the formation of glycomicelles, films, vesicles, or nanoparticles for sensing, diagnosing, or for biomacromolecule transport [28]. New developments will certainly be arising in the near future from the intimate merging of multivalency, supramolecular chemistry, and nanotechnology through the cyclodextrin connection.

17.6 Glycofullerenes and Sweet Carbon Nanotubes

Fullerene can be considered as a very attractive core with a rigid spherical shape for a unique globular carbohydrate multivalent presentation. However, its low solubility in polar solvents is one of the major drawbacks to address the preparation of biological active compounds. On the other hand, fullerenes allow the possibility to functionalize its three-dimensional (3D) structures in different positions of the C60 cage in a controlled fashion. Initially, the approaches reported in the literature to synthesize fullerene derivatives bearing carbohydrates were based on a monoaddition pattern [80,81] on the fullerene sphere or on a "one side" multi-addition pattern [82]. In all these cases, fullerene derivatives showed an amphiphilic character, with the fullerene being the lipophilic part of the structures. Recently, it has been shown that this amphiphilic character of sugar fullerene conjugates can be avoided by obtaining hexakis-adducts of [60]fullerene in which the C60 sphere could be completely surrounded by the sugar moieties in a T-symmetrical octahedral addition pattern [30]. The conjugation of sugar residues to an alkyne-substituted Bingel–Hirsch hexakis-adduct was performed by employing the CuAAC methodology. So, the first example of a globular polytopic ligand was obtained and named glycofulleres or "sugar balls" [30].

To increase the valency of these systems, a divalent dendron with mannose was used to obtain the first glycodendrofullerene with 24 sugar moieties on the periphery of the C60 [83]. These compounds were water soluble and interacted with Concanavalin A in a multivalent manner, demonstrating the accessibility of the sugars on the fullerene surface to be recognized by a lectin. Moreover, the compounds showed a good stability and low toxicity, which confirmed the appropriate features to be used in cellular assays.

Globular glycofullerenes with higher valency were prepared by a convergent strategy, combining trimannosylated glycodendrons with a Bingel–Hirsch hexakis-adduct [60] fullerene to obtain glycodendrofullerenes with 36 mannoses. Also, the biological activity of these compounds in a cellular infection model was introduced for the first time, providing important information about the potential use of these new glycomimetics as viral inhibitors. These glycodendrofullerenes act as antiviral agents in a DC-SIGN dependent Ebola pseudotype infection model [84]. Using a glycodendrofullerene with 36 mannoses, including a long spacer, a remarkable increase of the inhibitory activity of this compound (IC_{50} value of 0.3 μM) was observed, probably due to an efficient interaction of the mannose ligand with DC-SIGN. This result highlights the importance of combining an adequate scaffold to achieve the multivalency (the spherical fullerene) with the adequate ligand accessibility and flexibility. The valency of the compound is an important factor to obtain good affinities in a carbohydrate–lectin interaction, but as it has been shown in this work, it is not the only factor to be taken into account. Based on these results, fullerenes appear as very attractive scaffolds for a globular multivalent presentation of sugars.

Other types of carbon nanostructures have been functionalized with sugars for carbohydrate multivalent presentations with potential applications in biosensing and biomedicine. The use of carbohydrate conjugates, such as glycoproteins, glycolipids, and glycans to modify carbon nanostructures, will largely improve the biocompatibility of carbon nanotubes (CNTs) and graphene, and thus, will open the way for the application of these systems in biosensing, bioimaging, and drug delivery [85–87]. The sweet carbon nanostructures could be prepared following two different approaches: covalent and non-covalent conjugation with carbohydrates. One of the most commonly used covalent functionalizations of CNTs involves reactions with carboxylic groups on CNT sidewalls and ends, generated by strong oxidation. Then, many approaches would be used to functionalize CNTs with glycoconjugates: direct amination, Staudinger ligation, etc. Besides the use of carboxylic groups on CNTs, azomethine ylide 1,3-dipolar cycloaddition has also been reported for glycoconjugation [88]. Galactose functionalized CNTs have been prepared for the capture of pathogenic *Escherichia coli* cells in solution through specific carbohydrate–lectin interactions [89]. Similarly, aggregation of *Bacillus anthracis* spores (anthrax) was achieved using mannose-coated CNTs for anti-bioterrorism applications [90]. The synthesis of hetero-functionalized CNT-based glycoconjugates, incorporating both galactose and mannose residues, has also attracted interest and their interaction with pathogenic *E. coli* [89] and *B. anthracis* were studied [90]. Moreover, the introduction of these two carbohydrate epitopes led to a better specificity in the recognition process of bacteria.

In contrast to covalent functionalization, non-covalent functionalization can preserve the intrinsic electronic properties of carbon nanostructures, while bringing the desired functionality [91]. Carbohydrates conjugated with pyrene or porphyrin functional tails can be attached to the graphitic side walls of the CNT surfaces through π–π stacking through the aromatic moieties [85]. Glycopolymers and glycolipids have been attached to CNTs by hydrophobic and π–π interactions [85]. These kinds of non-covalent glyconanotubes were recently applied in the preparation of glycoconjugate/carbon nanotube field-effect transistor (NTFET) devices. Also, multivalent glycodendrons with a pyrene hydrophobic tag were used for the non-covalent functionalization of CNTs [92]. The cytotoxicity of CNTs towards live cells was drastically decreased offering potential applications in targeted drug delivery systems or for the design of biosensors [86].

17.7 Metallic Glyconanoparticles

Metallic glyconanoparticles (GNPs) are polyvalent systems based on metallic cores, such as gold, iron oxide, or semiconductor metals coated with carbohydrate molecules. The gold polyvalent systems are based on self-assembled monolayers (SAMs) of thiol-ending glycoconjugates bound to gold clusters through gold–sulfur (Au–S) bonds. GNPs present a multivalent array of carbohydrates and provide a glycocalix-like surface with globular carbohydrate display and a chemically well-defined composition. There are different methodologies to prepare metallic and, in particular, gold nanoparticles. The synthetic method based on a one-step protocol described by Brust *et al.* allows the preparation of GNPs with a varying percentage of carbohydrates in a controlled fashion ("direct" method) [93]. But the direct synthetic method presents limitations (control over the core size or sensitivity to reaction conditions of ligands) that can be overcome by multistep strategies generally involving chemical modification of previously prepared nanoparticles. One of these strategies is based on the so-called ligand place exchange (LPE) reaction [94] on pre-formed gold nanoparticles with thiol-ending ligands, in order to displace the protection shell (citrates, phosphines, amines, or thiols) on the gold surface by insertion of other ligands. The direct method allows the straightforward preparation of multifunctional nanoparticles, incorporating a different carbohydrate density and other molecules of interest (peptides, fluorescent probes, nucleic acids, magnetic elements, radionuclides, etc.). The first synthetic carbohydrate vaccine based on gold nanoparticles was obtained following this methodology [95]. Gold nanoparticles were coated with a tetrasaccharide (TetraPn), a repeating unit of the *Streptococcus pneumonia* type 14 capsular polysaccharide (Pn14PS), a conjugate of the T-Helper ovalbumin ($OVA_{323-339}$), and 5-mercaptopentyl β-D-glucopyranoside (Glc) in variable ratios. These nanoparticles triggered specific anti-Pn14PS immunoglobulin G (IgG) antibodies, proving that gold nanoparticles have great potential as carriers for the development of a huge diversity of fully synthetic carbohydrate-based vaccines. Metallic glyconanoparticles are very versatile multivalent systems; by changing the organic shell and/or the composition of the metallic nucleus, it is possible to obtain GNPs with magnetic and luminescent properties. Preparation of sugar-coated iron oxide and semiconductor quantum dots (QDs) nanoparticles, as well as their application as probes in magnetic resonance imaging (MRI) and optical imaging, has been recently revised [96]. New methodologies to convert gold nanoparticles into magnetic ones have been developed in the past decade. One of them consists of the introduction of a gadolinium (Gd) complex in the organic shell to obtain paramagnetic glyconanoparticles. A second strategy implies the modification of gold cores by introducing magnetic elements (iron, cobalt, or manganese) in order to produce superparamagnetic GNPs [32]. Paramagnetic Gd-based GNPs with different sugars and 5% of thiol-ending tetraazocyclododecane triacetic (DO_3A) Gd(III)-chelating molecules have been prepared and tested as MRI positive contrast agents in mouse glioma. Another recent example was the preparation of glycoferrite nanoparticles (superparamagnetic water-soluble core-shell XFe_2O_4 (where X = iron (Fe), manganese (Mn), or cobalt (Co)) nanoparticles) with different carbohydrates and carboxylic linkers and functionalized with a fluorescent dye (Texas Red). Additionally, protein G was conjugated to attach well oriented antibodies to obtain immuno-fluorescent-magnetic glycoferrites [97,98]. These multimodal GNPs have allowed for the specific labeling of peripheral blood mononuclear cells (PBMCs) present in blood in a percentage below 2% by fluorescent microscopy and by magnetic resonance imaging [97]. The study of their uptake, the intracellular fate by confocal microscopy [99], and the *in vivo* tracking of the migration of endogenous neural progenitors in response to an induced tumor have been carried out by MRI.

17.8 Conclusions

The need of carbohydrate multivalent systems as tools to investigate the role of carbohydrates in many relevant biological processes has led to the synthesis of a plethora of compounds based on different scaffolds. It is evident that the lack of a rational design for these multivalent systems due to the difficulty, in most of the cases, to define the spatial location in the nature of ligands and the corresponding receptors, prevents the preparation of ideal systems. Therefore, research in this topic is still required to achieve successful compounds that can be applied in biomedicine in a near future.

Acknowledgments

We would like to thank the financial support by MINECO of Spain CTQ2011-23410 and the EU RTN CARMUSYS (PITN-GA-2008-213592). JJR thanks the CSIC for a JAE-Doc contract.

References

1. Varki, A. (1993) *Glycobiology* **3**, 97–130.
2. Kamerling, J. P. (2007) *Comprehesive Glycoscience: From Chemistry to System Biology* (Volumes 3 and 4) (Amsterdam).
3. Werz, D. B. and Seeberger, P.H. (2007) In *Chemical Biology: From Small Molecules to Systems Biology and Drug Design*. Schreiber, S. L., T. M. K., Wess, G. Eds. Wiley-VCH (Weinheim). pp. 668–691.
4. Varki, A. C., R., Esko, J., Hart, G., Marth, J. Eds (1999) *Essentials of Glycobiology* **1999**. Cold Spring Harbor Laboratory Press: Cold Spring Harbor (New York)
5. Bertozzi, C. R., Kiessling, L. L. (2001) *Science* **291**, 2357–2364.
6. Gabius, H. J., Andre, S., Jimenez-Barbero, J., Romero, A., Solis, D. (2011) *Trends Biochem. Sci.* **36**, 298–313.
7. Gupta, G., Surolia, A. (2012) *Adv. Exp. Med. Biol.* **749**, 1–13.
8. Weis, W. I., Taylor, M. E., Drickamer, K. (1998) The C-type lectin superfamily in the immune system. *Immunol. Rev.* **163**, 19–34.
9. Sharon, N., Lis, H. (2004) *Glycobiology* **14**, 53R–62R.
10. Ghazarian, H., Idoni, B., Oppenheimer, S. B. (2011) *Acta Histochem.* **113**, 236–247.
11. Lee, Y. C., Lee, R. T. (1995) *Acc. Chem. Res.* **28**, 321–327.
12. Chabre, Y. M., Roy, R. (2010) in *Advances in Carbohydrate Chemistry and Biochemistry* **63**, Horton, D. Ed. Elsevier Academic Press Inc. (San Diego). pp. 165–393.
13. Mammen, M., Choi, S. K., Whitesides, G. M. (1998) *Angew. Chem. Int. Edit.* **37**, 2755–2794.
14. Culf, A. S., Cuperlovic-Culf, M., Ouellette, R. J. (2006) *Omics* **10**, 289–310.
15. Turnbull, J. E., Field, R. A. (2007) *Nat. Chem. Biol.* **3**, 74–77.
16. Bernardi, A., Jimenez-Barbero, J., Casnati, A., De Castro, C., Darbre, T., Fieschi, F., Finne, J., Funken, H., Jaeger, K. E., Lahmann, M., Lindhorst, T. K., Marradi, M., Messner, P., Molinaro, A., Murphy, P. V., Nativi, C., Oscarson, S., Penades, S., Peri, F., Pieters, R. J., Renaudet, O., Reymond, J. L., Richichi, B., Rojo, J., Sansone, F., Schaffer, C., Turnbull, W. B., Velasco-Torrijos, T., Vidal, S., Vincent, S., Wennekes, T., Zuilhof, H., Imberty, A. (2013) *Chem. Soc. Rev.* **42**, 4709–4727.
17. Davis, B. G. (2002) *Chem. Rev.* **102**, 579–602.
18. Payne, R. J., Wong, C. H. (2010) *Chem. Commun.* **46**, 21–43.
19. Westerlind, U. (2012) *Beilstein J. Org. Chem.* **8**, 804–818.
20. Miura, Y. (2012) *Polym. J.* **44**, 679–689.

21. Eissa, A. M., Cameron, N. R. (2013) in *Bio-Synthetic Polymer Conjugates* **253**. Schlaad, H. Ed. Springer-Verlag (Berlin). pp. 71–114.
22. Baldini, L., Casnati, A., Sansone, F., Ungaro, R. (2007) *Chem. Soc. Rev.* **36**, 254–266.
23. Dondoni, A., Marra, A. (2010) *Chem. Rev.* **110**, 4949–4977.
24. Li, Y., Cheng, Y., Xu, T. (2007) *Curr. Drug Discov. Tech.* **4**, 246–254.
25. Chabre, Y. M., Roy, R. (2008) *Curr. Top. Med. Chem.* **8**, 1237–1285.
26. Fulton, D. A., Stoddart, J. F. (2001) *Bioconjugate Chem.* **12**, 655–672.
27. Gomez-Garcia, M., Benito, J. M., Butera, A. P., Mellet, C. O., Fernandez, J. M. G., Blanco, J. L. J. (2012) *J. Org. Chem.* **77**, 1273–1288.
28. Martinez, A., Ortiz Mellet, C., Garcia Fernandez, J. M. (2013) *Chem. Soc. Rev.* **42**, 4746–4773.
29. Galan, M. C., Dumy, P., Renaudet, O. (2013) *Chem. Soc. Rev.* **42**, 4599–4612.
30. Nierengarten, J. F., Iehl, J., Oerthel, V., Holler, M., Illescas, B. M., Munoz, A., Martin, N., Rojo, J., Sanchez-Navarro, M., Cecioni, S., Vidal, S., Buffet, K., Durka, M., Vincent, S. P. (2010) *Chem. Commun.* **46**, 3860–3862.
31. El-Boubbou, K., Huang, X. F. (2011) *Curr. Med. Chem.* **18**, 2060–2078.
32. Marradi, M., Chiodo, F., Garcia, I., Penades, S. (2013) *Chem. Soc. Rev.* **42**, 4728–4745.
33. Wong, S. Y. C. (1995) *Curr. Opin. Struct. Biol.* **5**, 599–604.
34. Yarema, K. J., Bertozzi, C. R. (1998) *Curr. Opin. Chem. Biol.* **2**, 49–61.
35. Davis, B. G. (1999) *J. Chem. Soc. –Perkin Transactions* **1**, 3215–3237.
36. Lloyd, R. C., Davis, B. G., Jones, J. B. (2000) *Bioorg. Med. Chem.* **8**, 1537–1544.
37. Rademacher, T. W., Parekh, R. B., Dwek, R. A. (1988) *Annu. Rev. Biochem.* **57**, 785–838.
38. Bill, R. M., Flitsch, S. L. (1996) *Chem. Bio.* **3**, 145–149.
39. Buskas, T., Ingale, S., Boons, G. J. (2006) *Glycobiology* **16**, 113R–136R.
40. Arsequell, G., Valencia, G. (1997) *Tetrahedron Asymm.* **8**, 2839–2876.
41. Arsequell, G., Valencia, G. (1999) *Tetrahedron Asymm.* **10**, 3045–3094.
42. Gamblin, D. P., Scanlan, E. M., Davis, B. G. (2009) *Chem. Rev.* **109**, 131–163.
43. Bang, D., Pentelute, B. L., Kent, S. B. H. (2006) *Angew. Chem. Int. Ed.* **45**, 3985–3988.
44. Dawson, P. E., Muir, T. W., Clarklewis, I., Kent, S. B. H. (1994) *Science* **266**, 776–779.

45. Shin, Y., Winans, K. A., Backes, B. J., Kent, S. B. H., Ellman, J. A., Bertozzi, C. R. (1999) *J. Am. Chem. Soc.* **121**, 11684–11689.
46. Macmillan, D. (2006) *Angew. Chem. Int. Ed.* **45**, 7668–7672.
47. Dawson, P. E., Kent, S. B. H. (2000) *Annu. Rev. Biochem.* **69**, 923–960.
48. Yeo, D. S. Y., Srinivasan, R., Chen, G. Y. J., Yao, S. Q. (2004) *Chem. — Eur. J.* **10**, 4664–4672.
49. Tornoe, C. W., Christensen, C., Meldal, M. (2002) *J. Org. Chem.* **67**, 3057–3064.
50. Rostovtsev, V. V., Green, L. G., Fokin, V. V., Sharpless, K. B. (2002) *Angew. Chem. Int. Ed.* **41**, 2598–2599.
51. Peri, F. (2013) *Chem. Soc. Rev.* **42**, 4543–4556.
52. Verez-Bencomo, V., Fernandez-Santana, V., Hardy, E., Toledo, M. E., Rodriguez, M. C., Heynngnezz, L., Rodriguez, A., Baly, A., Herrera, L., Izquierdo, M., Villar, A., Valdes, Y., Cosme, K., Deler, M. L., Montane, M., Garcia, E., Ramos, A., Aguilar, A., Medina, E., Torano, G., Sosa, I., Hernandez, I., Martinez, R., Muzachio, A., Carmenates, A., Costa, L., Cardoso, F., Campa, C., Diaz, M., Roy, R. (2004) *Science* **305**, 522–525.
53. Ribeiro-Viana, R., Sanchez-Navarro, M., Luczkowiak, J., Koeppe, J. R., Delgado, R., Rojo, J., Davis, B. G. (2012) *Nat. Commun.* **3**, 1–8.
54. Kozlovska, T. M., Cielēns, I., Dreilinša, D., Dišlers, A., Baumanis, V., Ose, V., Pumpēns, P. (1993) *Gene* **137**, 133–137.
55. Renaudet, O., Dumy, P. (2003) *Org. Lett.* **5**, 243–246.
56. Krauss, I. J., Joyce, J. G., Finnefrock, A. C., Song, H. C., Dudkin, V. Y., Geng, X., Warren, J. D., Chastain, M., Shiver, J. W., Danishefsky, S. J. (2007) *J. Am. Chem. Soc.* **129**, 11042–11044.
57. Kiessling, L. L., Gestwicki, J. E., Strong, L. E. (2006) *Angew. Chem. Int. Ed.* **45**, 2348–2368.
58. Spain, S. G., Gibson, M. I., Cameron, N. R. (2007) *J. Polym. Sci. Part A: Polym. Chem.* **45**, 2059–2072.
59. Ting, S. R. S., Chen, G., Stenzel, M. H. (2010) *Polym. Chem.* **1**, 1392–1412.
60. Slavin, S., Burns, J., Haddleton, D. M., Becer, C. R. (2011) *Eur. Polym. J.* **47**, 435–446.
61. Kiessling, L. L., Gestwicki, J. E., Strong, L. E. (2000) *Curr. Opin. Chem. Biol.* **4**, 696–703.
62. Ouchi, M., Terashima, T., Sawamoto, M. (2009) *Chem. Rev.* **109**, 4963–5050.
63. Moad, G., Rizzardo, E., Thang, S. H. (2006) *Aust. J. Chem.* **59**, 669–692.
64. Hawker, C. J., Bosman, A. W., Harth, E. (2001) *Chem. Rev.* **101**, 3661–3688.

65. Kiessling, L. L., Grim, J. C. (2013) *Chem. Soc. Rev.* **42**, 4476–4491.
66. Moad, G., Rizzardo, E., Thang, S. H. (2008) *Accounts Chem. Res.* **41**, 1133–1142.
67. Kolb, H. C., Finn, M. G., Sharpless, K. B. (2001) *Angew. Chem. Int. Ed.* **40**, 2004–2021.
68. Kempe, K., Krieg, A., Becer, C. R., Schubert, U. S. (2012) *Chem. Soc. Rev.* **41**, 176–191.
69. Mammen, M., Choi, S.-K., Whitesides, G. M. (1998) *Angew. Chem. Int. Ed.* **37**, 2754–2794.
70. De Clercq, E. (2006) *Nat. Rev. Drug. Discov.* **5**, 1015–1025.
71. Reina, J. J., Bernardi, A. (2012) *Mini-Rev. Med. Chem.* **12**, 1434–1442.
72. Sansone, F., Casnati, A. (2013) *Chem. Soc. Rev.* **42**, 4623–4639.
73. Geraci, C., Consoli, G. M. L., Galante, E., Bousquet, E., Pappalardo, M., Spadaro, A. (2008) *Bioconjugate Chem.* **19**, 751–758.
74. Bernardes, G. J. L., Kikkeri, R., Maglinao, M., Laurino, P., Collot, M., Hong, S. Y., Lepenies, B., Seeberger, P. H. (2010) *Org. Biomol. Chem.* **8**, 4987–4996.
75. Challa, R., Ahuja, A., Ali, J., Khar, R. K. (2005) *AAPS PharmSciTech* **6**, E329–E357.
76. Naguib, M. (2007) *Anesthesia and Analgesia* **104**, 575–581.
77. Bom, A., Bradley, M., Cameron, K., Clark, J. K., van Egmond, J., Feilden, H., MacLean, E. J., Muir, A. W., Palin, R., Rees, D. C., Zhang, M. Q. (2002) *Angew. Chem. Int. Ed.* **41**, 265–270.
78. Gabor, F., Bogner, E., Weissenboeck, A., Wirth, M. (2004) *Adv. Drug Delivery Rev.* **56**, 459–480.
79. Benito, J. M., Gomez-Garcia, M., Mellet, C. O., Baussanne, I., Defaye, J., Fernandez, J. M. G. (2004) *J. Am. Chem. Soc.* **126**, 10355–10363.
80. Nishida, Y., Mizuno, A., Kato, H., Yashiro, A., Ohtake, T., Kobayashi, K. (2004) *Chem. Biodiv.* **1**, 1452–1464.
81. Enes, R. F., Tome, A. C., Cavaleiro, J. A. S., El-Agamey, A., McGarvey, D. J. (2005) *Tetrahedron* **61**, 11873–11881.
82. Isobe, H., Cho, K., Solin, N., Werz, D. B., Seeberger, P. H., Nakamura, E. (2007) *Org. Lett.* **9**, 4611–4614.
83. Sanchez–Navarro, M., Munoz, A., Illescas, B. M., Rojo, J., Martin, N. (2011) *Chem. — Eur. J.* **17**, 766–769.
84. Luczkowiak, J., Munoz, A., Sanchez-Navarro, M., Ribeiro-Viana, R., Ginieis, A., Illescas, B. M., Martin, N., Delgado, R., Rojo, J. (2013) *Biomacromolecules* **14**, 431–437.
85. Chen, Y., Star, A., Vidal, S. (2013) *Chem. Soc. Rev.* **42**, 4532–4542.

86. Gorityala, B. K., Ma, J. M., Wang, X., Chen, P., Liu, X. W. (2010) *Chem. Soc. Rev.* **39**, 2925–2934.
87. Bandaru, N. M., Voelcker, N. H. (2012) *J. Mat. Chem.* **22**, 8748–8758.
88. Hong, S. Y., Tobias, G., Al-Jamal, K. T., Ballesteros, B., Ali-Boucetta, H., Lozano-Perez, S., Nellist, P. D., Sim, R. B., Finucane, C., Mather, S. J., Green, M. L. H., Kostarelos, K., Davis, B. G. (2010) *Nat. Mat.* **9**, 485–490.
89. Gu, L. R., Elkin, T., Jiang, X. P., Li, H. P., Lin, Y., Qu, L. W., Tzeng, T. R. J., Joseph, R., Sun, Y. P. (2005) *Chem. Commun.* 874–876.
90. Wang, H. F., Gu, L. R., Lin, Y., Lu, F. S., Meziani, M. J., Luo, P. G. J., Wang, W., Cao, L., Sun, Y. P. (2006) *J. Am. Chem. Soc.* **128**, 13364–13365.
91. Zhao, Y. L., Stoddart, J. F. (2009) *Acc. Chem. Res.* **42**, 1161–1171.
92. Chen, Y. A., Star, A., Vidal, S. (2013) *Chem. Soc. Rev.* **42**, 4532–4542.
93. Brust, M., Walker, M., Bethell, D., Schiffrin, D. J., Whyman, R. (1994) *Chem. Commun.* 801–802.
94. Hostetler, M. J., Templeton, A. C., Murray, R. W. (1999) *Langmuir* **15**, 3782–3789.
95. Safari, D., Marradi, M., Chiodo, F., Dekker, H. A. T., Shan, Y. L., Adamo, R., Oscarson, S., Rijkers, G. T., Lahmann, M., Kamerling, J. P., Penades, S., Snippe, H. (2012) *Nanomedicine* **7**, 651–662.
96. Marradi, M., García, I., Penadés, S. (2011) in *Progress in Molecular Biology and Translational Science* **104**. Antonio, V. Ed. Academic Press (London). pp. 141–173.
97. Gallo, J., Garcia, I., Genicio, N., Padro, D., Penades, S. (2011) *Biomaterials* **32**, 9818–9825.
98. Garcia, I., Gallo, J., Genicio, N., Padro, D., Penades, S. (2011) *Bioconjugate Chem.* **22**, 264–273.
99. Gallo, J., Genicio, N., Penades, S. (2012) *Adv. Healthcare Mater.* **1**, 302–307.

Chapter 18
Antitubercular Drugs Based on Carbohydrate Derivatives

Vishwanath Gaitonde and Steven J. Sucheck

Department of Chemistry, The University of Toledo, 2801 W. Bancroft Street, Toledo, Ohio 43606, USA

18.1 *Mycobacterium tuberculosis*

18.1.1 *Introduction*

Discovered in 1882 by Dr. Robert Koch, *Mycobacterium tuberculosis* (*MTB*) is the causative agent of tuberculosis (TB). TB is one of the leading causes of death due to an infectious disease. In 2012, 8.6 million people developed TB, and of these, 1.3 million people died, which demonstrates the danger of this pathogenic bacillus [1]. Streptomycin, an aminoglycoside identified in the forties, was the first antibiotic to be recognized as a validated anti-TB drug demonstrating potent inhibition of *MTB* with comparatively low toxicity [2–4]. The rapid development of drug resistant strains led to the introduction of assorted anti-TB drugs, such as para-aminosalicylic acid, isoniazid, pyrazinamide, ethambutol, rifampin, etc., as first-line TB treatments, subsequently followed by second-line treatment for multi drug resistant (MDR) and extensive drug resistant (XDR) strains [5–15].

18.1.2 *Carbohydrates as anti-tuberculosis drugs*

Carbohydrates are ubiquitous compounds, most commonly known for providing energy for working and stockpiled as glycogen. Chemical and biological inquiries over the last few decades have established the association of carbohydrates with various pathogenic conditions and have identified glycan(s) as untapped tools in drug discovery [16–19]. An excellent review on the synthesis of carbohydrate-based therapeutics has been reported by Cipolla and coworkers [20]. Development of the carbohydrate-based drug discovery can be categorized by one of the following approaches: (i) producing native glycans with desired biological activity; (ii) structural modifications via glycomimetic approaches; (iii) generating glyco-scaffolds for achieving carbohydrate diversity; and (iv) developing glyco-fused therapeutics [20].

18.1.3 *Cell wall of MTB*

The cell wall of *MTB* is a sturdy structure essential for the organism's viability in the host [21,22]. The cell wall consists of an outer and an inner layer (Fig. 18.1). The inner layer consists of a plasma membrane associated with membrane proteins. The key cell wall components consist of a covalently attached complex of mycolic acids, D-arabinan, and D-galactan, known as the mycolyl arabinogalactan (mAG). The arabinogalactan (AG) portion of the mAG is made up of attached furanoses, and the arabinose and galactose residues, linked in a linear alternating pattern, consisting of 30 units each. The mAG glycosidic linkages are formed via the β-(1→5) and β-(1→6) of the galactan fragment, and the α-(1→3), α-(1→5), and β-(1→2) of the arabinan residues. The mycolic acid portion caps the arabinan domain, completing the mAG [23, 24]. The mAG is further covalently linked to peptidoglycan (PG) via a phosphoryl-*N*-acetylglucosaminosyl-rhamnose linker, and is commonly referred to as the mycolyl arabinogalactan peptidoglycan (mAGP) complex [25–28]. The PG layer consists of alternating units made of *N*-acetylglucosamine (GlcNAc) and modified glycosylmuramic acid [26, 29, 30].

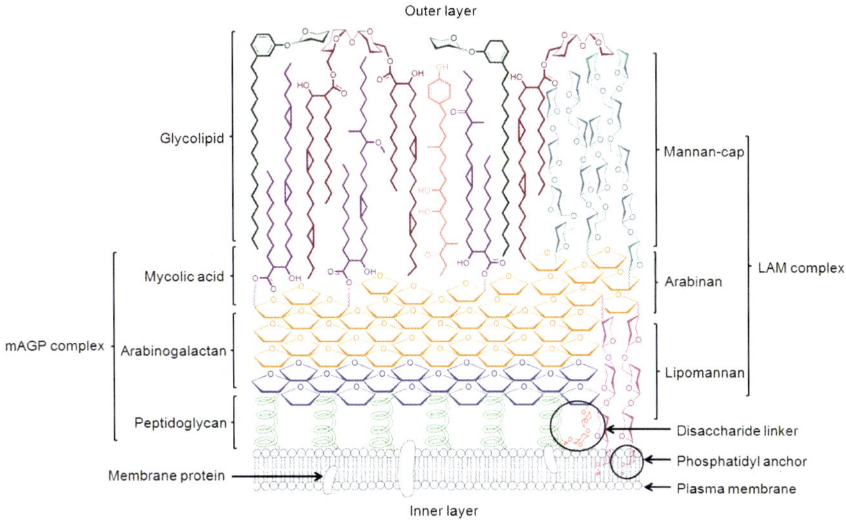

Figure 18.1 *Mycobacterium tuberculosis* cell wall.

The arabinofuranosyl (Ara*f*) residues are placed at the non-reducing end of the AG domain, two-thirds of which are esterified with mycolic acid (alpha-, methoxy-, and keto mycolates), of the outer membrane [31–34]. These attached mycolic acid moieties consist of long β-hydroxy chains with a short α-alkyl side chain. The alpha-mycolic acid makes up 70% of the mycolyl fatty acid residues, with methoxy- and keto-mycolic acid forming ~10–15% each [35,36]. The mycolic acids chains are further complexed with free glycolipids like trehalose monomycolate (TMM), trehalose dimycolate (TDM, cord factor), sulfolipids, phosphotidylinositol mannoside (PIM), phenolic glycolipids (PGLs), phthiocerol dimycocerosate (PDIM), and lipoarabinomannan (LAM) [29,37]. The integrity of these lipids with mycolic acids aligned perpendicular to the plasma membrane implicitly provides a resilient lipid barrier [38].

The LAM, also known as a glycosylphosphoinositide (GPI), is made up of three domains. These are the PIM anchor, and the D-mannan and D-arabinan polysaccharide backbones [39]. LAMs are synthesized using mannosyltransferases (ManTs), via the addition of mannose residues on to phosphoinositol (PI) to form PIM and lipomannan (LM) [40, 41]. Glycosylation of the LM segment with arabinan residues produces the LAM complex. The terminal arabinan residues of the LAM are further glycosylated with a mannose, the "mannose capping domain" giving rise to the mannosylated LAM (ManLAM) motif [39]. The LAM is non-covalently attached to the plasma membrane via an acylated PI moiety rooted in the membrane [38,42–45]. LAM represents an important component of the cell wall and is known to possess immunomodulatory activity, and together with the LM, PIM and ManLAM, plays a crucial role in cell wall integrity [46–49]. There is an ongoing quest to understand the mechanism of the ManLAM as an immunosuppressant [50–57]. Recent studies have demonstrated that the mannan core branching of the LM and/or LAM capping is relevant to mycobacterial pathogenesis and immunity [58]. The detailed structure and description of the biosynthesis of the LAM has been extensively reviewed [39, 46, 59]. Briefly, the biosynthetic pathway, from PI→PIM→LM→LAM→ManLAM, is catalyzed by a discrete set of enzymes, and improved understanding of these processes is growing with the assistance of designed substrates, inhibitors, and new enzyme assays.

18.2 Glycans in *Mycobacterium tuberculosis*

18.2.1 *Carbohydrate processing enzymes*

Glycans play a central role in various biological processes by promoting cell–cell interaction, recognition, adhesion, division, and tissue development in complex organisms [60–62]. The organization of saccharides to generate the complex array of known glycans has overwhelmed researchers. This is evident from theoretical calculations suggesting the possibility of 10^{12} isomers of linear and branched hexasaccharides [63]. Gaining insight into the subtle modifications of these precise glycan assemblies in *MTB* could lead to groundbreaking progress in TB treatments. The syntheses of these glycans are primarily orchestrated by two major enzyme families: the glycoside hydrolases (GHs) and the glycosyltransferases (GTs).

18.2.2 *Glycoside hydrolases (GHs)*

GHs are group of catabolic enzymes which hydrolyze the glycosidic bond to cleave saccharides from glycoconjugates. Depending on the cleavage pattern, these enzymes are classified as exoglycosidases (enzymes that hydrolyze the sugar residues from the non-reducing end) or endoglycosidases (enzymes that hydrolyze the sugar residues from within the oligo- or polysaccharide chain) [64–66]. The current update reported by the Carbohydrate-Active enZyme database (CAZy; http://www.cazy.org) lists 133 families of GHs, categorized based on the amino acid sequence and three-dimensional (3D) fold [67–70].

Koshland in 1953 was the first to describe that GHs follow either a retaining or an inverting mechanism, and this finding has been significant in terms of identifying other glycosidases [71]. The mechanisms of GHs have been excellently reviewed [72–74]. The three well understood mechanisms are provided in Fig. 18.2. Generally, it is observed that both the retaining (Fig. 18.2.1) and inverting (Fig. 18.2.2) enzyme have two amino acid residues in the active site, one acting as general acid and the other as a nucleophilic base [75]. The presence of an acetamido (*N*-acetyl) group at the C-2 position of the glycan (Fig. 18.2.3) acts as a catalytic nucleophile and the reaction proceeds via an intermediate oxazolinium ion, for the β-*N*-acetylhexosaminidases [72].

In addition to these glycosidase mechanisms, elimination and/or hydration mechanisms are also known [76]. For example, the GH4 family of glycoside hydrolyses undergoes both elimination and hydration, involving the active site bound nicotinamide adenine dinucleotide (NAD^+) as an oxidant and typically, Mn^{2+} counterbalancing the negative charge developed during the reaction [76–78].

18.2.3 *Glycosyltransferase (GT)*

GTs transport carbohydrate moieties from an activated donor to an acceptor substrate with overall retention or inversion of configuration at the anomeric center to produce α- or β- glycosylated

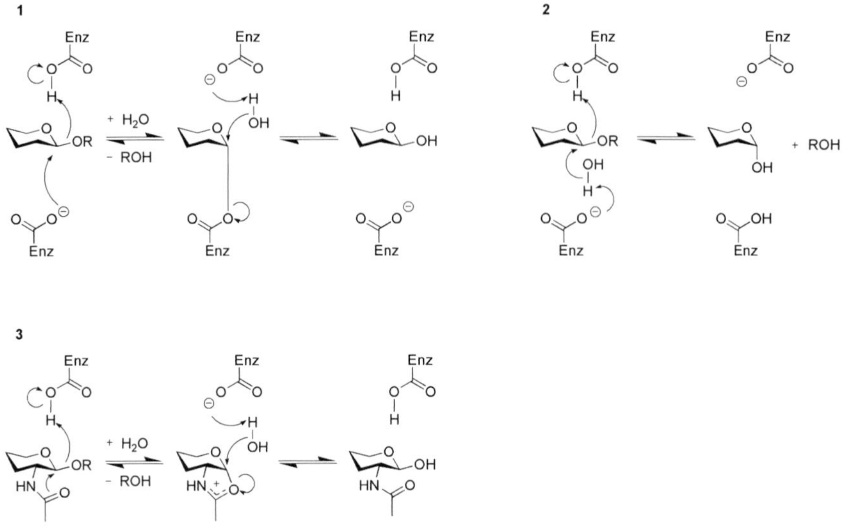

Figure 18.2 Mechanism of glycosidase, 1. Retaining mechanism, 2. Inverting mechanism, and 3. Retaining β-*N*-acetylhexosaminidases.

moiety. The current update reported by the CAZy database lists 94 families of GTs, categorized as either GT-A, GT-B, or GT-C sub-families based on the amino acid sequence and 3D fold [67,79–82].

The GT-A and GT-B protein topology consists of two β/α/β Rosmann domains [82, 83]. GT-A is understood to catalyze the transfer of activated sugar nucleotides from the donor to an acceptor with the help of divalent-metal ion co-ordinated as asparagine (Asp)–X–Asp, with the divalent-metal stabilizing the negatively charged phosphate [83–85]. GT-B catalyzes the same set of transfer reactions in which the negatively charged phosphate is stabilized by positively charged active-site residues [64,65,83]. GT-A and GT-B enzymes display both inverting and retaining mechanisms, whereas only the inverting GT mechanism is expected with GT-C folds [80,83]. Mechanistically the inverting GTs are understood to catalyze S_N2 type reactions, in which there is a nucleophilic attack from the acceptor substrate followed by a concurrent departure of the

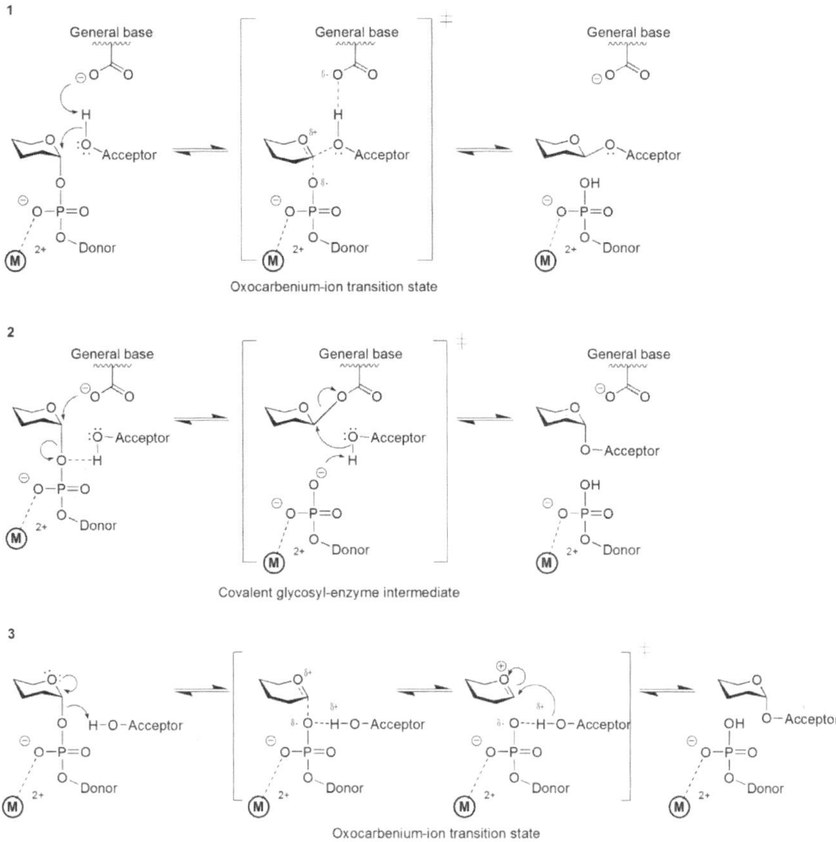

Figure 18.3 Mechanisms of glycosyltransferase, 1. Inverting mechanism via single displacement (S_N2-like), 2. Retaining mechanism via double displacement ($S_N2 \times 2$), and 3. retaining mechanism via S_Ni-type mechanism.

leaving group (Fig. 18.3.1). The retaining GTs are not well understood, but are believed to use a double-displacement mechanism, with an intermittent glycosyl–enzyme intermediate (Fig. 18.3.2) [83,86,87]. Alternatively, the retaining GTs via S_Ni-type mechanisms (Fig. 18.3.3) are also anticipated and backed with relevant experimental studies like the design and synthesis of glycomimetic inhibitory studies, transition state mimicry, kinetic analysis, quantum mechanics, and other computational approaches [87–91].

18.3 Targeting Glycosyltransferase Activity

18.3.1 *Galactosyltransferase inhibitors (GalT) as anti-tuberculosis drug targets*

The growing understanding of the biochemical pathways used in synthesis of the *MTB* cell wall creates a venue for developing carbohydrate-based inhibitors as anti-TB drugs. There are some notable contributions in this direction. For example, *MTB* galactosyltransferases (GalTs) (GlfT1 and GlfT2) are involved in the biosynthesis of the arabinogalactan cell wall of the organism. There are examples for generating substrates for investigating GlfT2 activity and studies related to the synthesis and activity of β-(1→4), β-(1→5), and β-(1→6) GalTs [92–95]. One recent study based on the use of synthetic uridine 5c-diphosphate galactofuranose (UDP-Gal*f*) analogs revealed critical structural features involved in substrate activity for GlfT2 [96]. The conversion of uridine 5c-diphosphate galactopyranose (UDP-Gal*p*) (Fig. 18.4.1) to UDP-Gal*f* (Fig. 18.4.2) is catalyzed by UDP galactopyranose mutase (UGM) [30, 97–100]. A set of *O*-galactopyranosides with

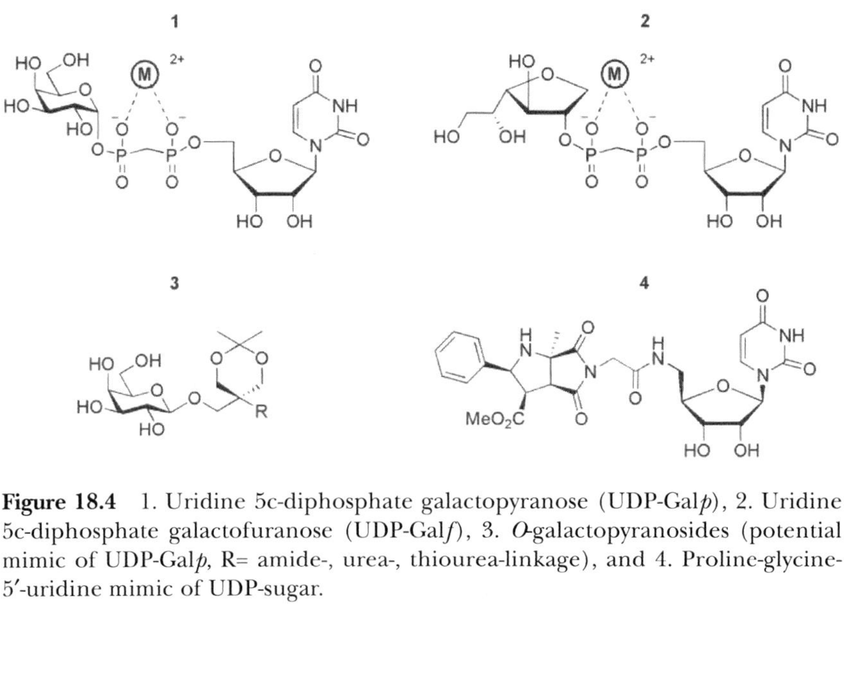

Figure 18.4 1. Uridine 5c-diphosphate galactopyranose (UDP-Gal*p*), 2. Uridine 5c-diphosphate galactofuranose (UDP-Gal*f*), 3. *O*-galactopyranosides (potential mimic of UDP-Gal*p*, R= amide-, urea-, thiourea-linkage), and 4. Proline-glycine-5′-uridine mimic of UDP-sugar.

functionalized R groups (amide-, urea-, and thiourea-linked derivatives) (Fig. 18.4.3) mimicking UDP-Gal*p* was synthesized and tested against the *MTB* $H_{37}Rv$ strain and displayed promising results with minimal inhibitory concentration (MIC) values in the 3 to 25 μg/mL range [98].

Analogous reports on derivatives of the sugar nucleotide UDP-Gal*f* showed prevention in the formation of mycobacterial galactan [101]. In yet another report, the synthesis of a library of transition state analogs was found to inhibit both of the glycosyltransferases MurG and *MTB* GlfT2 [102]. MurG catalyzes the transfer of *N*-acetyl-D-glucosamine to a lipid linked disaccharide of the bacterial PG biosynthesis [102,103]. A set of functionalized prolines attached to 5′-uridine via an amino acid (glycine, proline, or sarcosine) or diester linker were prepared and tested against *Escherichia coli* MurG and *MTB* GlfT2. One of the analogs (Fig. 18.4.4) displayed an inhibition of 80% at 1 mM concentration against *MTB* GlfT2 [102].

An interesting report on GlfT2 has suggested a single active site for catalyzing both β-(1→5) and β-(1→6) galactofuranosyl residues. Two octyl glycosides, the trisaccharide acceptors β-D-Gal*f*-(1→6)-β-D-Gal*f*-(1→5)-β-D-Gal*f*-$O(CH_2)_7CH_3$ (Fig. 18.5.1) and β-D-Gal*f*-(1→5)-β-D-Gal*f*-(1→6)-β-D-Gal*f*-$O(CH_2)_7CH_3$ (Fig. 18.5.2), were used for saturation-transfer difference (STD) nuclear magnetic resonance (NMR) spectroscopic studies [104].

1

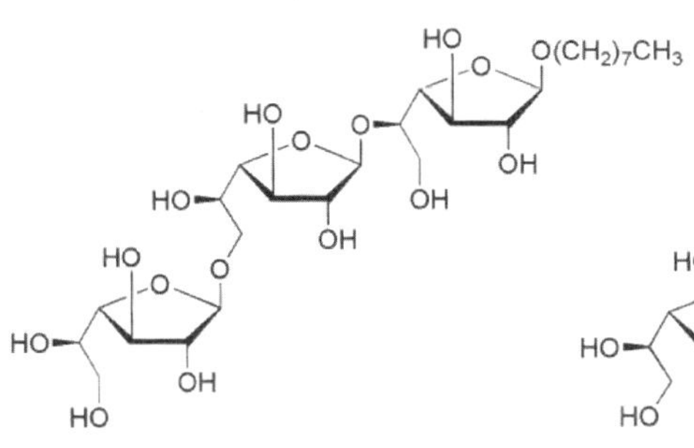

2

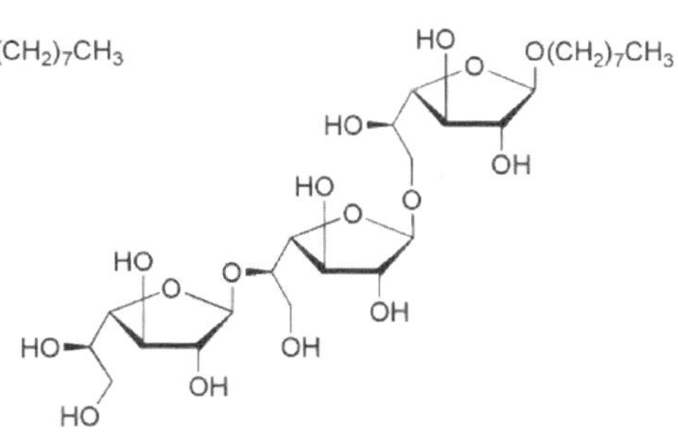

Figure 18.5 Trisaccharide acceptor substrates used in STD-NMR studies.

Contemporary work with sugar–amino acid–nucleoside (SAAN) analogs of D-galactose and L-arabinose is reported by Lowary and coworkers [105]. The D-galactopyranose and L-arabinofuranose-based motifs were synthesized, by conjugating with basic amino acids (lysine, glutamine, tryptophan, and histidine) to mimic a glycosyl nucleotide (Fig. 18.6) [105]. These targets were further tested against GlfT2. The histidine conjugate (Fig. 18.6.1) and tryptophan conjugate (Fig. 18.6.2) analogs of D-galactopyranose demonstrated inhibition of 79% and 75%, at 4 mM, respectively [105].

18.3.2 *Rhamnosyltransferase as an anti-tuberculosis drug target*

The mycobacterial cell wall PG is covalently attached to the mAG via a disaccharide phosphodiester linker α-L-Rha-(1→3)-α-D-GlcNAc-(1→phosphate) (L-Rha-D-GlcNAc-P) (Fig. 18.7.1) [32,106–108]. The L-Rha-D-GlcNAc-P connects the PG to the reducing end of mAG at the linear galactofuran residue. Detailed investigation in the AG domain has identified the crucial role of the L-Rha residue present in the L-Rha-D-GlcNAc-P linker [109, 110]. These structural findings present a unique target opportunity for antitubercular drug development [28,110]. The disaccharide phosphodiester linker, L-Rha-D-GlcNAc-P, is produced in a reaction (Fig. 18.7.2) catalyzed by an α-3-L-rhamnosyltransferase called WbbL 1 [32,106,111].

1

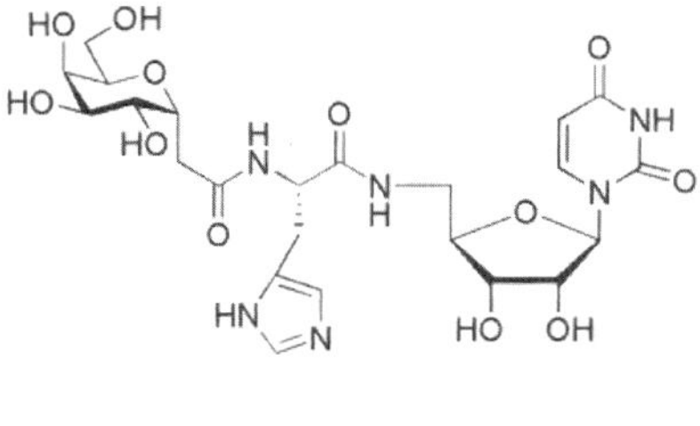

2

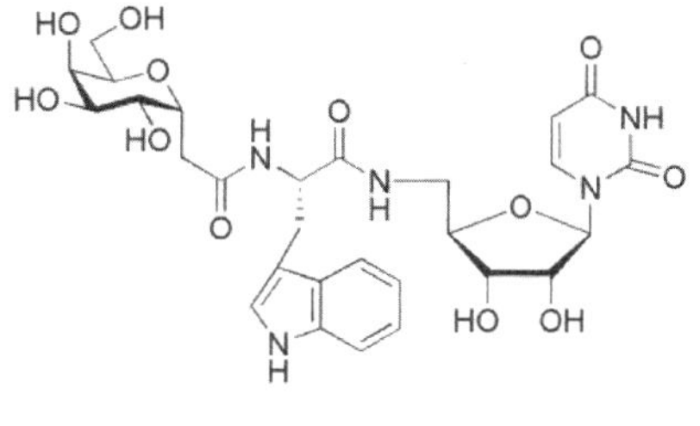

Figure 18.6 SAAN analog 1. histidine derivative, and 2. tryptophan derivative inhibition at 4 mM against GlfT2.

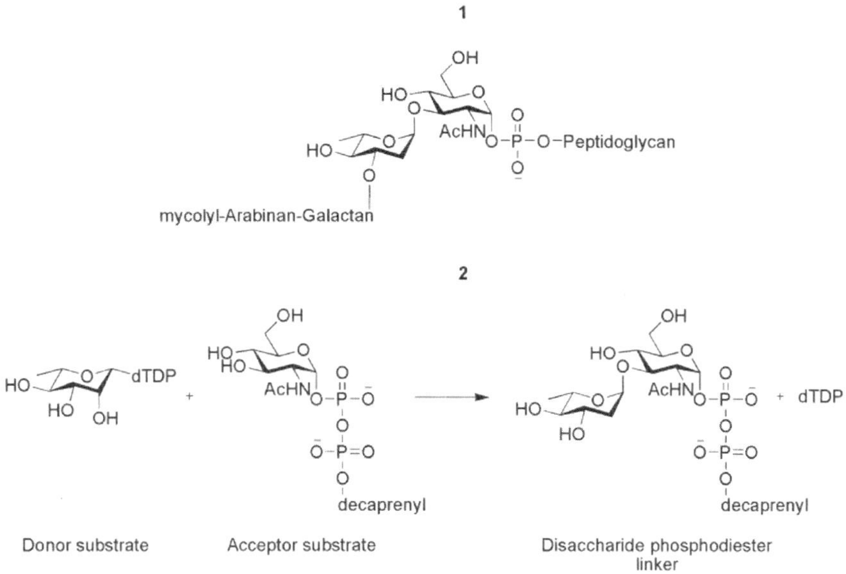

Figure 18.7 1. Disaccharide phosphodiester linker joining the mAG with the peptidoglycan, and 2. Reaction catalyzed by WbbL 1, producing L-Rha-D-GlcNAc-P.

Briefly, WbbL 1 catalyzes the glycosylation between the donor thymidine diphosphate dTDP-Rha and the α-D-GlcNAc acceptor substrate, and is established to be an essential step in the biogenesis of the bacterial cell wall. Interestingly, the absence of WbbL 1 homologues in the human populations provides a degree of leverage to synthesize non-toxic inhibitors [32,106,111,112]. Limited efforts have been directed in developing inhibitors of this L-rhamnose processing enzyme; however, inhibitors of rhamnosidases are known and may serve a good starting point [113–115].

Recently, a report on accessing iminosugar L-rhamnomimetics as effective inhibitors of mycobacterial WbbL 1 has appeared [112]. A library of α- and β-pseudoanomers of donor substrate, the dTDP-β-L-Rhamnose (dTDP-β-L-Rha), based on the L-rhamnose-aza-*C*-glycoside were synthesized via a divergent synthetic route (Fig. 18.8) [112]. The accessed compounds were then tested for inhibitory activity in *Mycobacterium smegmatis* mc^2 155. At 1 mM, compounds from the β-series showed a moderate inhibition and no

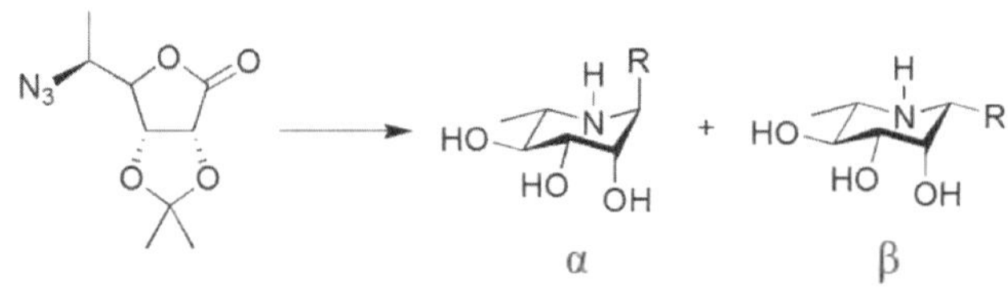

Entry β	R	Inhibition (%)
1	H	39
2	Me	25
3	Et	32
4	Butyl	19
5	Octyl	26
6	Naphthyl	25
7	$MeOPhCH_2$	23
8	$HOCH_2$	30
9	MeOPh	11

Figure 18.8 Screened library of β-pseudoanomers of L-rhamnose-aza-*C*-glycoside.

inhibition was displayed for the compounds from α-series [95,106, 112,116]. The selective inhibition exhibited by the β-configured mimetics is interpreted as mimics of dTDP-β-L-Rha, the donor substrate, or the β-rhamnosyl configuration in the reaction transition state [112].

18.3.3 *Arabinosyltransferases (AraTs) as anti-tuberculosis drug targets*

Arabinans are essential components of the mycobacterial cell wall, and these structures are unique to the bacteria. Enzymes which incorporate and polymerize the arabinan, the arabinofuranosyltransferases (Aft), are potential targets for an anti-tuberculosis drug-development [32, 117–119]. The arabinofuranose (Ara*f*) motif, is a biologically rare sugar. Ara*f*s are present in the cell wall in two distinct frameworks; namely, the AG and the LAM, with around

60–70 residues of Ara*f* [120–123]. The decoration of Ara*f* residues on the AG and the LAM is accomplished by the AraTs encoded by the Emb proteins; namely, EmbA, EmbB, and EmbC [124–129]. EmbA and EmbB are anticipated to work in concert as a heterodimer and overall the Emb proteins are considered to be excellent drug targets [126,130,131]. Structural studies have revealed that the D-arabinan motif is comprised of three major domains: (i) A linear 5-linked α-D-Ara*f* domain; (ii) a branched 3,5-linked α-D-Ara*f* domain modified with 5-linked α-D-Ara*f* residues at both branched positions; and (iii) the non-reducing termini consisting of 3,5-linked α-D-Ara*f* residues modified at both branching sites with the disaccharide β-D-Ara*f*-(1→2)-α-D-Ara*f* [32,34,129]. The role played by EmbA and EmbB in the AG synthesis, and EmbC in the LAM synthesis, respectively, are explained via genetic disruption experiments in *M. smegmatis* [119,127,129,132]. These studies elucidate the contributions of EmbA and EmbB in the formation of the terminal hexaarabinofuranosyl (Ara_6) (Fig. 18.9) assembly by catalyzing the transfer of the β-D-Ara*f*-(1→2)-α-D-Ara*f* disaccharide to the AG [119,129]. Analogous studies on the inactivation of

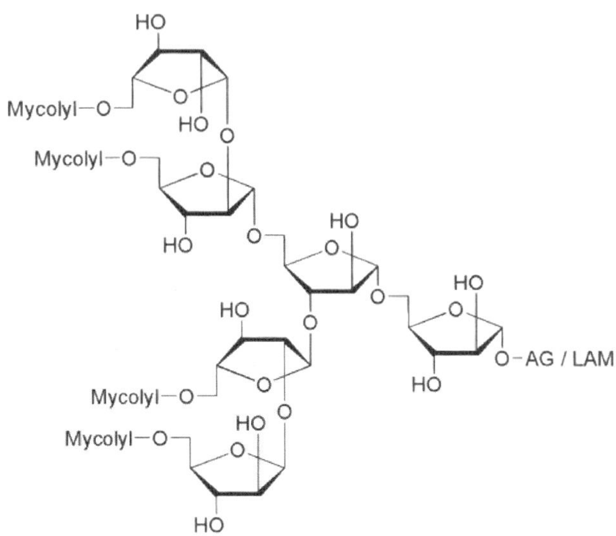

Figure 18.9 Terminating residue of arabinofuranosyl hexasaccharide.

the *embC* gene was shown to terminate LAM synthesis, and restoration was established on complementation with the *embC* wild-type gene [132].

In addition to these known AraTs (EmbA, EmbB, and EmbC), newly discovered AraTs, AftA (*Rv3792*), AftB (*Rv3805c*), AftC (*Rv2773*), AftD (*Rv0236c*) and recently identified AftE, are involved in arabinan biosynthesis. These AraTs are hinged to the AG and LAM biogenesis in the *MTB*, providing novel anti-tuberculosis drug targets [117,118,133–136]. Disruption of the *aftA* ortholog in *Corynebacterium glutamicum* showed a complete disappearance of arabinose residues in the cell wall. AftA is believed to catalyze the transfer of the first Ara*f* residue on to the galactan domain of the AG complex, as a priming enzyme using β-D-arabinofuranosyl-1-monophosphoryldecaprenol (DPA) (Fig. 18.10) as the sugar donor [117,121,123,137]. Analogous studies, involving deletion of the *aftB* ortholog in *C. glutamicum* demonstrated decline in the cell wall mycolic acid residues. This was a result of partial loss of the mycolation site, revealing the responsibility of AftB for addition of terminal β-(1→2)-linked Ara*f* residues [118]. Determination of the role of *MSMEG2785* in *M. smegmatis* and *Rv2673* in the *MTB* ortholog was accomplished by deletion of the *MSMEG2785*. The result was the absence of AG α-(1→3)-linked Ara*f* residues. This study elucidates the role of AftC as a branching enzyme on the synthetic linear α-(1→5) linked Ara_5 neoglycolipid acceptor [133]. Akin to AftC branching activity, AftD is anticipated to participate in the α-1,3-branching of

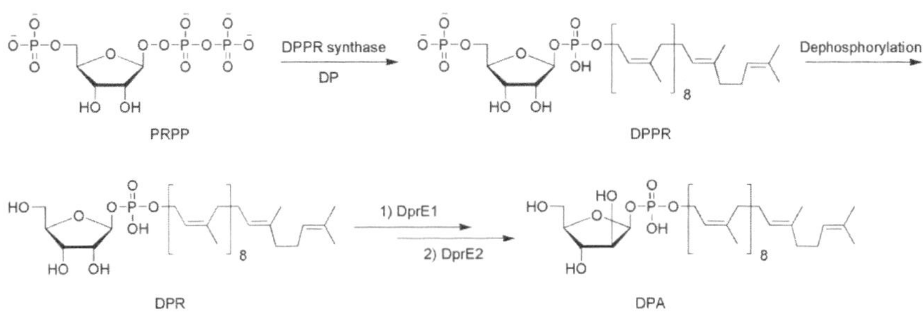

Figure 18.10 DPA biogenesis.

the non-reducing termini of the arabinan domain of the AG and LAM [134]. Studies relating to the *NCgl2096* gene deletion from *C. glutamicum* have identified a novel arabinosyltransferase, AftE, responsible for α-(1→2) AraT activity and showed its involvement in the biosynthesis of singular arabinans of LAM [135]. The above discussion highlights the significance of the arabinan in the cell wall biogenesis.

The polymerized region of Ara*f* residues in mycobacteria are synthesized by the only known precursor, DPA, and not a sugar nucleotide [137–144]. DPA is produced from 5-phosphoribose-1-diphosphate (PRPP) over a four-step process (Fig. 18.10) [140,145, 146]. Addition of decaprenyl phosphate (DP) to PRPP, via 5-phospho-ribosyltransferase (DPPR synthase, *Rv3806c*), generates decaprenylphosphoryl 5-phosphoribose (DPPR). DPPR is dephosphorylated to yield decaprenylphosphoryl ribose (DPR). DPR is subsequently converted to DPA via a two-step epimerization reaction. First, DPR is oxidized by decaprenylphosphoryl-β-D-ribose oxidase (DprE1, *Rv3790*) to decaprenylphosphoryl-X. Then it is subsequently reduced by decaprenylphosphoryl-D-2-keto erythro pentose reductase (DprE2, *Rv3791*) to generate DPA [138,140,147]. Studies related to the blocking of arabinan synthesis has revealed identification of decaprenylphosphoryl-β-D-ribose-2′-epimerase (DprE) involved in DPA synthesis [147,148]. Insight in the DPA biogenesis, the sole precursor of Ara, has aided by generating substrates and inhibitors of AraTs. This includes DPA-based analogs which are being investigated for activity as donor substrates for AraTs [123,149–154].

Research on the synthesis and testing of modified Ara*f* fragments as inhibitors of AraTs has been ongoing. One of the early reports involved the synthesis of di-, tri-, and tetrasachharide fragments of the terminating Ara_6 residue (Fig. 18.9) [155]. The target fragments showed conversion to a larger oligosaccharide via AraTs. Consequently, over a decade, acceptors and inhibitors of AraTs have been developed. Some of these include: (i) Replacing C-3 hydroxyl of the arabinofuranosyl hexasaccharide fragments with the amino, azido, methoxy, and C-5 groups modified to *N*-alkyl, amide, and sulphonamide derivatives; (ii) symmetrical *C*- and pseudo-symmetrical *O*-linked disaccharide motifs; (iii) deoxygenated α-(1→5) arabinofuranose disaccharide analogs; (iv) dansyl modified fluorescence active disaccharide acceptors and photoaffinity probes utilized in studying enzyme activity; (v) octyl-glycosides as oligosaccharide fragments; (vi) *n*-octyl-5-(α-D-arabinofuranosyl)-β-D-galactofuranoside derivatives; (vii) imino sugar-oligoarabinofuranoside hybrids; (viii) *C*-disaccharide analogs; and (ix) total synthesis of identified residues of 18- and 22-arabinan motifs and carbasugars analogs [48,156–174].

18.3.4 *Mannosyltransferase (ManT) as anti-tuberculosis drug targets*

As described previously, the biogenesis of LAM plays an important role in cell wall integrity and influences the pathogenesis of *MTB* [175,176]. Targeting LAM biosynthesis is considered as an important anti-tuberculosis drug approach [177]. The α-D-mannose-(1→2)-α-D-mannose-(1→6)-D-*myo*-inositol (PIM_2) (Fig. 18.11.1) is an intermediate in LAM biosynthesis and is implicated in the host immune

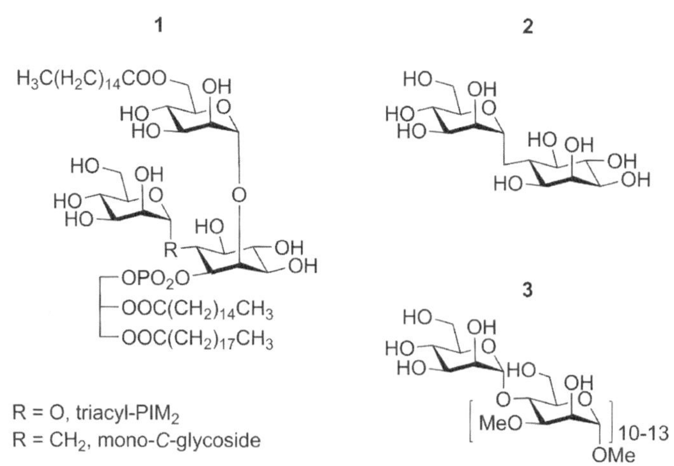

Figure 18.11 1. PIM_2, 2. *C*-glycoside analog of pseudotrisaccharide, and 3. Methylmannose polysaccharide.

response [177–180]. There was a recent report on the synthesis of a *C*-glycoside analog of the α-D-mannose-(1→6)-D-*myo*-inositol (Fig. 18.11.2) fragment of the pseudotrisaccharide [178]. The *C*-glycoside analog was studied as a hydrolytically stable substrate mimetic [181,182].

Polymethylated polysaccharides (PMPS) are unique carbohydrates. Two known PMPS in mycobacteria are the 3-*O*-methylmannose polysaccharides (MMPs) (Fig. 18.11.3) and 6-*O*-methylglucose lipopolysaccharides (MGLPs, not shown) [183–185]. Proposed MMP biosynthesis follows elongation with alternate mannosylation via α-(1→4)-ManT, and methylation via 3-*O*-methyltransferase (OMT) [183, 185]. Recently, an investigation on specificity of the ManT involved in MMPs biosynthesis was reported [186]. Two series of α-(1→4)-linked 3-*O*-methyl-mannopyranosides with either 3-*O*-methyl or free mannopyranose terminating residues were synthesized and tested as acceptors of the α-(1→4)-ManT [186]. The result suggests that ManT could produce linear MMPs independent of OMT and contradicts the joint activity between ManT and OMT [186].

Efforts to develop inhibitors and substrates for α-(1,6)-ManT involved in LM and LAM biosynthesis have also been reported [187–194]. For example, a triazole-linked α-1,6-*C*-oligomannoside up to hexadecamer is reported [192]. The key chemistry involves the use of a Cu(I)-catalyzed azide–alkyne cycloaddition (CuAAC) approach to attain the target oligomers (Fig. 18.12.1) [192]. The resistance from hydrolytic degradation is attributed to the anomeric carbon–carbon bonds and triazole rings as an interglycosidic link. These compounds revealed variable inhibitory activity against α-(1,6)-ManT, and maximal activity of IC_{50} with values of 0.14–0.22 mM displayed with hexamannoside and octamannoside. The study also indicated efficient binding of the triazole analog with the polyprenolphosphomannose (PPM)-dependent α-(1,6)-ManT with the length of $n = 4$ and 6.

It was previously demonstrated that the C-6 deoxy D-mannose analog hinders α-(1,6)-ManT activity, suggesting that mannose transfer takes place at the C-6 position of the acceptor substrate [195]. To confirm this finding, a triazole-linked 1,6-dimannoside

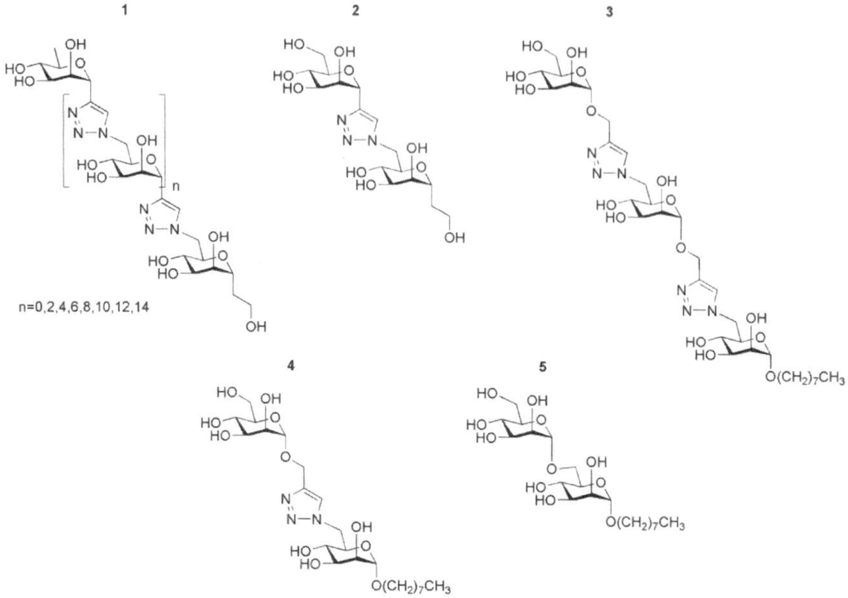

Figure 18.12 1. A series of triazole-linked *C*-(1,6)-oligomannoside, 2. triazole-linked 1,6-dimannoside acceptor, 3. triazole-linked oligomannoside (three mannose units), 4. triazole-linked oligomannoside (two mannose units), and 5. α-Man*p*-(1→6)-α-Man*p* Octyl disaccharide.

acceptor substrate (Fig. 18.12.2) was accessed and tested on membranes extracted from *M. smegmatis* [192]. This compound (Fig. 18.12.2) was found to be a poor acceptor substrate, up to 4 mM, as well as an inadequate inhibitor of α-(1,6)-ManT with an IC_{50} value of 0.96 mM. It was then shown that the presence of a C-6 deoxy position leads to complete blocking age of the ManT activity; however, but this structural attribute is not vital for inhibition [192].

Further, in order to investigate the elongation of the compounds, acting as ManT acceptors, similar triazole-linked substrates with a free hydroxyl group at the non-reducing termini (C-6, –OH) were synthesized (Figs. 18.12.3; Fig. 18.12.4). Compounds 18.12.3 and 18.12.4 were synthesized and tested for ManT acceptor activity and compared against compound 18.12.5, known for its ManT acceptor activity, and has the minimum structural requirements as synthetic acceptor (Fig. 18.12.5) [195,196]. These octyl glycosides act as substrate analogs for GTs in the mycobacteria [121,197–199]. They were then tested in an *in vitro* assay monitoring ManT acceptor activity in the mycobacterial membrane extract at 4 mM concentration [195,196]. A comparative study of the acceptor ability of the triazole linked compounds revealed the optimum size of acceptor and spatial orientation. Two structural requirements, (i) a hydrophobic aglycone, and (ii) a free hydroxyl group at the non-reducing end of the mannose residue, were identified as necessary [196].

There have been notable contributions in developing α-(1,6)-ManT inhibitors and substrates [187, 188, 193]. The α-(1→6)-mannopyranose Man*p* (1→6)-α-Man*p* motif found in the mannan core (Fig. 18.13.1) of the LAM was identified as minimum acceptor motif for enzymes synthesizing the LAM mannan structure [187, 188,193,200]. These studies also revealed substrate specificity, optimal length, and key interactions with the α-(1,6)-ManT involved in the LAM biosynthesis. Epimeric- (*manno*-, *talo*- and *gluco*-) and 6-deoxyamino disaccharide analogs of Man*p*-(1→6)-α-Man*p* Octyl glycoside (Fig. 18.13.2) were synthesized and evaluated as PPM-dependent α-(1→6)-ManT substrates and inhibitors in *M. smegmatis* [193]. This study identified a prerequisite of a *manno*-configuration. In comparison with the *talo*- and *gluco*- configurations which

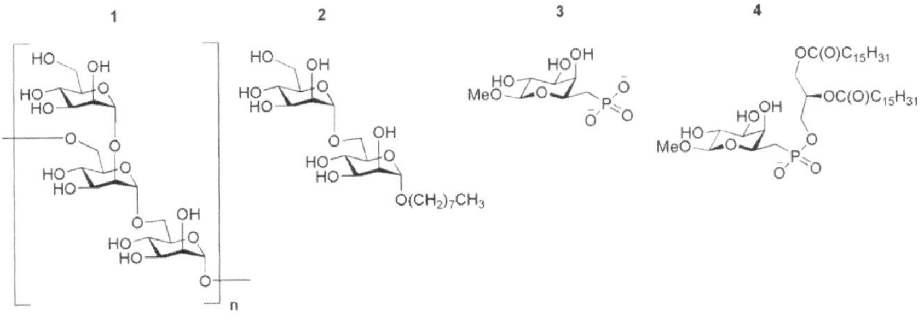

Figure 18.13 1. Mannan core found in the LAM complex, [39,48] 2. α-Man*p*-(1→6)-α-Man*p* Octyl glycoside (minimum acceptor) [187,188], 3. and 4. D-galactose-phosphonate analogs of *myo*-inositol-1-phosphate and PI.

displayed loss of activity. A 6-deoxyamino analog was a weak inhibitor and showed 50% inhibition of activity at ~1 mM concentration [193]. Other analogs of the Man*p*-(1→6)-α-Man*p* Octyl glycosides, such as deoxy, methoxy, fluoro, and amino functionalities were also synthesized and evaluated as substrates and inhibitors of PPM-dependent α-(1→6)-ManT [188,190,191,194].

PimA is another key ManT involved in catalyzing the generation of PIM from PI. In one study, galactose-derived phosphonate analogs (Fig. 18.13.3; Fig. 18.13.4) of *myo*-inositol-1-phosphate (IP), and phosphatidylinositol (PI) respectively, were synthesized and tested as substrates and inhibitors of PimA activity in *M. smegmatis* mc^2 155 [201]. Compound 18.13.3 showed potent inhibitory activity against PimA, which catalyzes the synthesis of PIM1 from PI at 0.01 mM concentration, and is a weak inhibitor of the ManT, which converts PIM1 to PIM2. Compound 18.13.4 displayed weak ManT inhibitory activity in the conversion of PIM1 to PIM2 [201].

18.3.5 *GlgE inhibitors as anti-tuberculosis drug targets*

GlgE is an α-1,4-glucan–phosphate maltosyltransferase, and a member of the glycoside hydrolase subfamily GH13_3 in the CAZy database, and a genetically validated tuberculosis drug target [202–205]. Transposon site hybridization analysis has identified genes essential for optimum growth in *MTB* [206–207]. These studies identified GlgE (*glgE, Rv1327c*) to be essential for *in vitro* growth [206]. GlgE catalyzes the biosynthesis of the α-1,4-glucan synthesis by transferring the donor maltose-1-phosphate (M1P) to the reducing end of the elongating α-1,4-glucan chain with an α-retaining mechanism (Fig. 18.14) [202,203,208]. Briefly, trehalose synthase (TreS) interconverts trehalose to maltose, which is further transformed into an activated donor, M1P, via phosphorylation by adenosine triphosphate (ATP)-dependent maltokinase, Pep2. M1P is then transferred onto the elongating α-1,4-glucan by the action of GlgE, generating the glucan polymer in the process [202].

Biochemical and genetic studies have revealed that inhibition of GlgE causes rapid death in *MTB* as a direct consequence of

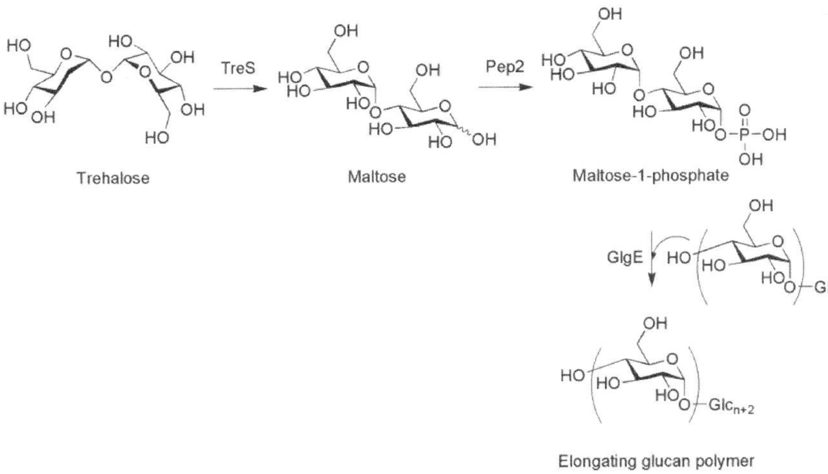

Figure 18.14 Overview of α-1,4-glucan elongation in MTB via TreS, Pep2 and GlgE.

self-poisoning due to the accumulation of M1P [202]. Analysis of the whole transcriptome revealed that the fatal nature of the accumulated M1P is the result of sparking a cellular stress responses in *MTB*; this causes deoxyribonucleic acid (DNA) damage due to reactive oxygen species [202,208–210]. These findings also report "synthetic lethality" demonstrated by simultaneous inactivation of TreS and the glucosyl transferase, *Rv3032*, leading to the death of *MTB* [202]. *Rv3032* elongates the α-1,4-glucan polymer using UDP-D-Glc as donor substrate [184].

Another study revealed structural insights in the binding site of the GlgE using GlgE isoform I from *Streptomyces coelicolor* [203]. This study reports important details, such as (i) Consistent structural features with respect to maltooligosaccharide acceptor extending at the non-reducing ends, and (ii) identical catalytic profiles and high similarities in the kinetic properties and well-conserved donor-binding sites [203].

These publications have led to the introduction of the first and only known inhibitor of the GlgE, a maltose-C-phosphonate (MCP) (Fig. 18.15.1) [211]. MCP is an isosteric non-hydrolysable mimic of

1 2

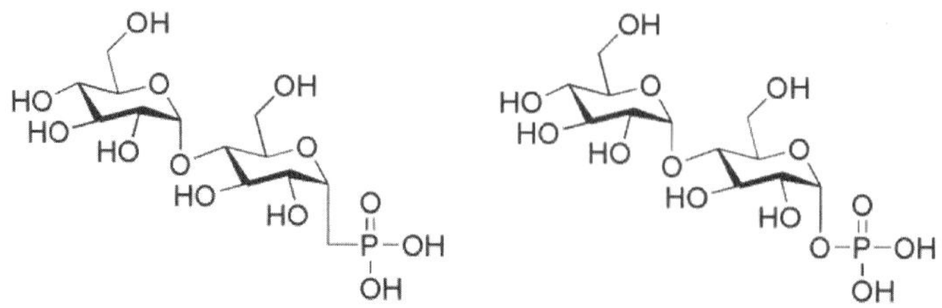

Figure 18.15 1. Maltose-C-phosphonate, and 2. Maltose-1-phosphate.

M1P (Fig. 18.15.2). MCP showed a moderate inhibition at IC_{50} = 230 ± 24 µM against the *MTB* GlgE [211,212]. A new poly-hydroxypyrolidine has also been recently reported to mimic the transition state of the reaction catalyzed by GlgE and inhibites both *MTB* GlgE and *S. coelicolor* GlgE1 [212].

18.4 *Mycobacterium tuberculosis* Antigen 85 Complex

18.4.1 *Antigen 85 (Ag85) as an anti-tuberculosis drug target*

The cell wall of mycobacteria presents a robust hydrophobic membrane made up of mycolic acids, which provide a line of defense against therapeutic drugs [31,35,213,214]. There are five different class mycolic acid moieties associated with *MTB*, namely the alpha-mycolic acid, and the *cis*- and *trans*-isomers of the methoxy and keto mycolic acids [35,215–217]. The hydrophobic membrane consists of a bilayer, with the inner layer consisting of mAG covalently attached to the PG, and the outer layer made of free glycolipids, TMM and TDM, (Fig. 18.1) [35,111,218]. The presence of TMD is associated with the pathogenesis of tuberculosis [215,219,220].

The Antigen 85 (Ag85) complex is made up of a triad of proteins (Ag85A, Ag85B, and Ag85C), each possessing mycolyltransferase activity. The transfer of mycolic acids is implied to be the final step of the cell envelope synthesis [221,222]. Ag85 A–C retains 68–80%

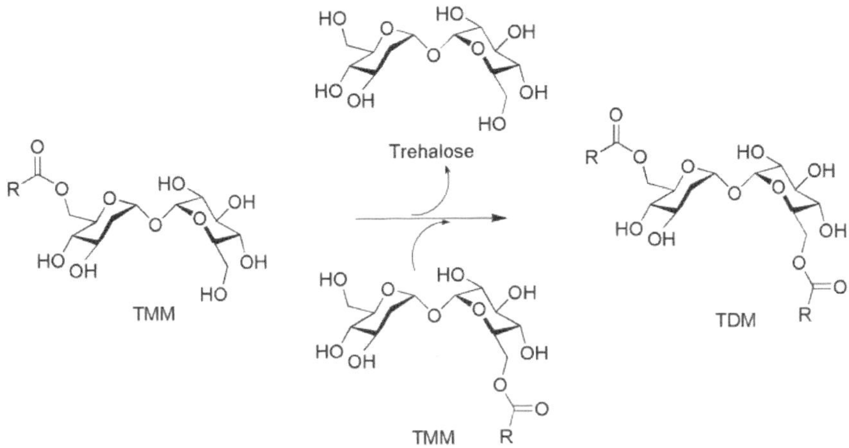

Figure 18.16 Transfer of a mycolic acid residue from one molecule of TMM to another, producing TDM (R = mycolic acid residue).

identical amino acid sequence homology, and catalyzes the transfer of mycolic acid residues from TMM to produce TDM (Fig. 18.16) [220–223].

Mechanistically, the two-step mycolyl transfer reaction is initiated by the nucleophilic attack by the serine residue in the enzyme binding pocket onto the mycolyl donor substrate, creating a covalent acyl–enzyme intermediate with the release of trehalose [220,221]. This intermediate collapses as an incoming TMM, the nucleophilic acyl acceptor, substrate attacks the intermediate to produce TDM. X-ray crystallographic studies have shown a high level of similarity between the active sites of Ag85s [221,223,224].

The critical role of the Ag85s in cell wall biosynthesis has been validated by *fbpA*, *fbpB* and *fbpC* gene knockout studies [221,222, 225–227]. Efforts to design inhibitors of Ag85s have been reported. Some of the earliest inhibitors were the substrate analogs 6-azido-6-deoxy-α,α-trehalose (ADT) and 6-amino-6-deoxy-α,α-trehalose [228]. ADT displayed complete inhibition of the growth of *Mycobacterium aurum* A+, a nonpathogenic strain of *MTB*, at a MIC of 200 μg/mL, and showed inhibition of mycolyltransferase activity *in vitro* [220,229]. Further studies revealed inhibition of the

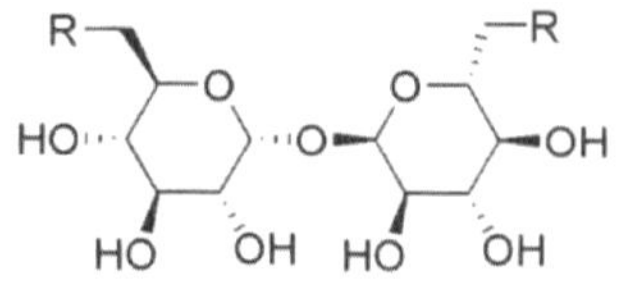

Entry	R	MIC (μg/ml)
1	$CH_3(CH_2)_7SO_2NH$-	16 - 32
2	$[CH_3(CH_2)_{11}]_2N$-	>128
3	$CH_3(CH_2)_7NH$-	4
4	$CH_3(CH_2)_{11}NH$-	8
5	$[CH_3(CH_2)_3CH(Et)(CH_2)]_2N$-	128
6	$CH_3(CH_2)_5NH$-	128
7	$CH_3(CH_2)_9NH$-	>1.3≤13
8	$CH_3(CH_2)_3CH(Et)CH_2NH$-	32

Figure 18.17 MIC values of 6,6′-dideoxytrehalose derivatives against *MTB* $H_{37}Ra$.

synthesis of TMM (44 ± 7%), TDM (87 ± 5%), and mycolic acids (62 ± 18%) at a concentration of 100 μg/mL [220]. One study reported the activity of derivatives of 6,6′-dideoxytrehalose (Fig. 18.17) against *MTB* $H_{37}Ra$ and *M. avium* [230]. Related inhibitor molecules with mono- or di-substituted hydrocarbon chains on a trehalose core generated a library of compounds and showed activity against *M. smegmatis* [231].

Other approaches to develop inhibitors of the Ag85 complex include the preparation of arabinofuranoside analogs based on docking studies [232]. Two compounds, 5-*S*-octyl-5-thio-α-D-arabinofuranoside (Fig. 18.18.1) and 5-*S*-octyl-5-thio-β-D-arabinofuranoside (Fig. 18.18.2), showed a MIC of 256 and 512 μg/mL against *M. smegmatis* ATCC 14468, respectively. Additionally, ester-, α-ketoester- and α-ketoamide derivatives of D-arabinose and D-trehalose based inhibitors were synthesized and tested against *MTB* Ag85C [233]. Among the generated library, methyl esters (Fig. 18.18.3) showed an inhibitory profile against the acyltransferase activity of Ag85C with an IC50 = 25 mM.

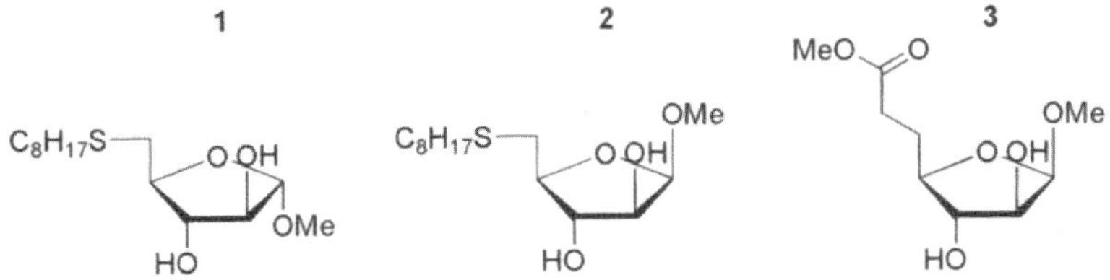

Figure 18.18 1. 5-*S*-octyl-5-thio-α-D-arabinofuranoside, 2. 5-*S*-octyl-5-thio-β-D arabinofuranoside, and 3. Methyl ester analog of an arabinofuranoside.

Research on developing inhibitors of Ag85, based on docking studies reported design, synthesis, testing, and X-ray analysis of thiophene-based compounds (Fig. 18.19.1; Fig. 18.19.2) [234]. The conjugates showed K_is in a range of 18.2 to 71.0 μM. The ethyl ester conjugate with methylene linker (Fig. 18.19.1(a)) showed the lowest K_i, with a value of 18.2 ± 1.8 μM, and was subsequently used to obtain crystal structures of the inhibitor bound in the active site of Ag85C. These compounds were tested for inhibition against *M. smegmatis* in a disk diffusion assay. Although no compound displayed a notable inhibitory profile, general conclusions were drawn, providing insight for generating second-generation compounds with improved inhibitory activity. It was noted that an ethyl ester modified thiophene conjugate (Fig. 18.19.1(a)) displayed better enzyme inhibition in comparison to the nitrile modified thiophene conjugate (Fig. 18.19.1(c)). Also, the length of the linker in Fig. 18.19.1(a) seemed to improve activity in comparison to the propylene linked compound (Fig. 18.19.1(b)) [234].

Lately, a study based on the design and synthesis of a library comprising of 24 compounds of trehalose analogs was reported [235]. Twenty two of these compounds consisted of trehalose and trehalose-modified analogs (1-methyl, 2-fluoro/iodo/deoxy/amino, 3-fluoro, 4-fluoro, 6-fluoro/bromo/phosphate/azido, fluorescein, and fluorobenzyl). The other compounds of the library were monosaccharides of the D-glucose and D-arabinose configuration. All analogs were screened as substrates for Ag85s via mass spectrometric analysis [235,236]. Striking substrate pliancy was observed as all the analogs were processed as substrates to a varying degree.

1 2

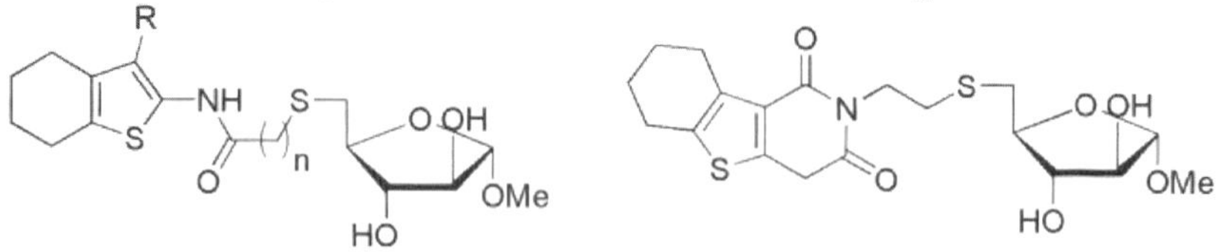

a: R = $COOC_2H_5$, n = 1 c: R = CN, n = 1
b: R = $COOC_2H_5$, n = 3 d: R = CN, n = 3

Figure 18.19 1. Thiophenyl-thio-arabinofuranoside conjugate (a, b = ethyl ester modified thiophene, and c, d = nitrile modified thiophene), and 2. Pyrimidinedione-thio-arabinofuranoside conjugate.

Furthermore, uptake of fluorescein–trehalose (FITC–trehalose) was assayed in growing *MTB* H37Rv. This result provided evidence of FITC–trehalose uptake [235,237,238].

18.5 Summary

Tuberculosis remains a major world health threat and takes the lives of millions every year. TB treatment options, while effective, are expensive and difficult to implement. Significant progress has been made in establishing the roles carbohydrate processing enzymes play in orchestrating the synthesis of glycans and glycoconjugates in *MTB*. Carbohydrate-based inhibitors and the preparation of substrate analogs for these enzymes have helped to characterize the respective enzyme active sites and activities. New drugs for the treatment of TB are expected to emerge as our understanding of the diverse roles these enzymes play within the cell begins to unfold. The development of lead inhibitors for essential and synthetically lethal enzyme combinations will inform researchers in their future efforts to develop potent carbohydrate-based inhibitors.

References

1. WHO. (2013) Global tuberculosis report. World Health Organization (Geneva).
2. Singh, B., Mitchison, D. (1954) *BMJ* 1, 130–132.

3. Woodruff, H. B. (2014) *Appl. Environ. Microbiol.* **80**, 2–8.
4. Schatz, A., Bugie E. S. W. (1944) *Proc. Soc. Exp. Biol. Med.* **55**, 66–69.
5. Crofton, J. (1960) *BMJ* **2**, 679–687.
6. Mitchison, D. A., Davies, G. R. (2008) *Open Infect. Dis. J.* **2**, 59–76.
7. Chiang, C. Y., Centis, R., Migliori, G. B. (2010) *Respirology* **15**, 413–432.
8. Mitchison, D. (2000) *Int. J. Tuberc. Lung Dis.* **4**, 796–806.
9. Udwadia, Z. F., Amale, R. A., Ajbani, K. K., Rodrigues, C. (2012) *Clin. Infect. Dis.* **54**, 579–581.
10. Bernheim, F. (1940) *Science* **92**, 204.
11. Lehmann, J. (1946) *Lancet* **247**, 15–16.
12. Maggi, N., Pasqualucci, C., Ballotta, R., Sensi, P. (1966) *Chemotherapy* **11**, 285–292.
13. Thomas, J., Baughn, C., Wilkinson, R., Shepherd, R. (1961) *Am. Rev. Respir. Dis.* **83**, 891–893.
14. Malone, L., Schurr, A., Lindh, H., McKenzie, D., Kiser, J., Williams, J. (1952) *Am. Rev. Tuberc.* **65**, 511–518.
15. Chorine, V. (1945) *C. R. Acad. Sci.* **220**, 150–151.
16. Ernst, B., Magnani, J. L. (2009) *Nat. Rev. Drug Discov.* **8**, 661–677.
17. Reina, J. J., Bernardi, A. (2012) *Mini Rev. Med. Chem.* **12**, 1434–1442.
18. Hudak, J. E., Bertozzi, C. R. (2014) *Chem. Biol.* **21**, 16–37.
19. Osborn, H. M., Evans, P. G., Gemmell, N., Osborne, S. D. (2004) *J. Pharm. Pharmacol.* **56**, 691–702.
20. Cipolla, L., Araújo, A. C., Bini, D., Gabrielli, L., Russo, L., Shaikh, N. (2010) *Expert Opin. Drug Discov.* **5**, 721–737.
21. Takayama, K., Kilburn, J. O. (1989) *Antimicrob. Agents Chemother.* **33**, 1493–1499.
22. Quemard, A., LaCave, C., Laneelle, G. (1991) *Antimicrob. Agents Chemother.* **35**, 1035–1039.
23. Lee, R., Brennan, P., Besra, G. (1996) *Curr. Top. Microbiol. Immunol.* **215**, 1.
24. Tam, P. H., Lowary, T. L. (2009) *Curr. Opin. Chem. Biol.* **13**, 618–625.
25. Kremer, L., Dover, L. G., Morehouse, C., Hitchin, P., Everett, M., Morris, H. R., Dell, A., Brennan, P. J., McNeil, M. R., Flaherty, C. (2001) *J. Biol. Chem.* **276**, 26430–26440.
26. Crick, D. C., Mahapatra, S., Brennan, P. J. (2001) *Glycobiology* **11**, 107R–118R.
27. Alderwick, L., Birch, H., Mishra, A., Eggeling, L., Besra, G. (2007) *Biochem. Soc. Trans.* **35**, 1325–1328.
28. Besra, G. S., Brennan, P. J. (1997) *J. Pharm. Pharmacol.* **49**, 25–30.
29. Brennan, P. J., Crick, D. C. (2007) *Curr. Top. Med. Chem.* **7**, 475–488.

30. Lowary, T. L. (2003) *Mini Rev. Med. Chem.* **3**, 689–702.
31. Dover, L. G., Alderwick, L. J., Brown, A. K., Futterer, K., Besra, G. S. (2007) *Curr. Mol. Med.* **7**, 247–276.
32. Daffe, M., Brennan, P. J., McNeil, M. (1990) *J. Biol. Chem.* **265**, 6734–6743.
33. Besra, G. S., Khoo, K. H., McNeil, M. R., Dell, A., Morris, H. R., Brennan, P. J. (1995) *Biochemistry* **34**, 4257–4266.
34. McNeil, M. R., Robuck, K. G., Harter, M., Brennan, P. J. (1994) *Glycobiology* **4**, 165–174.
35. Takayama, K., Wang, C., Besra, G. S. (2005) *Clin. Microbiol. Rev.* **18**, 81–101.
36. Qureshi, N., Takayama, K., Jordi, H. C., Schnoes, H. K. (1978) *J. Biol. Chem.* **253**, 5411–5417.
37. Kaur, D., Guerin, M. E., Škovierová, H., Brennan, P. J., Jackson, M. (2009) *Adv. Appl. Microbiol.* **69**, 23–78.
38. Favrot, L., Ronning, D. R. (2012) *Expert Rev. Anti. Infect. Ther.* **10**, 1023–1036.
39. Nigou, J., Gilleron, M., Puzo, G. (2003) *Biochimie* **85**, 153–166.
40. Kaur, D., Berg, S., Dinadayala, P., Gicquel, B., Chatterjee, D., McNeil, M. R., Vissa, V. D., Crick, D. C., Jackson, M., Brennan, P. J. (2006) *Proc. Natl. Acad. Sci. USA* **103**, 13664–13669.
41. Besra, G. S., Morehouse, C. B., Rittner, C. M., Waechter, C. J., Brennan, P. J. (1997) *J. Biol. Chem.* **272**, 18460–18466.
42. Tam, P. H., Lowary, T. L. (2010) *Carbohydr. Chem.* **36**, 38–63.
43. Korduláková, J., Gilleron, M., Mikušová, K. N., Puzo, G., Brennan, P. J., Gicquel, B., Jackson, M. (2002) *J. Biol. Chem.* **277**, 31335–31344.
44. Morita, Y., Patterson, J., Billman-Jacobe, H., McConville, M. (2004) *Biochem. J.* **378**, 589–597.
45. Hunter, S. W., Brennan, P. (1990) *J. Biol. Chem.* **265**, 9272–9279.
46. Mishra, A. K., Driessen, N. N., Appelmelk, B. J., Besra, G. S. (2011) *FEMS Microbiol. Rev.* **35**, 1126–1157.
47. Strohmeier, G. R., Fenton, M. J. (1999) *Microbes Infect.* **1**, 709–717.
48. Briken, V., Porcelli, S. A., Besra, G. S., Kremer, L. (2004) *Mol. Microbiol.* **53**, 391–403.
49. Fukuda, T., Matsumura, T., Ato, M., Hamasaki, M., Nishiuchi, Y., Murakami, Y., Maeda, Y., Yoshimori, T., Matsumoto, S., Kobayashi, K. (2013) *mBio* **4**, e00472–00012.
50. Wu, T., Guo, S., Wang, J., Li, L., Xu, L., Liu, P., Ma, S., Zhang, J., Xu, L., Luo, Y. (2011) *Cell. Immunol.* **272**, 94–101.
51. Nigou, J., Zelle-Rieser, C., Gilleron, M., Thurnher, M., Puzo, G. (2001) *J. Immunol.* **166**, 7477–7485.

52. Chan, J., Fan, X., Hunter, S., Brennan, P., Bloom, B. (1991) *Infect. Immun.* **59**, 1755–1761.
53. Sibley, L. D., Hunter, S., Brennan, P., Krahenbuhl, J. (1988) *Infect. Immun.* **56**, 1232–1236.
54. Hmama, Z., Sendide, K., Talal, A., Garcia, R., Dobos, K., Reiner, N. E. (2004) *J. Cell Sci.* **117**, 2131–2140.
55. Fratti, R. A., Chua, J., Vergne, I., Deretic, V. (2003) *Proc. Natl. Acad. Sci. USA* **100**, 5437–5442.
56. Vergne, I., Chua, J., Deretic, V. (2003) *J. Exp. Med.* **198**, 653–659.
57. Kang, P. B., Azad, A. K., Torrelles, J. B., Kaufman, T. M., Beharka, A., Tibesar, E., DesJardin, L. E., Schlesinger, L. S. (2005) *J. Exp. Med.* **202**, 987–999.
58. Stoop, E. J., Mishra, A. K., Driessen, N. N., Stempvoort, G., Bouchier, P., Verboom, T., Leeuwen, L. M., Sparrius, M., Raadsen, S. A., Zon, M. (2013) *Cell. Microbiol.* **15**, 2093–2108.
59. Ludwiczak, P., Brando, T., Monsarrat, B., Puzo, G. (2001) *Anal. Chem.* **73**, 2323–2330.
60. Weikkolainen, K., Aitio, O., Blomqvist, M., Natunen, J., Helin, J. (2007) *Glycoconjugate J.* **24**, 157–165.
61. Dwek, R. A. (1996) *Chemical Reviews* **96**, 683–720.
62. Varki, A. (1993) *Glycobiology* **3**, 97–130.
63. Laine, R. A. (1994) *Glycobiology* **4**, 759–767.
64. Gloster, T. M., Vocadlo, D. J. (2012) *Nat. Chem. Biol.* **8**, 683–694.
65. Gloster, T. (2012) *Biochem. Soc. Trans.* **40**, 913.
66. Ganem, B. (1996) *Accounts Chem. Res.* **29**, 340–347.
67. Lombard, V., Ramulu, H. G., Drula, E., Coutinho, P. M., Henrissat, B. (2013) *Nucleic Acids Res.* **42**, D490–D495.
68. Davies, G. J., Gloster, T. M., Henrissat, B. (2005) *Curr. Opin. Chem. Biol.* **15**, 637–645.
69. Henrissat, B. (1991) *Biochem. J.* **280**, 309–316.
70. Henrissat, B., Davies, G. (1997) *Curr. Opin. Struct. Biol.* **7**, 637–644.
71. Koshland, D. E. (1953) *Biol. Rev.* **28**, 416–436.
72. Zechel, D. L., Withers, S. G. (2000) *Acc. Chem. Res.* **33**, 11–18.
73. Rye, C. S., Withers, S. G. (2000) *Curr. Opin. Chem. Biol.* **4**, 573–580.
74. Vocadlo, D. J., Davies, G. J. (2008) *Curr. Opin. Chem. Biol.* **12**, 539–555.
75. Davies, G., Henrissat, B. (1995) *Structure* **3**, 853–859.
76. Jongkees, S. A., Withers, S. G. (2013) *Accounts Chem. Res.*
77. Yip, V. L., Withers, S. G. (2006) *Curr. Opin. Chem. Biol.* **10**, 147–155.

78. Yip, V. L., Varrot, A., Davies, G. J., Rajan, S. S., Yang, X., Thompson, J., Anderson, W. F., Withers, S. G. (2004) *J. Am. Chem. Soc.* **126**, 8354–8355.
79. Campbell, J., Davies, G., Bulone, V., Henrissat, B. (1997) *Biochem. J.* **326**, 929.
80. Coutinho, P. M., Deleury, E., Davies, G. J., Henrissat, B. (2003) *J. Mol. Biol.* **328**, 307–317.
81. Ünligil, U. M., Rini, J. M. (2000) *Curr. Opin. Struct. Biol.* **10**, 510–517.
82. Liu, J., Mushegian, A. (2003) *Protein Sci.* **12**, 1418–1431.
83. Lairson, L., Henrissat, B., Davies, G., Withers, S. (2008) *Annu. Rev. Biochem.* **77**, 521–555.
84. Wiggins, C. A. R., Munro, S. (1998) *Proc. Natl. Acad. Sci. USA* **95**, 7945–7950.
85. Breton, C., Bettler, E., Joziasse, D. H., Geremia, R. A., Imberty, A. (1998) *J. Biochem.* **123**, 1000–1009.
86. Soya, N., Fang, Y., Palcic, M. M., Klassen, J. S. (2011) *Glycobiology* **21**, 547–552.
87. Gómez, H., Polyak, I., Thiel, W., Lluch, J. M., Masgrau, L. (2012) *J. Am. Chem. Soc.* **134**, 4743–4752.
88. Lee, S. S., Hong, S. Y., Errey, J. C., Izumi, A., Davies, G. J., Davis, B. G. (2011) *Nat. Chem. Biol.* **7**, 631–638.
89. Breton, C., Fournel-Gigleux, S., Palcic, M. M. (2012) *Curr. Opin. Struct. Biol.* **22**, 540–549.
90. Tvaroška, I. (2006) *ACS Symp. Ser.* **930**, 285–301.
91. Errey, J., Lee, S., Gibson, R., MartinezFleites, C., Barry, C., Jung, P. J., O'Sullivan, A., Davis, B., Davies, G. (2010) *Angew. Chem. Int. Ed.* **49**, 1234–1237.
92. Pathak, A. K., Besra, G. S., Crick, D., Maddry, J. A., Morehouse, C. B., Suling, W. J., Reynolds, R. C. (1999) *Bioorg. Med. chem.* **7**, 2407–2413.
93. Snapper, S., Melton, R., Mustafa, S., Kieser, T., WR Jr., J. (1990) *Mol. Microbiol.* **4**, 1911–1919.
94. Splain, R. A., Kiessling, L. L. (2010) *Bioorg. Med. Chem.* **18**, 3753–3759.
95. Pathak, A. K., Pathak, V., Seitz, L., Maddry, J. A., Gurcha, S. S., Besra, G. S., Suling, W. J., Reynolds, R. C. (2001) *Bioorg. Med. Chem.* **9**, 3129–3143.
96. Poulin, M. B., Zhou, R., Lowary, T. L. (2012) *Org. Biomol. Chem.* **10**, 4074–4087.
97. Weston, A., Stern, R., Lee, R., Nassau, P., Monsey, D., Martin, S., Scherman, M., Besra, G., Duncan, K., McNeil, M. (1998) *Tuber. Lung Dis.* **78**, 123–131.

98. Mugunthan, G., Sriram, D., Yogeeswari, P., Kartha, K. R. (2012) *J. Carbohydr. Chem.* **31**, 553–570.
99. Richards, M. R., Lowary, T. L. (2009) *ChemBioChem* **10**, 1920–1938.
100. Borrelli, S., Zandberg, W. F., Mohan, S., Ko, M., Martinez-Gutierrez, F., Partha, S. K., Sanders, D. A., Av-Gay, Y., Pinto, B. M. (2010) *Int. J. Antimicrob. Agents* **36**, 364–368.
101. Peltier, P., Beláňová, M., Dianišková, P., Zhou, R., Zheng, R. B., Pearcey, J. A., Joe, M., Brennan, P. J., Nugier-Chauvin, C., Ferrières, V. (2010) *Chem. Biol.* **17**, 1356–1366.
102. Trunkfield, A. E., Gurcha, S. S., Besra, G. S., Bugg, T. D. (2010) *Bioorg. Med. Chem.* **18**, 2651–2663.
103. Bupp, K., van Heijenoort, J. (1993) *J. Bacteriol.* **175**, 1841–1843.
104. Szczepina, M. G., Zheng, R. B., Completo, G. C., Lowary, T. L., Pinto, B. M. (2009) *ChemBioChem* **10**, 2052–2059.
105. Vembaiyan, K., Pearcey, J. A., Bhasin, M., Lowary, T. L., Zou, W. (2011) *Bioorg. Med. Chem.* **19**, 58–66.
106. Mills, J. A., Motichka, K., Jucker, M., Wu, H. P., Uhlik, B. C., Stern, R. J., Scherman, M. S., Vissa, V. D., Pan, F., Kundu, M. (2004) *J. Biol. Chem.* **279**, 43540–43546.
107. Brennan, P. (2003) *Tuberculosis* **83**, 91–97.
108. Pan, F., Jackson, M., Ma, Y., McNeil, M. (2001) *J. Bacteriol.* **183**, 3991–3998.
109. Ma, Y., Mills, J. A., Belisle, J. T., Vissa, V., Howell, M., Bowlin, K., Scherman, M. S., McNeil, M. (1997) *Microbiology* **143**, 937–945.
110. McNeil, M., Daffe, M., Brennan, P. J. (1990) *J. Biol. Chem.* **265**, 18200–18206.
111. McNeil, M., Brennan, P. (1991) *Res. Microbiol.* **142**, 451–463.
112. Lucas, R., Balbuena, P., Errey, J. C., Squire, M. A., Gurcha, S. S., McNeil, M., Besra, G. S., Davis, B. G. (2008) *ChemBioChem* **9**, 2197–2199.
113. Fairbanks, A. J., Carpenter, N. C., Fleet, G. W., Ramsden, N. G., Cenci de Bello, I., Winchester, B. G., Al-Daher, S. S., Nagahashi, G. (1992) *Tetrahedron* **48**, 3365–3376.
114. Davis, B. G., Brandstetter, T. W., Hackett, L., Winchester, B. G., Nash, R. J., Watson, A. A., Griffiths, R. C., Smith, C., Fleet, G. W. (1999) *Tetrahedron* **55**, 4489–4500.
115. Davis, B. G., Nash, R. J., Watson, A. A., Smith, C., Fleet, G. W. (1999) *Tetrahedron* **55**, 4501–4520.
116. Mikušová, K. N., Yagi, T., Stern, R., McNeil, M. R., Besra, G. S., Crick, D. C., Brennan, P. J. (2000) *J. Biol. Chem.* **275**, 33890–33897.

117. Alderwick, L. J., Seidel, M., Sahm, H., Besra, G. S., Eggeling, L. (2006) *J. Biol. Chem.* **281**, 15653–15661.
118. Seidel, M., Alderwick, L. J., Birch, H. L., Sahm, H., Eggeling, L., Besra, G. S. (2007) *J. Biol. Chem.* **282**, 14729–14740.
119. Khasnobis, S., Zhang, J., Angala, S. K., Amin, A. G., McNeil, M. R., Crick, D. C., Chatterjee, D. (2006) *Chem. Biol.* **13**, 787–795.
120. Zhang, J., Khoo, K. H., Wu, S. W., Chatterjee, D. (2007) *J. Am. Chem. Soc.* **129**, 9650–9662.
121. Lee, R. E., Brennan, P. J., Besra, G. S. (1997) *Glycobiology* **7**, 1121–1128.
122. Brennan, P. J., Nikaido, H. (1995) *Annu. Rev. Biochem.* **64**, 29–63.
123. Lee, R., Mikusova, K., Brennan, P., Besra, G. (1995) *J. Am. Chem. Soc.* **117**, 11829–11832.
124. McNeil, M., Daffe, M., Brennan, P. (1991) *J. Biol. Chem.* **266**, 13217–13223.
125. Khoo, K. H., Tang, J. B., Chatterjee, D. (2001) *J. Biol. Chem.* **276**, 3863–3871.
126. Amin, A. G., Goude, R., Shi, L., Zhang, J., Chatterjee, D., Parish, T. (2008) *Microbiology* **154**, 240–248.
127. Belanger, A. E., Besra, G. S., Ford, M. E., Mikusová, K., Belisle, J. T., Brennan, P. J., Inamine, J. M. (1996) *Proc. Natl. Acad. Sci. USA* **93**, 11919–11924.
128. Cole, S., Brosch, R., Parkhill, J., Garnier, T., Churcher, C., Harris, D., Gordon, S., Eiglmeier, K., Gas, S., Barry, C. R. (1998) *Nature* **393**, 537–544.
129. Escuyer, V. E., Lety, M. A., Torrelles, J. B., Khoo, K. H., Tang, J. B., Rithner, C. D., Frehel, C., McNeil, M. R., Brennan, P. J., Chatterjee, D. (2001) *J. Biol. Chem.* **276**, 48854–48862.
130. Khoo, K. H., Douglas, E., Azadi, P., Inamine, J. M., Besra, G. S., Mikušová, K., Brennan, P. J., Chatterjee, D. (1996) *J. Biol. Chem.* **271**, 28682–28690.
131. Sreevatsan, S., Pan, X., Zhang, Y., Kreiswirth, B. N., Musser, J. M. (1997) *Antimicrob. Agents Chemother.* **41**, 636–640.
132. Zhang, N., Torrelles, J. B., McNeil, M. R., Escuyer, V. E., Khoo, K. H., Brennan, P. J., Chatterjee, D. (2003) *Mol. Microbiol.* **50**, 69–76.
133. Birch, H. L., Alderwick, L. J., Bhatt, A., Rittmann, D., Krumbach, K., Singh, A., Bai, Y., Lowary, T. L., Eggeling, L., Besra, G. S. (2008) *Mol. Microbiol.* **69**, 1191–1206.
134. Škovierová, H., Larrouy-Maumus, G., Zhang, J., Kaur, D., Barilone, N., Korduláková, J., Gilleron, M., Guadagnini, S., Belanová, M., Prevost, M. C. (2009) *Glycobiology* **19**, 1235–1247.

135. Mishra, A. K., Alves, J. E., Krumbach, K., Nigou, J., Castro, A. G., Geurtsen, J., Eggeling, L., Saraiva, M., Besra, G. S. (2012) *J. Biol. Chem.* **287**, 44173–44183.
136. Birch, H. L., Alderwick, L. J., Appelmelk, B. J., Maaskant, J., Bhatt, A., Singh, A., Nigou, J., Eggeling, L., Geurtsen, J., Besra, G. S. (2010) *Proc. Natl. Acad. Sci. USA* **107**, 2634–2639.
137. Wolucka, B. A., McNeil, M. R., De Hoffmann, E., Chojnacki, T., Brennan, P. J. (1994) *J. Biol. Chem.* **269**, 23328–23335.
138. Huang, H., Scherman, M. S., D'Haeze, W., Vereecke, D., Holsters, M., Crick, D. C., McNeil, M. R. (2005) *J. Biol. Chem.* **280**, 24539–24543.
139. Huang, H., Berg, S., Spencer, J. S., Vereecke, D., D'Haeze, W., Holsters, M., McNeil, M. R. (2008) *Microbiology* **154**, 736–743.
140. Mikušová, K., Huang, H., Yagi, T., Holsters, M., Vereecke, D., D'Haeze, W., Scherman, M. S., Brennan, P. J., McNeil, M. R., Crick, D. C. (2005) *J. Bacteriol.* **187**, 8020–8025.
141. Klutts, J. S., Hatanaka, K., Pan, Y., Elbein, A. D. (2002) *Arch. Biochem. Biophys.* **398**, 229–239.
142. Scherman, M., Weston, A., Duncan, K., Whittington, A., Upton, R., Deng, L., Comber, R., Friedrich, J. D., McNeil, M. (1995) *J. Bacteriol.* **177**, 7125–7130.
143. Scherman, M. S., Kalbe-Bournonville, L., Bush, D., Xin, Y., Deng, L., McNeil, M. (1996) *J. Biol. Chem.* **271**, 29652–29658.
144. Xin, Y., Lee, R. E., Scherman, M. S., Khoo, K. H., Besra, G. S., Brennan, P. J., McNeil, M. (1997) *Biochim. Biophys. Acta* **1335**, 231–234.
145. Wolucka, B. A. (2008) *FEBS J.* **275**, 2691–2711.
146. Barry, C. E., Crick, D. C., McNeil, M. R. (2007) *Infect. Disord. Drug Targets* **7**, 182–202.
147. Makarov, V., Manina, G., Mikusova, K., Möllmann, U., Ryabova, O., Saint-Joanis, B., Dhar, N., Pasca, M. R., Buroni, S., Lucarelli, A. P. (2009) *Science* **324**, 801–804.
148. Christophe, T., Jackson, M., Jeon, H. K., Fenistein, D., Contreras-Dominguez, M., Kim, J., Genovesio, A., Carralot, J. P., Ewann, F., Kim, E. H. (2009) *PLoS Pathog.* **5**, e1000645.
149. Lee, R., Brennan, P., Besra, G. (1998) *Bioorg. Med. Chem. Lett.* **8**, 951–954.
150. McGurk, P., Chang, G. X., Lowary, T. L., McNeil, M., Field, R. A. (2001) *Tetrahedron Lett.* **42**, 2231–2234.
151. Centrone, C. A., Lowary, T. L. (2002) *J. Org. Chem.* **67**, 8862–8870.

152. Joe, M., Lowary, T. L. (2006) *Carbohydr. Res.* **341**, 2723–2730.
153. Bosco, M., Bisseret, P., Constant, P., Eustache, J. (2007) *Tetrahedron Lett.* **48**, 153–157.
154. Zhang, J., Angala, S. K., Pramanik, P. K., Li, K., Crick, D. C., Liav, A., Jozwiak, A., Swiezewska, E., Jackson, M., Chatterjee, D. (2011) *ACS Chem. Biol.* **6**, 819–828.
155. Ayers, J. D., Lowary, T. L., Morehouse, C. B., Besra, G. S. (1998) *Bioorg. Med. Chem. Lett.* **8**, 437–442.
156. Ishiwata, A., Ito, Y. (2011) *J. Am. Chem. Soc.* **133**, 2275–2291.
157. Callam, C. S., Lowary, T. L. (2001) *J. Org. Chem.* **66**, 8961–8972.
158. Yin, H., D'Souz, F. W., Lowary, T. L. (2002) *J. Org. Chem.* **67**, 892–903.
159. Cociorva, O. M., Gurcha, S. S., Besra, G. S., Lowary, T. L. (2005) *Bioorg. Med. Chem.* **13**, 1369–1379.
160. D'Souz, F. W., Lowary, T. L. (2000) *Org. Lett.* **2**, 1493–1495.
161. Shi, L., Berg, S., Lee, A., Spencer, J. S., Zhang, J., Vissa, V., McNeil, M. R., Khoo, K. H., Chatterjee, D. (2006) *J. Biol. Chem.* **281**, 19512–19526.
162. Joe, M., Bai, Y., Nacario, R. C., Lowary, T. L. (2007) *J. Am. Chem. Soc.* **129**, 9885–9901.
163. Chang, G. X., Lowary, T. L. (2006) *Tetrahedron Lett.* **47**, 4561–4564.
164. Marotte, K., Ayad, T., Génisson, Y., Besra, G. S., Baltas, M., Prandi, J. (2003) *Eur. J. Org. Chem.* **2003**, 2557–2565.
165. Pathak, A. K., Pathak, V., Bansal, N., Maddry, J. A., Reynolds, R. C. (2001) *Tetrahedron Lett.* **42**, 979–982.
166. Han, J., Gadikota, R. R., McCarren, P. R., Lowary, T. L. (2003) *Carbohydr. Res.* **338**, 581–588.
167. Pathak, A. K., Pathak, V., Gurcha, S. S., Besra, G. S., Reynolds, R. C. (2002) *Bioorg. Med. Chem. Lett.* **12**, 2749–2752.
168. Pathak, A. K., Pathak, V., Seitz, L., Gurcha, S. S., Besra, G. S., Riordan, J. M., Reynolds, R. C. (2007) *Bioorg. Med. Chem.* **15**, 5629–5650.
169. Pathak, A. K., Pathak, V., Maddry, J. A., Suling, W. J., Gurcha, S. S., Besra, G. S., Reynolds, R. C. (2001) *Bioorg. Med. Chem.* **9**, 3145–3151.
170. Cociorva, O. M., Lowary, T. L. (2004) *Carbohydr. Res.* **339**, 853–865.
171. Pathak, A. K., Pathak, V., Kulshrestha, M., Kinnaird, D., Suling, W. J., Gurcha, S., Besra, G. S., Reynolds, R. C. (2003) *Tetrahedron* **59**, 10239–10248.
172. Pathak, A. K., Pathak, V., Riordan, J. R., Suling, W. J., Gurcha, S. S., Besra, G. S., Reynolds, R. C. (2007) *Bioorg. Med. Chem. Lett.* **17**, 4527–4530.

173. Pathak, A. K., Pathak, V., Suling, W. J., Riordan, J. R., Gurcha, S. S., Besra, G. S., Reynolds, R. C. (2009) *Bioorg. Med. Chem.* **17**, 872–881.
174. Pathak, A. K., Pathak, V., Suling, W. J., Gurcha, S. S., Morehouse, C. B., Besra, G. S., Maddry, J. A., Reynolds, R. C. (2002) *Bioorg. Med. Chem.* **10**, 923–928.
175. Chatterjee, D., Khoo, K. H. (1998) *Glycobiology* **8**, 113–120.
176. Schaeffer, M. L., Khoo, K.-H., Besra, G. S., Chatterjee, D., Brennan, P. J., Belisle, J. T., Inamine, J. M. (1999) *J. Biol. Chem.* **274**, 31625–31631.
177. Vergne, I., Daffé, M. (1998) *Front. Biosci.* **3**, 865–876.
178. Hans, S., Altiti, A., Mootoo, D. R. (2013) *Org. Biomol. Chem.* **11**, 6952–6959.
179. Parlane, N. A., Compton, B. J., Hayman, C. M., Painter, G. F., Basaraba, R. J., Heiser, A., Buddle, B. M. (2012) *Vaccine* **30**, 580–588.
180. Denis, M., Ainge, G. D., Larsen, D. S., Severn, W. B., Painter, G. F. (2009) *Immunopharmacol. Immunotoxicol.* **31**, 577–582.
181. J. C. Errey, S. S. Lee, R. P. Gibson, C. M. Fleites, C. S. Barry, P. M. J. Jung, A. C. O'Sullivan, Davis, B. G., Davies, A. G. J. (2010) *Angew. Chem. Int. Ed.* **49**, 1234–1237.
182. Waldscheck, B., Streiff, M., Notz, W., Kinzy, W., Schmidt, R. R. (2001) *Angew. Chem. Int. Ed.* **40**, 4007–4011.
183. Weisman, L., Ballou, C. (1984) *J. Biol. Chem.* **259**, 3457–3463.
184. Jackson, M., Brennan, P. J. (2009) *J. Biol. Chem.* **284**, 1949–1953.
185. Weisman, L., Ballou, C. (1984) *J. Biol. Chem.* **259**, 3464–3469.
186. Xia, L., Zheng, R. B., Lowary, T. L. (2012) *ChemBioChem* **13**, 1139–1151.
187. Brown, J. R., Field, R. A., Barker, A., Guy, M., Grewal, R., Khoo, K. H., Brennan, P. J., Besra, G. S., Chatterjee, D. (2001) *Bioorg. Med. Chem.* **9**, 815–824.
188. Tam, P. H., Besra, G. S., Lowary, T. L. (2008) *ChemBioChem* **9**, 267–278.
189. Watt, J. A., Williams, S. J. (2005) *Org. Biomol. Chem.* **3**, 1982–1992.
190. Subramaniam, V., Gurcha, S. S., Besra, G. S., Lowary, T. L. (2005) *Tetrahedron: Asymmetry* **16**, 553–567.
191. Subramaniam, V., Gurcha, S. S., Besra, G. S., Lowary, T. L. (2005) *Bioorg. Med. Chem.* **13**, 1083–1094.
192. Lo Conte, M., Marra, A., Chambery, A., Gurcha, S. S., Besra, G. S., Dondoni, A. (2010) *J. Org. Chem.* **75**, 6326–6336.
193. Tam, P. H., Lowary, T. L. (2010) *Org. Biomol. Chem.* **8**, 181–192.
194. Tam, P. H., Lowary, T. L. (2007) *Carbohydr. Res.* **342**, 1741–1772.

195. Poláková, M., Beláňová, M., Petruš, L., Mikušová, K. (2010) *Carbohydr. Res.* **345**, 1339–1347.
196. Poláková, M., Beláňová, M., Mikušová, K. N., Lattová, E., Perreault, H. L. N. (2011) *Bioconjugate Chem.* **22**, 289–298.
197. Lowary, T. L., Hindsgaul, O. (1994) *Carbohydr. Res.* **251**, 33–67.
198. Rose, N. L., Completo, G. C., Lin, S. J., McNeil, M., Palcic, M. M., Lowary, T. L. (2006) *J. Am. Chem. Soc.* **128**, 6721–6729.
199. Palcic, M. M., Heerze, L. D., Pierce, M., Hindsgaul, O. (1988) *Glycoconjugate J.* **5**, 49–63.
200. Yokoyama, K., Ballou, C. (1989) *J. Biol. Chem.* **264**, 21621–21628.
201. Dinev, Z., Gannon, C. T., Egan, C., Watt, J. A., McConville, M. J., Williams, S. J. (2007) *Org. Biomol. Chem.* **5**, 952–959.
202. Kalscheuer, R., Syson, K., Veeraraghavan, U., Weinrick, B., Biermann, K. E., Liu, Z., Sacchettini, J. C., Besra, G., Bornemann, S., Jacobs Jr., W. R. (2010) *Nat. Chem. Biol.* **6**, 376–384.
203. Syson, K., Stevenson, C. E., Rejzek, M., Fairhurst, S. A., Nair, A., Bruton, C. J., Field, R. A., Chater, K. F., Lawson, D. M., Bornemann, S. (2011) *J. Biol. Chem.* **286**, 38298–38310.
204. Stam, M. R., Danchin, E. G., Rancurel, C., Coutinho, P. M., Henrissat, B. (2006) *Protein Eng. Des. Sel.* **19**, 555–562.
205. Elbein, A. D., Pastuszak, I., Tackett, A. J., Wilson, T., Pan, Y. T. (2010) *J. Biol. Chem.* **285**, 9803–9812.
206. Sassetti, C. M., Boyd, D. H., Rubin, E. J. (2003) *Mol. Microbiol.* **48**, 77–84.
207. Sassetti, C. M., Rubin, E. J. (2003) *Proc. Natl. Acad. Sci. USA* **100**, 12989–12994.
208. Kalscheuer, R., Jacobs Jr., W. R. (2010) *Drug News Perspect.* **23**, 619–624.
209. Dahl, J. L., Kraus, C. N., Boshoff, H. I., Doan, B., Foley, K., Avarbock, D., Kaplan, G., Mizrahi, V., Rubin, H., Barry, C. E. (2003) *Proc. Natl. Acad. Sci. USA* **100**, 10026–10031.
210. Barea, F., Bonatto, D. (2008) *Mutat. Res.* **642**, 43–56.
211. Veleti, S. K., Lindenberger, J. J., Ronning, D. R., Sucheck, S. J. (2014) *Bioorg. Med. Chem.* **22**, 1401–1411.
212. Veleti, S. K., Lindenberger, J. J., Thanna, S., Ronning, D. R., Sucheck, S. J. (2014) *J. Org. Chem.* **79**, 9444–9450.
213. Jarlier, V., Nikaido, H. (1994) *FEMS Microbiol. Lett.* **123**, 11–18.
214. Yuan, Y., Lee, R. E., Besra, G. S., Belisle, J. T., Barry, C. E. (1995) *Proc. Natl. Acad. Sci. USA* **92**, 6630–6634.

215. Yuan, Y., Crane, D. C., Musser, J. M., Sreevatsan, S., Barry, C. E. (1997) *J. Biol. Chem.* **272**, 10041–10049.
216. Yuan, Y., Barry, C. E. (1996) *Proc. Natl. Acad. Sci. USA* **93**, 12828–12833.
217. Barry III, C. E., Lee, R. E., Mdluli, K., Sampson, A. E., Schroeder, B. G., Slayden, R. A., Yuan, Y. (1998) *Prog. Lipid Res.* **37**, 143–179.
218. Liu, J., Barry, C. E., Besra, G. S., Nikaido, H. (1996) *J. Biol. Chem.* **271**, 29545–29551.
219. Chatterjee, D. (1997) *Curr. Opin. Chem. Biol.* **1**, 579–588.
220. Belisle, J. T., Vissa, V. D., Sievert, T., Takayama, K., Brennan, P. J., Besra, G. S. (1997) *Science* **276**, 1420–1422.
221. Ronning, D. R., Klabunde, T., Besra, G. S., Vissa, V. D., Belisle, J. T., Sacchettini, J. C. (2000) *Nat. Struct. Biol.* **7**, 141–146.
222. Jackson, M., Raynaud, C., Lanéelle, M. A., Guilhot, C., Laurent-Winter, C., Ensergueix, D., Gicquel, B., Daffé, M. (1999) *Mol. Microbiol.* **31**, 1573–1587.
223. Ronning, D. R., Vissa, V., Besra, G. S., Belisle, J. T., Sacchettini, J. C. (2004) *J. Biol. Chem.* **279**, 36771–36777.
224. Anderson, D. H., Harth, G., Horwitz, M. A., Eisenberg, D. (2001) *J. Mol. Biol.* **307**, 671–681.
225. Nguyen, L., Chinnapapagari, S., Thompson, C. J. (2005) *J. Bacteriol.* **187**, 6603–6611.
226. Armitige, L. Y., Jagannath, C., Wanger, A. R., Norris, S. J. (2000) *Infect. Immun.* **68**, 767–778.
227. Matsunaga, I., Naka, T., Talekar, R. S., McConnell, M. J., Katoh, K., Nakao, H., Otsuka, A., Behar, S. M., Yano, I., Moody, D. B. (2008) *J. Biol. Chem.* **283**, 28835–28841.
228. Hanessian, S., Lavallee, P. (1972) *J. Antibiotics* **25**, 683–684.
229. Chung, G., Aktar, Z., Jackson, S., Duncan, K. (1995) *Antimicrob. Agents Chemother.* **39**, 2235–2238.
230. Rose, J. D., Maddry, J. A., Comber, R. N., Suling, W. J., Wilson, L. N., Reynolds, R. C. (2002) *Carbohydr. Res.* **337**, 105–120.
231. Wang, J., Elchert, B., Hui, Y., Takemoto, J. Y., Bensaci, M., Wennergren, J., Chang, H., Rai, R., Chang, C.-W. T. (2004) *Bioorg. Med. Chem.* **12**, 6397–6413.
232. Sanki, A. K., Boucau, J., Srivastava, P., Adams, S. S., Ronning, D. R., Sucheck, S. J. (2008) *Bioorg. Med. Chem.* **16**, 5672–5682.
233. Sanki, A. K., Boucau, J., Umesiri, F. E., Ronning, D. R., Sucheck, S. J. (2009) *Mol. Biosyst.* **5**, 945–956.

234. Ibrahim, D. A., Boucau, J., Lajiness, D. H., Veleti, S. K., Trabbic, K. R., Adams, S. S., Ronning, D. R., Sucheck, S. J. (2012) *Bioconjugate Chem.* **23**, 2403–2416.

235. Backus, K. M., Boshoff, H. I., Barry, C. S., Boutureira, O., Patel, M. K., D'Hooge, F., Lee, S. S., Via, L. E., Tahlan, K., Barry III, C. E., Davies, B. G. (2011) *Nat. Chem. Biol.* **7**, 228–235.

236. Yang, M., Proctor, M. R., Bolam, D. N., Errey, J. C., Field, R. A., Gilbert, H. J., Davis, B. G. (2005) *J. Am. Chem. Soc.* **127**, 9336–9337.

237. Favrot, L., Grzegorzewicz, A. E., Lajiness, D. H., Marvin, R. K., Boucau, J., Isailovic, D., Jackson, M., Ronning, D. R. (2013) *Nat. Commun.* **4**, 2748.

238. Favrot, L., Lajiness, D. H., Ronning, D. R. (2014) *J. Biol. Chem.* **289**, 25031–250340.

Part 5:
Conclusion

Chapter 19

Carbohydrate-Based Drugs on the Market: Overview and Future Directions

Matteo Gentili and Oscar Francesconi

Department of Chemistry Ugo Schiff, University of Florence, Polo Scientifico e Tecnologico, via della Lastruccia 13 50019 Sesto Fiorentino, Firenze, Italy

19.1 Introduction

The importance of carbohydrates in cellular physiology has been well known to biologists and biochemists for over two centuries. Nevertheless, these essential molecules have been living in the shadow of proteins and nucleic acids until the last decades, when their pivotal role in cell to cell and cell to pathogen interactions has started to emerge. This renaissance in glycobiology has led to an increasing interest towards the pathophysiological mechanisms in which carbohydrates take part, with the ultimate aim of discovering novel spots for therapeutic intervention. Although current research in structural biology has successfully unraveled many cellular processes in which carbohydrates are directly involved, drug discovery has not moved at the same speed in producing new carbohydrate-based drugs. This is not surprising, considering the complex chemical nature of sugars; indeed, carbohydrates are challenging and synthetically demanding substrates, which, in many cases, require complex multistep syntheses, making it difficult to create libraries of compounds to be tested in a classical pharmaceutical approach. Besides, they are usually endowed with poor pharmacokinetic properties, being unable to be passively absorbed at an intestinal level, but at the same time, possessing a high renal clearance, causing low plasmatic half-life if administered parenterally. Despite these drawbacks, several drugs, containing carbohydrate epitopes essential for their activity, are present on the market, often representing the first choice for the treatment of a broad range of pathologies, including bacterial, viral and micotic infections, tumors, diabetes, blood coagulation disorders, and epilepsy.

19.2 Drugs on the Market: The Role of Carbohydrates

Glycobiology is an interdisciplinary field of research, which was born from the interplay between the chemistry developed on carbohydrates and their biochemistry within the cell. Only recently, the development of new and technologically advanced techniques has allowed the comprehension of the multiple subtle roles in which carbohydrates are involved. For this reason, the first carbohydrate related drugs to be introduced on the market were not carbohydrate-based drugs, but rather, they can be defined as carbohydrate-containing drugs. In some cases, the saccharidic moiety has been exploited only to modulate the pharmacokinetics of the drug; in this case, the sugar does not take part in the interaction with the pharmacological target. On the other side, a broad range of carbohydrate-containing compounds that owe their biological activity to the presence of specific saccharidic residues are present on the therapeutic panorama. Historically, medicinal chemistry has mainly taken advantage of those carbohydrate containing drugs that could be easily encountered and isolated from natural sources, such as fungi or bacteria, thus overcoming the issue of problematic synthesis. This has allowed a fast introduction on the market of different kinds of chemotherapeutics starting from the forties, that were further chemically modified in order to improve their effectiveness or reduce their side effects. The very first glycosidic-based antibiotic to be approved was streptomycin in 1944, which is still a valuable weapon to treat tuberculosis; other related aminoglycosidic antibiotics followed, such as neomycin (1949) and gentamycin (1963) (Fig. 19.1). It was only in the late eighties that it was discovered that aminoglycosides owe their potent bactericidal activity to the ability of selectively binding the 16S ribosomal ribonucleic acid (rRNA) subunit of the 30S bacterial ribosome [1]. These aminosugars, whose amino groups can be protonated in the cellular environment, find strong electrostatic interactions with the negatively charged phosphate backbones of RNA [2], which are eventually reinforced by the contribution of hydrogen bonds between the amino and hydroxyl groups of both RNA and aminoglycosides [3]; the result of this is a strong complex that produces errors in ribosomal translational processes, leading to a lethal accumulation of truncated or nonfunctional proteins inside the cell. Interestingly, the fact that these drugs, due to their nephro- and ototoxicity, have been largely replaced in the past by other classes of better-tolerated antibacterials, has in part preserved them from bacterial resistance; this fact has now induced physicians to reconsider their use to overcome infections carried on by multidrug resistant bacteria [4,5]. In addition to this, aminoglycosidic derivatives have recently proven to be effective in hindering human immunodeficiency virus (HIV) infection by targeting different crucial steps in viral replication [6] and inducing the production in humans of θ-defensins, a class of peptides which inhibits the virus entry [7].

The importance of the saccharidic residue in the drug mechanism of action is well represented by the anthracyclines antibiotics (Fig. 19.1). The discovery of daunorubicin in the sixties as a natural product of *Streptomyces peucetius var. caesius* by Italian [8] and French researchers opened the way to this leading class of antineoplastics, widely used for the treatment of different types of tumors, including leukemias, lymphomas, and breast, uterine, and ovarian cancers [9]. All anthracyclines are characterized by the presence of an aminosugar residue linked to an anthraquinonic aglycone.

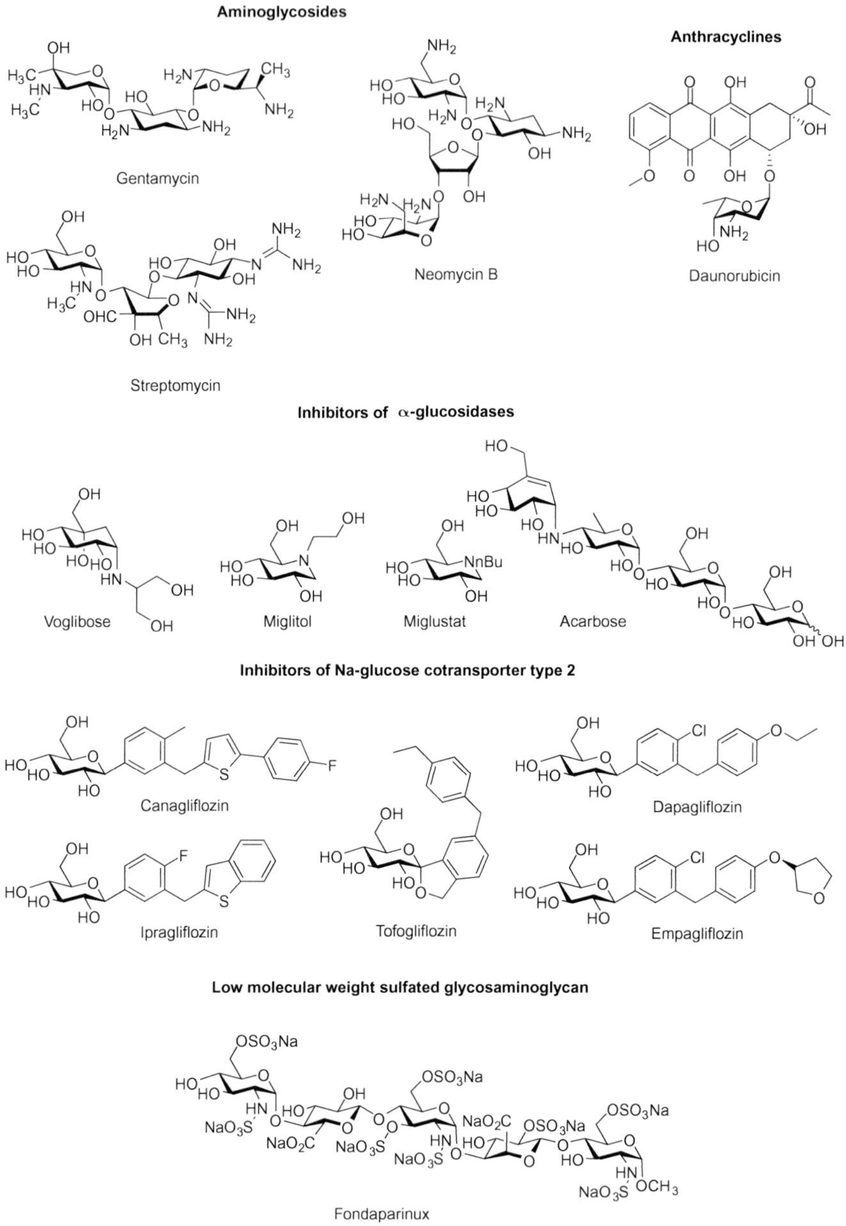

Figure 19.1 Carbohydrate-derived drugs on market.

The sugar is responsible for the interaction with the minor groove of deoxyribonucleic acid (DNA), while the intercalation of the anthraquinonic moiety in the double helix prevents the activity of DNA topoisomerase II, an essential nuclear enzyme which regulates DNA topology [10]. A number of chemical modifications at the sugar residue, such as the variation of the type of glycoside [11] and the implementation with disaccharides [12], exert an impact on the drug potency and can be useful in modulating side effects.

The first designed carbohydrate-based drugs were introduced on the market only in the nineties and belong to the family of oral drugs for the treatment of types I and II diabetes. Voglibose (Glustat, Takeda) [13], miglitol (Glyset, Pfizer) [14], and acarbose (Glucobay, Bayer) [15] are saccharidic competitive inhibitors of α-glucosidases present in the brush border of the small intestine (Fig. 19.1). These enzymes are responsible for the hydrolysis of oligosaccharides to monosaccharides, which can then be absorbed by the enterocytes. By binding to the catalytic site of glucosidases, these drugs prevent their action, thus resulting in diminished sugar absorption, causing a hypoglycemic effect. The efficacy of these drugs in controlling post-prandial glycemia has rendered them one of the most used classes of anti-diabetics, also due to their good tolerability. Chemically related to these compounds, miglustat (Zavesca, Actelion) was introduced in 2002 to treat type I Gaucher's disease [16], and it was the first drug to be approved for the treatment of Niemann–Pick type C disease in 2010 (Fig. 19.1) [17]. An interesting step forward in the treatment of diabetes type II using carbohydrate-based compounds was the introduction on the market of canagliflozin in April of 2013 (Invokana, Janssen Pharmaceuticals), followed by dapagliflozin (Fargixa, Bristol-Myers Squibbs) in January of 2014 (Fig. 19.1). These drugs act on the sodium–glucose cotransporter type 2 (SGLT-2) blocking the active reabsorption of glucose in the proximal convoluted tubule of the nephron in the kidney; this results in an excretion of glucose with urine, thus lowering plasma glucose levels without inducing increased insulin secretion or sensitivity, hypoglycemia, or weight gain [18,19]. This interesting new class is rapidly emerging, with more analogs awaiting the Food and Drug Administration (FDA) approval (ipragliflozin (Astellas), tofogliflozin (Chugai Pharmaceuticals) and empagliflozin (Boheringer Ingelheim)) (Fig. 19.1) [20].

The discovery of heparin in 1916 and its introduction in therapy in the thirties has represented a major progress in the therapy of coagulation-related dysfunctions. Since then, many sulfated glycosaminoglycans of lower molecular weight than heparin have been approved and are available on the market; usually, they are preferred to heparin for their better pharmacokinetic profile, as they can be administered subcutaneously instead of intravenously (Fig. 19.1). Although heparin is almost hitting the century of life, new opportunities are emerging for its use, such as in anti-inflammatory and anti-metastatic drugs [21].

Due to their structural complexity and to the fact that their bound conformations can be very different from their lowest energy ones, carbohydrate-based drugs have definitely taken advantage from the growing success of structure-based lead optimization techniques from the end of last century, as witnessed by the elegant example of the development of the antivirals oseltamivir (Tamiflu, Gilead/Roche) and zanamivir (Relenza, GlaxoSmithKline) (Fig. 19.2). As the last step for its replication, in order to leave the host cell, influenza virus needs to cleave a residue of sialic acid (Neu5Ac) to which it is attached. Viral neuraminidase (NA) is responsible for this task; therefore, its inhibition prevents the spreading of the infection. In 1994, the 2,3-didehydro analog of Neu5Ac was found to be a much more potent inhibitor (inhibitory constant (K_i) = 4 μM) than the natural substrate, in that it mimics the transition state in the hydrolysis of the Neu5Ac glycosidic bond [22]. The elucidated X-ray structure of the complex between NA and this compound was the starting point for its rational modification; the introduction in position 4 of the glycosidic ring of an aminic (oseltamivir) or guanidinic (zanamivir) residue dramatically improved the affinity for the enzyme, with a K_i value of 10^{-8} M and 10^{-10} M, respectively. Indeed, both of these groups form a salt bridge with glutamine (Glu)119 in the active site of the enzyme, the latter additionally interacting with Glu227. Despite their good potency, these compounds lacked good pharmacokinetics, so they

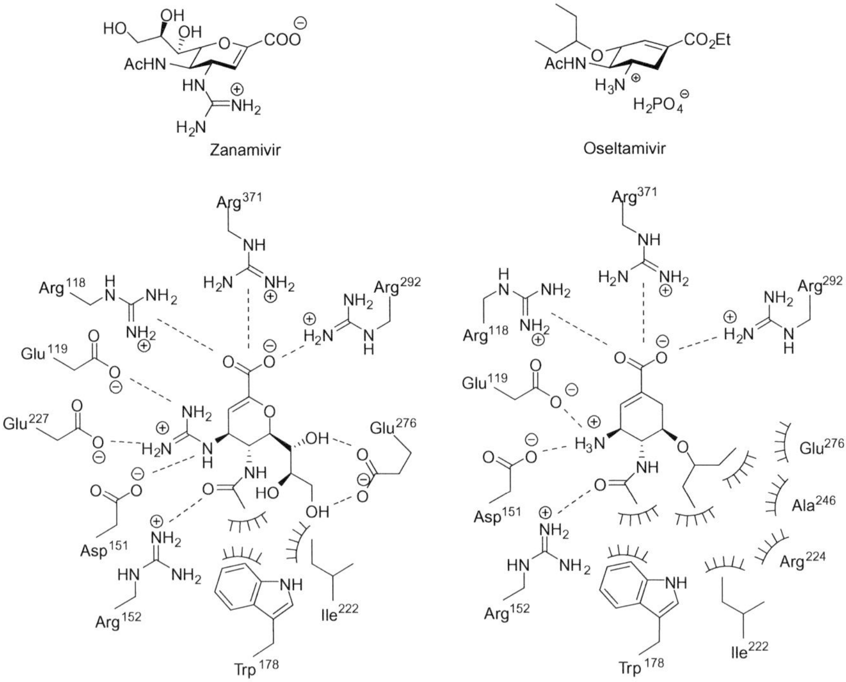

Figure 19.2 Schematic representation of the interactions in the binding site of neuraminidase by zanamivir (left) and oseltamivir (right).

were further modified by the elimination of the pyranosidic oxygen with the aim of increasing metabolic stability and to facilitate the chemical optimization of both pharmacological and biological activities.

Again, X-ray structures of the complex between Neu5Ac and its analogs bound to NA suggested that the hydrogen (H)-bonding interaction of the glycerol chain at the C_7 with Glu276 was only marginal, so this residue could be substituted with an alkyl group to instead take advantage of the presence of a lipophylic surface created by the hydrocarbon side chains of Glu276, alanine (Ala)246, arginine (Arg)224, and isoleucine (Ile)222 (Fig. 19.2). Therefore, the replacement of the glycerol moiety with a 3-pentyl chain finally produced oseltamivir, whose potency was enhanced by 150-fold (IC_{50} 150 nM) and that can be administered orally [23,24].

19.3 Carbohydrate-Based Vaccines

The observation that the capsular polysaccharide from *Pneumococcus* was able to stimulate the production of specific antibodies against this pathogen dates back to the twenties [25]. From this early discovery, the role of surface glycans as highly specific antigens has been deeply investigated and has led to the development of several carbohydrate-based vaccines. However, microbial polysaccharides are poorly immunogenic and can only elicit a short-term response, without developing any immunological memory; this issue was addressed by the conjugation of the saccharidic epitope to a carrier protein, a system that is able to stimulate a long lasting protection. While at present a variety of carbohydratic vaccines against different bacteria are available (see Table 19.1), only recently has research focused its attention on the development of vaccines based on viral- and tumor-associated antigens, among which carbohydrates are of preeminent importance. In Table 19.2, we wish to report a selection of some promising vaccines under clinical or preclinical evaluation for these pathologies.

Table 19.1 Licensed carbohydrate-based vaccines in US and EU.

Indication	Vaccines	Manufacturer (Trade name)
Haemophilus influenzae type b	Glycoconjugugate, polysaccharide with TT	Sanofi Pasteur (ActHIB, Hexacima, Hexyon, Hexavac); GlaxoSmithKline (Hiberix, Infarix Hexa)
	DT, TT and acellular pertussis adsorbed, inactivated poliovirus and Hib–TT conjugate vaccine	Sanofi Pasteur (Pentacel)
	Hib conjugate (meningococcal protein conjugate)	Merck & Co (PedvaxHIB)
	Hib conjugate (meningococcal protein conjugate) and hepatitis B (recombinant) vaccine	Merck & Co (Comvax)

(Continued)

Table 19.1 *(Continued)*

Indication	**Vaccines**	**Manufacturer (Trade name)**
Neisseria meningitidis A, C, Y and W-135	Glycoconjugate, meningococcal polysaccharide with DT	Sanofi Pasteur (Menactra)
	Meningococcal polysaccharide	Sanofi Pasteur (Menomune-A/C/Y/W-135)
	Glycoconjugate, meningococcal polysaccharide with CRM197	Novartis V&D (Menveo)
	Glycoconjugate, meningococcal polysaccharide with TT	GlaxoSmithKline (Nimenrix)
Salmonella typhi	Vi capsular polysaccharide	Sanofi Pasteur (TYPHIm Vi)
Streptococcus pneumoniae 4, 6B, 9V, 14, 18C, 19F and 23F	Pneumococcal polysaccharide 7-valent–CRM197 conjugate	Wyeth Pharmaceuticals (Prevnar)
Streptococcus pneumoniae 1, 4, 5, 6B, 7F, 9V, 14, 18C, 19F, and 23F	Pneumococcal polysaccharide, 10-valent	GlaxoSmithKline (Synflorix)
Streptococcus pneumoniae 1, 3, 4, 5, 6A, 6B, 7F, 9V, 14, 18C, 19A, 19F and 23F	Pneumococcal polysaccharide 13-valent–CRM197 conjugate	Wyeth Pharmaceuticals (Prevnar 13)
Streptococcus pneumoniae 1, 2, 3, 4, 5, 6B, 7F, 8, 9N, 9V, 10A, 11A, 12F, 14, 15B, 17F, 18C, 19F, 19A, 20, 22F, 23F and 33F	Pneumococcal polysaccharide, 23-valent	Merck & Co (Pneumovax 23)

This table reports vaccines authorized for immunization and distribution in the United States. (*Source*: Food and Drug Administration website) and in the European Union. (*Source*: European Medicines Agency website). This table should not be considered inclusive of all authorized carbohydrate-based vaccines currently available. Hib, *H. influenzae* type b; CRM197, a non-toxic mutant of diphtheria toxin; TT, tetanus toxoid; DT, diphtheria toxoid; Vi, capsular polysaccharide of *S. typhi*.

Table 19.2 Selected examples of carbohydrate vaccines under development.

Indication	Vaccines	Stage of development
Streptococcus pneumoniae	Glycoconjugate vaccine (1, 3, 4, 5, 6A, 6B, 7F, 9V, 14, 18C, 19A, 19F and 23F) conjugated to CRM197	Phase III[a]
	Glycoconjugate multivalent vaccine with novel adjuvants (e.g. MPL)	Phase I[a]
	Glycoconjugate vaccine (1, 3, 4, 5, 6B, 7F, 9V, 14, 18C, 19F, 23F) linked to either tetanus or diphtheria toxoid carrier	Phase III[a]
Group A *Streptococcus spp.*	Glycoconjugate group A polysaccharide with tetanus toxoid	Preclinical[a]
Group B *Streptococcus spp.*	Glycoconjugate vaccines of type Ia, lb, II, III, and V polysaccharides linked to carrier proteins	Phase II[a]
Haemophilus influenzae type B	Glycoconjugate of Hib PRP with CRM197	Phase III[a]
	Glycoconjugate of Hib PRP with diphtheria toxoid	Phase III[a]
	Glycoconjugate of Hib PRP with tetanus toxoid	Phase III[a]
	Glycoconjugate of Hib PRP with meningo-coccal type B outer membrane protein	Phase III[a]
Neisseria meningitidis group A	Glycoconjugate with tetanus toxoid	Preclinical[a]
Neisseria meningitidis group B	Outer membrane vesicles (OMV), high MW proteins, and C polysaccharide	Phase III[a]
	Glycoconjugate modified polysaccharide with recombinant PorB protein	Preclinical[a]
	De-*N*-acetyl sialic acid polysaccharide derivative-TT	Preclinical[a]
Neisseria meningitidis group B	Glycoconjugate with tetanus toxoid	Phase III[a]
Pseudomonas aeruginosa	Purified bacterial proteins, including flagellar Ag, LPS-O, porins, several inactivated bacterial toxins, and high MW polysaccharide antigen and glycoconjugate	Phase I[a]

(Continued)

Table 19.2 *(Continued)*

Indication	Vaccines	Stage of development
Salmonella typhi	Vi conjugate vaccine	Phase III[a]
HIV	V1V2Man_5GlcNAc glycopeptide	Preclinical [41]
	N332 ($Man_{8/9}GlcNAc_2$)	Preclinical [42]
Ovarian cancer	Globo H–STn–Tn–TF–GM_2–KLH unimolecular pentavalent conjugate	Phase I [43]
Breast cancer	Globo-H-DT plus α-galactosylceramide C34 as adjuvant	Preclinical [44]
	Globo-H-KLH plus QS-21 as adjuvant	Phase III [45]
Mammary cancer	PAM^3CysSK^4-peptide T_{HELPER} epitope-aberrant glycosilated MUC1 peptide	Preclinical [46]
Prostate cancer	Unimolecular hexavalent conjugates (Globo H–GM_2–$Lewis^y$–sTn–TF–Tn–R)	Preclinical [47]
	Globo H–GM_2–$Lewis^y$–MUC1–32(aa)–TF(c)–Tn(c)–KLH conjugate vaccine plus QS-21 as adjuvant	Phase II [48]

[a] Adapted from The Jordan Report Accelerated Development of Vaccines 2012. Hib, *H. influenzae* type b; (c), cluster; CRM197, a non-toxic mutant of diphtheria toxin; Globo H, globohexaosylceramide; KLH, keyhole limpet haemocyanin; MUC1-32(aa), mucin 1, 32 amino acids long; sTn, sialyl 2-6-α-N-acetylgalactosamine; GM_2, a tumor-associated ganglioside; $Lewis^y$, a tumor-associated surface antigen; TF, Thomsen–Friedenreich; Tn, 2-6-α-*N*-acetylgalactosamine; TT, tetanus toxoid; DT, diphtheria toxoid; Vi, capsular polysaccharide of *S. typhi*; QS-21, a *Quillaja saponaria* saponin.

19.4 Lectin Inhibitors: Drugs of the Future

Lectins are a class of proteins dedicated to carbohydrate recognition. Many years have passed from the discovery of ricin, the first identified lectin in 1888, but advances in the understanding of these proteins have been made only in the last decades. They have been found to be involved in a variety of physiological and pathological processes, making them promising new targets for therapeutical intervention. Thus, the inactivation of lectin's function by the use of competitive antagonists has become a field of increasing importance. In particular, extracellular lectins, which recognize terminal carbohydrate epitopes of other cells and pathogens, account for most of the biological targets that are currently under investigation for drug discovery. Since the carbohydrate–lectin interaction is one of the weakest in nature, with values in the millimolar range, chemical modification of the natural ligand in order to get higher binding values has been a widely employed approach that has led to new molecules endowed with increased inhibition abilities, with the aim of developing good lead candidates to be optimized to improve their drug-like characteristics.

19.5 Endogenous Lectins

In this context, selectin antagonists have been widely developed as candidate drugs. Selectins are calcium (Ca^{2+}) dependent lectins that recognize a common carbohydrate domain shared by sialyl Lewis$^{a/x}$. They cover different pathophysiological roles, mainly cell adhesion and migration from the bloodstream; these processes are majorly involved in inflammatory diseases (such as asthma, colitis, arthritis, and psoriasis) and cancer, since tumor cells use the selectin pathways to migrate from blood vessels to healthy tissues [26]. Therefore, selectin antagonists could be used to prevent tumor spreading and to control vaso-occlusive events due to selectin-mediated blood cell aggregation, such as in sickle cell disease [27]. A snapshot of the most promising carbohydrate-based selectin antagonists is reported in Table 19.3.

In particular bimosiamose (Revotar) [28] has been revealed to be highly effective in the treatment of chronic obstructive pulmonary disease (COPD) [29] and has completed phase II clinical trials for the treatment of psoriatic disorders, while GMI-1070 (Glycomimetics), which displays a >50-fold higher selectivity for E-selectin than for P- or L-selectin, is currently in phase II clinical trials for a separate indication, sickle-cell crisis [30].

A plethora of glycomimetics of lectin's natural substrates has been synthesized over the years, several times relying on the availability of the crystal structures of these proteins with their inhibitors. Dendritic cell-specific intercellular adhesion molecule-3-grabbing non-integrin (DC-SIGN) is probably the most investigated

Table 19.3 Carbohydrate-based selectin antagonists in preclinical and clinical trials.

Name and structure	Disease	Institution	Status
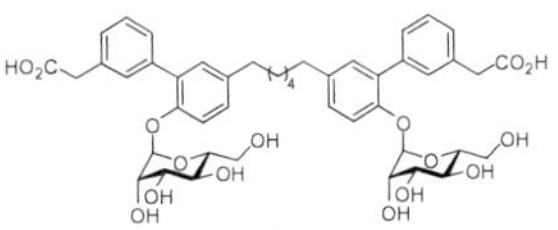 Bimosiamose	Asthma, COPD and psoriasis	Revotar	Phase II completed [28]
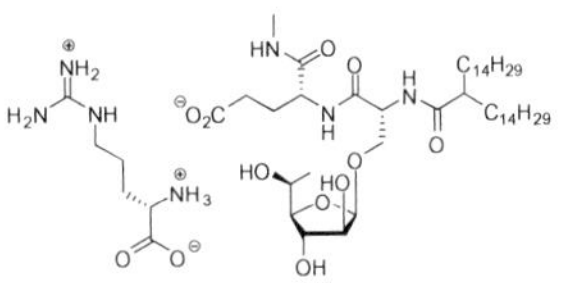 OJ-R9188	Allergic dermatitis	Nippon Organon	Preclinical [49]
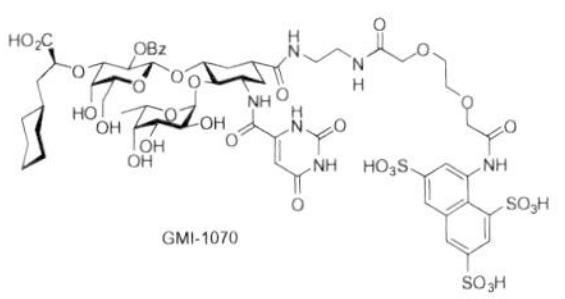 GMI-1070	Sickle cell crisis	GlycoMimetics	Phase II [30]
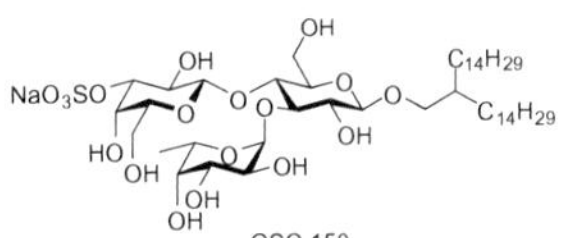 GSC-150	Metastatic cancer	Kanebo	Preclinical [50]

endogenous lectin, for its involvement in many infective processes carried on by viruses, bacteria, fungi, and protozoans, and in particular, in its role in HIV infection [31]. Compound (1) represents the most potent monovalent DC-SIGN inhibitor reported to date, displaying a IC_{50} value of 40 uM (Fig. 19.3) [32]. In this case, the authors have taken advantage of the presence of hydrophobic residues in the proximity of the mannose binding site to reinforce the interaction. As a future perspective, a strategic chemical modification of such small molecules will hopefully be envisaged to modulate their drug-like characteristics.

Myelin associated glycoprotein (MAG) represents an interesting target for drug development as well; this lectin binds to Neu5Ac containing glycosides and it is exclusively expressed by the glial cells

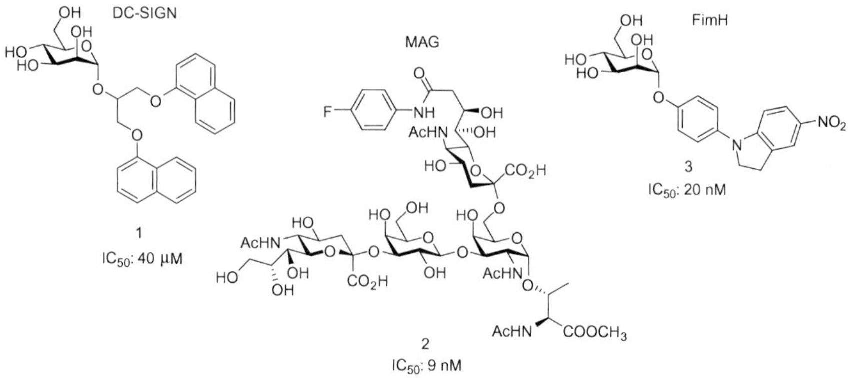

Figure 19.3 Rationally developed lead compounds as bacterial lectins antagonists and corresponding IC_{50}.

in the central nervous system (CNS), where it acts as an inhibitor protein which blocks axon growth, thus preventing the regeneration of the CNS after injury [33]. Targeting this protein could promote neuronal regeneration and the recovery of neurological functionality [34]. Compound (2) is reported to bind MAG with a median inhibitory concentration (IC_{50}) of 9 nM, 10,000-fold the affinity for the natural ligand (Fig. 19.3) [35]. Docking of this molecule to MAG suggested, among the others, a stacking interaction between the p-fluorophenyl residue and Arg118 that could account for the high stability of the complex.

19.6 Bacterial Lectins

Apart from endogenous lectins, bacterial lectins are responsible for the adhesion of the microorganism to the host tissues, enabling the growth of the infection. Targeting these lectins is emerging as a new strategy to fight bacterial infections, as multidrug resistant strains of different pathogens continue to appear [36]. Resistance is the product of the bactericidal stress exerted by conventional antibiotics; while killing the enemy, antibiotics also select those microorganisms that can neutralize them. On the contrary, drugs that are able to prevent the spreading of the infection without killing the pathogen, such as anti-adhesive agents, would be less likely to induce the development of resistant strains. In this context, different classes of bacterial lectins possessing different selectivities are being investigated as molecular targets.

Galactose (PA-IL) and fucose (PA-IIL) binding lectins are both virulence factors of *Pseudomonas aeruginosa*, which can be targeted to fight opportunistic infections carried on by this pathogen, especially on immunocompromised subjects. GMI-1051 (Glycomimetics) is currently under evaluation as an inhibitor of both of these lectins [37].

FimH is a lectin expressed on bacterial fimbriae, where it recognizes α-mannosidic residues exposed on the host tissues. Bacteria responsible for urinary tract infections, especially *Escherichia coli*, take advantage of this interaction to stick to the bladder and urethral epithelium and resist to shear stress and aggression by the immune system [38].

Recent advances in the field have been made by the rational design of new FimH antagonists on the basis of the crystal structure of the lectin–inhibitor complex [39].

Aryl α-mannosides have turned out to be promising leads: the π–π stacking interaction between the aromatic aglycone and the so-called "tyrosine gate" of FimH in its carbohydrate recognition domain is fundamental to achieve high affinities, which are further increased by the ortho substitution of the phenyl ring. Despite not being the most potent of its series, compound (3) (Fig. 19.3) was considered to be the best candidate for further development; this molecule displays very good drug-like characteristics, including lipophylicity/hydrophylicity balance (log $D_{7.4}$), plasma protein binding (PPB) and predicted oral bioavailability (using parallel artificial membranc permeability assays (PAMPAs)); these features resulted in an improved renal elimination profile, granting an *in vivo* urinary concentration above the minimum inhibitory concentration for the adhesion of the pathogen ($T > MIC_{adhesion}$) for more than eight hours [40]. Mice infected with uropathogenic *E. coli* were treated with a single intravenous administration of a 1 mg/kg dose, which resulted in a bacterial count reduction comparable to ciprofloxacin, a reference drug very commonly used to treat urinary infections. All these encouraging results testify the potential of this new non-bactericidal approach to treat infections.

19.7 Conclusions

Carbohydrates are proving to be promising drug leads, with the ability to cover several therapeutic niches; their potential has not been completely understood yet and it is probable that the future holds new and challenging discoveries in this field. The peculiar chemical features of glycosides, such as structural complexity and diversity, which were once considered as drawbacks from a pharmaceutical point of view, are now seen as valuable opportunities to create a variety of structures with an incredible biological and therapeutic potential. The language of carbohydrates requires a huge effort to be learned, but is meant to be one the most spoken in the future of drug development.

References

1. Moazed, D., Noller, H. F. (1987) *Nature* **327**, 389–394.
2. Wang, H., Tor, Y. (1997) *J. Am. Chem. Soc.* **119**, 8734–8735.
3. Jiang, L. C., Patel, D. J. (1998) *Nat. Struct. Biol.* **5**, 769–774.
4. Durante-Mangoni, E., Grammatikos, A., Utili, R., Falagas, M. E., *Inter. J. Antimicrob. Agents*, 2009. **33**(3), 201–205.
5. Falagas, M. E., Grammatikos, A. P., Michalopoulos, A. (2008) *Exp. Rev. Anti-infect. Ther.* **6**, 593–600.
6. Lapidot, A., Berchanski, A., Borkow, G. (2008) *FEBS Journal* **275**, 5236–5257.
7. Venkataraman, N., Cole, A. L., Ruchala, P., Waring, A. J., Lehrer, R. I., Stuchlik, O., Pohl, J., Cole, A. M. (2009) *PLoS Biol.* **7**, e1000095.
8. Di Marco, A., Gaetani, M., Orezzi, P., Scarpinato, B. M., Silvestrini, R., Soldati, M., Dasdia, T., Valentini, L. (1964) *Nature* **201**, 706–707.
9. Minotti, G., Menna, P., Salvatorelli, E., Cairo, G., Gianni, L. (2004) *Pharmacol. Rev.* **56**, 185–229.
10. Pommier, Y. (2012) *ACS Chem. Biol.* **8**, 82–95.
11. Křen, V., Řezanka, T. (2008) *FEMS Microbiol. Rev.* **32**, 858–889.

12. Arcamone, F., Animati, F., Bigioni, M., Capranico, G., Caserini, C., Cipollone, A., De Cesare, M., Ettorre, A., Guano, F., Manzini, S., Monteagudo, E., Pratesi, G., Salvatore, C., Supino, R., Zunino, F. (1999) *Biochem. Pharmacol.* **57**, 1133–1139.
13. Chen, X. L., Zheng, Y. G., Shen, Y. C. (2006) *Curr. Med. Chem.* **13**, 109–116.
14. Scott, L. J., Spencer, C. M. (2000) *Drugs* **59**, 521–549.
15. Truscheit, E., Frommer, W., Junge, B. (1981) *Angew. Chem. Int. Ed. in English* **20**, 744–761.
16. Dulsat, C., Mealy, N. (2009) *Drug. Future* **34**, 147–149.
17. Lyseng-Williamson, K. A. (2014) *Drugs* **74**, 61–74.
18. Wright, E. M., Loo, D. D. F., Hirayama, B. A. (2011) *Physiol. Rev.* **91**, 733–794.
19. Chao, E. C. (2014) *Clinical Diabetes* **32**, 4–11.
20. Sheridan, C. (2012) *Nat. Biotechnol.* **30**, 899–900.
21. Lever, R., Page, C. R. (2002) *Nat. Rev. Drug Discov.* **1**, 140–148.
22. Taylor, N. R., Von Itzstein, M. (1994) *J. Med. Chem.* **37**, 616–624.
23. Von Itzstein, M., Wu, W. Y., Kok, G. B., Pegg, M. S., Dyason, J. C., Jin, B., Van Phan, T., Smythe, M. L., White, H. F., Oliver, S. W., Colman, P. M., Varghese, J. N., Ryan, D. M., Woods, J. M., Bethell, R. C., Hotham, V. J., Cameronf, J. M., Penn, C. R. (1993) *Nature* **363**, 418–423.
24. Kim, C. U., Lew, W., Williams, M. A., Liu, H., Zhang, L., Swaminathan, S., Bischofberger, N., Chen, M. S., Mendel, D. B., Tai, C. Y., Laver, W. G., Stevens, R. C. (1997) *J. Am. Chem. Soc.* **119**, 681–690.
25. Heidelberger, M., Avery, O. T. (1923) *J. Exp. Med.* **38**, 73–79.
26. Barthel, S. R., Gavino, J. D., Descheny, L., Dimitroff, C. J. (2007) *Expert Opin. Ther. Tar.* **11**, 1473–1491.
27. Turhan, A., Weiss, L. A., Mohandas, N., Coller, B. S., Frenette, P. S. (2002) *Proceedings of the National Academy of Sciences of the United States of America* **99**, 3047–3051.
28. Kogan, T. P., Dupre, B., Bui, H., McAbee, K. L., Kassir, J. M., Scott, I. L., Hu, X., Vanderslice, P., Beck, P. J., Dixon, R. A. F. (1998) *J. Med. Chem.* **41**, 1099–1111.
29. Watz, H., Bock, D., Meyer, M., Schierhorn, K., Vollhardt, K., Woischwill, C., Pedersen, F., Kirsten, A., Beeh, K.-M., Meyer-Sabellek, W., Magnussen, H., Beier, J. (2013) *Pulm. Pharmacol. Ther.* **26,** 265–270.
30. Chang, J., Patton, J. T., Sarkar, A., Ernst, B., Magnani, J. L., Frenette, P. S. (2010) *Blood* **116**, 1779–1786.

31. Van Kooyk, Y., Geijtenbeek, T. B. H. (2003) *Nat. Rev. Immunol.* **3**, 697–709.
32. Tomašić, T., Hajsek, D., Svajger, U., Luzar, J., Obermajer, N., Petit-Haertlein, I., Fieschi, F., Anderluh, M. (2014) *Eur. J. Med. Chem.* **75**, 308–326.
33. Kelm, S., Pelz, A., Schauer, R., Filbin, M. T., Tang, S., Debellard, M. E., Schnaar, R. L., Mahoney, J. A., Hartnell, A., Bradfield, P., Crocker, P. R. (1994) *Curr. Biol.* **4**, 965–972.
34. Vyas, A. A., Blixt, O., Paulson, J. C., Schnaar, R. L. (2005) *J. Bio. Chem.* **280**, 16305–16310.
35. Zeng, Y., Rademacher, C., Nycholat, C. M., Futakawa, S., Lemme, K., Ernst, B., Paulson, J. C. (2011) *Bioorgan. Med. Chem. Letters* **21**, 5045–5049.
36. Ofek, I., Hasy, D. L., Sharon, N. (2003) *FEMS Immunol. Med. Mic.* **38**, 181–191.
37. Smith, T., Sarkar, A., Patton, J., Magnani, J. L. (2006) *Glycobiology* **16**, 1142–1143.
38. Hung, C. S., Bouckaert, J., Hung, D., Pinkner, J., Widberg, C., DeFusco, A., Auguste, C. G., Strouse, R., Langermann, S., Waksman, G., Hultgren, S. J. (2002) *Mol. Microbiol.* **44**, 903–915.
39. Han, Z., Pinkner, J. S., Ford, B., Obermann, R., Nolan, W., Wildman, S. A., Hobbs, D., Ellenberger, T., Cusumano, C. K., Hultgren, S. J., Janetka, J. W. (2010) *J. Med. Chem.* **53**, 4779–4792.
40. Jiang, X., Abgottspon, D., Kleeb, S., Rabbani, S., Scharenberg, M., Wittwer, M., Haug, M., Schwardt, O., Ernst, B. (2012) *J. Med. Chem.* **55**, 4700–4713.
41. Alam, S. M., Dennison, S. M., Aussedat, B., Vohra, Y., Park, P. K., Fernandez-Tejada, A., Stewart, S., Jaeger, F. H., Anasti, K., Blinn, J. H., Kepler, T. B., Bonsignori, M., Liao, H.-X., Sodroski, J. G., Danishefsky, S. J., Haynes, B. F. (2013) *Proceedings of the National Academy of Sciences* **110**, 18214–18219.
42. Pejchal, R., Doores, K. J., Walker, L. M., Khayat, R., Huang, P.-S., Wang, S.-K., Stanfield, R. L., Julien, J.-P., Ramos, A., Crispin, M., Depetris, R., Katpally, U., Marozsan, A., Cupo, A., Maloveste, S., Liu, Y., McBride, R., Ito, Y., Sanders, R. W., Ogohara, C., Paulson, J. C., Feizi, T., Scanlan, C. N., Wong, C.-H., Moore, J. P., Olson, W. C., Ward, A. B., Poignard, P., Schief, W. R., Burton, D. R., Wilson, I. A. (2011) *Science* **334**, 1097–1103.

43. Wilson, R. M., Danishefsky, S. J. (2013) *J. Am. Chem. Soc.* **135**, 14462–14472.
44. Huang, Y.-L., Hung, J.-T., Cheung, S. K. C., Lee, H.-Y., Chu, K.-C., Li, S.-T., Lin, Y.-C., Ren, C.-T., Cheng, T.-J. R., Hsu, T.-L., Yu, A. L., Wu, C.-Y., Wong, C.-H. (2013) *Proceedings of the National Academy of Sciences* **110**, 2517–2522.
45. Gilewski, T., Ragupathi, G., Bhuta, S., Williams, L. J., Musselli, C., Zhang, X.-F., Bencsath, K. P., Panageas, K. S., Chin, J., Hudis, C. A., Norton, L., Houghton, A. N., Livingston, P. O., Danishefsky, S. J. (2001) *Proceedings of the National Academy of Sciences* **98**, 3270–3275.
46. Lakshminarayanan, V., Thompson, P., Wolfert, M. A., Buskas, T., Bradley, J. M., Pathangey, L. B., Madsen, C. S., Cohen, P. A., Gendler, S. J., Boons, G. J. (2012) *Proceedings of the National Academy of Sciences of the United States of America* **109**, 261–266.
47. Ragupathi, G., Koide, F., Livingston, P. O., Cho, Y. S., Endo, A., Wan, Q., Spassova, M. K., Keding, S. J., Allen, J., Ouerfelli, O., Wilson, R. M., Danishefsky, S. J. (2006) *J. Am. Chem. Soc.* **128**, 2715–2725.
48. Slovin, S., Ragupathi, G., Fernandez, C., Diani, M., Jefferson, M., Wilton, A., Kelly, W. K., Morris, M., Solit, D., Clausen, H., Livingston, P., Scher, H. (2007) *Cancer Immunol. Immun.* **56**, 1921–1930.
49. Ikegami-Kuzuhara, A., Yoshinaka, T., Ohmoto, H., Inoue, Y., Saito, T. (2001) *Brit. J. Pharmacol.* **134**, 1498–1504.
50. Wada, Y., Saito, T., Matsuda, N., Ohmoto, H., Yoshino, K., Ohashi, M., Kondo, H., Ishida, H., Kiso, M., Hasegawa, A. (1996) *J. Med. Chem.* **39**, 2055–2059.

Index